# Student Solutions Manual

# Elementary Algebra

## FIFTH EDITION

## Ron Larson
The Pennsylvania State University
The Behrend College

Prepared by

## Carolyn F. Neptune
Johnson County Community College

BROOKS/COLE
CENGAGE Learning

Australia • Brazil • Japan • Korea • Mexico • Singapore • Spain • United Kingdom • United States

**BROOKS/COLE**
CENGAGE Learning

For product information and technology assistance, contact us at
**Cengage Learning Academic Resource Center,
1-800-423-0563**

For permission to use material from this text or product, submit all requests online at **www.cengage.com/permissions**
Further permissions questions can be emailed to
**permissionrequest@cengage.com**

ISBN-13: 978-0-547-14011-7
ISBN-10: 0-547-14011-8

**Brooks/Cole**
10 Davis Drive
Belmont, CA 94002-3098
USA

Cengage Learning products are re
Canada by Nelson Education, Ltd.

For your course and learning solut
**academic.cengage.com**

Purchase any of our products at yo
store or at our preferred online sto:
**www.ichapters.com**

Printed in the United States of America
1 2 3 4 5 6 7 13 12 11 10 09

# PREFACE

This *Student Solutions Manual* is a supplement to *Elementary Algebra*, Fifth Edition, by Ron Larson. The *Student Solutions Manual* includes solutions to the odd-numbered exercises in the text, including chapter reviews. In addition, this book contains the solutions to all exercises, even-numbered as well as odd-numbered, for the mid-chapter quizzes, chapter tests, and cumulative tests.

I have written detailed workouts of the problems. The algebraic steps are clearly shown and explanatory comments are included where appropriate. There are usually several "correct" ways to arrive at a solution to a problem in mathematics. Therefore you shouldn't be concerned if you have approached problems differently than I have.

I want to express my appreciation to the many people at Cengage Learning and Larson Texts, Inc. who made significant contributions to this project. Their efforts and cooperation made my work much easier. I am most grateful to my husband, Harold, for his unfailing patience, support, and understanding. His enthusiastic encouragement of my endeavors contributes immeasurably to their success.

I have made every effort to eliminate errors from this book, but I would sincerely appreciate and welcome your comments regarding corrections or suggested improvements for this supplement. You may contact me at the address shown below.

I hope you find the *Student Solutions Manual* to be a helpful supplement as you use the textbook, and I wish you well in your study of algebra.

Carolyn F. Neptune
Johnson County Community College
12345 College Boulevard
Overland Park, KS 66210
cneptune@jccc.edu

# CONTENTS

# C H A P T E R   1
## The Real Number System

# CHAPTER 1
## The Real Number System

### Section 1.1   Real Numbers: Order and Absolute Value

1. (a) natural numbers: $\left\{20, \frac{9}{3}\right\}$

   (b) integers: $\left\{-3, 20, \frac{9}{3}\right\}$

   (c) rational numbers: $\left\{-3, 20, -\frac{3}{2}, \frac{9}{3}, 4.5\right\}$

   (d) irrational numbers: $\{\pi\}$

3. (a) natural numbers: $\left\{\frac{8}{4}\right\}$

   (b) integers: $\left\{\frac{8}{4}\right\}$

   (c) rational numbers: $\left\{-\frac{5}{2}, 6.5, -4.5, \frac{8}{4}, \frac{3}{4}\right\}$

   (d) irrational numbers: $\left\{\sqrt{13}\right\}$

5.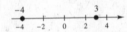

7.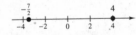

9. $3 > -4$ because 3 lies to the *right* of $-4$.

11. $4 > -\frac{7}{2}$ because 4 lies to the *right* of $-\frac{7}{2}$.

13. $0 > -\frac{7}{16}$ because 0 lies to the *right* of $-\frac{7}{16}$.

    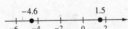

15. $\frac{9}{16} < \frac{5}{8}$ because $\frac{9}{16}$ lies to the *left* of $\frac{5}{8}$.

    Note: $\frac{5}{8} = \frac{10}{16}$

17. $-4.6 < 1.5$ because $-4.6$ lies to the *left* of 1.5.

19. $-6.58 > -7.66$ because $-6.58$ lies to the *right* of $-7.66$.

21. 2; The distance between 2 and zero is 2.

23. 8; The distance between $-8$ and 0 is 8.

25. The opposite of 3 is $-3$.

    The distance from 3 to 0 is 3, and the distance from $-3$ to 0 is also 3.

27. The opposite of $-3.8$ is 3.8.

    The distance from $-3.8$ to 0 is 3.8. and the distance from 3.8 to 0 is also 3.8.

29. The opposite of $\frac{5}{2}$ is $-\frac{5}{2}$.

    The distance from $\frac{5}{2}$ to 0 is $\frac{5}{2}$, and the distance from $-\frac{5}{2}$ to 0 is also $\frac{5}{2}$.

31. $\left|\frac{5}{2}\right| = \frac{5}{2}$

    The distance of $\frac{5}{2}$ from 0 is $\frac{5}{2}$.

33. $|-3| = 3$

    The distance of $-3$ from 0 is 3.

35. $|10| = 10$ because the distance between 10 and 0 is 10.

37. $|-3| = 3$ because the distance between $-3$ and 0 is 3.

39. $|-3.4| = 3.4$ because the distance between $-3.4$ and 0 is 3.4.

41. $\left|-\frac{7}{2}\right| = \frac{7}{2}$ because the distance between $-\frac{7}{2}$ and 0 is $\frac{7}{2}$.

43. $-|4.09| = -4.09$

    Note: $|4.09| = 4.09$

**45.** $-\left|-23.6\right| = -23.6$

Note: $\left|-23.6\right| = 23.6$

**47.** $\left|0\right| = 0$ because the distance between 0 and 0 is 0.

**49.** $\left|-16\right| = \left|16\right|$ because $\left|-16\right| = 16$ and $\left|16\right| = 16.$

**51.** $\left|-4\right| > \left|3\right|$ because $\left|-4\right| = 4$, $\left|3\right| = 3$ and 4 is greater than 3.

**53.** $\left|\frac{3}{16}\right| < \left|\frac{3}{2}\right|$ because $\left|\frac{3}{16}\right| = \frac{3}{16}$, $\left|\frac{3}{2}\right| = \frac{3}{2}$, and $\frac{3}{16} < \frac{3}{2}$.

Note: $\frac{3}{2} = \frac{24}{16}$, and $\frac{3}{16} < \frac{24}{16}$

**55.** $-\left|48.5\right| < \left|-48.5\right|$ because $-\left|-48.5\right| = -48.5,$
$\left|48.5\right| = 48.5$, and $-48.5$ is less than 48.5.

**57.** $\left|-\pi\right| > -\left|-2\pi\right|$ because $\left|-\pi\right| = \pi$, $-\left|-2\pi\right| = -2\pi$, and
$\pi > -2\pi.$

**59.**

**61.**

**63.** The number 12 units to the *right* of 8 is 20 because
$8 + 12 = 20.$

The number 12 units to the *left* of 8 is $-4$ because
$8 - 12 = -4.$

**65.** The number 6 units to the *right* of 21.3 is 27.3 because
$21.3 + 6 = 27.3.$

The number 6 units to the *left* of 21.3 is 15.3 because
$21.3 - 6 = 15.3.$

**67.** The number 3.5 units to the *right* of $-2$ is 1.5 because
$-2 + 3.5 = 1.5.$

The number 3.5 units to the *left* of $-2$ is $-5.5$ because
$-2 - 3.5 = -5.5.$

**69.** Sample answers: $-3, -100, -\frac{4}{1}$

**71.** Sample answers: $\sqrt{2}, \pi, -3\sqrt{3}$

**73.** Sample answers: $-7, 1, 341$

**75.** Sample answers: $\frac{1}{2}, 10, 20\frac{1}{5}$

**77.** Sample answers: $-1, -10, -100$

**79.** $-15$

The distance from $-15$ to 0 is 15: $\left|-15\right| = 15.$

The distance from 10 to 0 is 10; $\left|10\right| = 10.$

So, $-15$ is farther from 0 because $\left|-15\right| > \left|10\right|$ or
$15 > 10.$

**81.** 0.37 is smaller.

$\frac{3}{8} = 0.375$, and 0.37 lies to the *left* of 0.375, so 0.37 is smaller.

**83.** True. Their absolute values are equal because their distances from zero on the real number line are equal.

$\left|a\right| = \left|-a\right|$

For $a \geq 0$, $\left|a\right| = a$ and $\left|-a\right| = a.$

For $a < 0$, $\left|a\right| = -a$ and $\left|-a\right| = -a.$

**85.** True. The real number line is a picture used to represent the real numbers. In this picture each point on the line corresponds to exactly one real number and each real number corresponds to exactly one point on the real number line.

**87.** False. A rational number is a ratio of two integers. Some rational numbers, such as $\frac{6}{2}$ and $-\frac{12}{3}$, are integers, but others are not. For example, the rational numbers $\frac{1}{2}, \frac{7}{4}$, and $-\frac{3}{8}$ are not integers.

# Section 1.2   Adding and Subtracting Integers

**1.**

$2 + 7 = 9$

**3.**

$-8 + \left(-3\right) = -11$

**5.**

$10 + \left(-3\right) = 7$

**7.**

$-6 + 4 = -2$

**9.** $6 + 10 = 16$

**11.** $14 + (-14) = 0$

**13.** $-45 + 45 = 0$

**15.** $14 + 13 = 27$

**17.** $-23 + (-4) = -27$

**19.** $18 + (-12) = 6$

**21.** $75 + 100 = 175$

**23.** $9 + (-14) = -5$

**25.** $10 + (-6) + 34 = 38$

**27.** $-15 + (-3) + 8 = -10$

**29.** $9 + (-18) + 4 = -5$

**31.** $16 + 2 + (-7) = 11$

**33.** $-13 + 12 + 4 = 3$

**35.** $75 + (-75) + (-15) = -15$

**37.** $803 + (-104) + (-613) + 214 = 300$

**39.** $312 + (-564) + (-100) = -352$

**41.** $-890 + (-90) + 62 = -918$

**43.** $21 - 18 = 3$

**45.** $51 - 25 = 26$

**47.** $1 - (-4) = 1 + 4 = 5$

**49.** $15 - (-10) = 15 + 10 = 25$

**51.** $18 - (-18) = 18 + 18 = 36$

**53.** $19 - (-31) = 50$

**55.** $27 - 57 = -30$

**57.** $61 - 85 = -24$

**59.** $22 - 131 = -109$

**61.** $2 - 11 = -9$

**63.** $13 - 24 = -11$

**65.** $-135 - (-114) = -21$

**67.** $-4 - (-4) = 0$

**69.** $-10 - (-4) = -6$

**71.** $-71 - 32 = -103$

**73.** $-210 - 400 = -610$

**75.** $-110 - (-30) = -80$

**77.** $-6 - 15 = -21$

**79.** $380 - (-120) = 500$

**81.** $-43 - (-22) = -43 + 22 = -21$

**83.** $-15$

$10 + (-15) = -5$

**85.** $-36$

$-12 - (-36) = 24$

**87.** $-1 + 3 - (-4) + 10 = 16$

**89.** $6 + 7 - 12 - 5 = -4$

**91.** $-(-5) + 7 - 18 + 4 = -2$

**93.** The result is $-10$.

*Keystrokes:* Scientific calculator: 3 $\boxed{+/-}$ $\boxed{-}$ 7 $\boxed{=}$

Graphing calculator: $\boxed{(-)}$ 3 $\boxed{-}$ 7 $\boxed{\text{ENTER}}$

**95.** The result is 18.

*Keystrokes:* Scientific calculator: 6 $\boxed{+}$ 5 $\boxed{-}$ $\boxed{(}$ 7 $\boxed{+/-}$ $\boxed{)}$ $\boxed{=}$

Graphing calculator: 6 $\boxed{+}$ 5 $\boxed{-}$ $\boxed{(}$ $\boxed{(-)}$ 7 $\boxed{)}$ $\boxed{\text{ENTER}}$

**97.** $-10 + 22 = 12$

The temperature at noon was 12°F.

**99.** $-847 + 385 = -462$

He rested at a distance of 462 meters down the canyon.

**101.** $362,000 - (-650,000) = 362,000 + 650,000$
$$= 1,012,000$$

The profit during the second six months was $1,012,000.

**103.** $2750 - 350 - 500 + 450 + 6.42 = 2356.42$

The balance was $2356.42.

**105.** $25 + 4 + 3 - 9 = 23$

The temperature after soccer practice was $23°C$.

**107.** (a) Factory sales in 2000 were $1825 million; in 2001 factory sales were $1972 million.

$1972 - 1825 = 147$

The change in factory sales of digital cameras from 2000 to 2001 was $147 million.

(b) Factory sales in 2004 were $4739 million; in 2005 factory sales were $7468 million.

$7468 - 4739 = 2729$

The change in factory sales of digital cameras from 2004 to 2005 was $2729 million.

**109.** (a) $3 + 2 = 5$

(b) To add two numbers with like signs, add their absolute values and attach the common sign to the result.

**111.** To add two negative numbers, add their absolute values and attach the negative sign.

**113.** No, it is not possible that the sum of two positive integers is a negative number. To add two positive integers, add their absolute values and attach the common sign, which is always the positive sign because the numbers being added are positive.

# Section 1.3   Multiplying and Dividing Integers

**1.** $3 \cdot 2 = 2 + 2 + 2 = 6$

**3.** $5 \times (-3) = (-3) + (-3) + (-3) + (-3) + (-3) = -15$

**5.** $5 \times 7 = 35$

**7.** $0 \cdot 4 = 0$

**9.** $2(-16) = -32$

**11.** $-9(4) = -36$

**13.** $230(-3) = -690$

**15.** $-7(-13) = 91$

**17.** $-200(-8) = 1600$

**19.** $3(-5)(6) = -90$

**21.** $-7(3)(-1) = 21$

**23.** $-2(-3)(-5) = -30$

**25.** $|(-3)4| = |-12| = 12$

**27.** $|3(-5)(6)| = |-90| = 90$

**29.** $|6(20)(4)| = |480| = 480$

**31.**
$$\begin{array}{r} 26 \\ \times\ 13 \\ \hline 78 \\ 260 \\ \hline 338 \end{array}$$

**33.**
$$\begin{array}{r} 14 \\ \times\ 24 \\ \hline 56 \\ 280 \\ \hline 336 \end{array}$$

So, $-14 \times 24 = -336$.

**35.**
$$\begin{array}{r} 63 \\ \times\ 75 \\ \hline 315 \\ 4410 \\ \hline 4725 \end{array}$$

So, $75(-63) = -4725$.

**37.**
$$\begin{array}{r} 13 \\ \times\ 20 \\ \hline 260 \end{array}$$

So, $-13(-20) = 260$.

**39.**
$$\begin{array}{r} 429 \\ \times\ 21 \\ \hline 429 \\ 8580 \\ \hline 9009 \end{array}$$

So, $-21(-429) = 9009$.

**41.** $27 \div 9 = 3$

**43.** $72 \div (-12) = -6$

**45.** $-28 \div 4 = -7$

**47.** $-56 \div (-8) = 7$

**49.** $\frac{8}{0}$ is undefined.

    Division by zero is undefined.

**51.** $\frac{0}{8} = 0$

**53.** $\frac{-81}{-3} = 27$

**55.** $\frac{6}{-1} = -6$

**57.** $\frac{-28}{4} = -7$

**59.** $-27 \div (-27) = 1$

**61.**

$$
\begin{array}{r}
32 \\
45\overline{)1440} \\
135\phantom{0} \\
\hline
90 \\
90 \\
\hline
\end{array}
$$

    So, $1440 \div 45 = 32$.

**63.**

$$
\begin{array}{r}
160 \\
9\overline{)1440} \\
9\phantom{00} \\
\hline
54\phantom{0} \\
54\phantom{0} \\
\hline
0 \\
0 \\
\hline
\end{array}
$$

    So, $1440 \div (-9) = -160$.

**65.**

$$
\begin{array}{r}
82 \\
16\overline{)1312} \\
128\phantom{0} \\
\hline
32 \\
32 \\
\hline
\end{array}
$$

    So, $-1312 \div 16 = -82$.

**67.**

$$
\begin{array}{r}
110 \\
25\overline{)2750} \\
25\phantom{00} \\
\hline
25\phantom{0} \\
25\phantom{0} \\
\hline
0 \\
0 \\
\hline
\end{array}
$$

    So, $2750 \div 25 = 110$.

**69.**

$$
\begin{array}{r}
331 \\
28\overline{)9268} \\
84\phantom{00} \\
\hline
86\phantom{0} \\
84\phantom{0} \\
\hline
28 \\
28 \\
\hline
\end{array}
$$

    So, $-9268 \div (-28) = 331$.

**71.** $\dfrac{44{,}290}{515} = 86$

**73.** $\dfrac{169{,}290}{162} = 1045$

**75.** 240 is composite; its prime factorization is
    $2 \cdot 2 \cdot 2 \cdot 2 \cdot 3 \cdot 5$.

**77.** 643 is prime; the divisibility tests yield no factors of 643.

    By testing the remaining primes less than or equal to $\sqrt{643} \approx 25$, you can conclude that 643 is a prime number.

**79.** 3911 is prime; the divisibility tests yield no factors of 3911. By testing the remaining primes less than or equal to $\sqrt{3911} \approx 63$, you can conclude that 3911 is a prime number.

**81.** 1281 is composite, its prime factorization is $3 \cdot 7 \cdot 61$.

**83.** 3555 is composite; its prime factorization is
    $3 \cdot 3 \cdot 5 \cdot 79$.

**85.** $12 = 2 \cdot 2 \cdot 3$

**87.** $561 = 3 \cdot 11 \cdot 17$

**89.** $210 = 2 \cdot 3 \cdot 5 \cdot 7$

**91.** $2535 = 3 \cdot 5 \cdot 13 \cdot 13$

**93.** $192 = 2 \cdot 2 \cdot 2 \cdot 2 \cdot 2 \cdot 2 \cdot 3$

**95.** Example: $1 \times (-7) = -7$ or $1 \times 364 = 364$

    Algebraic description: If $a$ is a real number, then $1 \times a = a$.

**97.** $12 \div 4 = 3$

**99.** $6 \cdot 0 = 0 = 0 \cdot 6$

**101.** $-1(0) = 0$

    $-2(0) = 0$

    Rule: The product of an integer and zero is 0.

**103.** $-0.29(4) = -1.16$

The total price change per share during the four days is −$1.16.

The negative value indicates that the price per share *decreased* by $1.16.

**105.** $250(12)(2) = 6000$

The homeowner saves $250 per month for 2 years, so $250 is saved 12 times a year for 2 years. After 2 years, the homeowner has saved $6000.

**111.** (a) $\frac{328}{4} = 82$; The average number of points scored per exam is 82.

(b)
```
      73   77   82      87   91
   ─●─┼─●─┼─┼─●─┼─┼─●─┼─┼─●─┼→
    72   76   80   84   88   92
```

(c) $(87 - 82) + (73 - 82) + (77 - 82) + (91 - 82) = 5 + (-9) + (-5) + (9) = 0$

The sum of these differences is 0. Explanations will vary.

**113.** $(9)(6)(11) = 594$

The volume of the rectangular solid is 594 cubic inches.

**115.** The only even prime number is 2. There are no other even prime numbers because every other even number is divisible by itself, by 1, and by 2; all other even numbers are composites because they have more than two factors.

**107.** $(160)(360) = 57,600$

The area of the football field is 57,600 square feet.

**109.** $\frac{45}{9} = 5$

If the space shuttle travels 45 miles in 9 seconds, its average speed is 5 miles per second.

**117.** An even integer has a factor of 2 so the product of this integer and any other integer will also have a factor of 2. Therefore, the product is even.

The product of two odd integers is odd.

**119.** There are seven other twin primes less than 100.

They are 5 and 7, 11 and 13, 17 and 19, 29 and 31, 41 and 43, 59 and 61, and 71 and 73.

**121.** (a)
```
 01  02  03  04  05  06  07  08  09  10
 11  12  13  14  15  16  17  18  19  20
 21  22  23  24  25  26  27  28  29  30
 31  32  33  34  35  36  37  38  39  40
 41  42  43  44  45  46  47  48  49  50
 51  52  53  54  55  56  57  58  59  60
 61  62  63  64  65  66  67  68  69  70
 71  72  73  74  75  76  77  78  79  80
 81  82  83  84  85  86  87  88  89  90
 91  92  93  94  95  96  97  98  99  100
```

(b) The remaining numbers are prime.

Explanations will vary. Here is one explanation: Every number on the list has a square root that is less than or equal to 10. The numbers 2, 3, 5, and 7 are the only prime numbers less than 10. Because all the multiples of these four prime numbers have been crossed out and the number 1 has been crossed out also, any number which remains on the list has no prime factors less than its square root. Thus, any number which remains on the list must be prime.

# Mid-Chapter Quiz for Chapter 1

**1.** $\frac{3}{16} < \frac{3}{8}$

**2.** $-2.5 > -4$

**3.** $-7 < 3$

```
      -7              3
   ─┼─●─┼─┼─┼─┼─●─┼→
   -8  -6  -4  -2  0   2   4
```

**4.** $2\pi > 6$

Note: $2\pi \approx 6.28$

```
            6 2π
   ─┼─────┼─●●─┼──→
    5      6      7
```

**5.** $-\left|-0.75\right| = -0.75$

Note: $\left|-0.75\right| = 0.75$

**6.** $\left|-\frac{17}{19}\right| = \frac{17}{19}$

**7.** $\left|\frac{7}{2}\right| = \left|-3.5\right|$

Note: $\left|\frac{7}{2}\right| = \frac{7}{2}$ or 3.5, and $\left|-3.5\right| = 3.5$.

**8.** $\left|\frac{3}{4}\right| > -\left|0.75\right|$

Note: $\left|\frac{3}{4}\right| = \frac{3}{4}$ or 0.75, and $-\left|0.75\right| = -0.75$,

$0.75 > -0.75$.

**9.** $-22 - (-13) = -22 + 13 = -9$

**10.** $\left|-54 + 26\right| = \left|28\right| = 28$

**11.** $52 + 47 = 99$

**12.** $-18 + (-35) = -53$

**13.** $-15 - 12 = -27$

**14.** $-35 - (-10) = -35 + 10 = -25$

**15.** $25 + (-75) = -50$

**16.** $72 - 134 = -62$

**17.** $12 + (-6) - 8 + 10 = 8$

**18.** $-9 - 17 + 36 + (-15) = -5$

**19.** $-6(10) = -60$

**20.** $-7(-13) = 91$

**21.** $\frac{-45}{-3} = 15$

**22.** $\frac{-24}{6} = -4$

**23.** 23 is prime

**24.** 91 is composite; its prime factorization is $7 \cdot 13$

**25.** 111 is composite; its prime factorization is $3 \cdot 37$

**26.** 144 is composite; its prime factorization is
$2 \cdot 2 \cdot 2 \cdot 2 \cdot 3 \cdot 3$

**27.** $513,200 + 136,500 + (-97,750) + (-101,500) = 450,450$

The total profit for the year $450,450.

**28.** $(8)(4)(4) = 128$

There are 128 cubic feet in a cord of wood.

**29.** $90 \div 6 = 15$

Each piece of rope is 15 feet long.

**30.** $738 - 550 + 189 - 10 = 367$

The balance at the end of the month was $367.

## Section 1.4    Operations with Rational Numbers

**1.** 5 is prime, and by prime factorization $10 = 2 \cdot 5$. So, the greatest common factor is 5.

**3.** By prime factorization, $20 = 2 \cdot 2 \cdot 5$ and
$45 = 3 \cdot 3 \cdot 5$.
So, the greatest common factor is 5.

**5.** By prime factorization, $45 = 3 \cdot 3 \cdot 5$ and
$90 = 2 \cdot 3 \cdot 3 \cdot 5$. So, the greatest common factor is
$3 \cdot 3 \cdot 5$ or 45.

**7.** By prime factorization, $18 = 2 \cdot 3 \cdot 3$,
$84 = 2 \cdot 2 \cdot 3 \cdot 7$, and $90 = 2 \cdot 3 \cdot 3 \cdot 5$. So, the
greatest common factor is $2 \cdot 3$, or 6.

**9.** By prime factorization, $240 = 2 \cdot 2 \cdot 2 \cdot 2 \cdot 3 \cdot 5$,
$300 = 2 \cdot 2 \cdot 3 \cdot 5 \cdot 5$, and $360 = 2 \cdot 2 \cdot 2 \cdot 3 \cdot 3 \cdot 5$.
So, the greatest common factor is $2 \cdot 2 \cdot 3 \cdot 5$ or 60.

**11.** By prime factorization, $134 = 2 \cdot 67$,
$225 = 3 \cdot 3 \cdot 5 \cdot 5$, $315 = 3 \cdot 3 \cdot 5 \cdot 7$, and
$945 = 3 \cdot 3 \cdot 3 \cdot 5 \cdot 7$. Because there are no common
prime factors, the greatest common factor is 1.

**13.** $\frac{2}{4} = \frac{(1)(\cancel{2})}{(2)(\cancel{2})} = \frac{1}{2}$

**15.** $\frac{12}{15} = \frac{(4)(\cancel{3})}{(5)(\cancel{3})} = \frac{4}{5}$

**17.** $\dfrac{60}{192} = \dfrac{(5)(\cancel{12})}{(16)(\cancel{12})} = \dfrac{5}{16}$

Note: This reducing could be done using several steps, such as the following.

$\dfrac{60}{192} = \dfrac{(30)(\cancel{2})}{(96)(\cancel{2})} = \dfrac{(10)(\cancel{3})}{(32)(\cancel{3})} = \dfrac{(5)(\cancel{2})}{(16)(\cancel{2})} = \dfrac{5}{16}$

**19.** $\dfrac{28}{350} = \dfrac{2(\cancel{14})}{25(\cancel{14})} = \dfrac{2}{25}$

**21.** $\dfrac{1}{5} + \dfrac{2}{5} = \dfrac{1+2}{5} = \dfrac{3}{5}$

**23.** $\dfrac{2}{10} + \dfrac{4}{10} = \dfrac{2+4}{10} = \dfrac{6}{10} = \dfrac{(3)(\cancel{2})}{(5)(\cancel{2})} = \dfrac{3}{5}$

**25.** $\dfrac{3}{8} = \dfrac{3(2)}{8(2)} = \dfrac{6}{16}$

**27.** Write the original fraction in simplest form, and then find an equivalent fraction with the indicated denominator.

$\dfrac{6}{15} = \dfrac{2(\cancel{3})}{5(\cancel{3})} = \dfrac{2}{5} = \dfrac{2(5)}{5(5)} = \dfrac{10}{25}$

You could do both steps at once by multiplying the numerator and the denominator by $\frac{5}{3}$.

$\dfrac{6}{15} = \dfrac{6\left(\frac{5}{3}\right)}{15\left(\frac{5}{3}\right)} = \dfrac{10}{25}$

**29.** $\dfrac{7}{15} + \dfrac{1}{15} = \dfrac{7+1}{15} = \dfrac{8}{15}$

**31.** $\dfrac{3}{2} + \dfrac{5}{2} = \dfrac{3+5}{2} = \dfrac{8}{2} = 4$

**33.** $\dfrac{9}{16} - \dfrac{3}{16} = \dfrac{9}{16} + \dfrac{-3}{16}$

$= \dfrac{9+(-3)}{16} = \dfrac{6}{16} = \dfrac{(3)(\cancel{2})}{(8)(\cancel{2})} = \dfrac{3}{8}$

Note: This problem can also be written as follows.

$\dfrac{9}{16} - \dfrac{3}{16} = \dfrac{9-3}{16} = \dfrac{6}{16} = \dfrac{(3)(\cancel{2})}{(8)(\cancel{2})} = \dfrac{3}{8}$

**35.** $-\dfrac{23}{11} + \dfrac{12}{11} = \dfrac{-23+12}{11} = \dfrac{-11}{11} = -1$

**37.** $\dfrac{3}{4} - \dfrac{5}{4} = \dfrac{3}{4} + \dfrac{-5}{4}$

$= \dfrac{3+(-5)}{4} = \dfrac{-2}{4} = -\dfrac{(1)(\cancel{2})}{(2)(\cancel{2})} = -\dfrac{1}{2}$

**39.** $\dfrac{7}{10} + \left(-\dfrac{3}{10}\right) = \dfrac{7+(-3)}{10} = \dfrac{4}{10} = \dfrac{\cancel{2}(2)}{\cancel{2}(5)} = \dfrac{2}{5}$

**41.** $\dfrac{2}{5} + \dfrac{4}{5} + \dfrac{1}{5} = \dfrac{2+4+1}{5} = \dfrac{7}{5}$

**43.** $\dfrac{1}{2} + \dfrac{1}{3} = \dfrac{1(3)}{2(3)} + \dfrac{1(2)}{3(2)} = \dfrac{3}{6} + \dfrac{2}{6} = \dfrac{3+2}{6} = \dfrac{5}{6}$

**45.** $\dfrac{3}{16} + \dfrac{3}{8} = \dfrac{3}{16} + \dfrac{3(2)}{8(2)} = \dfrac{3}{16} + \dfrac{6}{16} = \dfrac{3+6}{16} = \dfrac{9}{16}$

**47.** $\dfrac{1}{4} - \dfrac{1}{3} = \dfrac{1(3)}{4(3)} - \dfrac{1(4)}{3(4)} = \dfrac{3}{12} - \dfrac{4}{12} = -\dfrac{1}{12}$

**49.** $-\dfrac{1}{8} - \dfrac{1}{6} = \dfrac{-1}{8} - \dfrac{1}{6}$

$= \dfrac{-1(3)}{8(3)} - \dfrac{1(4)}{6(4)}$

$= \dfrac{-3}{24} - \dfrac{4}{24}$

$= \dfrac{-3-4}{24}$

$= \dfrac{-7}{24} = -\dfrac{7}{24}$

**51.** $4 + \dfrac{8}{3} = \dfrac{4(3)}{1(3)} + \dfrac{8}{3}$

$= \dfrac{12}{3} + \dfrac{8}{3}$

$= \dfrac{12+8}{3}$

$= \dfrac{20}{3}$

**53.** $-\dfrac{3}{8} - \dfrac{1}{12} = -\dfrac{3(3)}{8(3)} - \dfrac{1(2)}{12(2)}$

$= -\dfrac{9}{24} - \dfrac{2}{24}$

$= \dfrac{-9-2}{24}$

$= -\dfrac{11}{24}$

**55.** $\dfrac{3}{4} - \dfrac{2}{5} = \dfrac{3(5)}{4(5)} - \dfrac{2(4)}{5(4)}$

$\quad = \dfrac{15}{20} - \dfrac{8}{20}$

$\quad = \dfrac{15 - 8}{20}$

$\quad = \dfrac{7}{20}$

**57.** $-\dfrac{5}{6} - \left(-\dfrac{3}{4}\right) = -\dfrac{5}{6} + \dfrac{3}{4}$

$\quad = -\dfrac{5(2)}{6(2)} + \dfrac{3(3)}{4(3)}$

$\quad = -\dfrac{10}{12} + \dfrac{9}{12}$

$\quad = -\dfrac{1}{12}$

**59.** $3\dfrac{1}{2} + 5\dfrac{2}{3} = \dfrac{3(2) + 1}{2} + \dfrac{5(3) + 2}{3}$

$\quad = \dfrac{7}{2} + \dfrac{17}{3}$

$\quad = \dfrac{7(3)}{2(3)} + \dfrac{17(2)}{3(2)}$

$\quad = \dfrac{21}{6} + \dfrac{34}{6}$

$\quad = \dfrac{21 + 34}{6}$

$\quad = \dfrac{55}{6}$

**61.** $1\dfrac{3}{16} - 2\dfrac{1}{4} = \dfrac{19}{16} - \dfrac{9}{4}$

$\quad = \dfrac{19}{16} - \dfrac{9(4)}{4(4)}$

$\quad = \dfrac{19}{16} - \dfrac{36}{16}$

$\quad = \dfrac{19 - 36}{16}$

$\quad = \dfrac{-17}{16}$

$\quad = -\dfrac{17}{16}$

**63.** $15 - 20\dfrac{1}{4} = \dfrac{15(4)}{1(4)} - \dfrac{20(4) + 1}{4}$

$\quad = \dfrac{60}{4} - \dfrac{81}{4}$

$\quad = \dfrac{60 - 81}{4}$

$\quad = \dfrac{-21}{4}$

$\quad = -\dfrac{21}{4}$

**65.** $-5\dfrac{1}{3} - 4\dfrac{5}{12} = -\dfrac{5(3) + 1}{3} - \dfrac{4(12) + 5}{12}$

$\quad = -\dfrac{16}{3} - \dfrac{53}{12}$

$\quad = -\dfrac{16(4)}{3(4)} - \dfrac{53}{12}$

$\quad = \dfrac{-64 - 53}{12}$

$\quad = \dfrac{-117}{12}$

$\quad = -\dfrac{39(\cancel{3})}{4(\cancel{3})}$

$\quad = -\dfrac{39}{4}$

**67.** $\dfrac{5}{12} - \dfrac{3}{8} + \dfrac{5}{4} = \dfrac{5(2)}{12(2)} - \dfrac{3(3)}{8(3)} + \dfrac{5(6)}{4(6)}$

$\quad = \dfrac{10}{24} - \dfrac{9}{24} + \dfrac{30}{24}$

$\quad = \dfrac{10 - 9 + 30}{24}$

$\quad = \dfrac{31}{24}$

**69.** $3 + \dfrac{12}{3} + \dfrac{1}{9} = \dfrac{3(9)}{1(9)} + \dfrac{12(3)}{3(3)} + \dfrac{1}{9}$

$\quad = \dfrac{27 + 36 + 1}{9}$

$\quad = \dfrac{64}{9}$

**71.** $2 - \dfrac{25}{6} - \dfrac{3}{4} = \dfrac{2(12)}{1(12)} - \dfrac{25(2)}{6(2)} - \dfrac{3(3)}{4(3)}$

$\quad = \dfrac{24}{12} - \dfrac{50}{12} - \dfrac{9}{12}$

$\quad = \dfrac{24 - 50 - 9}{12}$

$\quad = -\dfrac{35}{12}$

**73.** $1 - \dfrac{3}{10} - \dfrac{2}{5} = \dfrac{1}{1} - \dfrac{3}{10} - \dfrac{2}{5}$

$\qquad = \dfrac{1(10)}{1(10)} - \dfrac{3}{10} - \dfrac{2(2)}{5(2)}$

$\qquad = \dfrac{10}{10} - \dfrac{3}{10} - \dfrac{4}{10} = \dfrac{10 - 3 - 4}{10} = \dfrac{3}{10}$

Note: This problem could also be worked in two steps.
First, add the two known fractions.

$\dfrac{3}{10} + \dfrac{2}{5} = \dfrac{3}{10} + \dfrac{2(2)}{5(2)} = \dfrac{3}{10} + \dfrac{4}{10} = \dfrac{3 + 4}{10} = \dfrac{7}{10}$

Then subtract the sum from 1.

$1 - \dfrac{7}{10} = \dfrac{1}{1} - \dfrac{7}{10}$

$\qquad = \dfrac{1(10)}{1(10)} - \dfrac{7}{10}$

$\qquad = \dfrac{10}{10} - \dfrac{7}{10} = \dfrac{10 - 7}{10} = \dfrac{3}{10}$

**75.** $1 - \dfrac{1}{3} - \dfrac{1}{4} - \dfrac{1}{5} = \dfrac{1(60)}{1(60)} - \dfrac{1(20)}{3(20)} - \dfrac{1(15)}{4(15)} - \dfrac{1(12)}{5(12)}$

$\qquad = \dfrac{60}{60} - \dfrac{20}{60} - \dfrac{15}{60} - \dfrac{12}{60}$

$\qquad = \dfrac{60 - 20 - 15 - 12}{60}$

$\qquad = \dfrac{13}{60}$

**77.** $\dfrac{1}{2} \cdot \dfrac{3}{4} = \dfrac{1 \cdot 3}{2 \cdot 4} = \dfrac{3}{8}$

**79.** $-\dfrac{2}{3} \cdot \dfrac{5}{7} = \dfrac{-2 \cdot 5}{3 \cdot 7} = \dfrac{-10}{21} = -\dfrac{10}{21}$

**81.** $\dfrac{2}{3}\left(-\dfrac{9}{16}\right) = -\dfrac{2 \cdot 9}{3 \cdot 16} = -\dfrac{(2)(3)(3)}{(3)(8)(2)} = -\dfrac{3}{8}$

Note: The reducing could also be written this way.

$\dfrac{2}{3}\left(-\dfrac{9}{16}\right) = -\dfrac{\overset{1}{\cancel{2}} \cdot \overset{3}{\cancel{9}}}{\underset{1}{\cancel{3}} \cdot \underset{8}{\cancel{16}}} = -\dfrac{3}{8}$

**83.** $-\dfrac{3}{4}\left(-\dfrac{4}{9}\right) = \dfrac{3 \cdot 4}{4 \cdot 9} = \dfrac{(3)(4)}{(4)(3)(3)} = \dfrac{1}{3}$

**85.** $\dfrac{5}{18}\left(\dfrac{3}{4}\right) = \dfrac{(5)(3)}{(18)(4)} = \dfrac{(5)(3)}{(3)(6)(4)} = \dfrac{5}{24}$

**87.** $\dfrac{11}{12}\left(-\dfrac{9}{44}\right) = -\dfrac{11(9)}{12(44)} = -\dfrac{(11)(3)(3)}{(4)(3)(4)(11)} = -\dfrac{3}{16}$

**89.** $-\dfrac{3}{11}\left(-\dfrac{11}{3}\right) = \dfrac{(-3)(-11)}{(11)(3)} = \dfrac{33}{33} = 1$

**91.** $9\left(\dfrac{4}{15}\right) = \dfrac{9}{1}\left(\dfrac{4}{15}\right) = \dfrac{(3)(3)(4)}{(3)(5)} = \dfrac{12}{5}$

**93.** $2\tfrac{3}{4} \cdot 3\tfrac{2}{3} = \left(\dfrac{11}{4}\right)\left(\dfrac{11}{3}\right) = \dfrac{121}{12}$

**95.** $-5\tfrac{2}{3} \cdot 4\tfrac{1}{2} = \left(-\dfrac{17}{3}\right)\left(\dfrac{9}{2}\right)$

$\qquad = -\dfrac{(17)(9)}{(3)(2)} = -\dfrac{(17)(3)(3)}{(3)(2)} = -\dfrac{51}{2}$

**97.** $-\dfrac{3}{2}\left(-\dfrac{15}{16}\right)\left(\dfrac{12}{25}\right) = \dfrac{3}{2} \cdot \dfrac{15}{16} \cdot \dfrac{12}{25} = \dfrac{3 \cdot 15 \cdot 12}{2 \cdot 16 \cdot 25}$

$\qquad = \dfrac{(3)(5)(3)(4)(3)}{(2)(4)(4)(5)(5)} = \dfrac{27}{40}$

Note: The reducing could also be written this way.

$-\dfrac{3}{2}\left(-\dfrac{5}{16}\right)\left(\dfrac{12}{25}\right) = \dfrac{3}{2} \cdot \dfrac{15}{16} \cdot \dfrac{12}{25} = \dfrac{3 \cdot \overset{3}{\cancel{15}} \cdot \overset{3}{\cancel{12}}}{2 \cdot \underset{4}{\cancel{16}} \cdot \underset{5}{\cancel{25}}} = \dfrac{27}{40}$

**99.** $6\left(\dfrac{3}{4}\right)\left(\dfrac{2}{9}\right) = \dfrac{6 \cdot 3 \cdot 2}{4 \cdot 9} = \dfrac{\cancel{3} \cdot \cancel{2} \cdot \cancel{3} \cdot \cancel{2}}{\cancel{2} \cdot \cancel{2} \cdot \cancel{3} \cdot \cancel{3}} = 1$

**101.** The reciprocal of 7 is $\tfrac{1}{7}$.

$7\left(\tfrac{1}{7}\right) = \dfrac{7}{1}\left(\dfrac{1}{7}\right) = \dfrac{7}{7} = 1$

**103.** The reciprocal of $\tfrac{4}{7}$ is $\tfrac{7}{4}$.

$\tfrac{4}{7}\left(\tfrac{7}{4}\right) = \dfrac{28}{28} = 1$

**105.** $\dfrac{3}{8} \div \dfrac{3}{4} = \dfrac{3}{8} \cdot \dfrac{4}{3} = \dfrac{3 \cdot 4}{8 \cdot 3} = \dfrac{(3)(1)(4)}{(2)(4)(3)} = \dfrac{1}{2}$

**107.** $-\dfrac{5}{12} \div \dfrac{45}{32} = -\dfrac{5}{12} \cdot \dfrac{32}{45}$

$\qquad = -\dfrac{5 \cdot 32}{12 \cdot 45} = -\dfrac{(5)(8)(4)}{(3)(4)(9)(5)} = -\dfrac{8}{27}$

**109.** $\dfrac{8}{3} \div \dfrac{8}{3} = \dfrac{8}{3} \cdot \dfrac{3}{8} = \dfrac{(8)(3)}{(3)(8)} = 1$

**111.** $\dfrac{3}{5} \div \dfrac{7}{5} = \left(\dfrac{3}{5}\right)\left(\dfrac{5}{7}\right) = \dfrac{(3)(5)}{(5)(7)} = \dfrac{3}{7}$

**113.** $-\dfrac{5}{6} \div \left(-\dfrac{8}{10}\right) = \dfrac{5}{6} \div \dfrac{8}{10}$

$= \dfrac{5}{6} \cdot \dfrac{10}{8}$

$= \dfrac{5 \cdot 10}{6 \cdot 8}$

$= \dfrac{5(\cancel{2})(5)}{\cancel{2}(3)(8)} = \dfrac{25}{24}$

**115.** $-10 \div \dfrac{1}{9} = -\dfrac{10}{1} \cdot \dfrac{9}{1} = -\dfrac{90}{1} = -90$

**117.** $0 \div (-21) = 0$

**119.** $\dfrac{3}{5} \div 0$ is undefined.

Division by zero is undefined.

**121.** $3\dfrac{3}{4} \div 1\dfrac{1}{2} = \dfrac{15}{4} \div \dfrac{3}{2}$

$= \dfrac{15}{4} \cdot \dfrac{2}{3}$

$= \dfrac{15(2)}{4(3)}$

$= \dfrac{5(\cancel{3})(\cancel{2})}{(2)(\cancel{2})(\cancel{3})}$

$= \dfrac{5}{2}$

**123.** $3\dfrac{1}{4} \div 2\dfrac{5}{8} = \dfrac{3(4)+1}{4} \div \dfrac{2(8)+5}{8}$

$= \dfrac{13}{4} \div \dfrac{21}{8}$

$= \dfrac{13}{4} \cdot \dfrac{8}{21}$

$= \dfrac{13(2)(\cancel{4})}{\cancel{4}(21)}$

$= \dfrac{26}{21}$

**125.** $\dfrac{1}{4} = 0.25$

$$
\begin{array}{r}
.25 \\
4\overline{)1.00} \\
\underline{8} \\
20 \\
\underline{20}
\end{array}
$$

**127.** $\dfrac{5}{16} = 0.3125$

$$
\begin{array}{r}
.3125 \\
16\overline{)5.0000} \\
\underline{48} \\
20 \\
\underline{16} \\
40 \\
\underline{32} \\
80 \\
\underline{80}
\end{array}
$$

**129.** $\dfrac{2}{9} = 0.\overline{2}$

$$
\begin{array}{r}
.22\ldots \\
9\overline{)2.00} \\
\underline{18} \\
20 \\
\underline{18} \\
2
\end{array}
$$

**131.** $\dfrac{5}{12} = 0.41\overline{6}$

$$
\begin{array}{r}
.4166\ldots \\
12\overline{)5.0000} \\
\underline{48} \\
20 \\
\underline{12} \\
80 \\
\underline{72} \\
80 \\
\underline{72} \\
8
\end{array}
$$

**133.** $\dfrac{4}{11} = 0.\overline{36}$

$$
\begin{array}{r}
.3636\ldots \\
11\overline{)4.0000} \\
\underline{33} \\
70 \\
\underline{66} \\
40 \\
\underline{33} \\
70 \\
\underline{66} \\
4
\end{array}
$$

**135.** $12.33 + 14.76 = 27.09$

**137.** $132.1 + (-25.45) = 106.65$

**139.** $1.21 + 4.06 - 3.00 = 2.27$

**141.** $-0.0005 - 2.01 + 0.111 = -1.8995 \approx -1.90$

**143.** $-6.3(9.05) \approx -57.02$

**145.** $-0.05(-85.95) = 4.2975 \approx 4.30$

**147.** $4.69 \div 0.12 \approx 39.08$ (rounded to two decimal places)

$$0.12)\overline{4.69} = 12)\overline{469} \quad \frac{39.0833...}{} \approx 39.08 \text{ (Rounded)}$$

$$\begin{array}{r} 36 \\ \hline 109 \\ 108 \\ \hline 100 \\ 96 \\ \hline 40 \\ 36 \\ \hline 40 \\ 36 \\ \hline 4 \end{array}$$

**149.** $1.062 \div (-2.1) = -0.51$

(rounded to two decimal places)

$$2.1)\overline{1.0620} \quad \frac{.505}{}$$

$$\begin{array}{r} 105 \\ \hline 12 \\ 0 \\ \hline 120 \\ 105 \\ \hline 15 \end{array}$$

**151.** $13,912.94 - 13,878.15 = 34.79$

The increase in the Dow Jones Industrial Average was 34.79 points.

**153.** $8\dfrac{3}{4} + 7\dfrac{1}{5} + 9\dfrac{3}{8} = \dfrac{35}{4} + \dfrac{36}{5} + \dfrac{75}{8}$

$$= \dfrac{35(10)}{4(10)} + \dfrac{36(8)}{5(8)} + \dfrac{75(5)}{8(5)}$$

$$= \dfrac{350}{40} + \dfrac{288}{40} + \dfrac{375}{40}$$

$$= \dfrac{1013}{40}$$

$$= 25\dfrac{13}{40}$$

So, $25\dfrac{13}{40}$ tons, or 25.325 tons, of feed were purchased during the first quarter of the year.

**155.** $\dfrac{2}{3} - \dfrac{5}{16} = \dfrac{2(16)}{3(16)} - \dfrac{5(3)}{16(3)}$

$$= \dfrac{32}{48} - \dfrac{15}{48}$$

$$= \dfrac{32 - 15}{48}$$

$$= \dfrac{17}{48}$$

So, $\dfrac{17}{48}$ of the work was completed during May.

**157.** $60 \div \dfrac{5}{4} = \dfrac{60}{1} \cdot \dfrac{4}{5}$

$$= \dfrac{60(4)}{5}$$

$$= \dfrac{12(\cancel{5})(4)}{\cancel{5}}$$

$$= 48$$

So, you can make 48 breadsticks.

**159.** The number of gallons needed to drive 12,000 miles in a car which gets 22.3 miles per gallon is

$$\dfrac{12,000}{22.3} \approx 538.117 \text{ gallons.}$$

At \$2.859 per gallon, the annual fuel cost is

$$(538.117)(2.859) \approx \$1538.48. \text{(Rounded)}$$

(Note: More accurate answers are obtained if you round your answer *only* after all calculations are done.)

**161.** The cost of the milk is $2(3.75) = \$7.50$ and the cost of the bread is $3(1.68) = \$5.04$. So, the total cost is $7.50 + 5.04 = \$12.54$. Your change is the difference $20 - 12.54 = \$7.46$.

**163.** (a) Two hundred times 23.63 is approximately 5000. Three hundred times 86.25 is approximately 26,000. So, the total cost of the stock is approximately \$31,000.

Note: Answers will vary.

(b) $200(23.63) + 300(86.25) = 30,601$

So, the actual total cost of the stocks is \$30,601.

**165.** $\dfrac{2.859 + 2.969 + 3.079}{3} = \dfrac{8.907}{3} = 2.969$

So, the average price per gallon is 2.969.

**167.** No, it is not true.

Here are two fractions with the same sign, and their sum is negative.

$$\dfrac{-3}{4} + \dfrac{-1}{8} = -\dfrac{7}{8}$$

**169.** If two fractions have the same sign, their product is positive. If two fractions have opposite signs, their product is negative.

**171.** $3 = \dfrac{3}{1} = \dfrac{3(4)}{1(4)} = \dfrac{12}{4}$

Using the diagram, you can count that the number of one-fourths in 3 is 12.

You can also divide 3 by $\dfrac{1}{4}$ to find this same result.

$\dfrac{3}{\frac{1}{4}} = \dfrac{3}{1} \div \dfrac{1}{4} = \dfrac{3}{1} \cdot \dfrac{4}{1} = 12$

**173.** First method—rounding after calculations are done

$5.24(3.03)(2.749) = 43.6464228 \approx 43.6$

Second method—rounding first

$5.24(3.03)(2.749) \approx 5.2(3.0)(2.7) \approx 42.12$

Rounding after the calculations are done produces the more accurate answer.

**175.** True.

The reciprocal of $a/b$ is $b/a$, $a \neq 0$.

Fractions are rational numbers.

**177.** False.

The product of these two positive fractions is less than either factor: $\frac{1}{2} \cdot \frac{1}{4} = \frac{1}{8}$

**179.** False.

For example, $6 > 0$ and $8 > 0$, but $6 - 8 < 0$ because $6 - 8 = -2$.

**181.** $\dfrac{3}{6} + \dfrac{4}{5} = \dfrac{3(5)}{6(5)} + \dfrac{4(6)}{5(6)} = \dfrac{15 + 24}{30} = \dfrac{39}{30} = \dfrac{\cancel{3}(13)}{\cancel{3}(10)} = \dfrac{13}{10}$

## Section 1.5   Exponents, Order of Operations, and Properties of Real Numbers

**1.** $2 \cdot 2 \cdot 2 \cdot 2 \cdot 2 = 2^5$

**3.** $(-5)(-5)(-5)(-5) = (-5)^4$

**5.** $\left(-\frac{1}{4}\right) \cdot \left(-\frac{1}{4}\right) = \left(-\frac{1}{4}\right)^2$

**7.** $-\left[(1.6) \cdot (1.6) \cdot (1.6) \cdot (1.6) \cdot (1.6)\right] = -(1.6)^5$

**9.** $(-3)^6 = (-3)(-3)(-3)(-3)(-3)(-3)$

**11.** $\left(\frac{3}{8}\right)^5 = \left(\frac{3}{8}\right)\left(\frac{3}{8}\right)\left(\frac{3}{8}\right)\left(\frac{3}{8}\right)\left(\frac{3}{8}\right)$

**13.** $\left(-\frac{1}{2}\right)^7 = \left(-\frac{1}{2}\right)\left(-\frac{1}{2}\right)\left(-\frac{1}{2}\right)\left(-\frac{1}{2}\right)\left(-\frac{1}{2}\right)\left(-\frac{1}{2}\right)\left(-\frac{1}{2}\right)$

**15.** $-(9.8)^3 = -\left[(9.8)(9.8)(9.8)\right]$

**17.** $3^2 = 3 \cdot 3 = 9$

**19.** $2^6 = (2)(2)(2)(2)(2)(2) = 64$

**21.** $\left(\frac{1}{4}\right)^3 = \left(\frac{1}{4}\right)\left(\frac{1}{4}\right)\left(\frac{1}{4}\right) = \frac{1}{64}$

**23.** $(-5)^3 = (-5)(-5)(-5) = -125$

**25.** $-4^2 = -(4 \cdot 4) = -16$

**27.** $(-1.2)^3 = (-1.2)(-1.2)(-1.2) = -1.728$

**29.** $4 - 6 + 10 = -2 + 10 = 8$

**31.** $5 - (8 - 15) = 5 - (-7) = 5 + 7 = 12$

**33.** $17 - |2 - (6 + 5)| = 17 - |2 - 11|$
$= 17 - |-9|$
$= 17 - 9$
$= 8$

**35.** $15 + 3 \cdot 4 = 15 + 12 = 27$

**37.** $25 - 32 \div 4 = 25 - 8 = 17$

**39.** $(16 - 5) \div (3 - 5) = 11 \div (-2) = -\frac{11}{2}$

**41.** $(10 - 16) \cdot (20 - 26) = (-6) \cdot (-6) = 36$

**43.** $(45 \div 10) \cdot 2 = (4.5)(2) = 9$

**45.** $\left[360 - (8 + 12)\right] \div 5 = \left[360 - 20\right] \div 5$
$= 340 \div 5$
$= 68$

**47.** $5 + (2^2 \cdot 3) = 5 + (4 \cdot 3) = 5 + 12 = 17$

**49.** $(-6)^2 - (48 \div 4^2) = 36 - (48 \div 16) = 36 - 3 = 33$

**51.** $\left(3 \cdot \dfrac{5}{9}\right) + 1 - \dfrac{1}{3} = \left(\dfrac{3}{1} \cdot \dfrac{5}{9}\right) + \dfrac{1}{1} - \dfrac{1}{3}$

$\qquad = \dfrac{(\cancel{3})(5)}{(\cancel{3})(3)} + \dfrac{1}{1} - \dfrac{1}{3}$

$\qquad = \dfrac{5}{3} + \dfrac{3}{3} - \dfrac{1}{3} = \dfrac{7}{3}$

**53.** $18\left(\dfrac{1}{2} + \dfrac{2}{3}\right) = 18\left(\dfrac{3}{6} + \dfrac{4}{6}\right)$

$\qquad = 18\left(\dfrac{3+4}{6}\right)$

$\qquad = \dfrac{18}{1}\left(\dfrac{7}{6}\right)$

$\qquad = \dfrac{3(\cancel{6})(7)}{1(\cancel{6})}$

$\qquad = 21$

**55.** $\dfrac{7}{25}\left(\dfrac{7}{16} - \dfrac{1}{8}\right) = \dfrac{7}{25}\left(\dfrac{7}{16} - \dfrac{(1)(2)}{(8)(2)}\right)$

$\qquad = \dfrac{7}{25}\left(\dfrac{7}{16} - \dfrac{2}{16}\right)$

$\qquad = \dfrac{7}{25}\left(\dfrac{5}{16}\right)$

$\qquad = \dfrac{(7)(\cancel{5})}{(\cancel{5})(5)(16)} = \dfrac{7}{80}$

**57.** $\dfrac{7}{3}\left(\dfrac{2}{3}\right) \div \dfrac{28}{15} = \dfrac{14}{9} \div \dfrac{28}{15}$

$\qquad = \dfrac{14}{9} \cdot \dfrac{15}{28}$

$\qquad = \dfrac{\cancel{14}(\cancel{3})(5)}{3(\cancel{3})(2)(\cancel{14})}$

$\qquad = \dfrac{5}{6}$

**59.** $\dfrac{3 + \left[15 \div (-3)\right]}{16} = \dfrac{3 + (-5)}{16} = \dfrac{-2}{16} = -\dfrac{1(\cancel{2})}{\cancel{2}(8)} = -\dfrac{1}{8}$

**61.** $\dfrac{1 - 3^2}{-2} = \dfrac{1 - 9}{-2} = \dfrac{-8}{-2} = 4$

**63.** $\dfrac{7^2 - 4^2}{0}$ is undefined.

Division by zero is undefined.

**65.** $\dfrac{0}{6^2 + 1} = \dfrac{0}{6 \cdot 6 + 1} = \dfrac{0}{36 + 1} = \dfrac{0}{37} = 0$

**67.** $\dfrac{5^2 + 12^2}{13} = \dfrac{25 + 144}{13} = \dfrac{169}{13} = 13$

**69.** $\dfrac{3 \cdot 6 - 4 \cdot 6}{5 + 1} = \dfrac{18 - 24}{6} = \dfrac{-6}{6} = -1$

**71.** $7 - \dfrac{4 + 6}{2^2 + 1} + 5 = 7 - \dfrac{4 + 6}{4 + 1} + 5$

$\qquad = 7 - \dfrac{10}{5} + 5$

$\qquad = 7 - 2 + 5$

$\qquad = 10$

**73.** $300\left(1 + \dfrac{0.1}{12}\right)^{24} \approx 366.12$

(Rounded to two decimal places)

**75.** $\dfrac{1.32 + 4(3.68)}{1.5} = \dfrac{1.32 + 14.72}{1.5} = \dfrac{16.04}{1.5} \approx 10.69$

(Rounded to two decimal places)

**77.** Commutative Property of Multiplication

**79.** Commutative Property of Addition

**81.** Distributive Identity Property

**83.** Additive Inverse Property

**85.** Associative Property of Addition

**87.** Distributive Property

**89.** Multiplicative Inverse Property

**91.** Additive Identity Property

**93.** Distributive Property

**95.** Associative Property of Multiplication

**97.** $18 + 5 = 5 + 18$

**99.** $10(-3) = (-3)10$

**101.** $6(19 + 2) = 6 \cdot 19 + 6 \cdot 2$

**103.** $3 \cdot 4 + 5 \cdot 4 = (3 + 5)4$

**105.** $18 + (12 + 9) = (18 + 12) + 9$

**107.** $12(3 \cdot 4) = (12 \cdot 3)4$

**109.** (a) Additive inverse: $-50$

(b) Multiplicative inverse: $\dfrac{1}{50}$

**111.** (a) Additive inverse: 1

(b) Multiplicative inverse: $-1$

**113.** (a) Additive inverse: $\frac{1}{2}$

(b) Multiplicative inverse: $(-2)$

**115.** (a) Additive inverse: $-0.2$

(b) Multiplicative inverse: $\frac{10}{2} = 5$

**121.** $7 \cdot 4 + 9 + 2 \cdot 4 = 7 \cdot 4 + 2 \cdot 4 + 9$     Commutative Property of Addition

$= (7 \cdot 4 + 2 \cdot 4) + 9$     Associative Property of Addition

$= (7 + 2)4 + 9$     Distributive Property

$= 9 \cdot 4 + 9$     Add.

$= 9(4 + 1)$     Distributive Property

$= 9(5)$     Add.

$= 45$     Multiply.

**123.** $\left(\frac{7}{9} + 6\right) + \frac{2}{9} = \frac{7}{9} + \left(6 + \frac{2}{9}\right)$     Associative Property of Addition

$= \frac{7}{9} + \left(\frac{2}{9} + 6\right)$     Commutative Property of Addition

$= \left(\frac{7}{9} + \frac{2}{9}\right) + 6$     Associative Property of Addition

$= 1 + 6$     Add.

$= 7$     Add.

**125.** $4 \cdot 4 \cdot 4 \cdot 4 = 4^4 = 256$

The truck can transport 256 propane tanks.

**127.** The total area of the figure is the *sum* of the area of the upper rectangle and the area of the lower rectangle. The upper rectangle has length 3 and width 3. Its area is $3 \cdot 3 = 9$. The lower rectangle has length 9 and width 3. Its area is $9 \cdot 3 = 27$. Therefore, the total area is $9 + 27 = 36$ square units.

Area $= 3 \cdot 3 + 9 \cdot 3 = 9 + 27 = 36$ square units

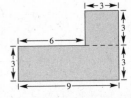

**129.** (a) $35.95 + 0.06(35.95) = 35.95(1 + 0.06)$

$= 35.95(1.06)$

(b) $35.95(1.06) = 38.107 \approx 38.11$

You must pay \$38.11.

**131.** (a) Area $= 30(30 - 8)$

(b) Area $= 30(30) - 30(8)$

(c) Area $= 660$ square feet

The area of the movie screen is 660 square feet.

**133.** Perimeter $= (8 - 2) + (3 + 11) + (2 \cdot 6 + 3)$

$= 6 + 14 + 15$

$= 35$

**117.** (a) $3(6 + 10) = 3(6) + 3(10) = 18 + 30 = 48$

(b) $3(6 + 10) = 3(16) = 48$

**119.** (a) $\frac{2}{3}(9 + 24) = \frac{2}{3} \cdot 9 + \frac{2}{3} \cdot 24 = 6 + 16 = 22$

(b) $\frac{2}{3}(9 + 24) = \frac{2}{3}(33) = 22$

**135.** No; the order in which these two activities are performed does affect the results.

**137.** No

$-6^2 = -(6)(6) = -36$

$(-6) = (-6)(-6) = 36$

So, $-6^2 \neq (-6)^2$.

**139.** $4 \cdot 6^2 = 4 \cdot 6 \cdot 6 = 144$

$24^2 = 24 \cdot 24 = 576$

So, $4 \cdot 6^2 \neq 24^2$.

Note: $24^2 = (4 \cdot 6)^2 = 4^2 \cdot 6^2$, *not* $4 \cdot 6^2$.

**141.** $4 - (6 - 2) = 4 - 4 = 0$

$4 - 6 - 2 = -2 - 2 = -4$

So, $4 - (6 - 2) \neq 4 - 6 - 2$.

**143.** $100 \div 2 \times 50 = 50 \times 50 = 2500$

So, $100 \div 2 \times 50 \neq 1$.

**145.** $5(7 + 3) = 5(7) + 5(3) = 35 + 15$

So, $5(7 + 3) \neq 5(7) + 3$. The 5 should be multiplied by *both* terms in the parentheses.

**147.** $\frac{8}{0}$ is undefined because division by 0 is undefined. So, $\frac{8}{0} \neq 0$.

**149.** The error was in the first step where the fraction was simplified incorrectly.

$-9 + \frac{9 + 20}{3(5)} - (-3) = -9 + \frac{29}{15} + 3$

$= -6 + \frac{29}{15}$

$= \frac{-90 + 29}{15}$

$= -\frac{61}{15}$

**151.** $(6 + 2) \cdot (5 + 3) = 8 \cdot 8 = 64$

$(6 + 2) \cdot 5 + 3 = 8 \cdot 5 + 3 = 40 + 3 = 43$

$6 + 2 \cdot 5 + 3 = 6 + 10 + 3 = 19$

$6 + 2 \cdot (5 + 3) = 6 + 2 \cdot 8 = 6 + 16 = 22$

## Review Exercises for Chapter 1

**1.** (a) natural numbers: $\{\sqrt{4}\}$

(b) integers: $\{-1, \sqrt{4}\}$

(c) rational numbers: $\{-1, 4.5, \frac{2}{5}, -\frac{1}{7}, \sqrt{4}\}$

(d) irrational numbers: $\{\sqrt{5}\}$

**3.** (a) natural numbers: $\{\frac{30}{2}, 2\}$

(b) integers: $\{\frac{30}{2}, 2\}$

(c) rational numbers: $\{\frac{30}{2}, 2, 1.5, -\frac{10}{7}\}$

(d) irrational numbers: $\{-\sqrt{3}, -\pi\}$

**153.** (a) Area of region I: $2(2)$ square units

Area of region II: $3(2)$ square units

Total area of rectangle:

$2(2) + 3(2) = 4 + 6 = 10$ square units

(b) Length of rectangle: $2 + 3$

Width of rectangle: 2 units

Area of rectangle: $(2 + 3)(2) = 5(2)$

$= 10$ square units

(c) $2(2) + 3(2) = (2 + 3)(2)$ (This is an application of the Distributive Property)

$= 5(2)$

$= 10$

**155.** One method: The area of the shaded rectangle can be found by multiplying its width of 11 and its height of $7 - 3$.

$11(7 - 3) = 11(4) = 44$

Another method: The area of the shaded rectangle can be found by finding the area of the entire largest rectangle and subtracting the area of the rectangle that is not shaded.

$11(7) - 11(3) = 77 - 33 = 44$

By the Distributive Property, the two expressions for area, $11(7 - 3)$ and $11(7) - 11(3)$, are equivalent.

**5.**

**7.**

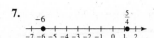

**9.**

**11.** $-\frac{1}{10} < 4$ because $-\frac{1}{10}$ lies to the *left* of 4.

**13.** $-3 > -7$ because $-3$ lies to the *right* of $-7$.

**15.** $5 > \frac{7}{2}$ because 5 lies to the *right* of $\frac{7}{2}$.

**17.** The smaller number is 0.6.

Note: $\frac{2}{3} = 0.666...$ and $0.666... > 0.6$.

**19.** The opposite of 152 is $-152$. The distance from 152 to 0 is 152 units, and the distance from $-152$ to 0 is also 152 units.

**21.** The opposite of $-\frac{7}{3}$ is $\frac{7}{3}$. The distance from $-\frac{7}{3}$ to 0 is $\frac{7}{3}$ units, and the distance from $\frac{7}{3}$ to 0 is also $\frac{7}{3}$ units.

**23.** $|-8.5| = 8.5$

**25.** $|3.4| = 3.4$

**27.** $-|-6.2| = -6.2$

Note: $|-6.2| = 6.2$

**29.** $-\left|\frac{8}{5}\right| = -\frac{8}{5}$

**31.** $|-84| = |84|$ because $|-84| = 84$ and $|84| = 84$.

**33.** $\left|\frac{5}{2}\right| > \left|\frac{8}{9}\right|$ because $\left|\frac{5}{2}\right| = \frac{5}{2}$, $\left|\frac{8}{9}\right| = \frac{8}{9}$, and $\frac{5}{2} > \frac{8}{9}$

**35.** $\left|\frac{3}{10}\right| > -\left|\frac{4}{5}\right|$ because $\left|\frac{3}{10}\right| = \frac{3}{10}$, $-\left|\frac{4}{5}\right| = -\frac{4}{5}$, and $\frac{3}{10} > -\frac{4}{5}$.

**37.** The number 7 units to the right of 5 is 12 because $5 + 7 = 12$.

The number 7 units to the left of 5 is $-2$ because $5 - 7 = -2$.

**39.** The number 5 units to the right of 2.6 is 7.6 because $2.6 + 5 = 7.6$.

The number 5 units to the left of 2.6 is $-2.4$ because $2.6 - 5 = -2.4$.

**41.** $4 + 3 = 7$

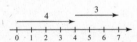

**43.** $-1 + (-4) = -5$

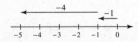

**45.** $16 + (-5) = 11$

**47.** $-125 + 30 = -95$

**49.** $(-13) + (-76) = -89$

**51.** $-10 + 21 + (-6) = 5$

**53.** $-17 + (-3) + (-9) = -29$

**55.** $95,000 - 64,400 + 51,800 = \$82,400$

The company's profit for the three months was $82,400.

**57.** The sum can be positive or negative. The sign is determined by the integer with the greater absolute value.

**59.** $28 - 7 = 21$

**61.** $8 - 15 = -7$

**63.** $14 - (-19) = 33$

**65.** $-18 - 4 = -22$

**67.** $-12 - (-7) - 4 = -9$

**69.** $613 - (-549) = 1162$

**71.** $1560 - 50 - 255 - 490 = 765$

The balance at the end of the month was $765.

**73.** $15 \times 3 = 45$

**75.** $-3 \cdot 24 = -72$

**77.** $6(-8) = -48$

**79.** $-5(-9) = 45$

**81.** $3(-6)(3) = -54$

**83.** $-4(-5)(-2) = -40$

**85.** $150 \times 12 \times 2 = 3600$

The total amount saved during the two years is $3600.

**87.** $72 \div 8 = 9$

**89.** $\dfrac{-72}{6} = -12$

**91.** $75 \div (-5) = -15$

**93.** $\dfrac{-52}{-4} = 13$

**95.** $0 \div 815 = 0$

**97.** $135 \div 0$ is undefined.

Division by zero is undefined.

**99.** $\frac{195}{3} = 65$

The average speed of the train is 65 miles per hour.

**101.** 137 is prime; the divisibility tests yield no factors of 137. By testing the remaining primes less than $\sqrt{137} \approx 11.7$, you can conclude that 137 is a prime number.

**103.** 839 is prime: the divisibility tests yield no factors of 839.

By testing the remaining primes less than or equal to $\sqrt{839} \approx 29$, you can conclude that 839 is a prime number.

**105.** 1764 is composite; its prime factorization is $1764 = 2 \cdot 2 \cdot 3 \cdot 3 \cdot 7 \cdot 7$.

**107.** $264 = 2 \cdot 2 \cdot 2 \cdot 3 \cdot 11$

**109.** $378 = 2 \cdot 3 \cdot 3 \cdot 3 \cdot 7$

**111.** $1612 = 2 \cdot 2 \cdot 13 \cdot 31$

**113.** $12 \times (-3) = -36$

**115.** $|-7| = 7$

**117.** By prime factorization, $54 = 2 \cdot 3 \cdot 3 \cdot 3$ and $90 = 2 \cdot 3 \cdot 3 \cdot 5$. So, the greatest common factor is $2 \cdot 3 \cdot 3$, or 18.

**119.** The first number, 2, is a prime number. By prime factorization, $6 = 2 \cdot 3$ and $9 = 3 \cdot 3$. The three numbers have no common factor other than 1. So the greatest common factor is 1.

**121.** By prime factorization, $63 = 3 \cdot 3 \cdot 7$, $84 = 2 \cdot 2 \cdot 3 \cdot 7$, and $441 = 3 \cdot 3 \cdot 7 \cdot 7$. So, the greatest common factor is $3 \cdot 7$, or 21.

**123.** $\frac{3}{12} = \frac{1(\cancel{3})}{2(2)(\cancel{3})} = \frac{1}{4}$

**125.** $\frac{30}{48} = \frac{5(\cancel{2})(\cancel{3})}{2(2)(2)(\cancel{2})(\cancel{3})} = \frac{5}{8}$

**127.** $\frac{2}{3} = \frac{2(5)}{3(5)} = \frac{10}{15}$

**129.** $\frac{6}{10} = \frac{3(\cancel{2})}{5(\cancel{2})} = \frac{3}{5} = \frac{3(5)}{5(5)} = \frac{15}{25}$

**131.** $\frac{3}{25} + \frac{7}{25} = \frac{3 + 7}{25} = \frac{10}{25} = \frac{(2)(\cancel{5})}{(5)(\cancel{5})} = \frac{2}{5}$

**133.** $\frac{27}{16} - \frac{15}{16} = \frac{27 - 15}{16} = \frac{12}{16} = \frac{(3)(\cancel{4})}{(4)(\cancel{4})} = \frac{3}{4}$

**135.** $\frac{3}{8} + \frac{1}{2} = \frac{3}{8} + \frac{1(4)}{2(4)} = \frac{3 + 4}{8} = \frac{7}{8}$

**137.** $-\frac{5}{9} + \frac{2}{3} = \frac{-5}{9} + \frac{2(3)}{3(3)} = \frac{-5}{9} + \frac{6}{9} = \frac{-5 + 6}{9} = \frac{1}{9}$

**139.** $-\frac{25}{32} + \left(-\frac{7}{24}\right) = -\frac{25(3)}{32(3)} - \frac{7(4)}{24(4)}$

$$= -\frac{75}{96} - \frac{28}{96} = \frac{-75 - 28}{96} = -\frac{103}{96}$$

**141.** $5 - \frac{15}{4} = \frac{5(4)}{1(4)} - \frac{15}{4}$

$$= \frac{20}{4} - \frac{15}{4}$$

$$= \frac{20 - 15}{4} = \frac{5}{4}$$

**143.** $5\frac{3}{4} - 3\frac{5}{8} = \frac{23}{4} - \frac{29}{8}$

$$= \frac{23(2)}{4(2)} - \frac{29}{8}$$

$$= \frac{46}{8} - \frac{29}{8} = \frac{46 - 29}{8} = \frac{17}{8}$$

**145.** $\frac{3}{8} + \frac{1}{2} + \frac{1}{8} + 1\frac{1}{4} + \frac{1}{2} = \frac{3 + 4 + 1 + 10 + 4}{8}$

$$= \frac{22}{8} = 2\frac{3}{4}$$

The total rainfall for the five days was $2\frac{3}{4}$ inches.

**147.** $\frac{5}{8} \cdot \frac{-2}{15} = \frac{5(-2)}{8 \cdot 15} = -\frac{(1)(\cancel{5})(\cancel{2})}{(4)(\cancel{2})(3)(\cancel{5})} = -\frac{1}{12}$

**149.** $35\left(\frac{1}{35}\right) = \frac{35}{1} \cdot \frac{1}{35} = \frac{35 \cdot 1}{1 \cdot 35} = \frac{(\cancel{35})(1)}{(1)(\cancel{35})} = 1$

**151.** $\frac{3}{8} \cdot \left(-\frac{2}{27}\right) = -\frac{3(2)}{8(27)} = -\frac{\cancel{3}(\cancel{2})}{\cancel{2}(4)(\cancel{3})(9)} = -\frac{1}{36}$

**153.** $\dfrac{5}{14} \div \dfrac{15}{28} = \dfrac{5}{14} \cdot \dfrac{28}{15} = \dfrac{5 \cdot 28}{14 \cdot 15} = \dfrac{(\cancel{5})(\cancel{14})(2)}{(\cancel{14})(\cancel{5})(3)} = \dfrac{2}{3}$

**155.** $-\dfrac{3}{4} \div \left(-\dfrac{7}{8}\right) = \dfrac{3}{4} \cdot \dfrac{8}{7} = \dfrac{3 \cdot 8}{4 \cdot 7} = \dfrac{(3)(2)(\cancel{4})}{(\cancel{4})(7)} = \dfrac{6}{7}$

**157.** $-\dfrac{5}{9} \div 0$ is undefined. Division by zero is undefined.

**159.** $-5 \cdot 0 = 0$

**161.** The average rate of snowfall is the quotient of the amount of snow that fell and the time.

$$\dfrac{6\frac{3}{4}}{8} = \dfrac{\frac{27}{4}}{8} = \dfrac{27}{4} \cdot \dfrac{1}{8} = \dfrac{(27)(1)}{(4)(8)} = \dfrac{27}{32}$$

The average rate of snowfall was $\dfrac{27}{32}$ inches per hour.

**163.** $\frac{5}{8} = 0.625$

```
     .625
8)5.000
   48
   ──
   20
   16
   ──
    40
    40
    ──
```

**165.** $\frac{8}{15} = 0.5\overline{3}$

```
      .533...
15)8.000
   75
   ──
   50
   45
   ──
    50
    45
    ──
     5
```

**167.** $4.89 + 0.76 = 5.65$

**169.** $3.815 - 5.19 \approx -1.38$

**171.** $(1.49)(-0.5) \approx -0.75$

**173.** $5.25 \div 0.25 = 21$

**175.** $600 + 200(8 - 4) = 600 + 200(4)$
$= 600 + 800$
$= 1400$

The cost to hire the DJ for an 8-hour event is \$1400.

**177.** $6 \cdot 6 \cdot 6 \cdot 6 \cdot 6 = 6^5$

**179.** $\left(\frac{6}{7}\right) \cdot \left(\frac{6}{7}\right) \cdot \left(\frac{6}{7}\right) \cdot \left(\frac{6}{7}\right) = \left(\frac{6}{7}\right)^4$

**181.** $(-7)^4 = (-7)(-7)(-7)(-7)$

**183.** $(1.25)^3 = (1.25) \cdot (1.25) \cdot (1.25)$

**185.** $2^4 = 16$

**187.** $\left(-\dfrac{3}{4}\right)^3 = -\dfrac{3^3}{4^3} = -\dfrac{27}{64}$

**189.** $-7^2 = -(7 \cdot 7) = -49$

**191.** $12 - 2 \cdot 3 = 12 - 6 = 6$

**193.** $18 \div 6 \cdot 7 = 3 \cdot 7 = 21$

**195.** $20 + \left(8^2 \div 2\right) = 20 + (64 \div 2) = 20 + 32 = 52$

**197.** $240 - \left(4^2 \cdot 5\right) = 240 - (16 \cdot 5) = 240 - 80 = 160$

**199.** $3^2(5 - 2)^2 = 9(3)^2 = 9(9) = 81$

**201.** $\dfrac{3}{4}\left(\dfrac{5}{6}\right) + 4 = \dfrac{3 \cdot 5}{4 \cdot 6} + 4$
$= \dfrac{(\cancel{3})(5)}{(4)(2)(\cancel{3})} + 4$
$= \dfrac{5}{8} + \dfrac{4}{1} = \dfrac{5}{8} + \dfrac{4(8)}{1(8)}$
$= \dfrac{5}{8} + \dfrac{32}{8}$
$= \dfrac{5 + 32}{8}$
$= \dfrac{37}{8}$

**203.** $122 - \left[45 - (32 + 8) - 23\right] = 122 - \left[45 - 40 - 23\right]$
$= 122 - \left[-18\right]$
$= 122 + 18$
$= 140$

**205.** $\dfrac{6 \cdot 4 - 36}{4} = \dfrac{24 - 36}{4} = \dfrac{-12}{4} = -3$

**207.** $\dfrac{54 - 4 \cdot 3}{6} = \dfrac{54 - 12}{6} = \dfrac{42}{6} = 7$

**209.** $\dfrac{78 - |-78|}{5} = \dfrac{78 - 78}{5} = \dfrac{0}{5} = 0$

**211.** $(5.8)^4 - (3.2)^5 = 1131.6496 - 335.54432$
$$= 796.10528 \approx 796.11$$

**213.** $\dfrac{3000}{(1.05)^{10}} \approx 1841.739761 \approx 1841.74$

**215.** (a) $25,000\left(\frac{3}{4}\right)^3 = 25,000 \cdot \frac{3}{4} \cdot \frac{3}{4} \cdot \frac{3}{4}$
$$= 10,546.875$$
$$\approx 10,546.88$$

The value of the car after 3 years is approximately $10,546.88.

(b) $25,000 - 10,546.88 = 14,453.12$

During the 3 years, the car depreciated $14,453.12.

**217.** Additive Inverse Property

**219.** Commutative Property of Multiplication

**221.** Multiplicative Identity Property

# Chapter Test for Chapter 1

**1.** (a) natural numbers:    $\{4\}$

(b) integers:    $\{4, -6, 0\}$

(c) rational numbers:    $\left\{4, -6, \frac{1}{2}, 0, \frac{7}{9}\right\}$

(d) Irrational numbers:    $\{\pi\}$

**2.** $-\frac{3}{5} > -|-2|$

**3.** $|-13| = 13$

**4.** $-|-6.8| = -6.8$

Note: $|-6.8| = 6.8$

**5.** $16 + (-20) = -4$

**6.** $-50 - (-60) = -50 + 60 = 10$

**7.** $7 + |-3| = 7 + 3 = 10$

**8.** $64 - (25 - 8) = 64 - 17 = 47$

**9.** $-5(32) = -160$

**10.** $\dfrac{-72}{-9} = 8$

**11.** $\dfrac{15(-6)}{3} = \dfrac{-90}{3} = -30$

**223.** Distributive Property

**225.** $-16 + 0 = -16$

**227.** $24 + 1 = 1 + 24$

**229.** One method:  The area of the shaded rectangle can be found by multiplying its height of 6 and its width of $18 - 5$.

$$6(18 - 5) = 6(13) = 78$$

Another method:  The area of the shaded rectangle can be found by finding the area of the entire largest rectangle and subtracting the area of the rectangle that is not shaded.

$$6(18) - 6(5) = 108 - 30 = 78$$

By the Distributive Property, the two expressions for area, $6(18 - 5)$ and $6(18) - 6(5)$, are equivalent.

**12.** $-\dfrac{(-2)(5)}{10} = -\dfrac{-10}{10} = -(-1) = 1$

**13.** $\dfrac{5}{6} - \dfrac{1}{8} = \dfrac{5(4)}{6(4)} - \dfrac{1(3)}{8(3)}$
$$= \dfrac{20}{24} - \dfrac{3}{24}$$
$$= \dfrac{20 - 3}{24}$$
$$= \dfrac{17}{24}$$

**14.** $-\dfrac{9}{50}\left(-\dfrac{20}{27}\right) = \dfrac{9(20)}{50(27)} = \dfrac{\cancel{9}(2)(\cancel{10})}{5(\cancel{10})(\cancel{9})(3)} = \dfrac{2}{15}$

**15.** $\dfrac{7}{16} \div \dfrac{21}{28} = \dfrac{7}{16} \cdot \dfrac{28}{21} = \dfrac{7 \cdot 28}{16 \cdot 21} = \dfrac{(\cancel{7})(7)(\cancel{4})}{(4)(\cancel{4})(3)(\cancel{7})} = \dfrac{7}{12}$

**16.** $\dfrac{-8.1}{0.3} = -\dfrac{8.1}{0.3} = -27$

$$0.3\overline{)8.1} = 3\overline{)81}\;^{27}$$
$$\phantom{0.3\overline{)8.1} = 3\overline{)}}\underline{6}$$
$$\phantom{0.3\overline{)8.1} = 3\overline{)}}21$$
$$\phantom{0.3\overline{)8.1} = 3\overline{)}}\underline{21}$$

**17.** $-(0.8)^2 = -0.64$

**18.** $35 - (50 \div 5^2) = 35 - (50 \div 25) = 35 - 2 = 33$

**19.**  $5(3 + 4)^2 - 10 = 5(7)^2 - 10$

$$= 5(49) - 10$$
$$= 245 - 10$$
$$= 235$$

**20.**  $18 - 7 \cdot 4 + 2^3 = 18 - 28 + 8 = -2$

**21.**  Distributive Property

**22.**  Multiplicative Inverse Property

**23.**  Associative Property of Addition

**24.**  Commutative Property of Multiplication

**25.**  $\dfrac{36}{162} = \dfrac{2(\not9)(2)}{2(\not9)(9)} = \dfrac{2}{9}$

**26.**  $216 = 2 \cdot 2 \cdot 2 \cdot 3 \cdot 3 \cdot 3$

**27.**  $\dfrac{1218}{21} = 58$

The average speed of the railway is 58 feet per second.

**28.**  Total cost:  $5(1.49) + 2(3.06) = 7.45 + 6.12 = 13.57$

Change:  $20 - 13.57 = 6.43$

You will receive $6.43 in change.

# CHAPTER 2
# Fundamentals of Algebra

# CHAPTER 2
## Fundamentals of Algebra

### Section 2.1   Writing and Evaluating Algebraic Expressions

1. $7.55w$

   The variable quantity is the number of hours worked, and this quantity is represented by the letter $w$. If the income is $7.55 per hour, the income earned is $7.55w$.

3. $3.79m$

   The variable quantity is the number of pounds of meat, and this quantity is represented by the letter $m$. If the cost per pound is $3.79, then the cost of the meat is $3.79m$.

5. Variable: $x$

7. Variables: $m, n$

9. Variable: $k$

11. $4x, 3$

13. $6x, -1$

15. $\frac{5}{3}, -3y^3$

17. $a^2, 4ab, b^2$

19. $3(x + 5), 10$

21. $15, \dfrac{5}{x}$

23. $\dfrac{3}{x + 2}, -3x, 4$

25. $14$

27. $-\frac{1}{3}$

29. $\frac{2}{5}$

31. $2\pi$

33. $3.06$

35. $y^5 = y \cdot y \cdot y \cdot y \cdot y$

37. $2^2 x^4 = 2 \cdot 2 \cdot x \cdot x \cdot x \cdot x$

39. $4y^2 z^3 = 4 \cdot y \cdot y \cdot z \cdot z \cdot z$

41. $\left(a^2\right)^3 = a^2 \cdot a^2 \cdot a^2 = a \cdot a \cdot a \cdot a \cdot a \cdot a$

43. $-4x^3 \cdot x^4 = -4 \cdot x \cdot x \cdot x \cdot x \cdot x \cdot x \cdot x$

45. $-9(ab)^3 = -9(ab)(ab)(ab) = -9 \cdot a \cdot a \cdot a \cdot b \cdot b \cdot b$

47. $(x + y)^2 = (x + y)(x + y)$

49. $\left(\dfrac{a}{3s}\right)^4 = \left(\dfrac{a}{3s}\right)\left(\dfrac{a}{3s}\right)\left(\dfrac{a}{3s}\right)\left(\dfrac{a}{3s}\right)$

51. $\left[2(a - b)^3\right]\left[2(a - b)^2\right] = 2(a - b)(a - b)(a - b) \cdot 2(a - b)(a - b) = 2 \cdot 2(a - b)(a - b)(a - b)(a - b)(a - b)$

53. $-2 \cdot u \cdot u \cdot u \cdot u = -2u^4$

55. $(2u) \cdot (2u) \cdot (2u) \cdot (2u) = (2u)^4$

    $(2$ is a factor of the base.$)$

57. $-a \cdot (-a) \cdot (-a) \cdot b \cdot b = (-a)^3 b^2$

59. $-3 \cdot (x - y) \cdot (x - y) \cdot (-3) \cdot (-3) = (-3)^3 (x - y)^2$

61. $\dfrac{x + y}{4} \cdot \dfrac{x + y}{4} \cdot \dfrac{x + y}{4} = \left(\dfrac{x + y}{4}\right)^3$

63. (a) When $x = \frac{1}{2}$, the value of $2x - 1$ is

   $2\left(\frac{1}{2}\right) - 1 = 1 - 1 = 0.$

   (b) When $x = -4$, the value of $2x - 1$ is

   $2(-4) - 1 = -8 - 1 = -9.$

65. (a) When $x = -2$, the value of $2x^2 - 5$ is

   $2(-2)^2 - 5 = 2(4) - 5 = 8 - 5 = 3.$

   (b) When $x = 3$, the value of $2x^2 - 5$ is

   $2(3)^2 - 5 = 2(9) - 5 = 18 - 5 = 13.$

**67.** (a) When $x = 4$ and $y = 3$, the value of $3x - 2y$ is $3(4) - 2(3) = 12 - 6 = 6$.

(b) When $x = \frac{2}{3}$ and $y = -1$, the value of $3x - 2y$ is $3\left(\frac{2}{3}\right) - 2(-1) = 2 - (-2) = 4$.

**69.** (a) When $x = 2$ and $y = 3$, the value of $|2x - 3y|$ is $|2(2) - 3(3)| = |4 - 9| = |-5| = 5$.

(b) When $x = -1$ and $y = 4$, the value of $|2x - 3y|$ is $|2(-1) - 3(4)| = |-2 - 12| = |-14| = 14$.

**71.** (a) When $x = 3$ and $y = 3$, the value of $x - 3(x - y)$ is $3 - 3(3 - 3) = 3 - 3(0) = 3 - 0 = 3$.

(b) When $x = 4$ and $y = -4$, the value of $x - 3(x - y)$ is

$$4 - 3\bigl(4 - (-4)\bigr) = 4 - 3(4 + 4) = 4 - 3(8) = 4 - 24 = -20.$$

**73.** (a) When $a = 2$ and $b = -3$, the value of $b^2 - 4ab$ is $(-3)^2 - 4(2)(-3) = 9 + 24 = 33$.

(b) When $a = 6$ and $b = -4$, the value of $b^2 - 4ab$ is $(-4)^2 - 4(6)(-4) = 16 + 96 = 112$.

**75.** (a) When $x = 4$ and $y = 2$, the value of $\dfrac{x - 2y}{x + 2y}$ is $\dfrac{4 - 2 \cdot 2}{4 + 2 \cdot 2} = \dfrac{4 - 4}{4 + 4} = \dfrac{0}{8} = 0$.

(b) When $x = 4$ and $y = -2$, the value of $\dfrac{x - 2y}{x + 2y}$ is undefined because $\dfrac{4 - 2(-2)}{4 + 2(-2)} = \dfrac{4 + 4}{4 - 4} = \dfrac{8}{0}$

and division by 0 is undefined.

**77.** (a) When $x = 0$ and $y = 5$, the value of $\dfrac{-y}{x^2 + y^2}$ is $\dfrac{-5}{0^2 + 5^2} = \dfrac{-5}{0 + 25} = \dfrac{-5}{25} = \dfrac{-1(\cancel{5})}{5(\cancel{5})} = -\dfrac{1}{5}$.

(b) When $x = 1$ and $y = -3$, the value of $\dfrac{-y}{x^2 + y^2}$ is $\dfrac{-(-3)}{(1)^2 + (-3)^2} = \dfrac{3}{1 + 9} = \dfrac{3}{10}$.

**79.** (a) When $x = 2$, $y = -1$, and $z = -1$, the value of $(x + 2y)(-3x - z)$ is

$$\bigl[2 + 2(-1)\bigr]\bigl[-3(2) - (-1)\bigr] = (2 - 2)(-6 + 1) = 0(-5) = 0.$$

(b) When $x = -3$, $y = 2$, and $z = -2$, the value of $(x + 2y)(-3x - z)$ is

$$\bigl[-3 + 2(2)\bigr]\bigl[-3(-3) - (-2)\bigr] = (-3 + 4)(9 + 2) = 1(11) = 11.$$

**81.** (a) When $b = 3$ and $h = 5$, the value of $\frac{1}{2}bh$ is $\frac{1}{2} \cdot 3 \cdot 5 = \frac{15}{2}$.

(b) When $b = 2$ and $h = 10$, the value of $\frac{1}{2}bh$ is $\frac{1}{2} \cdot 2 \cdot 10 = 10$.

**83.** (a) When $l = 4$, $w = 2$, and $h = 9$, the value of $lwh$ is $4(2)(9) = 72$.

(b) When $l = 100$, $w = 0.8$, and $h = 4$, the value of $lwh$ is $100(0.8)(4) = 320$.

**85.** (a)

| $x$ | $-1$ | $0$ | $1$ | $2$ | $3$ | $4$ |
|---|---|---|---|---|---|---|
| $\frac{2}{3}x + 4$ | $\frac{10}{3}$ | $4$ | $\frac{14}{3}$ | $\frac{16}{3}$ | $6$ | $\frac{20}{3}$ |

When $x = -1$, $3x - 2 = 3(-1) - 2 = -3 - 2 = -5$.

When $x = 0$, $3x - 2 = 3 \cdot 0 - 2 = 0 - 2 = -2$.

When $x = 1$, $3x - 2 = 3 \cdot 1 - 2 = 3 - 2 = 1$.

When $x = 2$, $3x - 2 = 3 \cdot 2 - 2 = 6 - 2 = 4$.

When $x = 3$, $3x - 2 = 3 \cdot 3 - 2 = 9 - 2 = 7$.

When $x = 4$, $3x - 2 = 3 \cdot 4 - 2 = 12 - 2 = 10$.

(b) For each one-unit increase in $x$, the value of the expression $3x - 2$ increases by 3.

(c) You might notice that 3 is the coefficient of $x$ in the expression $3x - 2$. In the expression $\frac{2}{3}x + 4$, the coefficient of $x$ is $\frac{2}{3}$. You might predict that the value of this expression would increase by $\frac{2}{3}$ for each one-unit increase in the value of $x$.

**87.** (a) $x + 6$

(b) When $x = 23$, the value of $x + 6 = 23 + 6 = 29$. You can jump 29 inches wearing the new shoes.

**89.** Area $= (n - 5)^2$

If $n = 8$, the value of $(n - 5)^2 = (8 - 5)^2 = 3^2 = 9$.

So, the area is 9 square units.

**91.** Area $= a(a + b)$

If $a = 5$ and $b = 4$, the value of

$a(a + b) = 5(5 + 4) = 5(9) = 45$. So, the area is 45 square units.

**93.** (a) $n = 10$

$$1 + 2 + 3 = 6 \text{ and } \frac{3(3 + 1)}{2} = \frac{3(4)}{2} = \frac{12}{2} = 6$$

(b) $n = 6$

$1 + 2 + 3 + 4 + 5 + 6 = 21$ and

$$\frac{6(6 + 1)}{2} = \frac{6(7)}{2} = \frac{42}{2} = 21$$

(c) $n = 10$

$1 + 2 + 3 + 4 + 5 + 6 + 7 + 8 + 9 + 10 = 55$

and $\dfrac{10(10 + 1)}{2} = \dfrac{10(11)}{2} = \dfrac{110}{2} = 55$

**95.** (a) 4, 5, 5.5, 5.75, 5.875, 5.938, 5.969

The value appears to be approaching 6.

(b) 9, 7.5, 6.75, 6.375, 6.188, 6.094, 6.047

The value appears to be approaching 6.

**97.** No, $3x$ is not a term of $4 - 3x$. The terms of this expression are 4 and $-3x$.

**99.** Formula in Exercise 93: $\dfrac{n(n + 1)}{2}$

For any natural number $n \geq 1$, both $n$ and $n + 1$ are positive and either $n$ or $n + 1$ is an even number. Therefore, the product $n(n + 1)$ is a positive even number. When a positive even number is divided by 2, the quotient is a positive integer—a natural number. So, this formula always yields a natural number.

Formula in Exercise 94: $\dfrac{n(n - 3)}{2}$

For any natural number $n \geq 4$, both $n$ and $n - 3$ are positive and either $n$ or $n - 3$ is an even number. Therefore, the product $n(n - 3)$ is a positive even number. When a positive even number is divided by 2, the quotient is a positive integer — a natural number. So, this formula always yields a natural number.

**101.** $10 - (-7) = 17$

**103.** $-5 + 10 - (-9) - 4 = 10$

**105.** $(-6)(-4) = 24$

**107.** $\dfrac{-144}{-12} = 12$

**109.** Commutative Property of Multiplication

**111.** Distributive Property

# Section 2.2 Simplifying Algebraic Expressions

**1.** Commutative Property of Addition

**3.** Associative Property of Multiplication

**5.** Additive Identity Property

**7.** Multiplicative Identity Property

**9.** Associative Property of Addition

**11.** Commutative Property of Multiplication

**13.** Distributive Property

**15.** Additive Inverse Property

**17.** Multiplicative Inverse Property

**19.** Distributive Property

**21.** Additive Inverse Property and Additive Identity Property

**23.** $(-5r)s = -5(rs)$

Associative Property of Multiplication

**25.** $v(2) = 2v$

Commutative Property of Multiplication

**27.** $5(t - 2) = 5(t) + 5(-2)$

Distributive Property

**29.** $(2z - 3) + [-(2z - 3)] = 0$

Additive Inverse Property

**31.** $-5x\left(-\dfrac{1}{5x}\right) = 1,\ x \neq 0$

Multiplicative Inverse Property

**33.** $12 + (8 - x) = (12 + 8) - x$

Associative Property of Addition

**35.** $2(16 + 8z) = 2(16) + 2(8z) = 32 + 16z$

**37.** $8(-3 + 5m) = 8(-3) + 8(5m) = -24 + 40m$

**39.** $10(9 - 6x) = 10(9) - 10(6x) = 90 - 60x$

**41.** $-8(2 + 5t) = -8(2) + (-8)(5t) = -16 - 40t$

**43.** $-5(2x - y) = -5(2x) - (-5)(y) = -10x + 5y$

**45.** $(x + 1)(8) = x(8) + 1(8) = 8x + 8$

**47.** $(4 - t)(-6) = 4(-6) - t(-6) = -24 + 6t$

**49.** $4(x + xy + y^2) = 4(x) + 4(xy) + 4(y^2)$
$$= 4x + 4xy + 4y^2$$

**51.** $3(x^2 + x) = 3(x^2) + 3(x) = 3x^2 + 3x$

**53.** $4(2y^2 - y) = 4 \cdot 2y^2 - 4 \cdot y = 8y^2 - 4y$

**55.** $-z(5 - 2z) = -z \cdot 5 - (-z)(2z) = -5z + 2z^2$

**57.** $-4y(3y - 4) = -4y(3y) - (-4y)(4) = -12y^2 + 16y$

**59.** $-(u - v) = (-1)(u - v) = (-1)(u) - (-1)(v) = -u + v$

**61.** $x(3x - 4y) = x(3x) - x(4y) = 3x^2 - 4xy$

**63.** The area of the rectangle on the left is $ab$. The area of the rectangle on the right is $ac$. The area of the entire rectangle can be written as $a(b + c)$ and as the sum of the two smaller rectangles $ab + ac$.

Using the Distributive Property, you can see that these expressions are equal.

$a(b + c) = ab + ac$

**65.** The area of the rectangle on the left is $2a$. The area of the rectangle on the right is $2(b - a)$. The area of the entire rectangle can be written as 2b and as the sum of the two smaller rectangles $2a + 2(b - a)$.

Using the Distributive Property, you can see that these expressions are equal.

$2a + 2(b - a) = 2a + 2b - 2a = 2b$

**67.** In this expression, $16t^3$ and $3t^3$ are like terms, and $4t$ and $-5t$ are like terms.

**69.** In this expression, $4rs^2$ and $12rs^2$ are like terms, and $-5$ and 1 are like terms.

**71.** In this expression, $4x^2y$ and $10x^2y$ are like terms, and $x^3$ and $3x^3$ are like terms.

**73.** $3y - 5y = (3 - 5)y = -2y$

**75.** $x + 5 - 3x = x - 3x + 5 = (1 - 3)x + 5 = -2x + 5$

**77.** $2x + 9x + 4 = (2 + 9)x + 4 = 11x + 4$

**79.** $5r + 6 - 2r + 1 = 5r - 2r + 6 + 1$
$$= (5 - 2)r + (6 + 1) = 3r + 7$$

**81.** $x^2 - 2xy + 4 + xy = x^2 - 2xy + xy + 4$
$$= x^2 + (-2 + 1)xy + 4$$
$$= x^2 + (-1)xy + 4$$
$$= x^2 - xy + 4$$

**83.** $5z - 5 + 10z + 2z + 16 = 5z + 10z + 2z - 5 + 16$
$$= (5 + 10 + 2)z + (-5 + 16)$$
$$= 17z + 11$$

**85.** $z^3 + 2z^2 + z + z^2 + 2z + 1 = z^3 + 2z^2 + z^2 + z + 2z + 1 = z^3 + (2 + 1)z^2 + (1 + 2)z + 1 = z^3 + 3z^2 + 3z + 1$

**87.** $2x^2y + 5xy^2 - 3x^2y + 4xy + 7xy^2 = (2x^2y - 3x^2y) + (5xy^2 + 7xy^2) + 4xy$
$$= (2 - 3)x^2y + (5 + 7)xy^2 + 4xy = -x^2y + 12xy^2 + 4xy$$

**89.** $3\left(\dfrac{1}{x}\right) - \dfrac{1}{x} + 8 = (3 - 1)\left(\dfrac{1}{x}\right) + 8$
$$= 2\left(\dfrac{1}{x}\right) + 8 \text{ or } \dfrac{2}{x} + 8$$

**91.** $5\left(\dfrac{1}{t}\right) - 3t + 6\left(\dfrac{1}{t}\right) - 2t = (5 + 6)\left(\dfrac{1}{t}\right) - (3 + 2)t$
$$= 11\left(\dfrac{1}{t}\right) - 5t \text{ or } \dfrac{11}{t} - 5t$$

**93.** False
$$3(x - 4) = 3 \cdot x - 3 \cdot 4 = 3x - 12$$
So, $3(x - 4) \neq 3x - 4$.

**95.** True
$$6x - 4x = (6 - 4)x = 2x$$
So, $6x - 4x = 2x$.

**97.** False
$$2 - (x + 4) = 2 - 1(x + 4)$$
$$= 2 - 1x - 1(4)$$
$$= 2 - x - 4$$
$$= -x + (2 - 4)$$
$$= -x - 2$$
So, $2 - (x + 4) \neq -2x - 8$.

**99.** $8(52) = 8(50 + 2) = 8(50) + 8(2) = 400 + 16 = 416$

**101.** $9(48) = 9(50 - 2) = 9(50) - 9(2) = 450 - 18 = 432$

**103.** $-4(56) = -4(60 - 4)$
$$= -4(60) - (-4)(4)$$
$$= -240 + 16$$
$$= -224$$

**105.** $5(7.02) = 5(7 + 0.02)$
$$= 5(7) + (5)(0.02)$$
$$= 35 + 0.1$$
$$= 35.1$$

**107.** $2(6x) = (2 \cdot 6)x = 12x$

**109.** $-(4x) = (-1 \cdot 4)x = -4x$

**111.** $(-2x)(-3x) = (-2)(-3)(x \cdot x) = 6x^2$

**113.** $(-5z)(2z^2) = (-5)(2)(z \cdot z^2) = -10z^3$

**115.** $\dfrac{18a}{5} \cdot \dfrac{15}{6} = \dfrac{18a(15)}{5(6)} = \dfrac{9\,\cancel{2}(a)\,\cancel{5}\,\cancel{3}}{\cancel{5}\,\cancel{3}\,\cancel{2}} = 9a$

**117.** $\left(-\dfrac{3x^2}{2}\right)\left(\dfrac{4x}{18}\right) = -\dfrac{\cancel{3}\,\cancel{2}\,\cancel{2}(x^2)(x)}{\cancel{2}\,\cancel{2}\,\cancel{3}(3)} = -\dfrac{x^3}{3}$

**119.** $(12xy^2)(-2x^3y^2) = 12(-2)(x \cdot x^3)(y^2 \cdot y^2) = -24x^4y^4$

**121.** $2(x - 2) + 4 = 2x - 4 + 4 = 2x$

**123.** $6(2s - 1) + s + 4 = 12s - 6 + s + 4$
$$= 12s + s - 6 + 4$$
$$= 13s - 2$$

**125.** $m - 3(m - 7) = m - 3m + 21$
$$= (1 - 3)m + 21$$
$$= -2m + 21$$

**127.** $-6(2 - 3x) + 10(5 - x) = -12 + 18x + 50 - 10x$
$$= 18x - 10x - 12 + 50$$
$$= 8x + 38$$

**129.** $\dfrac{2}{3}(12x + 15) + 16 = 8x + 10 + 16 = 8x + 26$

**131.** $3 - 2[6 + (4 - x)] = 3 - 2[6 + 4 - x]$
$$= 3 - 2[10 - x]$$
$$= 3 - 20 + 2x$$
$$= 2x - 17 \text{ or } -17 + 2x$$

Note: This expression may also be simplified as follows.
$3 - 2[6 + (4 - x)] = 3 - 2[6 + 4 - x]$
$$= 3 - 12 - 8 + 2x$$
$$= 2x - 17$$

**133.** $7x(2 - x) - 4x = 14x - 7x^2 - 4x$
$$= -7x^2 + 10x$$

**135.** $4x^2 + x(5 - x) - 3 = 4x^2 + 5x - x^2 - 3$
$$= 4x^2 - x^2 + 5x - 3$$
$$= 3x^2 + 5x - 3$$

**137.** $-3t(4 - t) + t(t + 1) = -12t + 3t^2 + t^2 + t$
$$= 4t^2 - 11t$$

**139.** $3t[4 - (t - 3)] + t(t + 5) = 3t[4 - t + 3] + t^2 + 5t$
$$= 12t - 3t^2 + 9t + t^2 + 5t$$
$$= -2t^2 + 26t$$

**141.** $\dfrac{2x}{3} - \dfrac{x}{3} = \dfrac{2}{3}x - \dfrac{1}{3}x = \left(\dfrac{2}{3} - \dfrac{1}{3}\right)x = \dfrac{1}{3}x \text{ or } \dfrac{x}{3}$

**143.** $\dfrac{3z}{8} + \dfrac{7z}{8} = \dfrac{3}{8}z + \dfrac{7}{8}z = \left(\dfrac{3}{8} + \dfrac{7}{8}\right)z = \dfrac{10}{8}z = \dfrac{5}{4}z \text{ or } \dfrac{5z}{4}$

**145.** $\dfrac{x}{3} - \dfrac{5x}{4} = \dfrac{1}{3}x - \dfrac{5}{4}x$
$$= \left(\dfrac{1}{3} - \dfrac{5}{4}\right)x$$
$$= \left(\dfrac{1(4)}{3(4)} - \dfrac{5(3)}{4(3)}\right)x$$
$$= \left(\dfrac{4}{12} - \dfrac{15}{12}\right)x$$
$$= -\dfrac{11}{12}x \text{ or } -\dfrac{11x}{12}$$

**147.** $\dfrac{3x}{10} - \dfrac{x}{15} + \dfrac{4x}{5} = \dfrac{3}{10}x - \dfrac{1}{15}x + \dfrac{4}{5}x$
$$= \left(\dfrac{3}{10} - \dfrac{1}{15} + \dfrac{4}{5}\right)x$$
$$= \left(\dfrac{3(3)}{10(3)} - \dfrac{1(2)}{15(2)} + \dfrac{4(6)}{5(6)}\right)x$$
$$= \left(\dfrac{9}{30} - \dfrac{2}{30} + \dfrac{24}{30}\right)x$$
$$= \dfrac{31}{30}x \text{ or } \dfrac{31x}{30}$$

**149.** $5 + (3x - 1) + (2x + 5) = 5 + 3x - 1 + 2x + 5$
$$= (3 + 2)x + (5 - 1 + 5)$$
$$= 5x + 9$$

**151.** (a) Perimeter: $2(3x) + 2(x + 7) = 6x + 2x + 14$
$$= 8x + 14$$

(b) Area: $3x(x + 7) = 3x^2 + 21x$

**153.** (a) $b_1 h + \frac{1}{2}(b_2 - b_1)h = b_1 h + \frac{1}{2}h(b_2 - b_1)$
$$= b_1 h + \tfrac{1}{2}hb_2 - \tfrac{1}{2}hb_1$$
$$= b_1 h + \tfrac{1}{2}b_2 h - \tfrac{1}{2}b_1 h$$
$$= \tfrac{1}{2}b_1 h + \tfrac{1}{2}b_2 h = \tfrac{1}{2}h(b_1 + b_2)$$

$b_1 h$ is the area of the rectangle marked in the figure. $\frac{1}{2}(b_2 - b_1)h$ is the area of the triangle marked in the figure. The sum of these two areas is the area of the trapezoid.

(b) Area $= \frac{1}{2}h(b_1 + b_2)$
$$= \tfrac{1}{2}(3)(7 + 12)$$
$$= \tfrac{1}{2}(3)(19) = \tfrac{57}{2} \text{ square units}$$

or

Area $= b_1 h + \frac{1}{2}(b_2 - b_1)h$
$$= 7(3) + \tfrac{1}{2}(12 - 7)(3)$$
$$= 21 + \tfrac{1}{2}(5)(3)$$
$$= 21 + \tfrac{15}{2}$$
$$= \tfrac{42}{2} + \tfrac{15}{2} = \tfrac{57}{2} \text{ square units}$$

**155.** Area: $\frac{1}{2}h(b_1 + b_2)$
$$\tfrac{1}{2}x[(x + 75) + (x + 25)] = \tfrac{1}{2}x(2x + 100)$$
$$= x^2 + 50x$$

**157.** $(6x)^4 = 6^4 x^4 = 1296 x^4$

So, $(6x)^4 \neq 6x^4$.

**159.** $-16x^2y^3$ and $7x^2y$ are not like terms because their variable factors are not alike.

Note: $x^2y^3 = x \cdot x \cdot y \cdot y \cdot y$ and $x^2y = x \cdot x \cdot y$

**161.** The error is in using the Distributive Property; $-3(x - 1) = -3x + 3$, not $-3x - 3$.

$$4x - 3(x - 1) = 4x - 3(x) + 3(1)$$
$$= 4x - 3x + 3$$
$$= x + 3$$

**163.** $0 - (-12) = 0 + 12 = 12$

**165.** $-12 - 2 + |-3| = -12 - 2 + 3 = -11$

# Mid-Chapter Quiz for Chapter 2

**1.** (a) When $x = 3$, the value of the expression $x^2 - 3x$ is
$$3^2 - 3(3) = 9 - 9 = 0.$$

(b) When $x = -2$, the value of the expression $x^2 - 3x$ is $(-2)^2 - 3(-2) = 4 + 6 = 10.$

**2.** (a) When $x = 5$ and $y = 3$, the value of the expression $\dfrac{x}{y - 3}$ is undefined;
$$\frac{5}{3 - 3} = \frac{5}{0} \text{ and division by zero is undefined.}$$

(b) When $x = 0$ and $y = -1$, the value of the expression $\dfrac{x}{y - 3}$ is $\dfrac{0}{-1 - 3} = \dfrac{0}{-4} = 0.$

**3.**
| Term | Coefficient |
|------|-------------|
| $4x^2$ | $4$ |
| $-2x$ | $-2$ |

**4.**
| Term | Coefficient |
|------|-------------|
| $5x$ | $5$ |
| $3y$ | $3$ |
| $-z$ | $-1$ |

**5.** $(-3y)(-3y)(-3y)(-3y) = (-3y)^4$

**6.** $2 \cdot (x - 3) \cdot (x - 3) \cdot 2 \cdot 2 = 2^3(x - 3)^2$

**7.** Associative Property of Multiplication

**8.** Distributive Property

**9.** Multiplicative Inverse Property

**10.** Commutative Property of Addition

**11.** $2x(3x - 1) = 2x(3x) - 2x(1) = 6x^2 - 2x$

**12.** $-6(2y + 3y^2 - 6) = -6(2y) + (-6)(3y^2) - (-6)(6)$
$$= -12y - 18y^2 + 36$$

**13.** $-4(-5y^2) = 20y^2$

**14.** $\dfrac{x}{3}\left(-\dfrac{3x}{5}\right) = -\dfrac{x(\cancel{3})(x)}{\cancel{3}(5)} = -\dfrac{x^2}{5}$

**15.** $(-3y)^2y^3 = (-3)^2y^2y^3 = 9y^5$

**16.** $\dfrac{2z^2}{3y} \cdot \dfrac{5z}{7} = \dfrac{2z^2(5z)}{3y(7)} = \dfrac{10z^3}{21y}$

**17.** $y^2 - 3xy + y + 7xy = y^2 + (-3 + 7)xy + y$
$$= y^2 + 4xy + y$$

**18.** $10\left(\dfrac{1}{u}\right) - 7\left(\dfrac{1}{u}\right) + 3u = (10 - 7)\left(\dfrac{1}{u}\right) + 3u = 3\left(\dfrac{1}{u}\right) + 3u$

**19.** $5(a - 2b) + 3(a + b) = 5a - 10b + 3a + 3b$
$$= (5 + 3)a + (-10 + 3)b$$
$$= 8a - 7b$$

**20.** $4x + 3[2 - 4(x + 6)] = 4x + 3[2 - 4x - 24]$
$$= 4x + 3[-22 - 4x]$$
$$= 4x - 66 - 12x$$
$$= (4 - 12)x - 66$$
$$= -8x - 66$$

**21.** $8 + (x + 6) + (3x + 1) = 8 + x + 6 + 3x + 1$
$$= (1 + 3)x + (8 + 6 + 1)$$
$$= 4x + 15$$

**167.** $\dfrac{5}{16} - \dfrac{3}{10} = \dfrac{5(5)}{16(5)} - \dfrac{3(8)}{10(8)} = \dfrac{25}{80} - \dfrac{24}{80} = \dfrac{25 - 24}{80} = \dfrac{1}{80}$

**169.** (a) When $x = 2$, the value of $3x - 2$ is
$$3(2) - 2 = 6 - 2 = 4.$$

(b) When $x = -1$, the value of $3x - 2$ is
$$3(-1) - 2 = -3 - 2 = -5.$$

**171.** (a) When $x = 1$ and $y = 5$, the value of $2y - x$ is
$$2(5) - 1 = 10 - 1 = 9.$$

(b) When $x = -6$ and $y = 3$, the value of $2y - x$ is
$$2(3) - (-6) = 6 + 6 = 12.$$

**22.** (a) $\dfrac{x}{6}$

    The variable quantity is the number of students in the class, and this quantity is represented by the letter $x$.

    If the students are divided into 6 teams, then the number of students on each team would be represented by $\dfrac{x}{6}$.

  (b) If there are 30 students in the class, then $x = 30$ and the number of students on each team is $\dfrac{30}{6}$, or 5.

    There are 5 students on each team.

## Section 2.3   Algebra and Problem Solving

**1.** (d)

**3.** (e)

**5.** (b)

**7.** $x + 5$

**9.** $x - 25$

**11.** $x - 6$

**13.** $2x$

**15.** $\dfrac{x}{3}$

**17.** $\dfrac{x}{50}$

**19.** $\dfrac{3}{10}x$ or $\dfrac{3x}{10}$ or $0.3x$

**21.** $3x - 5$

**23.** $3(x - 5)$

**25.** $\dfrac{x}{5} + 15$

**27.** $|x + 4|$

**29.** $x^2 + 1$

**31.** A number decreased by 10

or

Ten less than a number.

**33.** A number is tripled and the product is increased by 2

or

The product of 3 and a number, increased by 2.

**35.** A number is multiplied by one-half and the product is decreased by 6

or

One-half a number decreased by 6.

**37.** A number is subtracted from 2 and the difference is multiplied by 3

or

Three times the difference of 2 and a number.

**39.** A number is increased by 1 and the sum is divided by 2

or

The sum of a number and 1, divided by 2.

**41.** One-half decreased by a number divided by 5

or

One-half decreased by the quotient of a number and 5.

**43.** The square of a number is increased by 5.

**45.** $x(x + 3) = x^2 + 3x$

**47.** $x - (25 + x) = x - 25 - x = x - x - 25 = -25$

**49.** $x^2 - x(2x) = x^2 - 2x^2 = -x^2$

**51.** $\dfrac{8(x + 24)}{2} = \dfrac{\cancel{2}(4)(x + 24)}{\cancel{2}} = 4(x + 24) = 4x + 96$

**53.** The amount of money is a product.

   *Verbal Model:*   $\boxed{\text{Value of dimes}}$ · $\boxed{\text{Number of dimes}}$

   *Labels:*        Value of dime $= 0.10$ (dollars)

                 Number of dimes $= d$

   *Algebraic Expression:*  $0.10d$ (dollars)

**55.** The amount of sales tax is a product.

*Verbal Model:* | Percent of sales tax | · | Amount of purchase |

*Labels:*  Percent of sales tax $= 0.06$ (in decimal form)

Amount of purchase $= L$ (dollars)

*Algebraic Expression:* $0.06L$ (dollars)

**57.** The travel time is a quotient.

*Verbal Model:* | Distance traveled | / | Average speed |

*Labels:*  Distance traveled $= 100$ (miles)

Average speed $= r$ (miles per hour)

*Algebraic Expression:* $\dfrac{100}{r}$ (hours)

**59.** The camping fee is a sum of products.

*Verbal Model:* | Fee per parent | · | Number of parents | + | Fee per child | · | Number of children |

*Labels:*  Fee per parent $= 15$ (dollars)

Number of parents $= m$

Fee per child $= 2$ (dollars)

Number of children $= n$

*Algebraic Expression:* $15m + 2n$ (dollars)

**61.** Guesses will vary.

$t = 10.2$ years

**63.** Guesses will vary.

$t = 11.9$ years

**65.** The value of the expression is 1 less than two times the number $n$.

When $n = 20$, the value of the expression is $2(20) - 1$

and $40 - 1 = 39$.

**67.** The differences in the last row are all 5's. So, the coefficient of $n$ must be 5. However, $5(0) - 0$,

$5(1) = 5$, $5(2) = 10$, $5(3) = 15$, etc. In each instance, the value of $an + b$ is 4 more than 5 times $n$. So, the expression $an + b$ must be $5n + 4$. This indicates that $a = 5$ and $b = 4$.

**69.**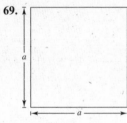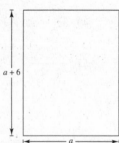

Perimeter of the square: $4a$ centimeters

Area of the square: $a^2$ square centimeters

Perimeter of the rectangle: $2(a) + 2(a + 6) = 2a + 2a + 12 = 4a + 12$ centimeters

Area of the rectangle: $a(a + 6) = a^2 + 6a$ square centimeters

**71.** Find the distance by multiplying the rate by the time.

$$1.15(5) = 5.75$$

In 5 seconds, the bubble will rise 5.75 feet.

**73.** The area of the trapezoid is the sum of the areas of the rectangle and the triangle.

Area of the rectangle: $2(2x)$

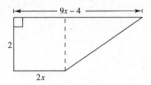

Area of the triangle: $\frac{1}{2}[(9x - 4) - 2x] \cdot 2$

$$2 \cdot 2x + \frac{1}{2}[(9x - 4) - 2x] \cdot 2 = 4x + \frac{[(9x - 4) - 2x]\cancel{2}}{\cancel{2}}$$

$$= 4x + [9x - 4 - 2x]$$

$$= 4x + 7x - 4$$

$$= 11x - 4$$

The area of the trapezoid is $11x - 4$.

**75.** Because there are 12 months in a year, the annual cost is 12 times the monthly cost. The annual cost in dollars is $12d$.

**77.** The perimeter is the sum of twice the length and twice the width.

$$2(1.5w) + 2(w) = 3w + 2w = 5w$$

The perimeter is $5w$.

**79.** The starting time is missing.

**81.** The amount of the weekly earnings is missing, as well as either the number of hours worked or the hourly wage.

**83.** a, b, and e.

Note:

(c) $n$ less than 4 would be equivalent to $4 - n$.

(d) The ratio of $n$ to 4 would be equivalent to $\frac{n}{4}$.

**85.** One possible interpretation: $\frac{5}{3n}$

Another possible interpretation: $\frac{5}{n} \cdot 3 = \frac{15}{n}$

The phrase "the quotient of 5 and a number" indicates the variable is in the denominator.

So, the expression $\frac{3n}{5}$ is not a possible interpretation.

**87.** $(-6)(-13) = 78$

**89.** $\left(-\frac{4}{3}\right)\left(-\frac{9}{16}\right) = \frac{\cancel{4}(\cancel{3})(3)}{\cancel{3}(\cancel{4})(4)} = \frac{3}{4}$

**91.** $\left|-\frac{5}{9}\right| + 2 = \frac{5}{9} + \frac{2(9)}{9} = \frac{5 + 18}{9} = \frac{23}{9}$

**93.** Commutative Property of Addition

**95.** Distributive Property

# Section 2.4   Introduction to Equations

**1.** (a)  $x = 0$

$$2(0) - 18 \overset{?}{=} 0$$

$$0 - 18 \overset{?}{=} 0$$

$$-18 \neq 0$$

0 is *not* a solution.

(b)  $x = 9$

$$2(9) - 18 \overset{?}{=} 0$$

$$18 - 18 \overset{?}{=} 0$$

$$0 = 0$$

9 *is* a solution.

**3.** (a)  $x = 2$

$$6(2) + 1 \overset{?}{=} -11$$

$$12 + 1 \overset{?}{=} -11$$

$$13 \neq -11$$

2 is *not* a solution.

(b)  $x = -2$

$$6(-2) + 1 \overset{?}{=} -11$$

$$-12 + 1 \overset{?}{=} -11$$

$$-11 = -11$$

$-2$ *is* a solution.

**5. (a)** $x = -1$

$$-1 + 5 \overset{?}{=} 2(-1)$$

$$4 \neq -2$$

$-1$ is *not* a solution.

**(b)** $x = 5$

$$5 + 5 \overset{?}{=} 2(5)$$

$$10 = 10$$

$5$ *is* a solution.

**7. (a)** $x = 1$

$$7(1) + 1 \overset{?}{=} 4(1 - 2)$$

$$7 + 1 \overset{?}{=} 4(-1)$$

$$8 \neq -4$$

$1$ is *not* a solution.

**(b)** $x = 12$

$$7(12) + 1 \overset{?}{=} 4(12 - 2)$$

$$84 + 1 \overset{?}{=} 4(10)$$

$$85 \neq 40$$

$12$ is *not* a solution.

**9. (a)** $x = \frac{3}{5}$

$$2\left(\tfrac{3}{5}\right) + 10 \overset{?}{=} 7\left(\tfrac{3}{5} + 1\right)$$

$$\tfrac{6}{5} + \tfrac{10}{1} \overset{?}{=} 7\left(\tfrac{3}{5} + \tfrac{5}{5}\right)$$

$$\tfrac{6}{5} + \tfrac{50}{5} \overset{?}{=} \tfrac{7}{1}\left(\tfrac{8}{5}\right)$$

$$\tfrac{56}{5} = \tfrac{56}{5}$$

$\frac{3}{5}$ *is* a solution.

**(b)** $x = -\frac{2}{3}$

$$2\left(-\tfrac{2}{3}\right) + 10 \overset{?}{=} 7\left(-\tfrac{2}{3} + 1\right)$$

$$-\tfrac{4}{3} + 10 \overset{?}{=} 7\left(\tfrac{1}{3}\right)$$

$$\tfrac{26}{3} \neq \tfrac{7}{3}$$

$-\frac{2}{3}$ is *not* a solution.

**11. (a)** $x = 3$

$$3^2 - 4 \overset{?}{=} 3 + 2$$

$$9 - 4 \overset{?}{=} 3 + 2$$

$$5 = 5$$

$3$ *is* a solution.

**(b)** $x = -2$

$$(-2)^2 - 4 \overset{?}{=} -2 + 2$$

$$4 - 4 \overset{?}{=} 0$$

$$0 = 0$$

$-2$ is a solution.

**13. (a)** $x = 0$

$$\frac{2}{0} - \frac{1}{0} \overset{?}{=} 1$$

$$\frac{2}{0} - \frac{1}{0} \neq 1$$

Division by 0 is not defined.

$0$ is *not* a solution.

**(b)** $x = \frac{1}{3}$

$$\frac{2}{1/3} - \frac{1}{1/3} \overset{?}{=} 1$$

$$\left(2 \div \tfrac{1}{3}\right) - \left(1 \div \tfrac{1}{3}\right) \overset{?}{=} 1$$

$$\left(\tfrac{2}{1} \cdot \tfrac{3}{1}\right) - \left(\tfrac{1}{1} \cdot \tfrac{3}{1}\right) \overset{?}{=} 1$$

$$6 - 3 \overset{?}{=} 1$$

$$3 \neq 1$$

$\frac{1}{3}$ is *not* a solution.

**15. (a)** $x = 3$

$$\frac{5}{3-1} + \frac{1}{3} \overset{?}{=} 5$$

$$\frac{5}{2} + \frac{1}{3} \overset{?}{=} 5$$

$$\frac{15}{6} + \frac{2}{6} \overset{?}{=} 5$$

$$\frac{17}{6} \neq 5$$

3 is *not* a solution.

**(b)** $x = \frac{1}{6}$

$$\frac{5}{(1/6)-1} + \frac{1}{1/6} \overset{?}{=} 5$$

$$\left[5 \div \left(\frac{1}{6} - 1\right)\right] + \left(1 \div \frac{1}{6}\right) \overset{?}{=} 5$$

$$\frac{5}{1} \div \left(-\frac{5}{6}\right) + \frac{1}{1} \div \frac{1}{6} \overset{?}{=} 5$$

$$\frac{5}{1} \cdot \frac{-6}{5} + \frac{1}{1} \cdot \frac{6}{1} \overset{?}{=} 5$$

$$\frac{-30}{5} + 6 \overset{?}{=} 5$$

$$-6 + 6 \overset{?}{=} 5$$

$$0 \neq 5$$

$\frac{1}{6}$ is *not* a solution.

**17. (a)** $x = -3.1$

$$-3.1 + 1.7 \overset{?}{=} 6.5$$

$$-1.4 \neq 6.5$$

−3.1 is *not* a solution.

**(b)** $x = 4.8$

$$4.8 + 1.7 \overset{?}{=} 6.5$$

$$6.5 = 6.5$$

4.8 *is* a solution.

**19. (a)** $x = 12.25$

$$40(12.25) - 490 \overset{?}{=} 0$$

$$0 = 0$$

12.25 *is* a solution.

**(b)** $x = -12.25$

$$40(-12.25) - 490 \overset{?}{=} 0$$

$$-980 \neq 0$$

−12.25 is *not* a solution.

**21. (a)** $x = \frac{5}{2}$

$$2\left(\frac{5}{2}\right)^2 - \frac{5}{2} - 10 \overset{?}{=} 0$$

$$0 = 0$$

$\frac{5}{2}$ *is* a solution.

**(b)** $x = -1.09$

$$2(-1.09)^2 - (-1.09) - 10 \overset{?}{=} 0$$

$$-6.5338 \neq 0$$

−1.09 is *not* a solution.

**23. (a)** $x = 0$

$$\frac{1}{0} \text{ is undefined.}$$

0 is *not* a solution.

**(b)** $x = -2$

$$\frac{1}{-2} - \frac{9}{-2-4} \overset{?}{=} 1$$

$$1 = 1$$

−2 *is* a solution.

**25. (a)** $x = \frac{6}{5}$

$$\left(\frac{6}{5}\right)^3 - 1.728 \overset{?}{=} 0$$

$$0 = 0$$

$\frac{6}{5}$ *is* a solution.

**(b)** $x = -\frac{6}{5}$

$$\left(-\frac{6}{5}\right)^3 - 1.728 \overset{?}{=} 0$$

$$-3.456 \neq 0$$

$-\frac{6}{5}$ is *not* a solution.

**27.**  $x - 8 = 3$  Original equation
$x - 8 + 8 = 3 + 8$  Add 8 to each side.
$x = 11$  Solution

**29.**  $\frac{2}{3}x = 12$  Original equation
$\frac{3}{2}\left(\frac{2}{3}x\right) = \frac{3}{2}(12)$  Multiply each side by $\frac{3}{2}$.
$x = 18$  Solution

**31.**  $5x + 12 = 22$  Original equation
$5x + 12 - 12 = 22 - 12$  Subtract 12 from each side.
$5x = 10$  Combine like terms.
$\frac{5x}{5} = \frac{10}{5}$  Divide each side by 5.
$x = 2$  Solution

**33.**

| | |
|---|---|
| $2(x - 1) = x + 3$ | Original equation |
| $2x - 2 = x + 3$ | Distributive Property |
| $2x - 2 - x = x + 3 - x$ | Subtract $x$ from each side. |
| $2x - x - 2 = x - x + 3$ | Group like terms. |
| $x - 2 = 3$ | Combine like terms. |
| $x - 2 + 2 = 3 + 2$ | Add 2 to each side. |
| $x = 5$ | Solution |

**35.**

| | |
|---|---|
| $\dfrac{x}{3} = x + 1$ | Original equation |
| $3\left(\dfrac{x}{3}\right) = 3(x + 1)$ | Multiply each side by 3. |
| $x = 3x + 3$ | Distributive Property |
| $x - 3x = 3x + 3 - 3x$ | Subtract $3x$ from each side. |
| $x - 3x = 3x - 3x + 3$ | Group like terms. |
| $-2x = 3$ | Combine like terms. |
| $\dfrac{-2x}{-2} = \dfrac{3}{-2}$ | Divide each side by $-2$. |
| $x = -\dfrac{3}{2}$ | Solution |

**37.** A number decreased by 6 is 32.

**39.** Twice a number increased by 5 is 21.

**41.** Ten times the difference of a number and 3 is 8 times the number.

**43.** The sum of a number and 1 is divided by 3, and the quotient is equal to 8.

**45.** $x + 12 = 45$

**47.** $\dfrac{x}{8} = 6$

**49.** $3x + 4 = 16$

**51.** $120 - 6x = 96$

**53.** $\dfrac{4}{x + 5} = 2$

**55.** $\dfrac{3 \text{ dollars}}{\cancel{\text{unit}}} \cdot \left(5 \cancel{\text{ units}}\right) = 15$ dollars

**57.** $\dfrac{50 \text{ pounds}}{\cancel{\text{foot}}} \cdot \left(3 \cancel{\text{ feet}}\right) = 150$ pounds

**59.** $\dfrac{5 \text{ feet}}{\cancel{\text{seconds}}} \cdot \dfrac{60 \cancel{\text{ seconds}}}{\cancel{\text{minute}}} \cdot \left(20 \cancel{\text{ minutes}}\right) = 6000$ feet

**61.** $\dfrac{100 \text{ centimeters}}{\cancel{\text{meter}}} \cdot \left(2.4 \cancel{\text{ meters}}\right) = 240$ centimeters

**63.**

$$x - 8 = 5$$
$$x - 8 + 8 = 5 + 8$$
$$x = 13$$

**65.** $3x = 30$

$$\dfrac{3x}{3} = \dfrac{30}{3}$$
$$x = 10$$

**Check:** $3(10) = 30$

**67.** *Verbal Model:*

$\boxed{\text{Original score}} + \boxed{\text{Additional points}} = \boxed{\text{Final score}}$

*Labels:*   Original score $= x$ (points)

Additional points $= 6$ (points)

Final score $= 94$ (points)

*Equation:*    $x + 6 = 94$

Note: There are other equivalent ways to write this equation. Here are some other possibilities:

$$94 - x = 6$$
$$94 = x + 6$$
$$x = 94 - 6$$

There are also equivalent ways of writing equations for the *other* exercises in this section.

**69.** *Verbal Model:* $\boxed{\begin{array}{c}\text{Amount}\\\text{saved}\end{array}} + \boxed{\begin{array}{c}\text{Additional}\\\text{savings}\\\text{needed}\end{array}} = \boxed{\begin{array}{c}\text{Computer}\\\text{cost}\end{array}}$

*Labels:*   Amount saved $= \$1044$

Additional savings needed $= x$ (dollars)

Computer cost $= \$1926$

*Equation:*    $1044 + x = 1926$

Note: Remember that there are other equivalent ways of writing this equation.

**71.** *Verbal Model:*

$\boxed{\text{Original price}} + \boxed{\text{Increase in price}} = \boxed{\text{Current price}}$

*Labels:*   Original price $= x$ (dollars)

Increase in price $= 45$ (dollars)

Current price $= 375$ (dollars)

*Equation:*    $x + 45 = 375$

**73.** *Verbal Model:*

$$\boxed{\dfrac{\text{Total yards for game}}{\text{Number of carries}}} = \boxed{\text{Average yards per carry}}$$

*Labels:*    Total yards for game = $x$ (yards)

Number of carries = 18

Average yards per carry = 4.5
(yards)

*Algebraic Equation:*  $\dfrac{x}{18} = 4.5$

Note: The equation could also be written as
$18(4.5) = x.$

**75.** *Verbal Model:*

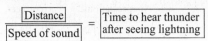

*Labels:*    Distance = $d$ (feet)

Speed of sound = 1100 (feet per second)

Time to hear thunder after seeing
lightning = 3 (seconds)

*Equation:*  $\dfrac{d}{1100} = 3$

**77.** *Verbal Model:*  $\boxed{\text{Volume}} = \boxed{\text{Length}} \cdot \boxed{\text{Width}} \cdot \boxed{\text{Height}}$

*Labels:*    Volume = 72 (cubic feet)

Length = 6 (feet)

Width = 4 (feet)

Height = $h$ (feet)

*Equation:*    $72 = (6)(4)(h)$

$72 = 24h$

**79.** *Verbal Model:*

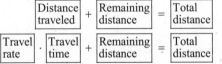

*Labels:*    Travel rate = $r$ (miles per hour)

Travel time = 3 (hours)

Remaining distance = 25 (miles)

Total distance = 160 (miles)

*Equation:*    $r \cdot 3 + 25 = 160$ or $3r + 25 = 160$

Note: Remember the formula $d = rt$; distance equals rate times time.

**81.** *Verbal Model:*    $\boxed{\substack{\text{Weeks} \\ \text{remaining}}} \cdot \boxed{\substack{\text{Average hours} \\ \text{per week}}} = \boxed{\substack{\text{Hours} \\ \text{for goal}}} - \boxed{\substack{\text{Hours} \\ \text{worked}}}$

*Labels:*    Weeks remaining = $15 - 8 = 7$ (weeks)

Average hours per week = $x$ (hours per week)

Number of hours for goal = 150 (hours)

Number of hours worked = 72 (hours)

*Algebraic Equation:*  $7x = 150 - 72$ or $7x = 78$

**83.** *Verbal model:*

$$\boxed{\text{Total depreciation}} = \boxed{3} \cdot \boxed{\text{Depreciation per year}} \text{ or}$$

$$\boxed{\text{Initial value}} - \boxed{\text{Final value}} = \boxed{3} \cdot \boxed{\text{Depreciation per year}}$$

*Labels:*   Initial value = \$750,000

Final value = \$75,000

Depreciation per year = $D$ (dollars)

*Equation:*   $750,000 - 75,000 = 3D$

Note: This equation could also be written as $\dfrac{750,000 - 75,000}{3} = D$ or as $750,000 - 3D = 75,000$.

**85.** *Verbal Model:*

$$\boxed{\text{Revenue from adults}} + \boxed{\text{Revenue from students}} = \boxed{\text{Total revenue}}$$

*Labels:*   Cost per adult ticket = 10 (dollars per person)

Number of adults = $a$ (persons)

Revenue from adults = $10a$ (dollars)

Cost per student ticket = 6 (dollars per person)

Number of students = $\dfrac{3}{4}a$ (persons)

Revenue from students = $6\left(\dfrac{3}{4}a\right)$ (dollars)

Total revenue = 986 (dollars)

*Algebraic Equation:*

$$10a + 6\left(\frac{3}{4}a\right) = 986$$

$$10a + \frac{3 \cdot \cancel{2} \cdot 3}{\cancel{2} \cdot 2}a = 986$$

$$\frac{10(2)}{2}a + \frac{9}{2}a = 986$$

$$\left(\frac{20}{2} + \frac{9}{2}\right)a = 986$$

$$\frac{29}{2}a = 986$$

**87.** No. There is only one value of $x$, $\dfrac{b}{a}$, that makes the equation true.

**89.** The total cost of a shipment of bulbs is \$840. Find the number of cases of bulbs if each case costs \$35.

**91.** $t^2 \cdot t^5 = t \cdot t \cdot t \cdot t \cdot t \cdot t \cdot t = t^7$

**93.** $6x + 9x = 15x$

**95.** $-(-8b) = 8b$

**97.** $x + 23$

**99.** $4x + 7$

# Review Exercises for Chapter 2

**1.** $60t$

The variable quantity is the number of hours traveled, and this quantity is represented by the letter $t$. If the average speed is 60 miles per hour, the distance traveled is $60t$.

**3.** Variable: $x$

**5.** Variables: $a$, $b$

**7.** Terms: $12y, y^2$

Coefficients: $12, 1$

**9.** Terms: $5x^2, -3xy, 10y^2$

Coefficients: $5, -3, 10$

**11.** Terms: $\dfrac{2y}{3}, -\dfrac{4x}{v}$

Coefficients: $\dfrac{2}{3}, -4$

**13.** $5z \cdot 5z \cdot 5z = (5z)^3 \text{ or } 5^3 z^3$

**15.** $(-3x) \cdot (-3x) \cdot (-3x) \cdot (-3x) \cdot (-3x) = (-3x)^5$

**17.** $(b - c) \cdot (b - c) \cdot 6 \cdot 6 = 6^2(b - c)^2$

**19.** (a) When $x = 0$, the value of $x^2 - 2x + 5$ is

$0^2 - 2 \cdot 0 + 5 = 0 - 0 + 5 = 5.$

(b) When $x = 2$, the value of $x^2 - 2x + 5$ is

$2^2 - 2 \cdot 2 + 5 = 4 - 4 + 5 = 5.$

**21.** (a) When $x = 2$ and $y = -1$, the value of

$x^2 - x(y + 1) = 2^2 - 2(-1 + 1)$

$= 4 - 2(0) = 4 - 0 = 4.$

(b) When $x = 1$ and $y = 2$, the value of

$x^2 - x(y + 1) = 1^2 - 1[2 + 1]$

$= 1 - 1(3) = 1 - 3 = -2.$

**23.** When $x = -5$ and $y = 3$, the value of

$\dfrac{x + 5}{y} = \dfrac{-5 + 5}{3} = \dfrac{0}{3} = 0.$

When $x = 2$ and $y = -1$, the value of

$\dfrac{x + 5}{y} = \dfrac{2 + 5}{-1} = \dfrac{7}{-1} = -7.$

**25.** (a) When $x = 1$, $y = 2$, and $z = 0$, the value of

$x^2 - 2y + z = (1)^2 - 2(2) + 0 = 1 - 4 + 0 = -3.$

(b) When $x = 2$, $y = -3$, and $z = -4$, the value of

$x^2 - 2y + z = (2)^2 - 2(-3) + (-4)$

$= 4 + 6 - 4 = 6.$

**27.** $xy \cdot \dfrac{1}{xy} = 1$

Multiplicative Inverse Property

**29.** $(x - y)(2) = 2(x - y)$

Commutative Property of Multiplication

**31.** $2x + (3y - z) = (2x + 3y) - z$

Associative Property of Addition

**33.** $3(m^2 n) = 3m^2(n)$

Associative Property of Multiplication

**35.** $(3x + 8) + \left[-(3x + 8)\right] = 0$

Additive Inverse Property

**37.** $4(x + 3y) = 4x + 4 \cdot 3y = 4x + 12y$

**39.** $-5(2u - 3v) = (-5)(2u) - (-5)(3v) = -10u + 15v$

**41.** $x(8x + 5y) = x(8x) + x(5y) = 8x^2 + 5xy$

**43.** $-(-a + 3b) = a - 3b$

Note: The sign of each term is changed.

The expression $-(-a + 3b)$ can be written as

$(-1)(-a + 3b) = (-1)(-a) + (-1)(3b) = a - 3b.$

**45.** $2(x + 3 - 2y) = 2(x) + 2(3) - 2(2y) = 2x + 6 - 4y$

**47.** In this expression, $3x$ and $2x$ are like terms.

**49.** In this expression, $10$ and $-2$ are like terms.

**51.** $3a - 5a = (3 - 5)a = -2a$

**53.** $3p - 4q + q + 8p = 3p + 8p - 4q + q$

$= (3 + 8)p + (-4 + 1)q$

$= 11p + (-3)q$

$= 11p - 3q$

**55.** $\frac{1}{4}s - 6t + \frac{7}{2}s + t = \frac{1}{4}s + \frac{7}{2}s - 6t + t$

$= \left(\frac{1}{4} + \frac{14}{4}\right)s + (-6 + 1)t$

$= \frac{15}{4}s - 5t$

**57.** $x^2 + 3xy - xy + 4 = x^2 + (3 - 1)xy + 4 = x^2 + 2xy + 4$

**59.** $5x - 5y + 3xy - 2x + 2y = 5x - 2x + 3xy - 5y + 2y = (5 - 2)x + 3xy + (-5 + 2)y = 3x + 3xy - 3y$

**61.** $5\left(1 + \dfrac{r}{n}\right)^2 - 2\left(1 + \dfrac{r}{n}\right)^2 = (5 - 2)\left(1 + \dfrac{r}{n}\right)^2 = 3\left(1 + \dfrac{r}{n}\right)^2$

**63.** $12(4t) = (12 \cdot 4)t = 48t$

**65.** $-5(-9x^2) = (-5)(-9)x^2 = 45x^2$

**67.** $(-6x)(2x^2) = (-6 \cdot 2)(x \cdot x^2) = -12x^3$

**69.** $\dfrac{12x}{5} \cdot \dfrac{10}{3} = \dfrac{12x \cdot 10}{5 \cdot 3} = \dfrac{4(3)x(5)(2)}{5(3)} = 8x$

**71.** $3(61) = 3(60 + 1) = 3(60) + 3(1) = 180 + 3 = 183$

**73.** $7(98) = 7(100 - 2)$

$\quad\quad = 7(100) - 7(2)$

$\quad\quad = 700 - 14$

$\quad\quad = 686$

**75.** $5(u - 4) + 10 = 5u - 5 \cdot 4 + 10$

$\quad\quad\quad\quad\quad\quad = 5u - 20 + 10$

$\quad\quad\quad\quad\quad\quad = 5u - 10$

**77.** $3s - (r - 2s) = 3s - r + 2s$

$\quad\quad\quad\quad\quad\quad = 3s + 2s - r$

$\quad\quad\quad\quad\quad\quad = 5s - r$

**79.** $-3(1 - 10z) + 2(1 - 10z) = (-3 + 2)(1 - 10z) = -1(1 - 10z) = -1 + 10z$

Note: the parentheses could be removed first.

$-3(1 - 10z) + 2(1 - 10z) = -3 \cdot 1 - (-3)(10z) + 2 \cdot 1 - 2(10z)$

$\quad\quad\quad\quad\quad = -3 + 30z + 2 - 20z = -3 + 2 + 30z - 20z$

$\quad\quad\quad\quad\quad = -1 + (30 - 20)z = -1 + 10z$

**81.** $\frac{1}{3}(42 - 18z) - 2(8 - 4z) = 14 - 6z - 16 + 8z$

$\quad\quad\quad\quad\quad\quad\quad\quad = -6z + 8z + 14 - 16$

$\quad\quad\quad\quad\quad\quad\quad\quad = (-6 + 8)z + 14 - 16$

$\quad\quad\quad\quad\quad\quad\quad\quad = 2z - 2$

**83.** $10 - [8(5 - x) + 2] = 10 - [40 - 8x + 2]$

$\quad\quad\quad\quad\quad\quad\quad = 10 - 40 + 8x - 2$

$\quad\quad\quad\quad\quad\quad\quad = 8x + 10 - 40 - 2$

$\quad\quad\quad\quad\quad\quad\quad = 8x - 32$

**85.** $2[x + 2(y - x)] = 2[x + 2y - 2x]$

$\quad\quad\quad\quad\quad\quad = 2x + 4y - 4x$

$\quad\quad\quad\quad\quad\quad = (2 - 4)x + 4y$

$\quad\quad\quad\quad\quad\quad = -2x + 4y$

**87.** $\dfrac{x}{4} - \dfrac{3x}{4} = \dfrac{1}{4}x - \dfrac{3}{4}x$

$\quad\quad\quad = \left(\dfrac{1}{4} - \dfrac{3}{4}\right)x$

$\quad\quad\quad = -\dfrac{2}{4}x$

$\quad\quad\quad = -\dfrac{1}{2}x \text{ or } -\dfrac{x}{2}$

**89.** $\dfrac{3z}{2} + \dfrac{z}{5} = \dfrac{3}{2}z + \dfrac{1}{5}z$

$\quad\quad\quad = \left(\dfrac{3}{2} + \dfrac{1}{5}\right)z$

$\quad\quad\quad = \left(\dfrac{3(5)}{2(5)} + \dfrac{1(2)}{5(2)}\right)z$

$\quad\quad\quad = \left(\dfrac{15}{10} + \dfrac{2}{10}\right)z$

$\quad\quad\quad = \dfrac{17}{10}z \text{ or } \dfrac{17z}{10}$

**91.** (a)  The perimeter is a sum of twice the length and twice the width.

$\quad\quad 2(x + 6) + 2(2x) = 2x + 12 + 4x = 6x + 12$

(b)  The area is the product of the length and the width.

$\quad\quad (x + 6)(2x) = x(2x) + 6(2x) = 2x^2 + 12x$

**93.** $(2n - 1) + (2n + 1) + (2n + 3) = 6n + 3$

**95.** The area of a rectangle is the product of the length and width.

$\quad\quad (16x)(4x) - 6x(x) = (16 \cdot 4)(x \cdot x) - 6(x \cdot x)$

$\quad\quad\quad\quad\quad\quad\quad\quad = 64x^2 - 6x^2 = 58x^2$

**97.** *Verbal Model:*

| Base pay per hour | + | Additional pay per unit | · | Number of units produced per hour |
|---|---|---|---|---|

*Algebraic Expression:* $8.25 + 0.60x$

**99.** $\frac{2}{3}x + 5$

**101.** $2x - 10$

**103.** $50 + 7x$

**105.** $\dfrac{x + 10}{8}$

**107.** $x^2 + 64$

**109.** The sum of a number and three *or* a number increased by three.

**115.** *Verbal Model:* $\boxed{\text{Monthly rent}} \cdot \boxed{\text{Number of months}}$

*Labels:* Monthly rent $= \$625$

Number of months $= n$

*Algebraic Expression:* $625n$

**117.** The value of the expression is 4 more than three times the number $n$.

When $n = 20$, the value of the expression is $3(20) + 4$

and $60 + 4 = 64$.

**119.** (a) $x = 3$

$5(3) + 6 \overset{?}{=} 36$

$15 + 6 \overset{?}{=} 36$

$21 \neq 36$

3 is *not* a solution.

(b) $x = 6$

$5(6) + 6 \overset{?}{=} 36$

$30 + 6 \overset{?}{=} 36$

$36 = 36$

6 *is* a solution.

**121.** (a) $x = -1$

$3(-1) - 12 \overset{?}{=} -1$

$-3 - 12 \overset{?}{=} -1$

$-15 \neq -1$

$-1$ is *not* a solution.

(b) $x = 6$

$3(6) - 12 \overset{?}{=} 6$

$18 - 12 \overset{?}{=} 6$

$6 = 6$

6 *is* a solution.

**111.** A number is decreased by two and the result is divided by three *or* two is subtracted from a number and the result is divided by three.

**113.** *Verbal Model:* $\boxed{\text{Commission rate}} \cdot \boxed{\text{Sales}}$

*Labels:* Commission rate $= 0.05$

Sales $= x$

*Algebraic Expression:* $0.05x$

**123.** (a) $x = \frac{2}{7}$

$4\left(2 - \frac{2}{7}\right) \overset{?}{=} 3\left(2 + \frac{2}{7}\right)$

$4\left(\frac{14}{7} - \frac{2}{7}\right) \overset{?}{=} 3\left(\frac{14}{7} + \frac{2}{7}\right)$

$4\left(\frac{12}{7}\right) \overset{?}{=} 3\left(\frac{16}{7}\right)$

$\frac{48}{7} = \frac{48}{7}$

$\frac{2}{7}$ *is* a solution.

(b) $x = -\frac{2}{3}$

$4\left(2 - \left(-\frac{2}{3}\right)\right) \overset{?}{=} 3\left(2 + \left(-\frac{2}{3}\right)\right)$

$4\left(\frac{6}{3} + \frac{2}{3}\right) \overset{?}{=} 3\left(\frac{6}{3} - \frac{2}{3}\right)$

$4\left(\frac{8}{3}\right) \overset{?}{=} 3\left(\frac{4}{3}\right)$

$\frac{32}{3} \neq 4$

$-\frac{2}{3}$ is *not* a solution.

**125.** (a) $x = -1$

$$\frac{4}{-1} - \frac{2}{-1} \overset{?}{=} 5$$

$$-4 - (-2) \overset{?}{=} 5$$

$$-4 + 2 \overset{?}{=} 5$$

$$-2 \neq 5$$

$-1$ is *not* a solution.

(b) $x = \dfrac{2}{5}$

$$\frac{4}{2/5} - \frac{2}{2/5} \overset{?}{=} 5$$

$$4\left(\frac{5}{2}\right) - 2\left(\frac{5}{2}\right) \overset{?}{=} 5$$

$$10 - 5 \overset{?}{=} 5$$

$$5 = 5$$

$\dfrac{2}{5}$ *is* a solution.

**127.** (a) $x = 3$

$$3(3 - 7) \overset{?}{=} -12$$

$$3(-4) \overset{?}{=} -12$$

$$-12 = -12$$

3 *is* a solution.

(b) $x = 4$

$$4(4 - 7) \overset{?}{=} -12$$

$$4(-3) \overset{?}{=} -12$$

$$-12 = -12$$

4 *is* a solution.

**129.**

| | |
|---|---|
| $-7x + 20 = -1$ | Original equation |
| $-7x + 20 - 20 = -1 - 20$ | Subtract 20 from each side. |
| $-7x = -21$ | Combine like terms. |
| $\dfrac{-7x}{-7} = \dfrac{-21}{-7}$ | Divide each side by $-7$. |
| $x = 3$ | Solution |

**131.**

| | |
|---|---|
| $x = -(x - 14)$ | Original equation |
| $x = -x + 14$ | Distributive Property |
| $x + x = -x + 14 + x$ | Add $x$ to each side. |
| $x + x = -x + x + 14$ | Group like terms. |
| $2x = 14$ | Combine like terms. |
| $\dfrac{2x}{2} = \dfrac{14}{2}$ | Divide each side by 2. |
| $x = 7$ | Solution |

**133.** *Verbal Model:*

$$\boxed{\text{Unknown number}} + \boxed{\text{Reciprocal of number}} = \boxed{\dfrac{37}{6}}$$

*Labels:*    Unknown number $= x$

Reciprocal $= \dfrac{1}{x}$

*Equation:*    $x + \dfrac{1}{x} = \dfrac{37}{6}$

**135.** *Verbal Model:*  $\boxed{\text{Area of rectangle}} - \boxed{\text{Area of triangle}} = \boxed{\text{Area of shaded region}}$

*Labels:*      Area of rectangle $= x(6)$ or $6x$ (square inches)

Area of triangle $= \frac{1}{2}(x)(6)$ or $3x$ (square inches)

Shaded area $= 24$ (square inches)

*Equation:*    $6x - \frac{1}{2}(x)(6) = 24$  or  $6x - 3x = 24$

This equation could also be simplified to  $3x = 24$.

Note: The area of a rectangle is the product of its length and width $(A = lw)$. The area of a triangle is one-half the product of its base and height $\left(A = \frac{1}{2}bh\right)$.

# Chapter Test for Chapter 2

**1.** Terms: $2x^2, -7xy, 3y^3$

Coefficients: $2, -7, 3$

**2.** $x \cdot (x + y) \cdot x \cdot (x + y) \cdot x = x^3(x + y)^2$

**3.** Associative Property of Multiplication

**4.** Commutative Property of Addition

**5.** Additive Inverse Property

**6.** Multiplicative Inverse Property

**7.** $3(x + 8) = 3x + 24$

**8.** $5(4r - s) = 20r - 5s$

**9.** $-y(3 - 2y) = -3y + 2y^2$

**10.** $-9(4 - 2x + x^2) = -36 + 18x - 9x^2$

**11.** $3b - 2a + a - 10b = (-2 + 1)a + (3 - 10)b$
$$= -a - 7b$$

**12.** $15(u - v) - 7(u - v) = (15 - 7)(u - v)$
$$= 8(u - v)$$
$$= 8u - 8v$$

Note: This problem can also be worked as follows.

$15(u - v) - 7(u - v) = 15u - 15v - 7u + 7v$
$$= (15 - 7)u + (-15 + 7)v$$
$$= 8u - 8v$$

**13.** $3z - (4 - z) = 3z - 4 + z = 3z + z - 4 = 4z - 4$

**14.** $2[10 - (t + 1)] = 2[10 - t - 1]$
$$= 2[10 - 1 - t]$$
$$= 2[9 - t]$$
$$= 18 - 2t$$

**15.** When $x = 2$, the value of the expression
$$x^3 - 2 = (2)^3 - 2 = 8 - 2 = 6.$$

**16.** When $x = 2$ and $y = -10$, the value of the expression
$$x^2 + 4(y + 2) = (2)^2 + 4(-10 + 2)$$
$$= 4 + 4(-8)$$
$$= 4 - 32$$
$$= -28.$$

**17.** When $a = 2$ and $b = 6$, the value of
$(a + 2b)/(3a - b)$ is undefined because

$$\frac{2 + 2(6)}{3(2) - 6} = \frac{2 + 12}{6 - 6} = \frac{14}{0}$$

and division by zero is undefined.

**18.** $\frac{1}{3}n - 4$ or $\frac{n}{3} - 4$

**19.** (a) *Perimeter:*    The perimeter is a sum of products.

*Verbal Model:*    2 Length + 2 Width

*Labels:*    Length = $2w - 4$
    Width = $w$

*Algebraic expression:*

$$2(2w - 4) + 2(w) = 4w - 8 + 2w = 6w - 8$$

*Area:*    The area is the product of the length and the width.

*Verbal Model:*    Length · Width

*Labels:*    Length = $2w - 4$
    Width = $w$

*Algebraic expression:*

$$(2w - 4)(w) = 2w(w) - 4(w) = 2w^2 - 4w$$

(b) *Perimeter:* When $w = 7$, the value of the expression $6w - 8 = 6(7) - 8 = 42 - 8 = 34$.

The perimeter is 34 units.

*Area:* When $w = 7$, the value of the expression

$$2w^2 - 4w = 2(7)^2 - 4(7)$$
$$= 2(49) - 28 = 98 - 28 = 70.$$

The area is 70 square units.

**20.** (a) The price is a sum of products.

*Verbal Model:*

25 Number of adults + 20 Number of children

*Labels:*    Number of adults = $m$
    Number of children = $n$

*Algebraic expression:*    $25m + 20n$

(b) If $m = 2$ and $n = 3$, the value of the expression
$$25m + 20n = 25(2) = 20(3) = 50 + 60 = 110.$$

It will cost $110 for two adults and three children to attend the concert.

**21.** (a) $6(3 - (-2)) - 5(2(-2) - 1) \overset{?}{=} 7$

$$6(3 + 2) - 5(-4 - 1) \overset{?}{=} 7$$

$$6(5) - 5(-5) \overset{?}{=} 7$$

$$30 + 25 \overset{?}{=} 7$$

$$55 \neq 7$$

$-2$ is *not* a solution.

(b) $6(3 - 1) - 5(2 \cdot 1 - 1) \overset{?}{=} 7$

$$6(2) - 5(2 - 1) \overset{?}{=} 7$$

$$12 - 5(1) \overset{?}{=} 7$$

$$12 - 5 \overset{?}{=} 7$$

$$7 = 7$$

$1$ *is* a solution.

# CHAPTER 3
## Linear Equations and Problem Solving

# CHAPTER 3
# Linear Equations and Problem Solving

## Section 3.1   Solving Linear Equations

**1.** $x = -6$

**3.** $x = 13$

**5.** $y = 4$

**7.** $z = -9$

**9.**

| | |
|---|---|
| $5x + 15 = 0$ | Original equation |
| $5x + 15 - 15 = 0 - 15$ | Subtract 15 from each side. |
| $5x = -15$ | Combine like terms. |
| $\dfrac{5x}{5} = \dfrac{-15}{5}$ | Divide each side by 5. |
| $x = -3$ | Simplify. |

**11.**

| | |
|---|---|
| $-2x - 8 = 0$ | Original equation |
| $-2x - 8 + 8 = 0 + 8$ | Add 8 to each side. |
| $-2x = 8$ | Combine like terms. |
| $\dfrac{-2x}{-2} = \dfrac{8}{-2}$ | Divide each side by $-2$. |
| $x = -4$ | Simplify. |

**13.** Addition

**15.** Multiplication

**17.**
$$8x - 16 = 0$$
$$8x - 16 + 16 = 0 + 16$$
$$8x = 16$$
$$\frac{8x}{8} = \frac{16}{8}$$
$$x = 2$$

**19.**
$$3x + 21 = 0$$
$$3x + 21 - 21 = 0 - 21$$
$$3x = -21$$
$$\frac{3x}{3} = \frac{-21}{3}$$
$$x = -7$$

**21.** $5x = 30$
$$\frac{5x}{5} = \frac{30}{5}$$
$$x = 6$$

**23.** $9x = -21$
$$\frac{9x}{9} = \frac{-21}{9}$$
$$x = -\frac{21}{9}$$
$$x = -\frac{7}{3}$$

**25.**
$$-8x + 4 = -20$$
$$-8x + 4 - 4 = -20 - 4$$
$$-8x = -24$$
$$\frac{-8x}{-8} = \frac{-24}{-8}$$
$$x = 3$$

**27.**
$$25x - 4 = 46$$
$$25x - 4 + 4 = 46 + 4$$
$$25x = 50$$
$$\frac{25x}{25} = \frac{50}{25}$$
$$x = 2$$

**29.**
$$10 - 4x = -6$$
$$10 - 10 - 4x = -6 - 10$$
$$-4x = -16$$
$$\frac{-4x}{-4} = \frac{-16}{-4}$$
$$x = 4$$

**31.**
$$6x - 4 = 0$$
$$6x - 4 + 4 = 0 + 4$$
$$6x = 4$$
$$\frac{6x}{6} = \frac{4}{6}$$
$$x = \frac{4}{6}$$
$$x = \frac{2}{3}$$

**33.**  $\dfrac{x}{3} = 10$

$3\left(\dfrac{x}{3}\right) = 3 \cdot 10$

$x = 30$

Note: $3\left(\dfrac{x}{3}\right) = \dfrac{3}{1} \cdot \dfrac{x}{3} = \dfrac{3x}{3} = x$

**35.**  $\dfrac{x}{2} + 1 = 0$

$\dfrac{x}{2} + 1 - 1 = 0 - 1$

$\dfrac{x}{2} = -1$

$2\left(\dfrac{x}{2}\right) = 2(-1)$

$x = -2$

**37.**  $x - \dfrac{1}{3} = \dfrac{4}{3}$

$x - \dfrac{1}{3} + \dfrac{1}{3} = \dfrac{4}{3} + \dfrac{1}{3}$

$x = \dfrac{5}{3}$

**39.**  $t - \dfrac{1}{3} = \dfrac{1}{2}$

$t - \dfrac{1}{3} + \dfrac{1}{3} = \dfrac{1}{2} + \dfrac{1}{3}$

$t = \dfrac{3}{6} + \dfrac{2}{6}$

$t = \dfrac{5}{6}$

**41.**  $3y - 2 = 2y$

$3y - 3y - 2 = 2y - 3y$

$-2 = -y$

$(-1)(-2) = (-1)(-y)$

$2 = y$

**43.**  $4 - 7x = 5x$

$4 - 7x + 7x = 5x + 7x$

$4 = 12x$

$\dfrac{4}{12} = \dfrac{12x}{12}$

$\dfrac{1}{3} = x$

**45.**  $4 - 5t = 16 + t$

$4 - 5t - t = 16 + t - t$

$4 - 6t = 16$

$4 - 4 - 6t = 16 - 4$

$-6t = 12$

$\dfrac{-6t}{-6} = \dfrac{12}{-6}$

$t = -2$

**47.**  $-3t + 5 = -3t$

$-3t + 3t + 5 = 3t + 3t$

$5 = 0$   (False)

The original equation has no solution.

**49.**  $15x - 3 = 15 - 3x$

$15x + 3x - 3 = 15 + 3x - 3x$

$18x - 3 = 15$

$18x - 3 + 3 = 15 + 3$

$18x = 18$

$\dfrac{18x}{18} = \dfrac{18}{18}$

$x = 1$

**51.**  $7a - 18 = 3a - 2$

$7a - 3a - 18 = 3a - 3a - 2$

$4a - 18 = -2$

$4a - 18 + 18 = -2 + 18$

$4a = 16$

$\dfrac{4a}{4} = \dfrac{16}{4}$

$a = 4$

**53.**  $7x + 9 = 3x + 1$

$7x - 3x + 9 = 3x - 3x + 1$

$4x + 9 = 1$

$4x + 9 - 9 = 1 - 9$

$4x = -8$

$\dfrac{4x}{4} = \dfrac{-8}{4}$

$x = -2$

**55.**  $4x - 6 = 4x - 6$

$4x - 4x - 6 = 4x - 4x - 6$

$-6 = -6$   Identity

The original equation has infinitely many solutions.

**57.**  $2x + 4 = -3(x - 2)$

$2x + 4 = -3x + 6$

$5x + 4 = 6$

$5x + 4 - 4 = 6 - 4$

$5x = 2$

$\dfrac{5x}{5} = \dfrac{2}{5}$

$x = \dfrac{2}{5}$

**59.**
$$5(3 - x) = x - 12$$
$$15 - 5x = x - 12$$
$$15 - 5x - x = x - x - 12$$
$$15 - 6x = -12$$
$$15 - 15 - 6x = -12 - 15$$
$$-6x = -27$$
$$\frac{-6x}{-6} = \frac{-27}{-6}$$
$$x = \frac{9}{2}$$

**61.**
$$2x = -3x$$
$$2x + 3x = -3x + 3x$$
$$5x = 0$$
$$\frac{5x}{5} = \frac{0}{5}$$
$$x = 0$$

**63.**
$$2x - 5 + 10x = 3$$
$$12x - 5 = 3$$
$$12x - 5 + 5 = 3 + 5$$
$$12x = 8$$
$$\frac{12x}{12} = \frac{8}{12}$$
$$x = \frac{2}{3}$$

**65.** $5t - 4 + 3t = 8t - 4$
$$8t - 4 = 8t - 4 \quad \text{Identity}$$
This equation has infinitely many solutions.

**67.**
$$2(y - 9) = -5y - 4$$
$$2y - 18 = -5y - 4$$
$$2y + 5y - 18 = -5y + 5y - 4$$
$$7y - 18 = -4$$
$$7y - 18 + 18 = -4 + 18$$
$$7y = 14$$
$$\frac{7y}{7} = \frac{14}{7}$$
$$y = 2$$

**69.**
$$3(2 - 7x) = 3(4 - 7x)$$
$$6 - 21x = 12 - 21x$$
$$6 - 21x + 21x = 12 - 21x + 21x$$
$$6 \neq 12$$
The original equation has no solution.

**71.** *Verbal Model:* $\boxed{\text{Perimeter}} = 3 \cdot \boxed{\text{Length of a side}}$

*Labels:* Perimeter $= 225$ (cm)

Length of a side $= x$ (cm)

*Equation:*
$$225 = 3x$$
$$\frac{225}{3} = \frac{3x}{3}$$
$$75 = x$$
The length of each side is 75 centimeters.

**73.** *Verbal Model:* $2\boxed{\text{Length}} + 2\boxed{\text{Width}} = \boxed{\text{Perimeter}}$

*Labels:* Width $= w$ (inches)

Length $= 2w$ (inches)

Perimeter $= 120$ (inches)

*Equation:*
$$2(2w) + 2w = 120$$
$$4w + 2w = 120$$
$$6w = 120$$
$$w = 20 \text{ and } 2w = 40$$

The width of the flag is 20 inches and the length is 40 inches.

**75.** Yes. Subtract the cost of parts from the total to find the cost of labor. Then divide the labor cost by 44 to find the number of hours spent on labor.

*Verbal Model:*  $\boxed{\text{Cost of parts}} + \boxed{\begin{array}{l}\text{Hourly cost}\\\text{of labor}\end{array}} \cdot \boxed{\begin{array}{l}\text{Hours of}\\\text{labor}\end{array}} = \boxed{\text{Total cost}}$

*Labels:*  Cost of parts = 285        (dollars)

Hourly cost of labor = 44    (dollars per hour)

Hours of labor = $t$        (hours)

Total cost = 384        (dollars)

*Equation:*
$$285 + 44t = 384$$
$$285 - 285 + 44t = 384 - 285$$
$$44t = 99$$
$$\frac{44t}{44} = \frac{99}{44}$$
$$t = \frac{9}{4}$$

The time spent on labor was $\frac{9}{4}$ hours or $2\frac{1}{4}$ hours. This answer could also be expressed as 2.25 hours or as 2 hours and 15 minutes.

**77.** *Verbal Model:*  $\boxed{\text{Total revenue}} = \boxed{\begin{array}{l}\text{Revenue from}\\\text{main floor seats}\end{array}} + \boxed{\begin{array}{l}\text{Revenue from}\\\text{balcony seats}\end{array}}$

*Labels:*  Total revenue = 5200            (dollars)

Price per main floor seat = 10        (dollars per seat)

Number of main floor seats = 400    (seats)

Price per balcony seat = 8            (dollars per seat)

Number of balcony seats = $x$        (seats)

*Equation:*
$$5200 = 400(10) + 8x$$
$$5200 = 4000 + 8x$$
$$5200 - 4000 = 4000 - 4000 + 8x$$
$$1200 = 8x$$
$$\frac{1200}{8} = \frac{8x}{8}$$
$$150 = x$$

There were 150 balcony seats sold.

**79.** *Verbal Model:*  $\boxed{\text{Total wages}} = \boxed{\text{Rate for job 1}} \cdot \boxed{\text{Time at job 1}} + \boxed{\text{Rate for job 2}} \cdot \boxed{\text{Time at job 2}}$

*Labels:*  Total wages = 425        (dollars)

Rate for job 1 = 9.25        (dollars per hour)

Time at job 1 = 40        (hours)

Rate for job 2 = 10.00        (dollars per hour)

Time at job 2 = $x$        (hours)

*Equation:*
$$425 = 9.25(40) + 10x$$
$$425 = 370 + 10x$$
$$425 - 370 = 370 - 370 + 10x$$
$$55 = 10x$$
$$\frac{55}{10} = \frac{10x}{10}$$
$$5.5 = x$$

You must tutor for 5.5 hours.

**81.** *Verbal Model:*   $5(\boxed{\text{Number}} + \boxed{16}) = \boxed{100}$

  *Label:*        Number $= x$

  *Equation:*
$$5(x + 16) = 100$$
$$5x + 80 = 100$$
$$5x + 80 - 80 = 100 - 80$$
$$5x = 20$$
$$\frac{5x}{5} = \frac{20}{5}$$
$$x = 4$$

The number is 4.

**83.** *Verbal Model:*   $\boxed{\begin{array}{l}\text{First consecutive}\\\text{odd integer}\end{array}} + \boxed{\begin{array}{l}\text{Second consecutive}\\\text{odd integer}\end{array}} = \boxed{72}$

  *Labels:*       First consecutive odd integer $= 2n + 1$

            Second consecutive odd integer $= 2n + 3$

  *Equation:*
$$(2n + 1) + (2n + 3) = 72$$
$$2n + 1 + 2n + 3 = 72$$
$$4n + 4 = 72$$
$$4n + 4 - 4 = 72 - 4$$
$$4n = 68$$
$$\frac{4n}{4} = \frac{68}{4}$$
$$n = 17$$

$2n + 1 = 2(17) + 1 = 34 + 1 = 35$

$2n + 3 = 2(17) + 3 = 34 + 3 = 37$

The integers are 35 and 37.

**85.** *Verbal Model:*   $\boxed{\begin{array}{l}\text{First consecutive}\\\text{odd integer}\end{array}} + \boxed{\begin{array}{l}\text{Second consecutive}\\\text{odd integer}\end{array}} + \boxed{\begin{array}{l}\text{Third consecutive}\\\text{odd integer}\end{array}} = \boxed{159}$

  *Labels:*       First consecutive integer $= 2n + 1$

            Second consecutive integer $= 2n + 3$

            Third consecutive integer $= 2n + 5$

  *Equation:*
$$(2n + 1) + (2n + 3) + (2n + 5) = 159$$
$$2n + 1 + 2n + 3 + 2n + 5 = 159$$
$$6n + 9 = 159$$
$$6n = 150$$
$$n = 25$$

$2n + 1 = 2(25) + 1 = 51$

$2n + 3 = 2(25) + 3 = 53$

$2n + 5 = 2(25) + 5 = 55$

The integers are 51, 53, and 55.

**87.** *Verbal Model:*   | Weight of red box | + 3 · | Weight of blue box | = 9 · | Weight of blue box |

  *Labels:*          Weight of red box $= R$    (ounces)

                     Weight of blue box $= 1$    (ounce)

  *Equation:*        $R + 3(1) = 9(1)$

                     $R + 3 = 9$

                     $R + 3 - 3 = 9 - 3$

                     $R = 6$

The red box weighs 6 ounces. If you remove three blue boxes from each side, the scale would still balance. The red box would balance the remaining six blue boxes, showing that the red box weighs 6 ounces. This illustrates the addition (or subtraction) property of equality.

**89.** False.

Multiplying each side of an equation by 0 yields $0 = 0$. For example, multiplying each side of the equation $3x = 9$ by 0, yields the equation $0 = 0$. The equation $3x = 9$ has one solution, $x = 3$; however, the equation $0 = 0$ is an identity and has infinitely many solutions. Because the original equation $3x = 9$ and the resulting equation $0 = 0$ have different solutions, they are not equivalent equations.

**91.** False.

An odd integer could be expressed as $2m + 1$ and an even integer could be expressed as $2n$. When they are added together, $(2m + 1) + 2n = 2m + 1 + 2n$ or $2m + 2n + 1$. This expression could be written as $2(m + n) + 1$; it represents an odd number because it is not a multiple of 2.

**93.**

**95.**

**97.** $x - 8 = -9$

  (a) $x = -1$

$$-1 - 8 \overset{?}{=} -9$$

$$-9 = -9$$

  $-1$ *is* a solution.

  (b) $x = 2$

$$2 - 8 \overset{?}{=} -9$$

$$-6 \neq -9$$

  2 *is not* a solution.

**99.** $2x - 1 = 3$

  (a) $x = -1$

$$2(-1) - 1 \overset{?}{=} 3$$

$$-2 - 1 \overset{?}{=} 3$$

$$-3 \neq 3$$

  $-1$ *is not* a solution.

  (b) $x = 2$

$$2(2) - 1 \overset{?}{=} 3$$

$$4 - 1 \overset{?}{=} 3$$

$$3 = 3$$

  2 *is* a solution.

**101.** $x + 4 = 2x$

  (a) $x = -1$

$$-1 + 4 \overset{?}{=} 2(-1)$$

$$3 \neq -2$$

  $-1$ *is not* a solution.

  (b) $x = 2$

$$2 + 4 \overset{?}{=} 2(2)$$

$$6 \neq 4$$

  2 *is not* a solution.

## Section 3.2  Equations That Reduce to Linear Form

1.  $2(y - 4) = 0$
    $2y - 8 = 0$
    $2y - 8 + 8 = 0 + 8$
    $2y = 8$
    $\dfrac{2y}{2} = \dfrac{8}{2}$
    $y = 4$

3.  $-5(t + 3) = 10$
    $-5t - 15 = 10$
    $-5t - 15 + 15 = 10 + 15$
    $-5t = 25$
    $\dfrac{-5t}{-5} = \dfrac{25}{-5}$
    $t = -5$

5.  $25(z - 2) = 60$
    $25z - 50 = 60$
    $25z - 50 + 50 = 60 + 50$
    $25z = 110$
    $\dfrac{25z}{25} = \dfrac{110}{25}$
    $z = \dfrac{22}{5}$

7.  $7(x + 5) = 49$
    $7x + 35 = 49$
    $7x + 35 - 35 = 49 - 35$
    $7x = 14$
    $\dfrac{7x}{7} = \dfrac{14}{7}$
    $x = 2$

9.  $4 - (z + 6) = 8$
    $4 - z - 6 = 8$
    $-z - 2 = 8$
    $-z - 2 + 2 = 8 + 2$
    $-z = 10$
    $(-1)(-z) = (-1)(10)$
    $z = -10$

11. $3 - (2x - 18) = 3$
    $3 - 2x + 18 = 3$
    $-2x + 21 = 3$
    $-2x + 21 - 21 = 3 - 21$
    $-2x = -18$
    $\dfrac{-2x}{-2} = \dfrac{-18}{-2}$
    $x = 9$

13. $12(x - 3) = 0$
    $12x - 36 = 0$
    $12x - 36 + 36 = 0 + 36$
    $12x = 36$
    $\dfrac{12x}{12} = \dfrac{36}{12}$
    $x = 3$

15. $5(x - 4) = 2(2x + 5)$
    $5x - 20 = 4x + 10$
    $5x - 4x - 20 = 4x - 4x + 10$
    $x - 20 = 10$
    $x - 20 + 20 = 10 + 20$
    $x = 30$

17. $3(2x - 1) = 3(2x + 5)$
    $6x - 3 = 6x + 15$
    $6x - 3 - 6x = 6x + 15 - 6x$
    $-3 = 15$  False

    This equation has no solution.

19. $-3(x + 4) = 4(x + 4)$
    $-3x - 12 = 4x + 16$
    $-3x - 4x - 12 = 4x - 4x + 16$
    $-7x - 12 = 16$
    $-7x - 12 + 12 = 16 + 12$
    $-7x = 28$
    $\dfrac{-7x}{-7} = \dfrac{28}{-7}$
    $x = -4$

21. $7 = 3(x + 2) - 3(x - 5)$
    $7 = 3x + 6 - 3x + 15$
    $7 = 21$    False

    This equation has no solution.

**23.** $7x - 2(x - 2) = 12$

$$7x - 2x + 4 = 12$$
$$5x + 4 = 12$$
$$5x + 4 - 4 = 12 - 4$$
$$5x = 8$$
$$\frac{5x}{5} = \frac{8}{5}$$
$$x = \frac{8}{5}$$

**25.** $4 - (y - 3) = 3(y + 1) - 4(1 - y)$

$$4 - y + 3 = 3y + 3 - 4 + 4y$$
$$-y + 7 = 7y - 1$$
$$-y - 7y + 7 = 7y - 7y - 1$$
$$-8y + 7 = -1$$
$$-8y + 7 - 7 = -1 - 7$$
$$-8y = -8$$
$$\frac{-8y}{-8} = \frac{-8}{-8}$$
$$y = 1$$

**27.** $-6(3 + x) + 2(3x + 5) = 0$

$$-18 - 6x + 6x + 10 = 0$$
$$-18 + 10 = 0 \quad \text{False}$$

This equation has no solution.

**29.** $2\big[(3x + 5) - 7\big] = 3(4x - 3)$

$$2\big[3x + 5 - 7\big] = 12x - 9$$
$$2(3x - 2) = 12x - 9$$
$$6x - 4 = 12x - 9$$
$$6x - 12x - 4 = 12x - 12x - 9$$
$$-6x - 4 = -9$$
$$-6x - 4 + 4 = -9 + 4$$
$$-6x = -5$$
$$\frac{-6x}{-6} = \frac{-5}{-6}$$
$$x = \frac{5}{6}$$

**31.** $4x + 3\big[x - 2(2x - 1)\big] = 4 - 3x$

$$4x + 3\big[x - 4x + 2\big] = 4 - 3x$$
$$4x + 3\big[-3x + 2\big] = 4 - 3x$$
$$4x - 9x + 6 = 4 - 3x$$
$$-5x + 6 = 4 - 3x$$
$$-5x + 3x + 6 = 4 - 3x + 3x$$
$$-2x + 6 = 4$$
$$-2x + 6 - 6 = 4 - 6$$
$$-2x = -2$$
$$\frac{-2x}{-2} = \frac{-2}{-2}$$
$$x = 1$$

**33.** $\dfrac{y}{5} = \dfrac{3}{5}$

$$5\left(\frac{y}{5}\right) = 5\left(\frac{3}{5}\right)$$
$$y = 3$$

**35.** $\dfrac{y}{5} = -\dfrac{3}{10}$

$$5\left(\frac{y}{5}\right) = 5\left(-\frac{3}{10}\right)$$
$$y = -\frac{3}{2}$$

**37.** $\dfrac{6x}{25} = \dfrac{3}{5}$

$$30x = 75 \quad \text{(Cross-multiply)}$$
$$\frac{30x}{30} = \frac{75}{30}$$
$$x = \frac{75}{30}$$
$$x = \frac{5}{2}$$

**39.** $\dfrac{5x}{4} + \dfrac{1}{2} = 0$

$$4\left(\frac{5x}{4} + \frac{1}{2}\right) = 4(0)$$
$$4\left(\frac{5x}{4}\right) + 4\left(\frac{1}{2}\right) = 0$$
$$5x + 2 = 0$$
$$5x + 2 - 2 = 0 - 2$$
$$5x = -2$$
$$\frac{5x}{5} = \frac{-2}{5}$$
$$x = -\frac{2}{5}$$

**41.**
$$\frac{x}{5} - \frac{1}{2} = 3$$
$$10\left(\frac{x}{5} - \frac{1}{2}\right) = 10(3)$$
$$10\left(\frac{x}{5}\right) - 10\left(\frac{1}{2}\right) = 30$$
$$2x - 5 = 30$$
$$2x - 5 + 5 = 30 + 5$$
$$2x = 35$$
$$\frac{2x}{2} = \frac{35}{2}$$
$$x = \frac{35}{2}$$

**43.**
$$\frac{x}{5} - \frac{x}{2} = 1$$
$$10\left(\frac{x}{5} - \frac{x}{2}\right) = 10(1)$$
$$10\left(\frac{x}{5}\right) - 10\left(\frac{x}{2}\right) = 10$$
$$2x - 5x = 10$$
$$-3x = 10$$
$$\frac{-3x}{-3} = \frac{10}{-3}$$
$$x = -\frac{10}{3}$$

**45.**
$$2s + \frac{3}{2} = 2s + 2$$
$$2s - 2s + \frac{3}{2} = 2s - 2s + 2$$
$$\frac{3}{2} = 2 \qquad \text{False}$$

This equation has no solution.

**47.**
$$3x + \frac{1}{4} = \frac{3}{4}$$
$$4\left(3x + \frac{1}{4}\right) = 4\left(\frac{3}{4}\right)$$
$$12x + 1 = 3$$
$$12x + 1 - 1 = 3 - 1$$
$$12x = 2$$
$$\frac{12x}{12} = \frac{2}{12}$$
$$x = \frac{1}{6}$$

**49.**
$$\frac{1}{5}x + 1 = \frac{3}{10}x - 4$$
$$10\left(\frac{1}{5}x + 1\right) = 10\left(\frac{3}{10}x - 4\right)$$
$$10\left(\frac{1}{5}x\right) + 10(1) = 10\left(\frac{3}{10}x\right) - 10(4)$$
$$2x + 10 = 3x - 40$$
$$2x - 3x + 10 = 3x - 3x - 40$$
$$-x + 10 = -40$$
$$-x + 10 - 10 = -40 - 10$$
$$-x = -50$$
$$-1(-x) = -1(-50)$$
$$x = 50$$

**51.**
$$\frac{2}{3}(z + 5) - \frac{1}{4}(z + 24) = 0$$
$$12\left[\frac{2}{3}(z + 5) - \frac{1}{4}(z + 24)\right] = 12(0)$$
$$12 \cdot \frac{2}{3}(z + 5) - 12 \cdot \frac{1}{4}(z + 24) = 0$$
$$8(z + 5) - 3(z + 24) = 0$$
$$8z + 40 - 3z - 72 = 0$$
$$5z - 32 = 0$$
$$5z - 32 + 32 = 0 + 32$$
$$5z = 32$$
$$\frac{5z}{5} = \frac{32}{5}$$
$$z = \frac{32}{5}$$

**53.**
$$\frac{100 - 4u}{3} = \frac{5u + 6}{4} + 6$$
$$12\left(\frac{100 - 4u}{3}\right) = 12\left(\frac{5u + 6}{4} + 6\right)$$
$$\frac{12}{1}\left(\frac{100 - 4u}{3}\right) = \frac{12}{1}\left(\frac{5u + 6}{4}\right) + 12(6)$$
$$4(100 - 4u) = 3(5u + 6) + 72$$
$$400 - 16u = 15u + 18 + 72$$
$$400 - 16u = 15u + 90$$
$$400 - 16u - 15u = 15u - 15u + 90$$
$$400 - 31u = 90$$
$$400 - 400 - 31u = 90 - 400$$
$$-31u = -310$$
$$\frac{-31u}{-31} = \frac{-310}{-31}$$
$$u = 10$$

**55.**
$$\left(\frac{2}{5}x - 1\right) + \left(\frac{1}{2}x\right) + \left(\frac{1}{10}x + 6\right) = 15$$

$$10\left(\frac{2}{5}x - 1\right) + 10\left(\frac{1}{2}x\right) + 10\left(\frac{1}{10}x + 6\right) = 10(15)$$

$$4x - 10 + 5x + x + 60 = 150$$

$$10x + 50 = 150$$

$$10x + 50 - 50 = 150 - 50$$

$$10x = 100$$

$$\frac{10x}{10} = \frac{100}{10}$$

$$x = 10$$

**57.**
$$\frac{t + 4}{6} = \frac{2}{3}$$

$$3(t + 4) = 12 \quad \text{(Cross-multiply)}$$

$$3t + 12 = 12$$

$$3t + 12 - 12 = 12 - 12$$

$$3t = 0$$

$$\frac{3t}{3} = \frac{0}{3}$$

$$t = 0$$

**59.**
$$\frac{x - 2}{5} = \frac{2}{3}$$

$$3(x - 2) = 5(2) \quad \text{(Cross-multiply)}$$

$$3x - 6 = 10$$

$$3x - 6 + 6 = 10 + 6$$

$$3x = 16$$

$$\frac{3x}{3} = \frac{16}{3}$$

$$x = \frac{16}{3}$$

**61.**
$$\frac{5x - 4}{4} = \frac{2}{3}$$

$$3(5x - 4) = 4(2) \quad \text{(Cross-multiply)}$$

$$15x - 12 = 8$$

$$15x - 12 + 12 = 8 + 12$$

$$15x = 20$$

$$\frac{15x}{15} = \frac{20}{15}$$

$$x = \frac{4}{3}$$

**63.**
$$\frac{x}{4} = \frac{1 - 2x}{3}$$

$$3(x) = 4(1 - 2x) \quad \text{(Cross-multiply)}$$

$$3x = 4 - 8x$$

$$3x + 8x = 4 - 8x + 8x$$

$$11x = 4$$

$$\frac{11x}{11} = \frac{4}{11}$$

$$x = \frac{4}{11}$$

**65.**
$$\frac{10 - x}{2} = \frac{x + 4}{5}$$

$$5(10 - x) = 2(x + 4) \quad \text{(Cross-multiply)}$$

$$50 - 5x = 2x + 8$$

$$50 - 5x - 2x = 2x - 2x + 8$$

$$50 - 7x = 8$$

$$50 - 50 - 7x = 8 - 50$$

$$-7x = -42$$

$$\frac{-7x}{-7} = \frac{-42}{-7}$$

$$x = 6$$

**67.** $3 + 0.03x = 5$

The power of ten that is needed to clear the decimals in this equation is 100 or $10^2$.

**69.** $1.205x - 0.003 = 0.5$

The power of ten that is needed to clear the decimals in this equation is 1000 or $10^3$.

**71.**
$$0.2x + 5 = 6$$

$$0.2x + 5 - 5 = 6 - 5$$

$$0.2x = 1$$

$$10(0.2x) = 10(1)$$

$$2x = 10$$

$$\frac{2x}{2} = \frac{10}{2}$$

$$x = 5 \text{ or } 5.00$$

**73.**
$$0.234x + 1 = 2.805$$

$$0.234x + 1 - 1 = 2.805 - 1$$

$$0.234x = 1.805$$

$$\frac{0.234x}{0.234} = \frac{1.805}{0.234}$$

$$x \approx 7.71$$

**75.**
$$0.42x - 0.4(x + 2.4) = 0.3(5)$$
$$0.42x - 0.4x - 0.96 = 1.5$$
$$0.02x - 0.96 = 1.5$$
$$100(0.02x - 0.96) = 100(1.5)$$
$$2x - 96 = 150$$
$$2x - 96 + 96 = 150 + 96$$
$$2x = 246$$
$$\frac{2x}{2} = \frac{246}{2}$$
$$x = 123$$

**77.**
$$\frac{x}{3.25} + 1 = 2.08$$
$$\frac{x}{3.25} + 1 - 1 = 2.08 - 1$$
$$\frac{x}{3.25} = 1.08$$
$$\frac{x}{3.25}(3.25) = 1.08(3.25)$$
$$x = 3.51$$

**79.**
$$\frac{x}{3.155} = 2.850$$
$$3.155\left(\frac{x}{3.155}\right) = 3.155(2.850)$$
$$x \approx 8.99$$

**81.**
$$\frac{t}{10} + \frac{t}{15} = 0.8$$
$$30\left(\frac{t}{10} + \frac{t}{15}\right) = 30(0.8)$$
$$30\left(\frac{t}{10}\right) + 30\left(\frac{t}{15}\right) = 24$$
$$3t + 2t = 24$$
$$5t = 24$$
$$\frac{5t}{5} = \frac{24}{5}$$
$$t = \frac{24}{5} \text{ or } 4.8 \text{ hours}$$

The required time is 4.8 hours.

**83.**
$$\frac{87 + 92 + 84 + x}{4} = 90$$
$$\frac{263 + x}{4} = 90$$
$$263 + x = 4(90)$$
$$263 + x = 360$$
$$263 + x - 263 = 360 - 263$$
$$x = 97$$

You must score 97 on the fourth exam.

**85.**
$$p_1x + p_2(a - x) = p_3a$$
$$0.1x + 0.3(100 - x) = 0.25(100)$$
$$0.1x + 30 - 0.3x = 25$$
$$-0.2x + 30 = 25$$
$$-0.2x + 30 - 30 = 25 - 30$$
$$-0.2x = -5$$
$$\frac{-0.2x}{-0.2} = \frac{-5}{-0.2}$$
$$x = 25$$

So, 25 quarts of the 10% solution must be used.

**87.**
$$N = 2.70t + 281.9$$
$$360 = 2.70t + 281.9$$
$$360 - 281.9 = 2.70t$$
$$78.1 = 2.70t$$
$$\frac{78.1}{2.70} = \frac{2.70t}{2.70}$$
$$28.9 \approx t$$

According to the model, the population will exceed 360 million during the year 2028.

**89.** Each brick is 8 inches long, so the $n$ bricks in each row have a combined length of $8n$ inches. There is $\frac{1}{2}$ inch of mortar between adjoining bricks and there are $n - 1$ mortar joints between the $n$ bricks; therefore, the mortar joints in each row have a combined length of $\frac{1}{2}(n - 1)$ inches. So, the $8n$ inches of bricks plus the $\frac{1}{2}(n - 1)$ inches of mortar joints equal the 93 inches of the width of the fireplace.

**91.** You could begin by dividing both sides of the equation by 3.
$$3(x - 7) = 15$$
$$\frac{3(x - 7)}{3} = \frac{15}{3}$$
$$x - 7 = 5$$
$$x = 12$$

**93.** Dividing each side of an equation by a variable factor assumes that the variable does not equal zero, and this assumption may yield a false solution.

**95.**
$$(-2x)^2 x^4 = (-2x)(-2x)x^4$$
$$= (-2)(-2)x \cdot x \cdot x \cdot x \cdot x \cdot x$$
$$= 4x^6$$

**97.** $5z^3(z^2) = 5 \cdot z \cdot z \cdot z \cdot z \cdot z = 5z^5$

**99.** $\dfrac{5x}{3} - \dfrac{2x}{3} - 4 = \dfrac{5}{3}x - \dfrac{2}{3}x - 4$

$\qquad\qquad\qquad = \left(\dfrac{5}{3} - \dfrac{2}{3}\right)x - 4$

$\qquad\qquad\qquad = \dfrac{3}{3}x - 4$

$\qquad\qquad\qquad = x - 4$

**105.** $\qquad 4 - 2x = 22$

$\qquad 4 - 2x - 4 = 22 - 4$

$\qquad\qquad -2x = 18$

$\qquad\qquad \dfrac{-2x}{-2} = \dfrac{18}{-2}$

$\qquad\qquad\quad x = -9$

**101.** $-y^2\left(y^2 + 4\right) + 6y^2 = -y^2 \cdot y^2 + \left(-y^2\right)4 + 6y^2$

$\qquad\qquad\qquad\qquad\quad = -y^4 - 4y^2 + 6y^2$

$\qquad\qquad\qquad\qquad\quad = -y^4 + 2y^2$

**103.** $\qquad 3x - 5 = 12$

$\qquad 3x - 5 + 5 = 12 + 5$

$\qquad\qquad\quad 3x = 17$

$\qquad\qquad\quad \dfrac{3x}{3} = \dfrac{17}{3}$

$\qquad\qquad\quad\ x = \dfrac{17}{3}$

# Section 3.3   Problem Solving with Percents

**1.** *Verbal Model:*   $\boxed{\text{Decimal}} \cdot \boxed{100\%} = \boxed{\text{Percent}}$

  *Label:*        Percent $= x$

  *Equation:*     $0.62(100\%) = x$

  $\qquad\qquad\qquad 62\% = x$

**3.** *Verbal Model:*   $\boxed{\text{Decimal}} \cdot \boxed{100\%} = \boxed{\text{Percent}}$

  *Label:*        Percent $= x$

  *Equation:*     $0.20(100\%) = x$

  $\qquad\qquad\qquad 20\% = x$

**5.** *Verbal Model:*   $\boxed{\text{Decimal}} \cdot \boxed{100\%} = \boxed{\text{Percent}}$

  *Label:*        Percent $= x$

  *Equation:*     $0.075(100\%) = x$

  $\qquad\qquad\qquad 7.5\% = x$

**7.** *Verbal Model:*   $\boxed{\text{Decimal}} \cdot \boxed{100\%} = \boxed{\text{Percent}}$

  *Label:*        Decimal $= x$

  *Equation:*     $2.38(100\%) = x$

  $\qquad\qquad\qquad 238\% = x$

**9.** *Verbal Model:*   $\boxed{\text{Fraction}} \cdot \boxed{100\%} = \boxed{\text{Percent}}$

  *Label:*        Percent $= x$

  *Equation:*     $\frac{4}{5}(100\%) = x$

  $\qquad\qquad\qquad 80\% = x$

**11.** *Verbal Model:*   $\boxed{\text{Fraction}} \cdot \boxed{100\%} = \boxed{\text{Percent}}$

  *Label:*        Percent $= x$

  *Equation:*     $\left(\frac{5}{4}\right)(100\%) = x$

  $\qquad\qquad\qquad 125\% = x$

**13.** *Verbal Model:*   $\boxed{\text{Fraction}} \cdot \boxed{100\%} = \boxed{\text{Percent}}$

  *Label:*        Percent $= x$

  *Equation:*     $\frac{5}{6}(100\%) = x$

  $\qquad\qquad\qquad 83\frac{1}{3}\% = x$

**15.** *Verbal Model:*   $\boxed{\text{Fraction}} \cdot \boxed{100\%} = \boxed{\text{Percent}}$

  *Label:*        Percent $= x$

  *Equation:*     $\frac{21}{20}(100\%) = x$

  $\qquad\qquad\qquad 105\% = x$

**17.** *Verbal Model:*   $\boxed{\text{Decimal}} \cdot \boxed{100\%} = \boxed{\text{Percent}}$

  *Label:*        Decimal $= x$

  *Equation:*     $x(100\%) = 12.5\%$

  $\qquad\qquad\qquad x = \dfrac{12.5\%}{100\%}$

  $\qquad\qquad\qquad x = 0.125$

**19.** *Verbal Model:* $\boxed{\text{Decimal}} \cdot \boxed{100\%} = \boxed{\text{Percent}}$

*Label:*        Decimal $= x$

*Equation:*     $x(100\%) = 125\%$

$$x = \frac{125\%}{100\%}$$

$$x = 1.25$$

**21.** *Verbal Model:* $\boxed{\text{Decimal}} \cdot \boxed{100\%} = \boxed{\text{Percent}}$

*Label:*        Decimal $= x$

*Equation:*     $x(100\%) = 8.5\%$

$$x = \frac{8.5\%}{100\%}$$

$$x = 0.085$$

**23.** *Verbal Model:* $\boxed{\text{Decimal}} \cdot \boxed{100\%} = \boxed{\text{Percent}}$

*Label:*        Decimal $= x$

*Equation:*     $x(100\%) = \frac{3}{4}\%$

$$x = \frac{\frac{3}{4}\%}{100\%}$$

$$x = \frac{0.75\%}{100\%}$$

$$x = 0.0075$$

**25.** *Verbal Model:* $\boxed{\text{Fraction}} \cdot \boxed{100\%} = \boxed{\text{Percent}}$

*Label:*        Fraction $= x$

*Equation:*     $x(100\%) = 37.5\%$

$$x = \frac{37.5\%}{100\%}$$

$$x = \frac{375}{1000}$$

$$x = \frac{3(125)}{8(125)}$$

$$x = \frac{3}{8}$$

**27.** *Verbal Model:* $\boxed{\text{Fraction}} \cdot \boxed{100\%} = \boxed{\text{Percent}}$

*Label:*        Fraction $= x$

*Equation:*     $x(100\%) = 130\%$

$$x = \frac{130\%}{100\%}$$

$$x = \frac{13}{10}$$

**29.** *Verbal Model:* $\boxed{\text{Fraction}} \cdot \boxed{100\%} = \boxed{\text{Percent}}$

*Label:*        Fraction $= x$

*Equation:*     $x(100\%) = 1.4\%$

$$x = \frac{1.4\%}{100\%}$$

$$x = \frac{14}{1000}$$

$$x = \frac{7}{500}$$

**31.** *Verbal Model:* $\boxed{\text{Fraction}} \cdot \boxed{100\%} = \boxed{\text{Percent}}$

*Label:*        Fraction $= x$

*Equation:*     $x(100\%) = \frac{1}{2}\%$

$$x = \frac{\frac{1}{2}\%}{100\%}$$

$$x = \frac{2\left(\frac{1}{2}\right)}{2(100)}$$

$$x = \frac{1}{200}$$

**33.**

| Percent | Parts out of 100 | Decimal | Fraction |
|---------|------------------|---------|----------|
| 40% | 40 | 0.4 | $\frac{2}{5}$ |

(a) 40% means 40 parts out of 100.

(b) *Verbal Model:* $\boxed{\text{Decimal}} \cdot \boxed{100\%} = \boxed{\text{Percent}}$

    *Label:*        Decimal $= x$

    *Equation:*     $x(100\%) = 40\%$

$$x = \frac{40\%}{100\%}$$

$$x = 0.4$$

(c) *Verbal Model:* $\boxed{\text{Fraction}} \cdot \boxed{100\%} = \boxed{\text{Percent}}$

    *Label:*        Fraction $= x$

    *Equation:*     $x(100\%) = 40\%$

$$x = \frac{40\%}{100\%}$$

$$x = \frac{2}{5}$$

**35.**

| Percent | Parts out of 100 | Decimal | Fraction |
|---------|------------------|---------|----------|
| 7.5% | 7.5 | 0.075 | $\frac{3}{40}$ |

(a) 7.5% means 7.5 parts out of 100.

(b) *Verbal Model:* ⬚Decimal · ⬚100% = ⬚Percent

    *Label:*        Decimal $= x$

    *Equation:*    $x(100\%) = 7.5\%$

$$x = \frac{7.5\%}{100\%}$$

$$x = 0.075$$

(c) *Verbal Model:* ⬚Fraction · ⬚100% = ⬚Percent

    *Label:*        Fraction $= x$

    *Equation:*    $x(100\%) = 7.5\%$

$$x = \frac{7.5\%}{100\%}$$

$$x = \frac{75}{1000}$$

$$x = \frac{3}{40}$$

**37.**

| Percent | Parts out of 100 | Decimal | Fraction |
|---------|------------------|---------|----------|
| 63% | 63 | 0.63 | $\frac{63}{100}$ |

(a) 63 parts out of 100 means 63%.

(b) *Verbal Model:* ⬚Decimal · ⬚100% = ⬚Percent

    *Label:*        Decimal $= x$

    *Equation:*    $x(100\%) = 63\%$

$$x = \frac{63\%}{100\%}$$

$$x = 0.63$$

(c) *Verbal Model:* ⬚Fraction · ⬚100% = ⬚Percent

    *Label:*        Fraction $= x$

    *Equation:*    $x(100\%) = 63\%$

$$x = \frac{63\%}{100\%}$$

$$x = \frac{63}{100}$$

**39.**

| Percent | Parts out of 100 | Decimal | Fraction |
|---------|------------------|---------|----------|
| 15.5% | 15.5 | 0.155 | $\frac{31}{200}$ |

(a) *Verbal Model:* ⬚Decimal · ⬚100% = ⬚Percent

    *Label:*        Percent $= x$

    *Equation:*    $(0.155)(100\%) = x$

$$15.5\% = x$$

(b) 15.5% means 15.5 parts out of 100.

(c) *Verbal Model:* ⬚Fraction · ⬚100% = ⬚Percent

    *Label:*        Fraction $= x$

    *Equation:*    $x(100\%) = 15.5\%$

$$x = \frac{15.5\%}{100\%}$$

$$x = \frac{155\%}{1000\%} = \frac{31}{200}$$

**41.**

| Percent | Parts out of 100 | Decimal | Fraction |
|---------|------------------|---------|----------|
| 60% | 60 | 0.6 | $\frac{3}{5}$ |

(a) *Verbal Model:* ⬚Fraction · ⬚100% = ⬚Percent

    *Label:*        Percent $= x$

    *Equation:*    $\frac{3}{5}(100\%) = x$

$$60\% = x$$

(b) 60% means 60 parts out of 100.

(c) *Verbal Model:* ⬚Decimal · ⬚100% = ⬚Percent

    *Label:*        Decimal $= x$

    *Equation:*    $x(100\%) = 60\%$

$$x = \frac{60\%}{100\%}$$

$$x = 0.6$$

**43.** $\frac{3}{8}$ of the figure is shaded.

$$\frac{3}{8}(100\%) = 37\frac{1}{2}\% \text{ or } 37.5\%$$

**45.** $\frac{150}{360}$ of the figure is shaded.

$$\frac{150}{360} = \frac{5}{12}$$

$$\frac{5}{12}(100\%) = 41\frac{2}{3}\% \approx 41.67\%$$

**47.** *Verbal Model:* $\boxed{\text{What number}} = \boxed{30\% \text{ of } 150}$ $(a = pb)$

*Label:* $a = $ unknown number

*Percent equation:* $a = 0.30(150)$

$a = 45$

Therefore, 45 is 30% of 150.

**49.** *Verbal Model:* $\boxed{\text{What number}} = \boxed{0.75\% \text{ of } 56}$ $(a = pb)$

*Label:* $a = $ unknown number

*Percent equation:* $a = 0.0075(56)$

$a = 0.42$

Note: $0.75\% = 0.0075$

Therefore, 0.42 is 0.75% of 56.

**51.** *Verbal Model:* $\boxed{903} = \boxed{43\% \text{ of what number}}$ $(a = pb)$

*Label:* $b = $ unknown number

*Percent equation:* $903 = 0.43b$

$\dfrac{903}{0.43} = b$

$2100 = b$

Therefore, 903 is 43% of 2100.

**53.** *Verbal Model:* $\boxed{594} = \boxed{450\% \text{ of what number}}$ $(a = pb)$

*Label:* $b = $ unknown number

*Percent equation:* $594 = 4.50b$

$\dfrac{594}{4.50} = b$

$132 = b$

Therefore, 594 is 450% of 132.

**55.** *Verbal Model:* $\boxed{576} = \boxed{\text{What per cent of } 800}$ $(a = pb)$

*Label:* $p = $ unknown percent

(in decimal form)

*Percent equation:* $576 = p(800)$

$\dfrac{576}{800} = p$

$0.72 = p$

Therefore, 576 is 72% of 800.

**57.** *Verbal Model:* $\boxed{22} = \boxed{\text{What percent of } 800}$ $(a = pb)$

*Label:* $p = $ unknown percent

(in decimal form)

*Percent equation:* $22 = p(800)$

$\dfrac{22}{800} = p$

$0.0275 = p$

Therefore, 22 is 2.75% of 800.

**59.** *Verbal Model:* $\boxed{\text{Selling price}} = \boxed{\text{Cost}} + \boxed{\text{Markup}}$

*Labels:* Selling price = $49.95

Cost = $26.97

Markup = $x$ (dollars)

*Equation:* $49.95 = 26.97 + x$

$49.95 - 26.97 = x$

$22.98 = x$

The markup is $22.98.

*Verbal Model:* $\boxed{\text{Markup}} = \boxed{\text{Markup rate}} \cdot \boxed{\text{Cost}}$

*Labels:* Markup = $22.98

Markup rate = $p$ $\left(\begin{array}{l}\text{percent in}\\\text{decimal form}\end{array}\right)$

Cost = $26.97

*Equation:* $22.98 = p(26.97)$

$\dfrac{22.98}{26.97} = p$

$0.852 \approx p$

The markup rate is approximately 85.2%.

**61.** *Verbal Model:* $\boxed{\text{Selling price}} = \boxed{\text{Cost}} + \boxed{\text{Markup}}$

*Labels:* Selling price = $74.38

Cost = $c$ (dollars)

Markup rate = 0.815

Markup = $0.815c$

*Equation:* $74.38 = c + 0.815c$

$74.38 = 1.815c$

$\dfrac{74.38}{1.815} = c$

$40.98 \approx c$

$0.815c \approx 0.815(40.98) \approx 33.40$

The cost is approximately $40.98 and the markup is $33.40.

**63.** *Verbal Model:*   $\boxed{\text{Selling price}} = \boxed{\text{Cost}} + \boxed{\text{Markup}}$

*Labels:*   Selling price = \$125.98

Cost = $c$

Markup = \$56.69

Markup rate = $p$ $\left(\begin{array}{l}\text{percent in} \\ \text{decimal form}\end{array}\right)$

*Equation:*   $125.98 = c + 56.69$

$125.98 - 56.69 = c$

$69.29 = c$

*Verbal Model:*   $\boxed{\text{Markup}} = \boxed{\text{Markup rate}} \cdot \boxed{\text{Cost}}$

*Equation:*   $56.69 = p(69.29)$

$\dfrac{56.69}{69.29} = p$

$0.818 \approx p$

The cost is \$69.29 and the markup rate is approximately 81.8%.

**65.** *Verbal Model:*   $\boxed{\text{Selling price}} = \boxed{\text{Cost}} + \boxed{\text{Markup}}$

*Labels:*   Selling price = \$15,900

Cost = $c$        (dollars)

Markup rate = $p$ $\left(\begin{array}{l}\text{percent in} \\ \text{decimal form}\end{array}\right)$

Markup = \$2650

*Equation:*   $15,900 = c + 2650$

$15,900 - 2650 = c$

$13,250 = c$

*Verbal Model:*   $\boxed{\text{Markup}} = \boxed{\text{Markup rate}} \cdot \boxed{\text{Cost}}$

*Equation:*   $2650 = p(13,250)$

$\dfrac{2650}{13,250} = p$

$0.2 = p$

The cost is \$13,250 and the markup rate is 20%.

**67.** *Verbal Model:*   $\boxed{\text{Selling price}} = \boxed{\text{Cost}} + \boxed{\text{Markup}}$

*Labels:*   Selling price = $x$        (dollars)

Cost = \$107.97

Markup rate = 0.852 $\left(\begin{array}{l}\text{percent in} \\ \text{decimal form}\end{array}\right)$

Markup = $0.852(107.97) \approx$ \$91.99

*Equation:* $x = 107.97 + 91.99 = 199.96$

The markup is \$91.99 and the selling price is \$199.96.

**69.** *Verbal Model:*   $\boxed{\text{Sale price}} = \boxed{\text{List price}} - \boxed{\text{Discount}}$

*Labels:*   Sale price = \$29.95

List price = \$39.95

Discount rate = $p$ $\left(\begin{array}{l}\text{percent in} \\ \text{decimal form}\end{array}\right)$

Discount = $p(39.95)$  (dollars)

*Equation:*   $29.95 = 39.95 - p(39.95)$

$29.95 - 39.95 = -p(39.95)$

$-10.00 = -p(39.95)$

$\dfrac{-10.00}{-39.95} = p$

$0.2503 \approx p$

The discount is \$39.95 - \$29.95 = \$10.00; the discount rate is approximately 25.03% (approximately 25%).

**71.** *Verbal Model:*   $\boxed{\text{Sale price}} = \boxed{\text{List price}} - \boxed{\text{Discount}}$

*Labels:*   Sale price = \$18.95

List price = $x$        (dollars)

Discount rate = 0.2 $\left(\begin{array}{l}\text{percent in} \\ \text{decimal form}\end{array}\right)$

Discount = $0.2x$        (dollars)

*Equation:*   $18.95 = x - 0.2x$

$18.95 = 0.8x$

$\dfrac{18.95}{0.8} = x$

$23.69 \approx x$

Note: $x - 0.2x = (1 - 0.2)x = 0.8x$

The list price is \$23.69 and the discount is \$23.69 - \$18.95 = \$4.74.

**73.** *Verbal Model:*   $\boxed{\text{Sale price}} = \boxed{\text{List price}} - \boxed{\text{Discount}}$

*Labels:*   Sale price = $x$        (dollars)

List price = \$189.99

Discount rate = $p$ $\left(\begin{array}{l}\text{percent in} \\ \text{decimal form}\end{array}\right)$

Discount = \$30.00

*Equation:*   $x = 189.99 - 30.00 = 159.99$

*Verbal Model:* $\boxed{\text{Discount}} = \boxed{\text{Discount rate}} \cdot \boxed{\text{List price}}$

*Equation:*   $30.00 = p(189.99)$

$\dfrac{30.00}{189.99} = p$

$0.158 \approx p$

The sale price is \$159.99 and the discount rate is approximately 15.8%.

**75.** *Verbal Model:* $\boxed{\text{Sale price}} = \boxed{\text{List price}} - \boxed{\text{Discount}}$

*Labels:*      Sale price $= x$ (dollars)

            List price $= \$119.96$

            Discount rate $= 0.50$ $\left(\begin{array}{l}\text{percent in}\\ \text{decimal form}\end{array}\right)$

            Discount $= 0.50(119.96)$

                    $\approx 59.98$ (dollars)

*Equation:*     $x = 119.96 - 59.98 = 59.98$

The sale price is \$59.98 and the discount is \$59.98.

**77.** *Verbal Model:* $\boxed{\text{Sale price}} = \boxed{\text{List price}} - \boxed{\text{Discount}}$

*Labels:*      Sale price $= \$695.00$

            List price $= x$ (dollars)

            Discount rate $= p$ $\left(\begin{array}{l}\text{percent in}\\ \text{decimal form}\end{array}\right)$

            Discount $= \$300$

*Equation:*         $695.00 = x - 300$

          $695.00 + 300 = x$

               $995 = x$

*Verbal Model:* $\boxed{\text{Discount}} = \boxed{\text{Discount rate}} \cdot \boxed{\text{List price}}$

*Equation:*      $300 = p(995)$

          $\dfrac{300}{995} = p$

          $0.302 \approx P$

The list price is \$995.00 and the discount rate is approximately 30.2%.

**79.** *Verbal Model:* $\boxed{\text{Rent}} = \boxed{\text{Percent of income}}$ $(a = pb)$

*Labels:*      $a =$ rent (dollars)

            $p = 0.17$ $\left(\begin{array}{l}\text{percent in}\\ \text{decimal form}\end{array}\right)$

            $b =$ income $= \$3200$

*Percent equation:* $a = 0.17(3200)$

                $a = 544$

The monthly rent payment is \$544.

**81.** *Verbal Model:* $\boxed{\begin{array}{l}\text{Annual retirement}\\ \text{fund contribution}\end{array}} = \boxed{\begin{array}{l}\text{Percent of annual}\\ \text{gross income}\end{array}}$

*Labels:*      Annual retirement

            fund contribution $= x$ (dollars)

            Percent $= 0.075$ (in decimal form)

            Gross income $= 45{,}800$ (dollars)

*Equation:*     $x = 0.075(45{,}800)$

             $x = 3435$

So, each year you put in \$3435.

**83.** *Verbal Model:*   $\boxed{\text{Commission}} = \boxed{\begin{array}{c}\text{Percent} \\ \text{(in decimal form)}\end{array}} \cdot \boxed{\begin{array}{c}\text{Sale} \\ \text{price}\end{array}}$

     *Labels:*        Commission = 14,506.50   (dollars)

                      Percent = $p$             (percent in decimal form)

                      Sale price = 152,700     (dollars)

     *Equation:*      $14,506.50 = p(152,700)$

$$\frac{14,506.50}{152,700} = p$$

$$0.095 = p$$

This is a commission of 9.5%.

**85.** *Verbal Model:*   $\boxed{\text{Increase in price}} = \boxed{\text{Percent of original price}}\;(a = pb)$

     *Labels:*        Increase in price = 3900 − 3750 = 150   (dollars)

                      Unknown percent = $p$        (in decimal form)

                      Original price = 3750        (dollars)

     *Equation:*      $150 = p(3750)$

$$\frac{150}{3750} = p$$

$$0.04 = p$$

The price of the lawn tractor increased by 4%.

**87.** *Verbal Model:*   $\boxed{\text{New salary}} = \boxed{\text{Percent (in decimal form)}} \cdot \boxed{\text{Old salary}} + \boxed{\text{Old salary}}$

     *Labels:*        New salary = $x$      (dollars)

                      Percent = 0.05      (percent in decimal form)

                      Old salary = 35,600    (dollars)

     *Equation:*      $x = 0.05(35,600) + 35,600$

                         $x = 1780 + 35,600$

                         $x = 37,380$

The new salary is $37,380.

**89.** *Verbal Model:*   $\boxed{\text{Points needed for a B}} = \boxed{\text{Percent of total points}}\;(a = pb)$

     *Labels:*        $a$ = Points needed for a B = 394 + 6 = 400

                      $p$ = 0.80   (percent in decimal form)

                      $b$ = total points in course

     *Equation:*      $400 = 0.80(b)$

$$\frac{400}{0.80} = b$$

$$500 = b$$

There were 500 possible points.

**91.** *Verbal Model:*    | Original price | $-$ | Percent of original price | $=$ | Sale price |

       *Labels:*    Original price $= x$    (dollars)

                   Percent $= 0.20$    (in decimal form)

                   Sale price $= 250$    (dollars)

       *Equation:*
$$x - 0.20x = 250$$
$$0.80x = 250$$
$$\frac{0.80x}{0.80} = \frac{250}{0.80}$$
$$x = 312.50$$

So, the original price of the coat was $312.50.

**93.** *Verbal Model:*    | Area of garage | $=$ | Percent of area of plot of land |    $(a = pb)$

       *Labels:*    $a =$ area of garage $= 24^2 = 576$    (square feet)

                   $p =$ percent in decimal form

                   $b =$ area of plot of land $= (650)(825) = 536{,}250$    (square feet)

       *Equation:*
$$576 = p(536{,}250)$$
$$\frac{576}{536{,}250} = p$$
$$0.00107 \approx p$$

The garage occupies approximately 0.107% of the plot of land.

**95. Media networks:**

*Verbal Model:*   $\boxed{\text{Revenue from media networks}} = \boxed{\text{Percent (in decimal form)}} \cdot \boxed{\text{Company revenue}}$

*Labels:*   Revenue from media networks $= x$   (billion dollars)

Percent $= 0.42$   (percent in decimal form)

Company revenue $= 35.5$   (billion dollars)

*Equation:*   $x = 0.42(35.5)$

$x = 14.91$

The revenue from media networks is $14.91 billion.

**Parks and resorts:**

*Verbal Model:*   $\boxed{\text{Revenue from parks and resorts}} = \boxed{\text{Percent (in decimal form)}} \cdot \boxed{\text{Company revenue}}$

*Labels:*   Revenue from parks and resorts $= x$   (billion dollars)

Percent $= 0.30$   (percent in decimal form)

Company revenue $= 35.5$   (billion dollars)

*Equation:*   $x = 0.30(35.5)$

$x = 10.65$

The revenue from parks and resorts is $10.65 billion.

**Studio entertainment:**

*Verbal Model:*   $\boxed{\text{Revenue from studio entertainment}} = \boxed{\text{Percent (in decimal form)}} \cdot \boxed{\text{Company revenue}}$

*Labels:*   Revenue from studio entertainment $= x$   (billion dollars)

Percent $= 0.21$   (percent in decimal form)

Company revenue $= 35.5$   (billion dollars)

*Equation:*   $x = 0.21(35.5)$

$x = 7.455$

The revenue from studio entertainment is $7.455 billion.

**Consumer products:**

*Verbal Model:*   $\boxed{\text{Revenue from consumer products}} = \boxed{\text{Percent (in decimal form)}} \cdot \boxed{\text{Company revenue}}$

*Labels:*   Revenue from consumer products $= x$   (billion dollars)

Percent $= 0.07$   (percent in decimal form)

Company revenue $= 35.5$   (billion dollars)

*Equation:*   $x = 0.07(35.5)$

$x = 2.485$

The revenue from consumer products is $2.485 billion.

**97. (a) Mathematicians and computer scientists in 2005:**

*Verbal Model:*   $\boxed{\text{Number of women}} = \boxed{\begin{array}{c}\text{Percent}\\ \text{(in decimal form)}\end{array}} \cdot \boxed{\begin{array}{c}\text{Total}\\ \text{number}\end{array}}$

*Labels:*   Number of women = 876,420   (people)

Percent = 0.27   (percent in decimal form)

Total number = $x$   (people)

*Equation:*   $876{,}420 = 0.27x$

$\dfrac{876{,}420}{0.27} = x$

$3{,}246{,}000 = x$

The total number of mathematicians and computer scientists in 2005 was 3,246,000.

**(b) Chemists in 1983:**

*Verbal Model:*   $\boxed{\text{Number of women}} = \boxed{\begin{array}{c}\text{Percent}\\ \text{(in decimal form)}\end{array}} \cdot \boxed{\begin{array}{c}\text{Total}\\ \text{number}\end{array}}$

*Labels:*   Number of women = 22,834   (people)

Percent = 0.233   (percent in decimal form)

Total number = $x$   (people)

*Equation:*   $22{,}834 = 0.233x$

$\dfrac{22{,}834}{0.233} = x$

$98{,}000 = x$

The total number of chemists in 1983 was 98,000.

**(c) Biologists in 2005:**

*Verbal Model:*   $\boxed{\text{Number of women}} = \boxed{\begin{array}{c}\text{Percent}\\ \text{(in decimal form)}\end{array}} \cdot \boxed{\begin{array}{c}\text{Total}\\ \text{number}\end{array}}$

*Labels:*   Number of women = 57,953   (people)

Percent = 0.487   (percent in decimal form)

Total number = $x$   (people)

*Equation:*   $57{,}953 = 0.487x$

$\dfrac{57{,}953}{0.487} = x$

$119{,}000 = x$

The total number of biologists in 2005 was 119,000.

**99.** If $a > b$ and $b \neq 0$, the percent is greater than 100%.

If $a < b$ and $b \neq 0$, the percent is less than 100%.

If $a = b$ and $b \neq 0$, the percent is equal to 100%.

**101.** False.

$1 = 100\%$ and $1\% = 0.01$, so 1 is not equal to 1%.

**103.** False.

$68\% = 0.68$. To find 68% of 50, multiply 0.68 times 50.

**105.** $8 - |-7 + 11| + (-4) = 8 - |4| - 4$

$= 8 - 4 - 4$

$= 0$

**107.** (a) When $x = 4$ and $y = 3$, the value of the expression is

$$x^2 - y^2 = 4^2 - 3^2 = 16 - 9 = 7.$$

(b) When $x = -5$ and $y = 3$, the value of the expression is

$$x^2 - y^2 = (-5)^2 - 3^2 = 25 - 9 = 16.$$

**109.** $4(2x - 5) = 4(2x) - 4(5) = 8x - 20$

**111.**
$$4(x + 3) = 0$$
$$4x + 12 = 0$$
$$4x + 12 - 12 = 0 - 12$$
$$4x = -12$$
$$\frac{4x}{4} = \frac{-12}{4}$$
$$x = -3$$

**113.**
$$22 - (z + 1) = 33$$
$$22 - z - 1 = 33$$
$$21 - z = 33$$
$$21 - 21 - z = 33 - 21$$
$$-z = 12$$
$$z = -12$$

# Mid-Chapter Quiz for Chapter 3

**1.**
$$74 - 12x = 2$$
$$74 - 74 - 12x = 2 - 74$$
$$-12x = -72$$
$$\frac{-12x}{-12} = \frac{-72}{-12}$$
$$x = 6$$

**2.**
$$10(y - 8) = 0$$
$$10y - 80 = 0$$
$$10y - 80 + 80 = 0 + 80$$
$$10y = 80$$
$$\frac{10y}{10} = \frac{80}{10}$$
$$y = 8$$

**3.**
$$3x + 1 = x + 20$$
$$3x - x + 1 = x - x + 20$$
$$2x + 1 = 20$$
$$2x + 1 - 1 = 20 - 1$$
$$2x = 19$$
$$\frac{2x}{2} = \frac{19}{2}$$
$$x = \frac{19}{2}$$

**4.**
$$6x + 8 = 8 - 2x$$
$$6x + 2x + 8 = 8 - 2x + 2x$$
$$8x + 8 = 8$$
$$8x + 8 - 8 = 8 - 8$$
$$8x = 0$$
$$\frac{8x}{8} = \frac{0}{8}$$
$$x = 0$$

**5.**
$$-10x + \frac{2}{3} = \frac{7}{3} - 5x$$
$$3\left(-10x + \frac{2}{3}\right) = 3\left(\frac{7}{3} - 5x\right)$$
$$-30x + 2 = 7 - 15x$$
$$-30x + 15x + 2 = 7 - 15x + 15x$$
$$-15x + 2 = 7$$
$$-15x + 2 - 2 = 7 - 2$$
$$-15x = 5$$
$$\frac{-15x}{-15} = \frac{5}{-15}$$
$$x = -\frac{1}{3}$$

**6.**
$$\frac{x}{5} + \frac{x}{7} = 1$$
$$35\left(\frac{x}{5} + \frac{x}{7}\right) = 35(1)$$
$$35\left(\frac{x}{5}\right) + 35\left(\frac{x}{7}\right) = 35(1)$$
$$7x + 5x = 35$$
$$12x = 35$$
$$\frac{12x}{12} = \frac{35}{12}$$
$$x = \frac{35}{12}$$

**7.**
$$\frac{9+x}{3} = 15$$
$$3\left(\frac{9+x}{3}\right) = 3(15)$$
$$9 + x = 45$$
$$9 - 9 + x = 45 - 9$$
$$x = 36$$

**8.**
$$3 - 5(4 - x) = -6$$
$$3 - 20 + 5x = -6$$
$$-17 + 5x = -6$$
$$-17 + 17 + 5x = -6 + 17$$
$$5x = 11$$
$$\frac{5x}{5} = \frac{11}{5}$$
$$x = \frac{11}{5}$$

**9.**
$$\frac{x+3}{6} = \frac{4}{3}$$
$$3(x + 3) = 24$$
$$3x + 9 = 24$$
$$3x + 9 - 9 = 24 - 9$$
$$3x = 15$$
$$x = \frac{15}{3}$$
$$x = 5$$

**10.**
$$\frac{x+7}{5} = \frac{x+9}{7}$$
$$7(x + 7) = 5(x + 9)$$
$$7x + 49 = 5x + 45$$
$$7x - 5x + 49 = 5x - 5x + 45$$
$$2x + 49 = 45$$
$$2x + 49 - 49 = 45 - 49$$
$$2x = -4$$
$$x = -2$$

**11.**
$$32.86 - 10.5x = 11.25$$
$$32.86 - 32.86 - 10.5x = 11.25 - 32.86$$
$$-10.5x = -21.61$$
$$\frac{-10.5x}{-10.5} = \frac{-21.61}{-10.5}$$
$$x \approx 2.06$$

**12.**
$$\frac{x}{5.45} + 3.2 = 12.6$$
$$\frac{x}{5.45} + 3.2 - 3.2 = 12.6 - 3.2$$
$$\frac{x}{5.45} = 9.4$$
$$5.45\left(\frac{x}{5.45}\right) = 5.45(9.4)$$
$$x = 51.23$$

**13.** *Verbal Model:* $\boxed{\text{What number}} = \boxed{62\% \text{ of } 25}$  $(a = pb)$

*Label:* $a$ = unknown number

*Equation:* $a = 0.62(25)$
$$a = 15.5$$

So, 62% of 25 is 15.5.

**14.** *Verbal Model:* $\boxed{\text{What number}} = \boxed{\frac{1}{2}\% \text{ of } 8400}$ $(a = pb)$

*Label:* $a$ = unknown number

*Equation:* $a = 0.005(8400)$
$$a = 42$$

So, 42 is $\frac{1}{2}$%, or 0.5%, of 8400.

**15.** *Verbal Model:* $\boxed{300} = \boxed{\text{What percent of } 150}$ $(a = pb)$

*Label:* $p$ = unknown percent

*Equation:* $300 = p(150)$
$$\frac{300}{150} = p$$
$$2 = p$$
$$200\% = p$$

So, 300 is 200% of 150.

**16.** *Verbal Model:* $\boxed{145.6} = \boxed{32\% \text{ of what number}}$ $(a = pb)$

*Label:* $b$ = unknown number

*Equation:* $145.6 = 0.32(b)$
$$\frac{145.6}{0.32} = b$$
$$455 = b$$

So, 145.6 is 32% of 455.

**17.** *Verbal Model:* 2 Length + 2 Width = Perimeter

    *Labels:*           Width = $w$      (meters)

                    Length = $\frac{3}{2}w$    (meters)

                    Perimeter = 60   (meters)

    *Equation:*        $2\left(\frac{3}{2}\right)w + 2w = 60$

                       $3w + 2w = 60$

                          $5w = 60$

                       $w = 12$ and $\frac{3}{2}w = 18$

The length of the rectangle is 18 meters and the width is 12 meters.

**18.** *Verbal Model:* $\boxed{\begin{array}{c}\text{Hours} \\ \text{at store}\end{array}} \cdot \boxed{\begin{array}{c}\text{Hourly earnings} \\ \text{at store}\end{array}} + \boxed{\begin{array}{c}\text{Hours of} \\ \text{baby-sitting}\end{array}} \cdot \boxed{\begin{array}{c}\text{Hourly earnings} \\ \text{baby-sitting}\end{array}} = \boxed{\begin{array}{c}\text{Total} \\ \text{earnings}\end{array}}$

    *Labels:*         Hours at store = 40           (hours)

                   Hourly earnings at store = 7.50    (dollars per hour)

                   Hours of baby-sitting = $x$       (hours)

                   Hourly earnings baby-sitting = 7   (dollars per hour)

                   Total earnings = 370         (dollars)

    *Equation:*        $40(7.50) + x(7) = 370$

                      $300 + 7x = 370$

             $300 - 300 + 7x = 370 - 300$

                         $7x = 70$

                         $\dfrac{7x}{7} = \dfrac{70}{7}$

                          $x = 10$

You must baby-sit for 10 hours.

**19.** *Verbal Model:* First area + Second area + Third area = Total area

    *Labels:*         First area = $x$           (square meters)

                   Second area = $2x$       (square meters)

                   Third area = $2(2x) = 4x$   (square meters)

                   Total area = 42         (square meters)

    *Equation:*        $x + 2x + 4x = 42$

                        $7x = 42$

                       $x = 6$ and $2x = 12, 4x = 24$

The area of the first subregion is 6 square meters, the area of the second subregion is 12 square meters, and the area of the third subregion is 24 square meters.

**20.** *Verbal Model:* $\frac{1}{3}\left(\boxed{\text{First score}} + \boxed{\text{Second score}} + \boxed{\text{Third score}}\right) = 0.90(100)$

*Labels:*  First score = 84  (points)

Second score = 93  (points)

Third score = $x$  (points)

*Equation:*  $\frac{1}{3}(84 + 93 + x) = 0.90(100)$

$\frac{1}{3}(84) + \frac{1}{3}(93) + \frac{1}{3}x = 90$

$28 + 31 + \frac{1}{3}x = 90$

$59 + \frac{1}{3}x = 90$

$\frac{1}{3}x = 31$

$3\left(\frac{1}{3}x\right) = 3(31)$

$x = 93$

You must score 93 on the third test to earn a 90% average.

**21.** *Verbal Model:* $\boxed{\text{New price}} = \boxed{\text{Percent (in decimal form)}} \cdot \boxed{\text{Old price}}$

*Labels:*  New price = 1099  (dollars)

Percent = 1.08  (percent in decimal form)

Old price = $x$  (dollars)

*Equation:*  $1099 = 1.08x$

$\frac{1099}{1.08} = x$

$1017.59 \approx x$

The price two years ago was approximately $1017.59.

**22.** *Verbal Model:* $\boxed{\begin{array}{l}\text{Endangered}\\\text{birds}\end{array}} = \boxed{\begin{array}{l}\text{Percent}\\\text{(in decimal form)}\end{array}} \cdot \boxed{\begin{array}{l}\text{Endangered}\\\text{species}\end{array}}$

*Labels:*  Endangered birds = 251  (endangered birds)

Percent = $p$  (percent in decimal form)

Endangered species = $599 + 325 + 232 + 78 + 85 + 251 = 1570$  (endangered species)

*Equation:*  $251 = p(1570)$

$\frac{251}{1570} = p$

$0.16 \approx p$

Birds were 16% of the total endangered wildlife and plant species.

**23.**
$$\frac{t}{4} + \frac{t}{12} = 1$$
$$12\left(\frac{t}{4} + \frac{t}{12}\right) = 12(1)$$
$$12\left(\frac{t}{4}\right) + 12\left(\frac{t}{12}\right) = 12$$
$$3t + t = 12$$
$$4t = 12$$
$$\frac{4t}{4} = \frac{12}{4}$$
$$t = 3$$

It will take 3 hours for the two people to paint the room.

## Section 3.4   Ratios and Proportions

**1.** $36$ to $9 = \frac{36}{9} = \frac{4}{1}$

**3.** $27$ to $54 = \frac{27}{54} = \frac{1}{2}$

**5.** $5\frac{2}{3}$ to $1\frac{1}{3} = \dfrac{\frac{17}{3}}{\frac{4}{3}} = \frac{17}{3} \cdot \frac{3}{4} = \frac{17(\cancel{3})}{\cancel{3}(4)} = \frac{17}{4}$

**7.** $14 : 21 = \frac{14}{21} = \frac{2}{3}$

**9.** $144 : 16 = \frac{144}{16} = \frac{9}{1}$

**11.** $3\frac{1}{5} : 5\frac{3}{10} = \dfrac{\frac{16}{5}}{\frac{53}{10}} = \frac{16}{5} \cdot \frac{10}{53} = \frac{16(\cancel{5})(2)}{\cancel{5}(53)} = \frac{32}{53}$

**13.** $\dfrac{42 \text{ inches}}{21 \text{ inches}} = \frac{42}{21} = \frac{2}{1}$

**15.** $\dfrac{40 \text{ dollars}}{60 \text{ dollars}} = \frac{40}{60} = \frac{2}{3}$

**17.** $\dfrac{1 \text{ quart}}{1 \text{ gallon}} = \dfrac{1 \text{ quart}}{4 \text{ quarts}} = \frac{1}{4}$

**19.** $\dfrac{7 \text{ nickels}}{3 \text{ quarters}} = \dfrac{7 \text{ nickels}}{15 \text{ nickels}} = \frac{7}{15}$

Note: This problem could also be done using cents as the common ratio.

$$\dfrac{7 \text{ nickels}}{3 \text{ quarters}} = \dfrac{35 \text{ cents}}{75 \text{ cents}} = \frac{35}{75} = \frac{7}{15}$$

**21.** $\dfrac{3 \text{ hours}}{90 \text{ minutes}} = \dfrac{180 \text{ minutes}}{90 \text{ minutes}} = \frac{180}{90} = \frac{2}{1}$

**23.** $\dfrac{75 \text{ centimeters}}{2 \text{ meters}} = \dfrac{75 \text{ centimeters}}{200 \text{ centimeters}} = \frac{75}{200} = \frac{3}{8}$

**25.** $\dfrac{60 \text{ milliliters}}{1 \text{ liter}} = \dfrac{60 \text{ milliliters}}{1000 \text{ milliliters}} = \frac{60}{1000} = \frac{3}{50}$

**27.** $\dfrac{2 \text{ kilometers}}{2500 \text{ meters}} = \dfrac{2000 \text{ meters}}{2500 \text{ meters}} = \frac{4}{5}$

**29.** $\dfrac{3000 \text{ pounds}}{5 \text{ tons}} = \dfrac{3000 \text{ pounds}}{10,000 \text{ pounds}} = \frac{3}{10}$

**31.** Verbal Model: $\boxed{\text{Unit price}} = \dfrac{\boxed{\text{Total price}}}{\boxed{\text{Total units}}}$

Unit price: $\dfrac{\$0.98}{20 \text{ oz}} = \$0.049$ per ounce

**33.** Verbal Model: $\boxed{\text{Unit price}} = \dfrac{\boxed{\text{Total price}}}{\boxed{\text{Total units}}}$

Total units: 1 pound + 4 ounces $= 1(16 \text{ oz}) + 4 \text{ oz}$
$$= 16 \text{ oz} + 4 \text{ oz}$$
$$= 20 \text{ oz}$$

Unit price: $\dfrac{\$1.46}{20 \text{ oz}} = \$0.073$ per ounce

**35.** (a) The unit price for the gallon (4 quart) container:

$$\text{Unit price} = \frac{\text{Total price}}{\text{Total units}} = \frac{3.49}{4 \text{ quarts}} \approx \$0.873 \text{ per quart}$$

(b) The unit price for the quart container:

$$\text{Unit price} = \frac{\text{Total price}}{\text{Total units}} = \frac{1.39}{1 \text{ quart}} = \$1.39 \text{ per quart}$$

The gallon container (a) has the lower unit price.

**37.** (a) The unit price for the 4-pound bag:

$$\text{Unit price} = \frac{\text{Total price}}{\text{Total units}} = \frac{1.89}{4 \text{ pounds}} = \$0.4725 \text{ per pound}$$

(b) The unit price for the 10-pound bag:

$$\text{Unit price} = \frac{\text{Total price}}{\text{Total units}} = \frac{4.49}{10 \text{ pounds}} = \$0.449 \text{ per pound}$$

The 10-pound bag (b) has the lower unit price.

**39.** (a) The unit price for the 2-liter bottle is

$$\text{Unit price} = \frac{\text{Total price}}{\text{Total units}} = \frac{\$1.09}{67.6 \text{ ounces}} \approx \$0.016 \text{ per ounce.}$$

(b) The unit price for the cans is

$$\text{Unit price} = \frac{\text{Total price}}{\text{Total units}} = \frac{\$1.69}{6(12) \text{ ounces}} = \frac{\$1.69}{72 \text{ ounces}} \approx \$0.023 \text{ per ounce.}$$

The 2-liter bottle (a) has the lower unit price.

**41.**
$$\frac{5}{3} = \frac{20}{y}$$

$5y = 60 \quad \text{(Cross-multiply)}$

$$y = \frac{60}{5}$$

$y = 12$

**43.**
$$\frac{5}{x} = \frac{3}{2}$$

$10 = 3x \quad \text{(Cross-multiply)}$

$$\frac{10}{3} = x$$

**45.**
$$\frac{z}{35} = \frac{5}{8}$$

$8z = (35)5 \quad \text{(Cross-multiply)}$

$8z = 175$

$$\frac{8z}{8} = \frac{175}{8}$$

$$z = \frac{175}{8}$$

**47.**
$$\frac{8}{3} = \frac{t}{6}$$

$$6\left(\frac{8}{3}\right) = 6\left(\frac{t}{6}\right)$$

$16 = t$

**49.**
$$\frac{0.5}{0.8} = \frac{n}{0.3}$$

$0.15 = 0.8n \quad \text{(Cross-multiply)}$

$$\frac{0.15}{0.8} = n$$

$$\frac{15}{80} = n$$

$$\frac{3}{16} = n$$

The answer could also be written as $n = 0.1875$.

**51.**
$$\frac{x+1}{5} = \frac{3}{10}$$
$$10(x+1) = 15 \quad \text{(Cross-multiply)}$$
$$10x + 10 = 15$$
$$10x + 10 - 10 = 15 - 10$$
$$10x = 5$$
$$\frac{10x}{10} = \frac{5}{10}$$
$$x = \frac{1}{2}$$

**53.**
$$\frac{x+6}{3} = \frac{x-5}{2}$$
$$2(x+6) = 3(x-5) \quad \text{(Cross-multiply)}$$
$$2x + 12 = 3x - 15$$
$$2x - 3x + 12 = 3x - 3x - 15$$
$$-x + 12 = -15$$
$$-x + 12 - 12 = -15 - 12$$
$$-x = -27$$
$$(-1)(-x) = (-1)(-27)$$
$$x = 27$$

**55.**
$$\frac{x+2}{8} = \frac{x-1}{3}$$
$$3(x+2) = 8(x-1)$$
$$3x + 6 = 8x - 8$$
$$3x - 8x + 6 = 8x - 8x - 8$$
$$-5x + 6 = -8$$
$$-5x + 6 - 6 = -8 - 6$$
$$-5x = -14$$
$$\frac{-5x}{-5} = \frac{-14}{-5}$$
$$x = \frac{14}{5}$$

**57.** $\dfrac{4 \text{ hours}}{6 \text{ hours}} = \dfrac{4}{6} = \dfrac{2}{3}$

**59.** $\dfrac{806 \text{ yards}}{217 \text{ yards}} = \dfrac{806}{217} = \dfrac{31(26)}{31(7)} = \dfrac{26}{7}$

**61.** $\dfrac{345 \text{ cubic centimeters}}{17.25 \text{ cubic centimeters}} = \dfrac{345}{17.25} = \dfrac{20}{1}$

**63.** $\dfrac{45 \text{ teeth}}{30 \text{ teeth}} = \dfrac{45}{30} = \dfrac{3}{2}$

**65.** $\dfrac{\pi(10)^2 \text{ square inches}}{\pi(7)^2 \text{ square inches}} = \dfrac{100\pi}{49\pi} = \dfrac{100}{49}$

**67.** *Verbal Model:*
$$\boxed{\frac{\text{Gallons for shorter trip}}{\text{Miles for shorter trip}}} = \boxed{\frac{\text{Gallons for longer trip}}{\text{Miles for longer trip}}}$$

*Labels:*   Gallons for shorter trip $= x$ (gallons)
Miles for shorter trip $= 400$ (miles)
Gallons for longer trip $= 20$ (gallons)
Miles for longer trip $= 500$ (miles)

*Proportion:*
$$\frac{x}{400} = \frac{20}{500}$$
$$400\left(\frac{x}{400}\right) = 400\left(\frac{20}{500}\right)$$
$$x = \frac{8000}{500}$$
$$x = 16$$

On a trip of 400 miles, 16 gallons of gas would be used.

**69.** *Verbal Model:*    $\boxed{\dfrac{\text{Blocks for smaller wall}}{\text{Length of smaller wall}}} = \boxed{\dfrac{\text{Blocks for larger wall}}{\text{Length of larger wall}}}$

      *Labels:*    Blocks for smaller wall = 100

                   Length of smaller wall = 16    (feet)

                   Blocks for larger wall = $x$

                   Length of larger wall = 40    (feet)

      *Proportion:*

$$\frac{100}{16} = \frac{x}{40}$$

$$40\left(\frac{100}{16}\right) = 40\left(\frac{x}{40}\right)$$

$$250 = x$$

So, 250 blocks are needed to build a 40-foot wall.

**71.** *Verbal Model:*    $\boxed{\dfrac{\text{Price of shirt}}{\text{Price of jeans}}} = \boxed{\dfrac{\text{Tax on shirt}}{\text{Tax on jeans}}}$

      *Labels:*    Price of shirt = 19.99    (dollars)

                   Price of jeans = 34.99    (dollars)

                   Tax on shirt = 1.20    (dollars)

                   Tax on jeans = $x$    (dollars)

      *Proportion:*

$$\frac{19.99}{34.99} = \frac{1.20}{x}$$

$$19.99x = 34.99(1.20) \qquad \text{(Cross-multiply)}$$

$$19.99x = 41.988$$

$$\frac{19.99x}{19.99} = \frac{41.988}{19.99}$$

$$x \approx 2.10$$

The tax on a pair of jeans is approximately $2.10.

**73.** *Verbal Model:*    $\boxed{\dfrac{\text{Poll voters for candidate}}{\text{Voters in poll}}} = \boxed{\dfrac{\text{Election voters for candidate}}{\text{Voters in election}}}$

      *Labels:*    Poll voters for candidate = 624

                   Voters in poll = 1100

                   Election voters for candidate = $x$

                   Voters in election = 40,000

      *Proportion:*

$$\frac{624}{1100} = \frac{x}{40,000}$$

$$40,000\left(\frac{624}{1100}\right) = 40,000\left(\frac{x}{40,000}\right)$$

$$22,691 \approx x$$

The candidate can expect 22,691 votes.

**75.** *Verbal Model:*   $\boxed{\dfrac{\text{Gallons in first tank}}{\text{Gallons in second tank}}} = \boxed{\dfrac{\text{Time to fill first tank}}{\text{Time to fill second tank}}}$

*Labels:*  Gallons in first tank $= 750$

Gallons in second tank $= 1000$

Time to fill first tank $= 35$    (minutes)

Time to fill second tank $= x$    (minutes)

*Proportion:*  $\dfrac{750}{1000} = \dfrac{35}{x}$

$750x = 35{,}000$

$x = \dfrac{35{,}000}{750}$

$x = 46\dfrac{2}{3}$

It would take $46\dfrac{2}{3}$ minutes to fill the second tank.

**77.** *Verbal Model:*   $\boxed{\dfrac{\text{Inches on map scale}}{\text{Miles represented on scale}}} = \boxed{\dfrac{\text{Inches between cities on map}}{\text{Miles between cities}}}$

*Labels:*  Inches on map scale $= 1\dfrac{1}{4}$

Miles represented on scale $= 80$

Inches between cities on map $= 6$

Miles between cities $= x$

*Proportion:*  $\dfrac{1.25}{80} = \dfrac{6}{x}$

$1.25x = 80(6)$    (Cross-multiply)

$1.25x = 480$

$\dfrac{1.25x}{1.25} = \dfrac{480}{1.25}$

$x = 384$

The distance between the cities is approximately 384 miles.

**79.** *Verbal Model:*   $\boxed{\text{Fresh water to salt ratio}} = \boxed{\dfrac{\text{Pounds of water}}{\text{Pounds of salt}}}$

*Labels:*  Fresh water to salt ratio $= 25{:}1$ or $\dfrac{25}{1}$

Amount of fresh water $= x$    (pounds)

Amount of salt $= \dfrac{1}{2}$    (pounds)

*Proportion:*  $\dfrac{25}{1} = \dfrac{x}{\frac{1}{2}}$

$25\left(\dfrac{1}{2}\right) = 1(x)$    (Cross-multiply)

$\dfrac{25}{2} = x$

The amount of fresh water required is $\dfrac{25}{2}$ or $12\dfrac{1}{2}$ pounds.

**81.** Corresponding sides of similar triangles are proportional.

$$\frac{1}{2} = \frac{x}{5}$$

$$5 = 2x \qquad \text{(Cross-multiply)}$$

$$\frac{5}{2} = \frac{2x}{2}$$

$$\frac{5}{2} = x$$

Note: There are several ways to set up this proportion. It could also be written as $\frac{2}{1} = \frac{5}{x}, \frac{2}{5} = \frac{1}{x},$ or $\frac{5}{2} = \frac{x}{1}$.

**83.** *Verbal Model:*
$$\boxed{\frac{\text{Height of smaller triangle}}{\text{Base of smaller triangle}}} = \boxed{\frac{\text{Height of larger triangle}}{\text{Base of larger triangle}}}$$

*Labels:*

Height of smaller triangle (person's height) = 6 (feet)

Base of smaller triangle (length of shadow) = $x$ (feet)

Height of larger triangle (height of streetlight) = 15 (feet)

Base of larger triangle = $10 + x$ (feet)

*Proportion:*

$$\frac{6}{x} = \frac{15}{10 + x}$$

$$6(10 + x) = 15x \qquad \text{(Cross-multiply)}$$

$$60 + 6x = 15x$$

$$60 + 6x - 6x = 15x - 6x$$

$$60 = 9x$$

$$\frac{60}{9} = \frac{9x}{9}$$

$$\frac{20}{3} = x$$

15 ft    6 ft    10 ft    x ft

The length of the shadow is $6\frac{2}{3}$ feet (or 6 feet, 8 inches).

**85.** *Verbal Model:*
$$\boxed{\frac{\text{New length}}{\text{Old length}}} = \boxed{\frac{\text{New percent}}{\text{Old percent}}}$$

*Labels:*

New length = 2 (inches)

Old length = 10 (inches)

New percent = $x$

Old percent = 100

*Proportion:*

$$\frac{2}{10} = \frac{x}{100}$$

$$100\left(\frac{2}{10}\right) = 100\left(\frac{x}{100}\right)$$

$$20 = x$$

The new photo needs to be 20% of the size of the original. (In other words, the original photo needs to be reduced by 80%.)

**87.** *Verbal Model:*
$$\boxed{\frac{\text{Price in 2006}}{\text{Price in 1979}}} = \boxed{\frac{\text{Index in 2006}}{\text{Index in 1979}}}$$

*Labels:*

Price in 2006 = $x$ (dollars)

Price in 1979 = 2875 (dollars)

Index in 2006 = 201.6

Index in 1979 = 72.6

*Proportion:*

$$\frac{x}{2875} = \frac{201.6}{72.6}$$

$$2875\left(\frac{x}{2875}\right) = 2875\left(\frac{201.6}{72.6}\right)$$

$$x \approx 7983$$

The 2006 price of the lawn tractor was approximately $7983.

**89.** *Verbal Model:*   $\dfrac{\boxed{\text{Price in 1975}}}{\boxed{\text{Price in 1996}}} = \dfrac{\boxed{\text{Index in 1975}}}{\boxed{\text{Index in 1996}}}$

*Labels:*        Price in 1975 $= x$       (dollars)

Price in 1996 $= 2.75$    (dollars)

Index in 1975 $= 53.8$

Index in 1996 $= 156.9$

*Proportion:*        $\dfrac{x}{2.75} = \dfrac{53.8}{156.9}$

$2.75\left(\dfrac{x}{2.75}\right) = 2.75\left(\dfrac{53.8}{156.9}\right)$

$x \approx 0.94$

The 1975 price of the gallon of milk was approximately $0.94.

**91.** No, the ratio of men to women is not enough information to tell you the total number of people in the class. To determine the total number of people in the class, you also need to know the number of men or the number of women in the class.

**103.** *Verbal Model:*    $\boxed{465} = \boxed{\text{What percent of 500}}$

*Label:*        $p =$ unknown percent (in decimal form)

*Percent equation:*    $465 = p(500)$

$\dfrac{465}{500} = p$

$0.93 = p$

So, 465 is 93% of 500.

**93.** Answers will vary.

**95.** $(-5)^3 + 3 = (-5)(-5)(-5) + 3 = -125 + 3 = -122$

**97.** $\dfrac{-\left|7 + 3^2\right|}{4} = \dfrac{-\left|16\right|}{4} = \dfrac{-16}{4} = -4$

**99.** $(8 \cdot 9) + (-4)^3 = 72 + (-4)(-4)(-4)$

$= 72 - 64$

$= 8$

**101.** *Verbal Model:*    $\boxed{\text{What number}} = \boxed{45\% \text{ of } 90}$

*Label:*        $a =$ unknown number

*Percent equation:*    $a = 0.45(90)$

$a = 40.5$

So, 45% of 90 is 40.5.

# Section 3.5   Geometric and Scientific Applications

**1.**  $A = \dfrac{1}{2}bh$

$2A = 2 \cdot \dfrac{1}{2}bh$

$2A = bh$

$\dfrac{2A}{b} = \dfrac{bh}{b}$

$\dfrac{2A}{b} = h$

**3.**    $A = P + Prt$

$A - P = P - P + Prt$

$A - P = Prt$

$\dfrac{A - P}{Pt} = \dfrac{Prt}{Pt}$

$\dfrac{A - P}{Pt} = r$

**5.**  $V = lwh$

$\dfrac{V}{wh} = \dfrac{lwh}{wh}$

$\dfrac{V}{wh} = l$

**7.**    $S = C + RC$

$S = C(1 + R)$

$\dfrac{S}{1 + R} = \dfrac{C(1 + R)}{1 + R}$

$\dfrac{S}{1 + R} = C$

**9.**    $F = \alpha\dfrac{m_1 m_2}{r^2}$

$Fr^2 = \alpha m_1 m_2$

$\dfrac{Fr^2}{\alpha m_1} = \dfrac{\alpha m_1 m_2}{\alpha m_1}$

$\dfrac{Fr^2}{\alpha m_1} = m_2$

**11.**
$$A = \frac{1}{2}(a + b)h$$
$$2A = 2 \cdot \frac{1}{2}(a + b)h$$
$$2A = (a + b)h$$
$$2A = ah + bh$$
$$2A - ah = bh$$
$$\frac{2A - ah}{h} = \frac{bh}{h}$$
$$\frac{2A - ah}{h} = b$$

**13.**
$$h = v_0 t + \frac{1}{2}at^2$$
$$h - v_0 t = \frac{1}{2}at^2$$
$$2(h - v_0 t) = 2 \cdot \frac{1}{2}at^2$$
$$2(h - v_0 t) = at^2$$
$$\frac{2(h - v_0 t)}{t^2} = a \text{ or } a = \frac{2h - 2v_0 t}{t^2}$$

**15.** $V = \pi r^2 h$
$$V = \pi(5)^2 4$$
$$V = \pi(25)4$$
$V = 100\pi$ cubic meters or approximately
314.2 cubic meters

**17.** $B = \dfrac{703w}{h^2}$
$$B = \frac{703(127)}{(61)^2}$$
$B \approx 24$ pounds per square inch

**19.** $I = Prt$
$I = (870)(0.038)(1.5)$ [Note: 18 months = 1.5 years]
$I = 49.59$
The interest is \$49.59.

**21.** $I = Prt$
$$54 = 450(r)(2)$$
$$54 = 900r$$
$$\frac{54}{900} = r$$
$$0.06 = r$$
The interest rate is 6%.

**23.** *Verbal Model:* $\boxed{\text{Distance}} = \boxed{\text{Rate}} \cdot \boxed{\text{Time}}$

*Labels:* Distance = $d$ (meters)

Rate = 4 (meters per minute)

Time = 12 (minutes)

*Equation:* $d = 4(12)$
$$d = 48$$
The distance is 48 meters.

**25.** *Verbal Model:* $\boxed{\text{Distance}} = \boxed{\text{Rate}} \cdot \boxed{\text{Time}}$

*Labels:*
Distance = 128 (kilometers)

Rate = 8 (kilometers per hour)

Time = $t$ (hours)

*Equation:* $128 = 8t$
$$\frac{128}{8} = t$$
$$16 = t$$
The time is 16 hours.

**27.** *Verbal Model:* $\boxed{\text{Distance}} = \boxed{\text{Rate}} \cdot \boxed{\text{Time}}$

*Labels:*
Distance = 2054 (meters)

Rate = $r$ (meters per second)

Time = 18 (seconds)

*Equation:* $2054 = r(18)$
$$2054 = 18r$$
$$\frac{2054}{18} = r$$
$$114.\overline{1} = r$$
The rate is $114.\overline{1}$ meters per second.

**29.** $A = \frac{1}{2}bh$
$$6 = \frac{1}{2}b(3)$$
$$2(6) = 2\left(\frac{1}{2}\right)(b)(3)$$
$$12 = 3b$$
$$\frac{12}{3} = b$$
$$4 = b$$
The length of the base is 4 inches.

**31.** $V = lwh$
$$3125 = 50(25)(h)$$
$$3125 = 1250h$$
$$\frac{3125}{1250} = h$$
$$2.5 = h$$
The depth of the pool is 2.5 meters.

**33.** (a)
$$P = 2l + 2w$$
$$66 = 2(18) + 2w$$
$$66 = 36 + 2w$$
$$66 - 36 = 36 + 2w - 36$$
$$30 = 2w$$
$$\frac{30}{2} = w$$
$$15 = w$$

The width of the floor is 15 feet.

(b) $A = lw$
$$A = 18(15)$$
$$A = 270$$

The area of the floor is 270 square feet.

The flooring costs \$12 per square foot, so the cost of the floor is $12(270) = \$3240$.

**35.** *Common formula:*   $I = Prt$

*Labels:*   $I = \$128.98$
   $P = \$1500$
   $r = $ annual interest rate
   $t = 1$      (year)

*Equation:*
$$I = Prt$$
$$128.98 = 1500(r)(1)$$
$$128.98 = 1500r$$
$$\frac{128.98}{1500} = r$$
$$0.086 \approx r$$

The annual interest rate is approximately 8.6%.

**37.** *Common formula:*   $A = P + Prt$

*Labels:*      $A = $ amount of payment  (dollars)
      $P = \$15,000$
      $r = 0.13$
      $t = \frac{1}{2}$               (year)

*Equation:*      $A = 15,000 + 15,000(0.13)\left(\frac{1}{2}\right)$
      $A = 15,000 + 975$
      $A = 15,975$

The amount of the payment is \$15,975.

**39.** *Verbal Model:*   $\boxed{\text{Distance}} = \boxed{\text{Rate}} \cdot \boxed{\text{Time}}$

*Labels:*      Distance $= 3000$   (miles)
      Rate $= r$      (miles per hour)
      Time $= 2.6$   (hours)

*Equation:*      $3000 = r(2.6)$
      $\dfrac{3000}{2.6} = r$
      $1154 \approx r$

The rate is approximately 1154 miles per hour.

**41.** *Verbal Model:*   $\boxed{\text{Distance}} = \boxed{\text{Rate}} \cdot \boxed{\text{Time}}$.

*Labels:*      Distance $= 3000$   (miles)
      Rate $= 17,500$   (miles per hour)
      Time $= t$      (hours)

*Equation:*      $3000 = 17,500t$
      $\dfrac{3000}{17,500} = t$
      $0.17 \approx t$

The time is approximately 0.17 hours (or approximately 10 minutes).

**43.** *Verbal Model:*  | Total cost of less expensive stamps | $+$ | Total cost of more expensive stamps | $=$ | Total cost of the 100 stamps |

*Labels:*

Unit price of less expensive stamp $= 0.27$     (dollars per stamp)

Unit price of more expensive stamp $= 0.42$     (dollars per stamp)

Number of less expensive stamps $= x$     (stamps)

Number of more expensive stamps $= 100 - x$     (stamps)

Total price of the stamps $= 35.10$     (dollars)

*Equation:*

$$0.27x + 0.42(100 - x) = 35.10$$
$$0.27x + 42 - 0.42x = 35.10$$
$$-0.15x + 42 = 35.10$$
$$-0.15x + 42 - 42 = 35.10 - 42$$
$$-0.15x = -6.90$$
$$\frac{-0.15x}{-0.15} = \frac{-6.90}{-0.15}$$
$$x = 46 \text{ and } 100 - x = 54$$

There are 46 of the \$0.27 stamps and 54 of the \$0.42 stamps.

**45.** *Verbal Model:*  | Amount of alcohol in Solution 1 | $+$ | Amount of alcohol in Solution 2 | $=$ | Amount of alcohol in final solution |

*Labels:*

Solution 1: Percent alcohol $= 0.10$; amount $= 25$     (gallons)

Solution 2: Percent alcohol $= 0.30$; amount $= x$     (gallons)

Final solution: Percent alcohol $= 0.25$; amount $= 25 + x$     (gallons)

*Equation:*

$$0.10(25) + 0.30(x) = 0.25(25 + x)$$
$$2.5 + 0.30x = 6.25 + 0.25x$$
$$2.5 + 0.30x - 0.25x = 6.25 + 0.25x - 0.25x$$
$$2.5 + 0.05x = 6.25$$
$$2.5 + 0.05x - 2.5 = 6.25 - 2.5$$
$$0.05x = 3.75$$
$$\frac{0.05x}{0.05} = \frac{3.75}{0.05}$$
$$x = 75 \text{ and } 25 + x = 100$$

There are 75 gallons of Solution 2 and 100 gallons of the final solution.

**47.** *Verbal Model:*

$$\boxed{\begin{array}{c}\text{Amount of alcohol} \\ \text{in Solution 1}\end{array}} + \boxed{\begin{array}{c}\text{Amount of alcohol} \\ \text{in Solution 2}\end{array}} = \boxed{\begin{array}{c}\text{Amount of alcohol} \\ \text{in final solution}\end{array}}$$

*Labels:*        Solution 1: Percent alcohol $= 0.15$; amount $= 5$      (quarts)

                 Solution 2: Percent alcohol $= 0.45$; amount $= x$      (quarts)

                 Final solution: Percent alcohol $= 0.30$; amount $= 5 + x$      (quarts)

*Equation:*

$$0.15(5) + 0.45(x) = 0.30(5 + x)$$
$$0.75 + 0.45x = 1.50 + 0.30x$$
$$0.75 + 0.45x - 0.30x = 1.50 + 0.30x - 0.30x$$
$$0.75 + 0.15x = 1.50$$
$$0.75 + 0.15x - 0.75 = 1.50 - 0.75$$
$$0.15x = 0.75$$
$$\frac{0.15x}{0.15} = \frac{0.75}{0.15}$$
$$x = 5 \quad \text{and} \quad 5 + x = 10$$

There are 5 quarts of Solution 2 and 10 quarts of the final solution.

**49.** *Verbal Model:*

$$\boxed{\begin{array}{c}\text{Amount of antifreeze in} \\ \text{first solution}\end{array}} + \boxed{\begin{array}{c}\text{Amount of antifreeze} \\ \text{in pure solution}\end{array}} = \boxed{\begin{array}{c}\text{Amount of antifreeze} \\ \text{in final solution}\end{array}}$$

*Labels:*        Original solution: percent antifreeze $= 0.30$, amount $= 4 - x$   (gallons)

                 Pure antifreeze: percent antifreeze $= 1.00$, amount $= x$       (gallons)

                 Final solution: percent antifreeze $= 0.50$, amount $= 4$        (gallons)

*Equation:*

$$0.30(4 - x) + 1.00(x) = 0.50(4)$$
$$1.2 - 0.3x + x = 2$$
$$1.2 + 0.7x = 2$$
$$0.7x = 0.8$$
$$x = \frac{0.8}{0.7} \approx 1.14$$

Approximately 1.14 gallons $\left(\text{or } 1\frac{1}{7} \text{ gallons}\right)$ must be withdrawn and replaced.

**51.** *Verbal Model:*    | Total interest | = | Interest from investment A | + | Interest from investment B |

*Labels:*    Total interest = \$500

Principal for investment A = $P$    (dollars)

Annual interest rate for investment A = 0.07

Time in investment A = 1    (year)

Interest from investment A = $0.07P$    (dollars)

Principal for investment B = $6000 - P$    (dollars)

Annual interest rate for investment B = 0.09

Time in investment B = 1    (year)

Interest from investment B = $0.09(6000 - P)$    (dollars)

*Equation:*

$$500 = 0.07P + 0.09(6000 - P)$$

$$500 = 0.07P + 540 - 0.09P$$

$$-40 = -0.02P$$

$$\frac{-40}{-0.02} = P$$

$$2000 = P \text{ and } 6000 - P = 4000$$

So, \$2000 was invested at 7% and \$4000 was invested at 9%.

**53.** *Verbal Model:*    | Work done | = | Portion done by you | + | Portion done by friend |

*Labels:*    Both persons: work done = 1 job, time = $t$    (hours)

Your work: rate = $\frac{1}{2}$ job per hour, time = $t$    (hours)

Friend's work: rate = $\frac{1}{3}$ job per hour, time = $t$    (hours)

*Equation:*

$$1 = \frac{1}{2}(t) + \frac{1}{3}(t)$$

$$1 = \left(\frac{1}{2} + \frac{1}{3}\right)t$$

$$1 = \left(\frac{3}{6} + \frac{2}{6}\right)t$$

$$1 = \frac{5}{6}t$$

$$\frac{6}{5}(1) = \frac{6}{5}\left(\frac{5}{6}t\right)$$

$$\frac{6}{5} = t$$

It would take $\frac{6}{5}$ hours (or 1 hour, 12 minutes) to mow the lawn.

**55.** *Verbal Model:*   $\boxed{\begin{array}{c}\text{Work}\\\text{done}\end{array}} = \boxed{\begin{array}{c}\text{Portion done}\\\text{by first worker}\end{array}} + \boxed{\begin{array}{c}\text{Portion done}\\\text{by second worker}\end{array}}$

*Labels:*   Both persons: work done $= 1$ task, time $= t$     (minutes)

First worker: rate $= \dfrac{1}{m}$ tasks per hour, time $= t$     (minutes)

Second worker: rate $= \dfrac{1}{9m}$ tasks per hour, time $= t$   (minutes)

*Equation:*
$$1 = \frac{1}{m}(t) + \frac{1}{9m}(t)$$
$$1 = \left(\frac{1}{m} + \frac{1}{9m}\right)t$$
$$1 = \left(\frac{9}{9m} + \frac{1}{9m}\right)t$$
$$1 = \frac{10}{9m}t$$
$$\frac{9m}{10}(1) = \frac{9m}{10}\left(\frac{10}{9m}t\right)$$
$$\frac{9}{10}m = t$$

Working together, the two people can complete the task in $\dfrac{9}{10}m$ minutes.

**57.** *Verbal Model:*   $\boxed{\text{Volume}} = \boxed{\text{Rate}} \cdot \boxed{\text{Time}}$

*Labels:*   Volume $= 1000$       (milliliters)

Rate $= r$       (milliliters per hour)

Time $= 8$       (hours)

*Equation:*
$$1000 = r(8)$$
$$\frac{1000}{8} = \frac{8r}{8}$$
$$125 = r$$

The solution flows through the tube at the rate of 125 milliliters per hour.

**59.** *Verbal Model:*   $\boxed{\begin{array}{c}\text{Cost of carnations}\\\text{per dozen}\end{array}} \cdot \boxed{\begin{array}{c}\text{Number of dozens}\\\text{of carnations}\end{array}} + \boxed{\begin{array}{c}\text{Cost of roses}\\\text{per dozen}\end{array}} \cdot \boxed{\begin{array}{c}\text{Number of dozens}\\\text{of roses}\end{array}} = \boxed{\begin{array}{c}\text{Total cost}\\\text{of flowers}\end{array}}$

*Labels:*   Cost of carnations per dozen $= 12$     (dollars per dozen)

Number of dozens of carnations $= x$     (dozens)

Cost of dozens of roses $= 18$     (dollars per dozen)

Number of dozens of roses $= 2x$     (dollars)

Total cost of flowers $= 384$     (dollars)

*Equation:*
$$12x + 18(2x) = 384$$
$$12x + 36x = 384$$
$$48x = 384$$
$$x = 8 \text{ and } 2x = 16$$

So, 8 dozen carnations and 16 dozen roses were ordered.

**61.** (a) *Verbal Model:* | Value of corn | + | Value of soybeans | = | Value of mixture |

   *Labels:*      Corn: price per ton $= \$125$, number of tons $= x$, value of corn $= \$125x$

   Soybeans: price per ton $= \$200$, number of tons $= 100 - x$, value of soybeans $= \$200(100 - x)$

   Mixture: price per ton $= m$ (dollars), number of tons $= 100$, value of mixture $= \$100m$

   *Equation:*    $125x + 200(100 - x) = 100m$

   $$\frac{125x + 200(100 - x)}{100} = m$$

| Corn weight, $x$ | Soybeans weight, $100 - x$ | Price/ton of the mixture, $m$ |
|---|---|---|
| 0 | 100 | $\dfrac{125(0) + 200(100)}{100} = \dfrac{20{,}000}{100} = \$200$ |
| 20 | 80 | $\dfrac{125(20) + 200(80)}{100} = \dfrac{18{,}500}{100} = \$185$ |
| 40 | 60 | $\dfrac{125(40) + 200(60)}{100} = \dfrac{17{,}000}{100} = \$170$ |
| 60 | 40 | $\dfrac{125(60) + 200(40)}{100} = \dfrac{15{,}500}{100} = \$155$ |
| 80 | 20 | $\dfrac{125(80) + 200(20)}{100} = \dfrac{14{,}000}{100} = \$140$ |
| 100 | 0 | $\dfrac{125(100) + 200(0)}{100} = \dfrac{12{,}500}{100} = \$125$ |

(b) As the number of tons of corn increases, the number of tons of soybeans *decreases* by the same amount.

(c) As the number of tons of corn increases, the price per ton of the mixture *decreases* and gets closer to the price per ton of the corn.

(d) If there were equal numbers of tons of corn and soybeans in the mixture, the price per ton of the mixture would be halfway between the prices of the two components. In other words, the price per ton of the mixture would be the *average of the two prices,* 125 and 200, which is

$$\frac{125 + 200}{2} = \frac{325}{2} = 162.50.$$

Note: This result can be verified with the equation above, using $x = 50$ and $100 - x = 50$.

$$m = \frac{125(50) + 200(50)}{100} = \$162.50$$

**63.** *Verbal Model:* | Perimeter | $= 2$ | Length | $+ 2$ | Width |

   *Labels:*      Perimeter $= 36$          (inches)

   Length $= w + 8$          (inches)

   Width $= w$          (inches)

   *Equation:*    $36 = 2(w + 8) + 2w$

   $36 = 2w + 16 + 2w$

   $36 = 4w + 16$

   $36 - 16 = 4w + 16 - 16$

   $20 = 4w$

   $\dfrac{20}{4} = \dfrac{4w}{4}$

   $5 = w$  and  $w + 8 = 13$

The length of the rectangle is 13 inches and the width is 5 inches.

**65.** *Verbal Model:*   $\boxed{\text{Perimeter}} = \boxed{\text{Side } a} + \boxed{\text{Side } b} + \boxed{\text{Side } c}$

*Labels:*       Perimeter = 83          (meters)

Side a = $x$          (meters)

Side b = $x$          (meters)

Side c = $4x - 7$          (meters)

*Equation:*       $83 = x + x + 4x - 7$

$83 = 6x - 7$

$83 + 7 = 6x - 7 + 7$

$90 = 6x$

$\dfrac{90}{6} = \dfrac{6x}{6}$

$15 = x$ and $4x - 7 = 4(15) - 7 = 53$

The lengths of the three sides of the triangle are 15 meters, 15 meters and 53 meters.

**67.** *Verbal Model:*   $\boxed{\begin{array}{c}\text{Faster car's}\\\text{distance}\end{array}} - \boxed{\begin{array}{c}\text{Slower car's}\\\text{distance}\end{array}} = \boxed{\begin{array}{c}\text{Distance}\\\text{between cars}\end{array}}$

*Labels:*       Faster car's rate = 52          (miles per hour)

Slower car's rate = 45          (miles per hour)

Time = 4          (hour)

Distance between cars = $x$     (miles)

*Equation:*       $52(4) - 45(4) = x$

$208 - 180 = x$

$28 = x$

The cars will be 28 miles apart.

**69.** *Verbal Model:*   $\boxed{\begin{array}{c}\text{Distance driven}\\\text{at higher speed}\end{array}} + \boxed{\begin{array}{c}\text{Distance driven}\\\text{at lower speed}\end{array}} = \boxed{\begin{array}{c}\text{Total}\\\text{distance}\end{array}}$

*Labels:*       Higher rate = 55          (miles per hour)

Lower rate = 48          (miles per hour)

Time at higher rate = $x$          (hours)

Time at lower rate = $4.25 - x$          (hours)

Total distance = 225          (miles)

*Equation:*       $55x + 48(4.25 - x) = 225$

$55x + 204 - 48x = 225$

$7x + 204 = 225$

$7x = 21$

$x = 3$

$4.25 - x = 4.25 - 3 = 1.25$

The 55 mph speed was maintained for 3 hours and the 48 mph speed was maintained for 1.25 hours (or 1 hour and 15 minutes).

**71.** To do this calculation mentally, you could divide 180 miles by 2 to obtain 90 miles, the distance you traveled in 2 hours. Then divide by 2 again to obtain 45 miles, the distance you traveled in 1 hour. Your average speed is 45 miles per hour.

$\text{Rate} = \dfrac{\text{Distance}}{\text{Time}} = \dfrac{180}{4} = 45$ miles per hour

**73.** Solving for $h$:

$$A = \frac{1}{2}(x + y)h$$

$$2A = 2 \cdot \frac{1}{2}(x + y)h$$

$$2A = (x + y)h$$

$$\frac{2A}{x + y} = \frac{(x + y)h}{x + y}$$

$$\frac{2A}{x + y} = h$$

You would use this result to find the height of a trapezoid.

Solving for $x$:

$$A = \frac{1}{2}(x + y)h$$

$$2A = 2 \cdot \frac{1}{2}(x + y)h$$

$$2A = (x + y)h$$

$$\frac{2A}{h} = \frac{(x + y)h}{h}$$

$$\frac{2A}{h} = x + y$$

$$\frac{2A}{h} - y = x + y - y$$

$$\frac{2A}{h} - y = x$$

You would use this result to find the base $x$ of a trapezoid.

Solving for $y$:

$$A = \frac{1}{2}(x + y)h$$

$$2A = 2 \cdot \frac{1}{2}(x + y)h$$

$$2A = (x + y)h$$

$$\frac{2A}{h} = \frac{(x + y)h}{h}$$

$$\frac{2A}{h} = x + y$$

$$\frac{2A}{h} - x = x + y - x$$

$$\frac{2A}{h} - x = y$$

You would use this result to find the base $y$ of a trapezoid.

## Section 3.6  Linear Inequalities

**1.** $x$ is greater than or equal to 3.

**75.** Yes, if the radius of a circle is doubled, the circumference is doubled.

$$c = 2\pi r$$

$$2\pi(2r) = 4\pi r = 2(2\pi r) = 2c$$

No, if the radius is doubled, the area is not doubled; the area is multiplied by 4.

$$A = \pi r^2$$

$$\pi(2r)^2 = \pi\left(4r^2\right) = 4\pi r^2 = 4A$$

**77.** (a)  Natural numbers: $\{7, 1\}$

(b)  Integers: $\{7, 1, -3\}$

(c)  Rational numbers: $\left\{1.8, \frac{1}{10}, 7, -2.75, 1, -3\right\}$

(d)  Irrational numbers: None

**79.** (a)  Natural numbers: $\{9\}$

(b)  Integers: $\{9, -6\}$

(c)  Rational numbers: $\left\{-2.2, 9, \frac{1}{3}, \frac{3}{5}, -6\right\}$

(d)  Irrational numbers: $\left\{\sqrt{13}\right\}$

**81.**    $\dfrac{1}{4} = \dfrac{y}{36}$

$1(36) = 4y$    (Cross-multiply)

$36 = 4y$

$\dfrac{36}{4} = \dfrac{4y}{4}$

$9 = y$

**83.**    $\dfrac{3}{2} = \dfrac{9}{x}$

$3x = 2(9)$    (Cross-multiply)

$3x = 18$

$\dfrac{3x}{3} = \dfrac{18}{3}$

$x = 6$

**85.**    $\dfrac{34}{x} = \dfrac{102}{48}$

$34(48) = 102x$    (Cross-multiply)

$1632 = 102x$

$\dfrac{1632}{102} = \dfrac{102x}{102}$

$16 = x$

**3.** $x$ is less than or equal to 10.

**5.** $y$ is less than $-9$.

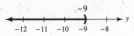

**7.** $z$ is greater than or equal to 5 and less than or equal to 10.

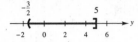

**9.** $y$ is greater than $-\frac{3}{2}$ and less than or equal to 5.

**11.** (f)

**12.** (b)

**13.** (d)

**14.** (c)

**15.** (a)

**16.** (e)

**17.** (a) $x = 3$

$$5(3) - 12 \overset{?}{>} 0$$

$$15 - 12 \overset{?}{>} 0$$

$$3 > 0$$

3 *is* a solution.

(b) $x = -3$

$$5(-3) - 12 \overset{?}{>} 0$$

$$-15 - 12 \overset{?}{>} 0$$

$$-27 \not> 0$$

$-3$ *is not* a solution.

(c) $x = \frac{5}{2}$

$$5\left(\frac{5}{2}\right) - 12 \overset{?}{>} 0$$

$$\frac{25}{2} - \frac{24}{2} \overset{?}{>} 0$$

$$\frac{1}{2} > 0$$

$\frac{5}{2}$ *is* a solution.

(d) $x = \frac{3}{2}$

$$5\left(\frac{3}{2}\right) - 12 \overset{?}{>} 0$$

$$\frac{15}{2} - \frac{24}{2} \overset{?}{>} 0$$

$$-\frac{9}{2} \not> 0$$

$\frac{3}{2}$ *is not* a solution.

**19.** (a) $x = 10$

$$3 - \frac{1}{2}(10) \overset{?}{>} 0$$

$$3 - 5 \overset{?}{>} 0$$

$$-2 \not> 0$$

10 *is not* a solution.

(b) $x = 6$

$$3 - \frac{1}{2}(6) \overset{?}{>} 0$$

$$3 - 3 \overset{?}{>} 0$$

$$0 > 0$$

6 *is not* solution.

(c) $x = -\frac{3}{4}$

$$3 - \frac{1}{2}\left(-\frac{3}{4}\right) \overset{?}{>} 0$$

$$3 + \frac{3}{8} \overset{?}{>} 0$$

$$3\frac{3}{8} > 0$$

$-\frac{3}{4}$ *is* solution.

(d) $x = 0$

$$3 - \frac{1}{2}(0) \overset{?}{>} 0$$

$$3 - 0 \overset{?}{>} 0$$

$$3 > 0$$

0 *is* a solution.

**21.** (a) $x = 4$

$$5(4 - 2) + 1 \overset{?}{<} 12$$

$$5(2) + 1 \overset{?}{<} 12$$

$$10 + 1 \overset{?}{<} 12$$

$$11 < 12$$

4 *is* a solution.

(b) $x = 1$

$$5(1 - 2) + 1 \overset{?}{<} 12$$

$$5(-1) + 1 \overset{?}{<} 12$$

$$-5 + 1 \overset{?}{<} 12$$

$$-4 < 12$$

1 *is* a solution.

(c) $x = 5$

$$5(5 - 2) + 1 \overset{?}{<} 12$$

$$5(3) + 1 \overset{?}{<} 12$$

$$15 + 1 \overset{?}{<} 12$$

$$16 \not< 12$$

5 *is not* a solution.

(d) $x = -3$

$$5(-3 - 2) + 1 \overset{?}{<} 12$$

$$5(-5) + 1 \overset{?}{<} 12$$

$$-25 + 1 \overset{?}{<} 12$$

$$-24 < 12$$

$-3$ *is* a solution.

**23.** (a) $x = 0$

$$15 - (0 + 8) \overset{?}{\geq} 13$$

$$15 - 8 \overset{?}{\geq} 13$$

$$7 \not\geq 13$$

0 *is not* a solution.

(b) $x = -6$

$$15 - (-6 + 8) \overset{?}{\geq} 13$$

$$15 - 2 \overset{?}{\geq} 13$$

$$13 \geq 13$$

$-6$ *is* a solution.

(c) $x = 2$

$$15 - (2 + 8) \overset{?}{\geq} 13$$

$$15 - 10 \overset{?}{\geq} 13$$

$$5 \not\geq 13$$

2 *is not* a solution.

(d) $x = -10$

$$15 - (-10 + 8) \overset{?}{\geq} 13$$

$$15 - (-2) \overset{?}{\geq} 13$$

$$17 \geq 13$$

$-10$ *is* a solution.

**25.** (a) $x = 1$

$$5(1) + 3 \overset{?}{\leq} 1 - 5$$

$$5 + 3 \overset{?}{\leq} -4$$

$$8 \not\leq -4$$

1 *is not* a solution.

(b) $x = -2$

$$5(-2) + 3 \overset{?}{\leq} -2 - 5$$

$$-10 + 3 \overset{?}{\leq} -7$$

$$-7 \leq -7$$

$-2$ *is* a solution.

(c) $x = -1$

$$5(-1) + 3 \overset{?}{\leq} -1 - 5$$

$$-5 + 3 \overset{?}{\leq} -6$$

$$-2 \not\leq -6$$

$-1$ *is not* a solution.

(d) $x = 2$

$$5(2) + 3 \overset{?}{\leq} 2 - 5$$

$$10 + 3 \overset{?}{\leq} -3$$

$$13 \not\leq -3$$

2 *is not* a solution

**27.**
$$t - 3 \geq 2$$
$$t - 3 + 3 \geq 2 + 3$$
$$t \geq 5$$

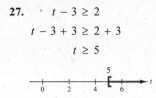

**29.**
$$x + 4 \leq 6$$
$$x + 4 - 4 \leq 6 - 4$$
$$x \leq 2$$

**31.** $4x < 12$
$$\frac{4x}{4} < \frac{12}{4}$$
$$x < 3$$

**33.** $-10x < 40$
$$\frac{-10x}{-10} > \frac{40}{-10} \quad (\text{Reverse inequality})$$
$$x > -4$$

**35.** $\frac{2}{3}x \leq 12$
$$3\left(\frac{2}{3}\right)x \leq 3(12)$$
$$2x < 36$$
$$\frac{2x}{2} \leq \frac{36}{2}$$
$$x \leq 18$$

**37.** $-\frac{5}{8}x \geq 10$
$$-\frac{8}{5}\left(-\frac{5}{8}x\right) \leq -\frac{8}{5}(10) \quad (\text{Reverse inequality})$$
$$x \leq -16$$

**39.** $3x \geq -\frac{4}{5}$
$$\frac{1}{3}(3x) \geq \frac{1}{3}\left(-\frac{4}{5}\right)$$
$$x \geq -\frac{4}{15}$$

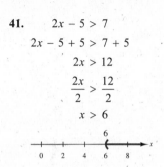

**41.**
$$2x - 5 > 7$$
$$2x - 5 + 5 > 7 + 5$$
$$2x > 12$$
$$\frac{2x}{2} > \frac{12}{2}$$
$$x > 6$$

**43.**
$$3x + 2 \leq 14$$
$$3x + 2 - 2 \leq 14 - 2$$
$$3x \leq 12$$
$$\frac{3x}{3} \leq \frac{12}{3}$$
$$x \leq 4$$

**45.**
$$4 - 2x < 3$$
$$4 - 4 - 2x < 3 - 4$$
$$-2x < -1$$
$$\frac{-2x}{-2} > \frac{-1}{-2} \quad (\text{Reverse inequality})$$
$$x > \frac{1}{2}$$

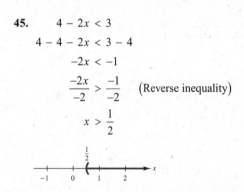

**47.**
$$12 - x > 4$$
$$12 - 12 - x > 4 - 12$$
$$-x > -8$$
$$\frac{-x}{-1} < \frac{-8}{-1} \quad (\text{Reverse inequality})$$
$$x < 8$$

**49.**
$$3x + 9 < 2x$$
$$3x - 2x + 9 < 2x - 2x$$
$$x + 9 < 0$$
$$x + 9 - 9 < 0 - 9$$
$$x < -9$$

**51.**
$$7t + 9 < 14 + 6t$$
$$7t + 9 - 6t < 14 + 6t - 6t$$
$$t + 9 < 14$$
$$t + 9 - 9 < 14 - 9$$
$$t < 5$$

**53.**
$$2x - 5 > -x + 6$$
$$2x + x - 5 > -x + x + 6$$
$$3x - 5 > 6$$
$$3x - 5 + 5 > 6 + 5$$
$$3x > 11$$
$$\frac{3x}{3} > \frac{11}{3}$$
$$x > \frac{11}{3}$$

**55.**
$$2(x + 7) > 12$$
$$2x + 14 > 12$$
$$2x + 14 - 14 > 12 - 14$$
$$2x > -2$$
$$\frac{2x}{2} > \frac{-2}{2}$$
$$x > -1$$

**57.**
$$-3(x + 11) \le 6$$
$$-3x - 33 \le 6$$
$$-3x - 33 + 33 \le 6 + 33$$
$$-3x \le 39$$
$$\frac{-3x}{-3} \ge \frac{39}{-3} \quad \text{(Reverse inequality)}$$
$$x \ge -13$$

**59.**
$$6 + \frac{2x}{3} < x + 7$$
$$3\left(6 + \frac{2x}{3}\right) < 3(x + 7)$$
$$18 + 2x < 3x + 21$$
$$18 + 2x - 3x < 3x + 21 - 3x$$
$$18 - x < 21$$
$$18 - x - 18 < 21 - 18$$
$$-x < 3$$
$$\frac{-x}{-1} > \frac{3}{-1} \quad \text{(Reverse inequality)}$$
$$x > -3$$

**61.**
$$-2(z + 1) \ge 3(z + 1)$$
$$-2z - 2 \ge 3z + 3$$
$$-2z - 2 + 2z \ge 3z + 3 + 2z$$
$$-2 \le 5z + 3$$
$$-2 - 3 \le 5z + 3 - 3$$
$$-5 \ge 5z$$
$$\frac{-5}{5} \ge \frac{5z}{5}$$
$$-1 \ge z \text{ or } z \le -1$$

**63.**
$$3(x + 1) \ge 2(x + 5)$$
$$3x + 3 \ge 2x + 10$$
$$3x - 2x + 3 \ge 2x - 2x + 10$$
$$x + 3 \ge 10$$
$$x + 3 - 3 \ge 10 - 3$$
$$x \ge 7$$

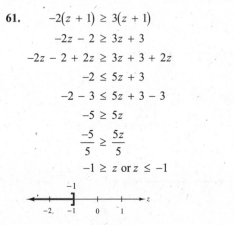

**65.**
$$10(1 - y) < -4(y - 2)$$
$$10 - 10y < -4y + 8$$
$$10 - 10y + 4y < -4y + 4y + 8$$
$$10 - 6y < 8$$
$$10 - 6y - 10 < 8 - 10$$
$$-6y < -2$$
$$\frac{-6y}{-6} > \frac{-2}{-6} \quad \text{(Reverse inequality)}$$
$$y > \frac{1}{3}$$

**67.**
$$\frac{x}{4} + \frac{1}{2} > 0$$
$$4\left(\frac{x}{4} + \frac{1}{2}\right) > 4(0)$$
$$4\left(\frac{x}{4}\right) + 4\left(\frac{1}{2}\right) > 0$$
$$x + 2 > 0$$
$$x + 2 - 2 > 0 - 2$$
$$x > -2$$

**69.**
$$\frac{x}{5} - \frac{x}{2} \le 1$$
$$10\left(\frac{x}{5} - \frac{x}{2}\right) \le 10(1)$$
$$10\left(\frac{x}{5}\right) - 10\left(\frac{x}{2}\right) \le 10$$
$$2x - 5x \le 10$$
$$-3x \le 10$$
$$\frac{-3x}{-3} \ge \frac{10}{-3} \quad \text{(Reverse inequality)}$$
$$x \ge -\frac{10}{3}$$

**71.** $x \le -1$

**73.** $y > -6$

**75.** $x \ge 4$

**77.** $0 < x \le 6$

**79.** Mars is farther from the sun than Mercury. (This illustrates the Transitive Property of Inequalities.)

**81.** *Verbal Model:* $\boxed{\text{Total points}} \ge \boxed{90\% \text{ of } 500}$

*Labels:*   Score for fifth test $= x$   (points)

Total points $= (93 + 88 + 91 + 82) + x$   (points)

*Inequality:*
$$(93 + 88 + 91 + 82) + x \ge 0.90(500)$$
$$354 + x \ge 450$$
$$354 + x - 354 \ge 450 - 354$$
$$x \ge 96$$

You must obtain at least 96 points on the fifth test to earn an A.

**83.** *Verbal Model:* $\boxed{\begin{array}{c}\text{Truck} \\ \text{weight}\end{array}} + \boxed{\begin{array}{c}\text{Bushel} \\ \text{weight}\end{array}} \cdot \boxed{\begin{array}{c}\text{Number} \\ \text{of bushels}\end{array}} \le \boxed{6000}$

*Labels:*   Truck weight $= 4350$   (pounds)

Bushel weight $= 48$   (pounds)

Number of bushels $= x$

*Inequality:*
$$4350 + 48x \le 6000$$
$$4350 + 48x - 4350 \le 6000 - 4350$$
$$48x \le 1650$$
$$\frac{48x}{48} \le \frac{1650}{48}$$
$$x \le 34.375$$

The truck can haul no more than 34 bushels; $0 \le x \le 34$.

**85.** *Verbal Model:*   $\boxed{\begin{array}{c}\text{Weekly cost for}\\\text{Company B}\end{array}} + 0.25 \cdot \boxed{\begin{array}{c}\text{Distance}\\\text{in miles}\end{array}} > \boxed{\begin{array}{c}\text{Weekly cost for}\\\text{Company A}\end{array}}$

*Labels:*  Weekly cost for Company B = 180   (dollars)

Distance in miles = $m$   (miles)

Weekly cost for Company A = 270   (dollars)

*Inequality:*
$$180 + 0.25m > 270$$
$$180 - 180 + 0.25m > 270 - 180$$
$$0.25m > 90$$
$$\frac{0.25m}{0.25} > \frac{90}{0.25}$$
$$m > 360$$

You must drive more than 360 miles for the cost of Company B to be more than the cost for Company A.

**87.** True

These two inequalities are equivalent. The second inequality can be obtained by subtracting 6 from both sides of the original inequality.

**89.** True

These two statements are equivalent.

**91.** True
$$x \overset{?}{<} x + 4$$
$$x - x \overset{?}{<} x - x + 4 \qquad \text{Subtract } x \text{ from each side.}$$
$$0 < 4$$

**93.** The error is the square bracket at 5; there should be a parenthesis at 5.

**95.** $2x + 7$

The sum of twice a number and 7

Two times a number plus 7

The sum of 7 and the product of 2 and a number

**97.** $-5(x + 7)$

Negative five times the sum of a number and 7

The product of negative five and the sum of a number and 7

**99.** $\dfrac{1}{3} + \dfrac{x}{2}$

One-third plus the quotient of a number and 2

The sum of one-third and the quotient of a number and 2

**101.** $d = rt$
$$d = 48(38)$$
$$d = 1824$$

The distance is 1824 feet.

**103.** $d = rt$
$$384 = 6t$$
$$\frac{384}{6} = \frac{6t}{6}$$
$$64 = t$$

The time is 64 minutes (or 1 hour, 4 minutes).

**105.** $d = rt$
$$240 = r(20)$$
$$\frac{240}{20} = \frac{r(20)}{20}$$
$$12 = r$$

The rate is 12 kilometers per hour.

# Review Exercises for Chapter 3

**1.**
$$2x - 10 = 0$$
$$2x - 10 + 10 = 0 + 10$$
$$2x = 10$$
$$\frac{2x}{2} = \frac{10}{2}$$
$$x = 5$$

**3.**
$$-3y - 12 = 0$$
$$-3y - 12 + 12 = 0 + 12$$
$$-3y = 12$$
$$\frac{-3y}{-3} = \frac{12}{-3}$$
$$y = -4$$

**5.**
$$5x - 3 = 0$$
$$5x - 3 + 3 = 0 + 3$$
$$5x = 3$$
$$\frac{5x}{5} = \frac{3}{5}$$
$$x = \frac{3}{5}$$

**7.**
$$x + 10 = 13$$
$$x + 10 - 10 = 13 - 10$$
$$x = 3$$

**9.**
$$5 - x = 2$$
$$5 - 5 - x = 2 - 5$$
$$-x = -3$$
$$-1(-x) = -1(-3)$$
$$x = 3$$

**11.** $10x = 50$
$$\frac{10x}{10} = \frac{50}{10}$$
$$x = 5$$

**13.**
$$8x + 7 = 39$$
$$8x + 7 - 7 = 39 - 7$$
$$8x = 32$$
$$\frac{8x}{8} = \frac{32}{8}$$
$$x = 4$$

**15.**
$$24 - 7x = 3$$
$$24 - 24 - 7x = 3 - 24$$
$$-7x = -21$$
$$\frac{-7x}{-7} = \frac{-21}{-7}$$
$$x = 3$$

**17.**
$$15x - 4 = 16$$
$$15x - 4 + 4 = 16 + 4$$
$$15x = 20$$
$$\frac{15x}{15} = \frac{20}{15}$$
$$x = \frac{20}{15}$$
$$x = \frac{4}{3}$$

**19.**
$$\frac{x}{5} = 4$$
$$5\left(\frac{x}{5}\right) = 5(4)$$
$$x = 20$$

**21.** *Verbal Model:* 
$$\boxed{\text{Hourly wage}} = \boxed{\text{Base hourly wage}} + \boxed{0.60} \cdot \boxed{\text{Number of units produced}}$$

*Labels:*    Hourly wage = 15.50    (dollars)

Base hourly wage = 8.30    (dollars)

Number of units produced = $x$

*Equation:*
$$15.50 = 8.30 + 0.60x$$
$$15.50 - 8.30 = 8.30 + 0.60x - 8.30$$
$$7.20 = 0.60x$$
$$\frac{7.20}{0.60} = \frac{0.60x}{0.60}$$
$$12 = x$$

You must produce 12 units in an hour.

**23.** *Verbal Model:* $2 \cdot \boxed{\text{Length}} + 2 \cdot \boxed{\text{Width}} = \boxed{\text{Perimeter}}$

*Labels:*  Width $= w$  (meters)

Length $= w + 30$  (meters)

Perimeter $= 260$  (meters)

*Equation:*  $2(w + 30) + 2w = 260$

$2w + 60 + 2w = 260$

$4w + 60 = 260$

$4w + 60 - 60 = 260 - 60$

$4w = 200$

$\dfrac{4w}{4} = \dfrac{200}{4}$

$w = 50 \text{ and } w + 30 = 80$

The length of the rectangle is 80 meters and the width is 50 meters.

**25.** $3x - 2(x + 5) = 10$

$3x - 2x - 10 = 10$

$x - 10 = 10$

$x - 10 + 10 = 10 + 10$

$x = 20$

**27.** $2(x + 3) = 6(x - 3)$

$2x + 6 = 6x - 18$

$2x - 6x + 6 = 6x - 6x - 18$

$-4x + 6 = -18$

$-4x + 6 - 6 = -18 - 6$

$-4x = -24$

$\dfrac{-4x}{-4} = \dfrac{-24}{-4}$

$x = 6$

**29.** $7 - \big[2(3x + 4) - 5\big] = x - 3$

$7 - \big[6x + 8 - 5\big] = x - 3$

$7 - \big[6x + 3\big] = x - 3$

$7 - 6x - 3 = x - 3$

$4 - 6x = x - 3$

$4 - 6x - x = x - x - 3$

$4 - 7x = -3$

$4 - 4 - 7x = -3 - 4$

$-7x = -7$

$\dfrac{-7x}{-7} = \dfrac{-7}{-7}$

$x = 1$

**31.** $\dfrac{2}{3}x - \dfrac{1}{6} = \dfrac{9}{2}$

$6\left(\dfrac{2}{3}x - \dfrac{1}{6}\right) = 6\left(\dfrac{9}{2}\right)$

$6\left(\dfrac{2}{3}x\right) - 6\left(\dfrac{1}{6}\right) = 27$

$4x - 1 = 27$

$4x - 1 + 1 = 27 + 1$

$4x = 28$

$\dfrac{4x}{4} = \dfrac{28}{4}$

$x = 7$

**33.** $\dfrac{x}{3} - \dfrac{1}{9} = 2$

$9\left(\dfrac{x}{3} - \dfrac{1}{9}\right) = 9(2)$

$9\left(\dfrac{x}{3}\right) - 9\left(\dfrac{1}{9}\right) = 18$

$3x - 1 = 18$

$3x - 1 + 1 = 18 + 1$

$3x = 19$

$\dfrac{3x}{3} = \dfrac{19}{3}$

$x = \dfrac{19}{3}$

**35.**
$$\frac{u}{10} + \frac{u}{5} = 6$$
$$10\left(\frac{u}{10} + \frac{u}{5}\right) = 10(6)$$
$$10\left(\frac{u}{10}\right) + 10\left(\frac{u}{5}\right) = 60$$
$$u + 2u = 60$$
$$3u = 60$$
$$\frac{3u}{3} = \frac{60}{3}$$
$$u = 20$$

**37.**
$$\frac{2x}{9} = \frac{2}{3}$$
$$2x(3) = 9(2)$$
$$6x = 18$$
$$\frac{6x}{6} = \frac{18}{6}$$
$$x = 3$$

**39.**
$$\frac{x+3}{5} = \frac{x+7}{12}$$
$$12(x+3) = 5(x+7)$$
$$12x + 36 = 5x + 35$$
$$12x + 36 - 5x = 5x + 35 - 5x$$
$$7x + 36 = 35$$
$$7x + 36 - 36 = 35 - 36$$
$$7x = -1$$
$$\frac{7x}{7} = \frac{-1}{7}$$
$$x = -\frac{1}{7}$$

**41.**
$$5.16x - 87.5 = 32.5$$
$$5.16x - 87.5 + 87.5 = 32.5 + 87.5$$
$$5.16x = 120$$
$$\frac{5.16x}{5.16} = \frac{120}{5.16}$$
$$x \approx 23.26$$

**43.**
$$\frac{x}{4.625} = 48.5$$
$$4.625\left(\frac{x}{4.625}\right) = 4.625(48.5)$$
$$x \approx 224.31$$

**45.**
$$\frac{t}{10} + \frac{t}{15} = 0.5$$
$$30\left(\frac{t}{10} + \frac{t}{15}\right) = 30(0.5)$$
$$3t + 2t = 15$$
$$5t = 15$$
$$\frac{5t}{5} = \frac{15}{5}$$
$$t = 3$$

It will take 3 hours for the two people to complete 50% of the task.

**47.**

| Percent | Parts out of 100 | Decimal | Fraction |
|---------|------------------|---------|----------|
| 60% | 60 | 0.60 | $\frac{3}{5}$ |

**Parts out of 100:** 60% means 60 parts out of 100.

**Decimal:**

*Verbal Model:*    $\boxed{\text{Decimal}} \cdot \boxed{100\%} = \boxed{\text{Percent}}$

*Label:*        Decimal $= x$

*Equation:*      $x(100\%) = 60\%$
$$x = \frac{60\%}{100\%}$$
$$x = 0.60$$

**Fraction:**

*Verbal Model:*    $\boxed{\text{Fraction}} \cdot \boxed{100\%} = \boxed{\text{Percent}}$

*Label:*        Fraction $= x$

*Equation:*      $x(100\%) = 60\%$
$$x = \frac{60\%}{100\%}$$
$$x = \frac{60}{100}$$
$$x = \frac{3}{5}$$

**49.**

| Percent | Parts out of 100 | Decimal | Fraction |
|---------|------------------|---------|----------|
| 80% | 80 | 0.80 | $\frac{4}{5}$ |

(a) *Verbal Model:*  $\boxed{\text{Fraction}} \cdot \boxed{100\%} = \boxed{\text{Percent}}$

    *Label:*         Percent $= x$

    *Equation:*     $\frac{4}{5}(100\%) = x$

                    $80\% = x$

(b) 80% means 80 parts out of 100.

(c) *Verbal Model:*  $\boxed{\text{Decimal}} \cdot \boxed{100\%} = \boxed{\text{Percent}}$

    *Label:*         Decimal $= x$

    *Equation:* $x(100\%) = 80\%$

$$x = \frac{80\%}{100\%}$$

$$x = 0.80$$

**51.**

| Percent | Parts out of 100 | Decimal | Fraction |
|---------|------------------|---------|----------|
| 20% | 20 | 0.20 | $\frac{1}{5}$ |

**Percent:**

*Verbal Model:* $\boxed{\text{Decimal}} \cdot \boxed{100\%} = \boxed{\text{Percent}}$

    *Label:*         Percent $= x$

    *Equation:*     $0.20(100\%) = x$

                   $20\% = x$

**Parts out of 100:**

20% means 20 parts out of 100.

**Fraction:**

*Verbal Model:* $\boxed{\text{Fraction}} \cdot \boxed{100\%} = \boxed{\text{Percent}}$

    *Label:*         Fraction $= x$

    *Equation:*     $x(100\%) = 20\%$

$$x = \frac{20\%}{100\%}$$

$$x = \frac{20}{100}$$

$$x = \frac{1}{5}$$

**55.** *Verbal Model:*     $\boxed{\text{What number}} = \boxed{125\% \text{ of } 16}$ $(a = pb)$

    *Label:*           $a = $ unknown number

    *Percent equation:*   $a = (1.25)(16)$

                      $a = 20$

So, 20 is 125% of 16.

**53.**

| Percent | Parts out of 100 | Decimal | Fraction |
|---------|------------------|---------|----------|
| 55% | 55 | 0.55 | $\frac{11}{20}$ |

**Percent:**

55 parts out of 100 means 55%.

**Decimal:**

*Verbal Model:*         $\boxed{\text{Decimal}} \cdot \boxed{100\%} = \boxed{\text{Percent}}$

*Label:*         Decimal $= x$

*Equation:*    $x(100\%) = 55\%$

$$x = \frac{55\%}{100\%}$$

$$x = \frac{55}{100}$$

$$x = 0.55$$

**Fraction:**

*Verbal Model:*         $\boxed{\text{Fraction}} \cdot \boxed{100\%} = \boxed{\text{Percent}}$

*Label:*         Fraction $= x$

*Equation:*    $x(100\%) = 55\%$

$$x = \frac{55\%}{100\%}$$

$$x = \frac{55}{100}$$

$$x = \frac{11}{20}$$

**57.** *Verbal Model:*    $\boxed{150} = \boxed{37\frac{1}{2}\% \text{ of what number}}$ $(a = pb)$

   *Label:*        $b = $ unknown number

   *Percent equation:*    $150 = 0.375b$

$$\frac{150}{0.375} = b$$

$$400 = b$$

So, 150 is $37\frac{1}{2}\%$ of 400.

**59.** *Verbal Model:*    $\boxed{150} = \boxed{\text{What percent of } 250}$ $(a = pb)$

   *Label:*        $p = $ unknown percent (in decimal form)

   *Percent equation:*    $150 = p(250)$

$$\frac{150}{250} = p$$

$$0.6 = p$$

So, 150 is 60% of 250.

**61.** *Verbal Model:*    $\boxed{\text{Selling price}} = \boxed{\text{Cost}} + \boxed{\text{Markup}}$

   *Labels:*        Selling price $= x$        (dollars)

            Cost $= 48$            (dollars)

            Markup rate $= 0.78$        (percent in decimal form)

            Markup $= (0.78)(48)$    (dollars)

   *Equation:*        $x = 48 + (0.78)(48)$

            $x = 48 + 37.44$

            $x = 85.44$

The selling price is $85.44.

**63.** (a) *Verbal Model:* $\boxed{\begin{array}{c}2005\\ \text{Sales}\end{array}} = \boxed{\begin{array}{c}\text{Percent}\\ \text{(in decimal form)}\end{array}} \cdot \boxed{\begin{array}{c}2004\\ \text{Sales}\end{array}} + \boxed{\begin{array}{c}2004\\ \text{Sales}\end{array}}$

*Label:* 2005 sales = 601.2    (millions of dollars)

Percent = $p$    (percent in decimal form)

2004 sales = 554.2    (millions of dollars)

*Equation:*
$$601.2 = p(554.2) + 554.2$$
$$601.2 - 554.2 = p(554.2) + 554.2 - 554.2$$
$$47 = p(554.2)$$
$$\frac{47}{554.2} = p$$
$$0.085 \approx p$$

The percent increase in sales from 2004 to 2005 was 8.5%.

(b) *Verbal Model:* $\boxed{\begin{array}{c}2005\\ \text{Sales}\end{array}} = \boxed{\begin{array}{c}2001\\ \text{Sales}\end{array}} + \boxed{\begin{array}{c}\text{Percent}\\ \text{(in decimal form)}\end{array}} \cdot \boxed{\begin{array}{c}2001\\ \text{Sales}\end{array}}$

*Label:* 2005 sales = 601.2    (millions of dollars)

Percent = $p$    (percent in decimal form)

2001 sales = 379.8    (millions of dollars)

*Equation:*
$$601.2 = 379.8 + p(379.8)$$
$$601.2 - 379.8 = 379.8 + p(379.8) - 379.8$$
$$221.4 = p(379.8)$$
$$\frac{221.4}{379.8} = p$$
$$0.583 \approx p$$

The percent increase in sales from 2001 to 2005 was 58.3%.

**65.** $\dfrac{18 \text{ inches}}{4 \text{ yards}} = \dfrac{18 \text{ inches}}{4(36) \text{ inches}} = \dfrac{18}{144} = \dfrac{1}{8}$

**67.** $\dfrac{2 \text{ hours}}{90 \text{ minutes}} = \dfrac{2(60) \text{ minutes}}{90 \text{ minutes}} = \dfrac{120}{90} = \dfrac{4}{3}$

**69.** $\dfrac{6 \text{ hours}}{7.5 \text{ hours}} = \dfrac{60}{75} = \dfrac{4(15)}{5(15)} = \dfrac{4}{5}$

**71.** (a) The unit price for the 18-ounce container:

Unit price $= \dfrac{\text{Total price}}{\text{Total units}} = \dfrac{0.89}{18} \approx \$0.049$ per ounce

(b) The unit price for the 24-ounce container:

Unit price $= \dfrac{\text{Total price}}{\text{Total units}} = \dfrac{1.12}{24} \approx \$0.047$ per ounce

The 24-ounce container (b) has the lower unit price.

**73.** $\dfrac{7}{16} = \dfrac{z}{8}$

$(7)(8) = 16z$

$56 = 16z$

$\dfrac{56}{16} = z$

$\dfrac{7}{2} = z$

**75.** $\dfrac{x+2}{4} = -\dfrac{1}{3}$

$12\left(\dfrac{x+2}{4}\right) = 12\left(-\dfrac{1}{3}\right)$

$3(x+2) = 4(-1)$

$3x + 6 = -4$

$3x + 6 - 6 = -4 - 6$

$3x = -10$

$\dfrac{3x}{3} = \dfrac{-10}{3}$

$x = -\dfrac{10}{3}$

**77.** $\dfrac{x-3}{2} = \dfrac{x+6}{5}$

$5(x-3) = 2(x+6)$

$5x - 15 = 2x + 12$

$5x - 2x - 15 = 2x + 12 - 2x$

$3x - 15 = 12$

$3x - 15 + 15 = 12 + 15$

$3x = 27$

$\dfrac{3x}{3} = \dfrac{27}{3}$

$x = 9$

**79.** *Verbal Model:*

$$\boxed{\dfrac{\text{Hours for shorter time}}{\text{Hours for longer time}}} = \boxed{\dfrac{\text{Cost for shorter time}}{\text{Cost for longer time}}}$$

*Labels:*

Hours for shorter time $= 2$     (hours)

Hours for longer time $= 3$     (hours)

Cost for shorter time $= x$     (dollars)

Cost for longer time $= 200$     (dollars)

*Equation:*

$\dfrac{2}{3} = \dfrac{x}{200}$

$2(200) = 3x$

$400 = 3x$

$\dfrac{400}{3} = \dfrac{3x}{3}$

$133.33 \approx x$

The band would charge \$133.33 to play for two hours.

**81.** *Verbal Model:* $\dfrac{\boxed{\text{Price in 2006}}}{\boxed{\text{Price in 1984}}} = \dfrac{\boxed{\text{Index in 2006}}}{\boxed{\text{Index in 1984}}}$

*Labels:*

Price in 2006 = $x$     (dollars)

Price in 1984 = 78     (dollars)

Index in 2006 = 201.6

Index in 1984 = 103.9

*Equation:* $\dfrac{x}{78} = \dfrac{201.6}{103.9}$

$$78\left(\dfrac{x}{78}\right) = 78\left(\dfrac{201.6}{103.9}\right)$$

$$x \approx 151$$

The 2006 price of the recliner chair was approximately \$151.

**83.** $P = 2l + 2w$

$P - 2l = 2w$

$\dfrac{P - 2l}{2} = \dfrac{2w}{2}$

$\dfrac{P - 2l}{2} = w$ or $w = \dfrac{P}{2} - l$

**85.** $z = \dfrac{x - m}{s}$

$s(z) = s\left(\dfrac{x - m}{s}\right)$

$sz = x - m$

$sz + m = x - m + m$

$sz + m = x$

**87.** *Verbal Model:* $\boxed{\text{Distance}} = \boxed{\text{Rate}} \cdot \boxed{\text{Time}}$

*Labels:*

Distance = $d$     (miles)

Rate = 65     (miles per hour)

Time = 8     (hours)

*Equation:* $d = 65(8)$

$d = 520$

The distance is 520 miles.

**89.** *Verbal Model:* $\boxed{\text{Distance}} = \boxed{\text{Rate}} \cdot \boxed{\text{Time}}$

*Labels:*

Distance = 855     (meters)

Rate = 5     (meters per minute)

Time = $t$     (minutes)

*Equation:* $855 = 5t$

$\dfrac{855}{5} = \dfrac{5t}{5}$

$171 = t$

The time is 171 minutes (or 2 hours, 51 minutes).

**91.** *Verbal Model:*   $\boxed{\text{Distance}} = \boxed{\text{Rate}} \cdot \boxed{\text{Time}}$

    *Labels:*

        Distance = 3000    (miles)

        Rate = $r$          (miles per hour)

        Time = 50        (hours)

    *Equation:*

$$3000 = r(50)$$

$$\frac{3000}{50} = \frac{r(50)}{50}$$

$$60 = r$$

The rate is 60 miles per hour.

**93.** *Verbal Model:*   $\boxed{\text{Distance}} = \boxed{\text{Rate}} \cdot \boxed{\text{Time}}$

    *Labels:*

        Distance = $d$    (miles)

        Rate = 475     (miles per hour)

        Time = $2\frac{1}{3}$     (hours)

    *Equation:*

$$d = 475\left(2\tfrac{1}{3}\right)$$

$$d = 475\left(\tfrac{7}{3}\right)$$

$$d = \tfrac{3325}{3}$$

$$d = 1108\tfrac{1}{3}$$

The distance is $1108\frac{1}{3}$ miles.

**95.** *Common formula:*   $P = 2l + 2w$

    *Labels:*

        $P = 112$            (feet)

        $l = $ length       (feet)

        $w = $ width $= l - 4$   (feet)

    *Equation:*

$$112 = 2l + 2(l - 4)$$

$$112 = 2l + 2l - 8$$

$$112 = 4l - 8$$

$$120 = 4l$$

$$\frac{120}{4} = l$$

$$30 = l$$

$$26 = l - 4$$

The length of the pool is 30 feet and the width is 26 feet.

**97.** *Verbal Model:*   $\boxed{\text{Interest}} = \boxed{\text{Principal}} \cdot \boxed{\text{Rate}} \cdot \boxed{\text{Time}}$

    *Labels:*

        Interest = $I$      (dollars)

        Principal = 1000   (dollars)

        Rate = 0.095     (percent in decimal form)

        Time = 5         (years)

    *Equation:*

$$I = 1000(0.095)(5)$$

$$I = 475$$

The total interest for the 5 years is $475.

**99.** *Verbal Model:*    | Value of dimes | + | Value of quarters | = | Total value |

*Labels:*    Mixed coins: total value = \$5.55, number of coins = 30

Dimes: value per coin = 0.10, number of coins = $x$

Quarters: value per coin = 0.25, number of coins = $30 - x$

*Equation:*    $0.10x + 0.25(30 - x) = 5.55$

$0.10x + 7.50 - 0.25x = 5.55$

$-0.15x + 7.50 = 5.55$

$-0.15x + 7.50 - 7.50 = 5.55 - 7.50$

$-0.15x = -1.95$

$x = \dfrac{-1.95}{-0.15}$

$x = 13 \text{ and } 30 - x = 17$

You have 13 dimes and 17 quarters.

**101.** *Verbal Model:*    | Work done | = | Portion done by first person | + | Portion done by second person |

*Labels:*    Both persons: work done = 1 complete task, time = $t$    (hours)

First person: rate = $\frac{1}{5}$ task per hour, time = $t$    (hours)

Second person: rate = $\frac{1}{6}$ task per hour, time = $t$    (hours)

*Equation:*    $1 = \left(\frac{1}{5}\right)t + \left(\frac{1}{6}\right)t$

$1 = \left(\frac{1}{5} + \frac{1}{6}\right)t$

$1 = \left(\frac{6}{30} + \frac{5}{30}\right)t$

$1 = \left(\frac{11}{30}\right)t$

$\frac{30}{11}(1) = \frac{30}{11}\left(\frac{11}{30}\right)t$

$\frac{30}{11} = t$

It would take $\frac{30}{11}$ hours (or approximately 2.7 hours) for both people to complete the task.

**103.** $x$ is less than 3.

**105.** $x$ is greater than or equal to 1 *and* less than 4.

**107.** (a)  $x = 3$

$7(3) - 10 \overset{?}{>} 0$

$21 - 10 \overset{?}{>} 0$

$11 > 0$

3 *is* a solution.

(b)  $x = \frac{1}{2}$

$7\left(\frac{1}{2}\right) - 10 \overset{?}{>} 0$

$\frac{7}{2} - \frac{20}{2} \overset{?}{>} 0$

$-\frac{13}{2} \not> 0$

$\frac{1}{2}$ *is not* a solution.

**109.** (a)  $x = 4$

$\frac{1}{4}(4) - 2 \overset{?}{<} 1$

$1 - 2 \overset{?}{<} 1$

$-1 < 1$

4 *is* a solution.

(b)  $x = 32$

$\frac{1}{4}(32) - 2 \overset{?}{<} 1$

$8 - 2 \overset{?}{<} 1$

$6 \not< 1$

32 *is not* a solution.

**111.** (a)  $x = 2$

$$3 - (2 - 4) \overset{?}{<} 0$$

$$3 - (-2) \overset{?}{<} 0$$

$$5 \not< 0$$

2 *is not* a solution.

(b)  $x = 9$

$$3 - (9 - 4) \overset{?}{<} 0$$

$$3 - 5 \overset{?}{<} 0$$

$$-2 < 0$$

9 *is* a solution.

**113.**  $x + 9 \geq 7$

$$x + 9 - 9 \geq 7 - 9$$

$$x \geq -2$$

**115.**  $3x - 8 < 1$

$$3x - 8 + 8 < 1 + 8$$

$$3x < 9$$

$$\frac{3x}{3} < \frac{9}{3}$$

$$x < 3$$

**117.**  $-11x \leq -22$

$$\frac{-11x}{-11} \geq \frac{-22}{-11} \quad \text{(Reverse inequality)}$$

$$x \geq 2$$

**119.**  $\frac{4}{5}x > 8$

$$5\left(\frac{4}{5}x\right) > 5(8)$$

$$4x > 40$$

$$\frac{4x}{4} > \frac{40}{4}$$

$$x > 10$$

**Note:** This could be done in one step.

$$\frac{4}{5}x > 8$$

$$\frac{5}{4}\left(\frac{4}{5}x\right) > \frac{5}{4}(8)$$

$$x > 10$$

**121.**  $14 - \frac{1}{2}t < 12$

$$2\left(14 - \frac{1}{2}t\right) < 2(12)$$

$$28 - t < 24$$

$$28 - 28 - t < 24 - 28$$

$$-t < -4$$

$$(-1)(-t) > (-1)(-4) \quad \text{(Reverse inequality)}$$

$$t > 4$$

**123.**  $3 - 3y \geq 2(4 + y)$

$$3 - 3y \geq 8 + 2y$$

$$3 - 3y - 2y \geq 8 + 2y - 2y$$

$$3 - 5y \geq 8$$

$$3 - 3 - 5y \geq 8 - 3$$

$$-5y \geq 5$$

$$\frac{-5y}{-5} \leq \frac{5}{-5} \quad \text{(Reverse inequality)}$$

$$y \leq -1$$

**125.**  $z \geq 10$

**127.**  $0 < y \leq 100$

**129.** *Verbal Model:*    | Weekly charge | + | Mileage charge | ≤ | Budget |

*Labels:*    Weekly charge = 184        (dollars)

Distance = $x$        (miles)

Mileage charge = $0.75x$    (dollars)

Budget = 250        (dollars)

*Inequality:*    $184 + 0.75x \leq 250$

$184 + 0.75x - 184 \leq 250 - 184$

$0.75x \leq 66$

$\dfrac{0.75x}{0.75} \leq \dfrac{66}{0.75}$

$x \leq 88$

To stay within the budget, you can drive no more than 88 miles.

# Chapter Test for Chapter 3

**1.**    $8x + 104 = 0$

$8x + 104 - 104 = 0 - 104$

$8x = -104$

$\dfrac{8x}{8} = \dfrac{-104}{8}$

$x = -13$

**2.**    $4x - 3 = 18$

$4x - 3 + 3 = 18 + 3$

$4x = 21$

$\dfrac{4x}{4} = \dfrac{21}{4}$

$x = \dfrac{21}{4}$

**3.**    $5 - 3x = -2x - 2$

$5 - 3x + 2x = -2x + 2x - 2$

$5 - x = -2$

$5 - 5 - x = -2 - 5$

$-x = -7$

$\dfrac{-x}{-1} = \dfrac{-7}{-1}$

$x = 7$

**4.**  $4 - (x - 3) = 5x + 1$

$4 - x + 3 = 5x + 1$

$7 - x = 5x + 1$

$7 - x - 5x = 5x + 1 - 5x$

$7 - 6x = 1$

$7 - 6x - 7 = 1 - 7$

$-6x = -6$

$\dfrac{-6x}{-6} = \dfrac{-6}{-6}$

$x = 1$

**5.**    $\dfrac{2}{3}x = \dfrac{1}{9} + x$

$9\left(\dfrac{2}{3}x\right) = 9\left(\dfrac{1}{9} + x\right)$

$6x = 1 + 9x$

$6x - 9x = 1 + 9x - 9x$

$-3x = 1$

$\dfrac{-3x}{-3} = \dfrac{1}{-3}$

$x = -\dfrac{1}{3}$

**6.**    $\dfrac{t + 2}{3} = \dfrac{2t}{9}$

$9(t + 2) = 3(2t)$

$9t + 18 = 6t$

$9t + 18 - 9t = 6t - 9t$

$18 = -3t$

$\dfrac{18}{-3} = \dfrac{-3t}{-3}$

$-6 = t$

**7.**
$$4.08(x + 10) = 9.50(x - 2)$$
$$4.08x + 40.8 = 9.50x - 19$$
$$4.08x - 9.50x + 40.8 = 9.50x - 9.50x - 19$$
$$-5.42x + 40.8 = -19$$
$$-5.42x + 40.8 - 40.8 = -19 - 40.8$$
$$-5.42x = -59.8$$
$$\frac{-5.42x}{-5.42} = \frac{-59.8}{-5.42}$$
$$x \approx 11.03$$

**8.** *Verbal Model:*  $\boxed{\text{Cost for parts}} + \boxed{\text{Labor cost per hour}} \cdot \boxed{\text{Hours of labor}} = \boxed{\text{Total bill}}$

*Labels:*
Cost for parts $= 62$   (dollars)
Labor cost per hour $= 32$   (dollars/hour)
Hours of labor $= x$   (hours)
Total bill $= 142$   (dollars)

*Equation:*
$$62 + 32x = 142$$
$$62 - 62 + 32x = 142 - 62$$
$$32x = 80$$
$$\frac{32x}{32} = \frac{80}{32}$$
$$x = 2\frac{1}{2}$$

So, $2\frac{1}{2}$ hours were spent repairing the oven.

**9.**

| Percent | Decimal | Fraction |
|---------|---------|----------|
| 31.25% | 0.3125 | $\frac{5}{16}$ |

**Percent:**

*Verbal Model:*  $\boxed{\text{Fraction}} \cdot \boxed{100\%} = \boxed{\text{Percent}}$

Label:  Percent $= x$

Equation:
$$\frac{5}{16}(100\%) = x$$
$$\frac{500\%}{16} = x$$
$$31.25\% = x \text{ or } x = 31\tfrac{1}{4}\%$$

**Decimal:**

*Verbal Model:*  $\boxed{\text{Decimal}} \cdot \boxed{100\%} = \boxed{\text{Percent}}$

Label:  Decimal $= x$

*Equation:*
$$x(100\%) = 31.25\%$$
$$x = \frac{31.25\%}{100\%}$$
$$x = 0.3125$$

**10.** *Verbal Model:*   $\boxed{324} = \boxed{27\% \text{ of what number}}\ (a = pb)$

*Label:*   $b =$ unknown number

*Equation:*   $324 = 0.27b$

$$\frac{324}{0.27} = b$$

$$1200 = b$$

So, 324 is 27% of 1200.

**11.** *Verbal Model:*   $\boxed{90} = \boxed{\text{What percent of } 250}\ (a = pb)$

*Label:*   $p =$ unknown percent (in decimal form)

*Percent equation:*   $90 = p(250)$

$$\frac{90}{250} = p$$

$$0.36 = p$$

So, 90 is 36% of 250.

**12.** $\dfrac{40 \text{ inches}}{2 \text{ yards}} = \dfrac{40 \cancel{\text{ inches}}}{2(36) \cancel{\text{ inches}}} = \dfrac{40}{72} = \dfrac{5}{9}$

**13.**   $\dfrac{2x}{3} = \dfrac{x+4}{5}$

$$5(2x) = 3(x+4)$$

$$10x = 3x + 12$$

$$10x - 3x = 3x + 12 - 3x$$

$$7x = 12$$

$$\frac{7x}{7} = \frac{12}{7}$$

$$x = \frac{12}{7}$$

**14.** $\dfrac{\text{Shorter side of big triangle}}{\text{Longer side of big triangle}} = \dfrac{\text{Shorter side of small triangle}}{\text{Longer side of small triangle}}$

$$\frac{x}{7} = \frac{4}{5.6}$$

$$5.6x = 7(4)$$

$$5.6x = 28$$

$$\frac{5.6x}{5.6} = \frac{28}{5.6}$$

$$x = 5$$

**15.** *Verbal Model:*   $\boxed{\text{Distance}} = \boxed{\text{Rate}} \cdot \boxed{\text{Time}}$

*Labels:*   Distance $= 264$   (miles)

Rate $= x$   (miles per hour)

Time $= 4$   (hours)

*Equation:*   $264 = x(4)$

$$\frac{264}{4} = \frac{4x}{4}$$

$$66 = x$$

Your average speed was 66 miles per hour.

**16.** *Verbal Model:*

$$\boxed{\begin{array}{c}\text{Work}\\\text{done}\end{array}} = \boxed{\begin{array}{c}\text{Portion done}\\\text{by you}\end{array}} + \boxed{\begin{array}{c}\text{Portion done}\\\text{by your friend}\end{array}}$$

*Labels:*       Both persons: work done = 1 task, time = $t$     (hours)

                   Your work: rate $= \frac{1}{9}$ task per hour, time = $t$     (hours)

                   Friend's work: rate $= \frac{1}{12}$ task per hour, time = $t$     (hours)

*Equation:*

$$1 = \left(\tfrac{1}{9}\right)t + \left(\tfrac{1}{12}\right)t$$

$$1 = \left(\tfrac{1}{9} + \tfrac{1}{12}\right)t$$

$$1 = \left(\tfrac{4}{36} + \tfrac{3}{36}\right)t$$

$$1 = \tfrac{7}{36}t$$

$$\tfrac{36}{7}(1) = \tfrac{36}{7}\left(\tfrac{7}{36}\right)t$$

$$\tfrac{36}{7} = t$$

It would take $\frac{36}{7}$ hours (or approximately 5.1 hours) to paint the building.

**17.**

$$a = pb + b$$

$$a = b(p + 1)$$

$$\frac{a}{p + 1} = \frac{b(p + 1)}{p + 1}$$

$$\frac{a}{p + 1} = b$$

**18.** *Common formula:*     $I = Prt$

     *Labels:*           $I = \$500$

                   $P = $ principal     (dollars)

                   $r = 0.08$         (percent in decimal form)

                   $t = 1$            (year)

     *Equation:*

$$500 = P(0.08)(1)$$

$$500 = P(0.08)$$

$$\frac{500}{0.08} = P$$

$$6250 = P$$

The required principal is $6250.

**19.**

$$x + 3 \leq 7$$

$$x + 3 - 3 \leq 7 - 3$$

$$x \leq 4$$

**20.** $-\dfrac{2x}{3} > 4$

$3\left(-\dfrac{2x}{3}\right) > 3(4)$

$-2x > 12$

$\dfrac{-2x}{-2} < \dfrac{12}{-2}$   (Reverse inequality)

$x < -6$

**21.** $21 - 3x \le 6$

$21 - 21 - 3x \le 6 - 21$

$-3x \le -15$

$\dfrac{-3x}{-3} \ge \dfrac{-15}{-3}$   (Reverse inequality)

$x \ge 5$

**22.** $3(6 + 2x) > 8$

$18 + 6x > 8$

$18 - 18 + 6x > 8 - 18$

$6x > -10$

$\dfrac{6x}{6} > \dfrac{-10}{6}$

$x > -\dfrac{5}{3}$

**23.** $-4(9 + 2x) \ge -40$

$-36 - 8x \ge -40$

$-36 + 36 - 8x \ge -40 + 36$

$-8x \ge -4$

$\dfrac{-8x}{-8} \le \dfrac{-4}{-8}$   (Reverse inequality)

$x \le \dfrac{1}{2}$

**24.** $-(3 + x) < 2(3x - 5)$

$-3 - x < 6x - 10$

$-3 - x - 6x < 6x - 6x - 10$

$-3 - 7x < -10$

$-3 + 3 - 7x < -10 + 3$

$-7x < -7$

$\dfrac{-7x}{-7} > \dfrac{-7}{-7}$   (Reverse inequality)

$x > 1$

# Cumulative Test for Chapters 1–3

**1.** $-\dfrac{3}{4} < \left|-\dfrac{7}{8}\right|$

Note: $\left|-\dfrac{7}{8}\right| = \dfrac{7}{8}$ and $-\dfrac{3}{4} < \dfrac{7}{8}$

**2.** $(-200)(2)(-3) = 1200$

**3.** $\dfrac{3}{8} - \dfrac{5}{6} = \dfrac{3 \cdot 3}{8 \cdot 3} - \dfrac{5 \cdot 4}{6 \cdot 4}$

$= \dfrac{9}{24} - \dfrac{20}{24} = \dfrac{9 - 20}{24} = -\dfrac{11}{24}$

**4.** $-\dfrac{2}{9} \div \dfrac{8}{75} = -\dfrac{2}{9} \cdot \dfrac{75}{8}$

$= -\dfrac{2 \cdot 75}{9 \cdot 8}$

$= \dfrac{(2)(3)(25)}{(3)(3)(2)(4)}$

$= -\dfrac{25}{12}$

**5.** $-(-2)^3 = -(-8) = 8$

**6.** $3 + 2(6) - 1 = 3 + 12 - 1 = 15 - 1 = 14$

**7.** $24 + 12 \div 3 = 24 + 4 = 28$

**8.** When $x = -2$ and $y = 3$, the expression

$$-3x - (2y)^2 = -3(-2) - (2 \cdot 3)^2$$
$$= 6 - 6^2$$
$$= 6 - 36$$
$$= -30.$$

**9.** When $x = -2$ and $y = 3$, the expression

$$\frac{5}{6}y + x^3 = \frac{5}{6}(3) + (-2)^3$$
$$= \frac{5}{2} + (-8)$$
$$= \frac{5}{2} - \frac{16}{2}$$
$$= \frac{5 - 16}{2}$$
$$= -\frac{11}{2}$$

**10.** $3 \cdot (x + y) \cdot (x + y) \cdot 3 \cdot 3 = 3^3(x + y)^2$

**11.** $-2x(x - 3) = -2x^2 + 6x$

**12.** Associative Property of Addition

**13.** $\left(3x^3\right)\left(5x^4\right) = 15x^7$

**14.** $2x^2 - 3x + 5x^2 - (2 + 3x)$
$$= 2x^2 - 3x + 5x^2 - 2 - 3x$$
$$= 7x^2 - 6x - 2$$

**15.** $4\left(x^2 + x\right) + 7\left(2x - x^2\right) = 4x^2 + 4x + 14x - 7x^2$
$$= -3x^2 + 18x$$

**16.** $12x - 3 = 7x + 27$
$$5x - 3 = 27$$
$$5x = 30$$
$$x = \frac{30}{5}$$
$$x = 6$$

**17.**
$$2x - \frac{5x}{4} = 13$$
$$4\left(2x - \frac{5x}{4}\right) = 4(13)$$
$$8x - 5x = 52$$
$$3x = 52$$
$$x = \frac{52}{3}$$

**18.**
$$5(x + 8) = -2x - 9$$
$$5x + 40 = -2x - 9$$
$$5x + 40 + 2x = -2x - 9 + 2x$$
$$7x + 40 = -9$$
$$7x + 40 - 40 = -9 - 40$$
$$7x = -49$$
$$\frac{7x}{7} = \frac{-49}{7}$$
$$x = -7$$

**19.**
$$-8(x + 5) \le 16$$
$$-8x - 40 \le 16$$
$$-8x - 40 + 40 \le 16 + 40$$
$$-8x \le 56$$
$$\frac{-8x}{-8} \ge \frac{56}{-8} \quad \text{(Reverse inequality)}$$
$$x \ge -7$$

**20.** To estimate the annual fuel cost, you could round the miles per gallon to 30 and round the fuel cost to $3.00 per gallon. You could determine how many gallons would be needed for the 15,000 miles and then multiply that result by the cost per gallon.

$$\frac{15,000 \text{ miles}}{1 \text{ year}} \cdot \frac{1 \text{ gallon}}{30 \text{ miles}} \cdot \frac{\$3.00}{1 \text{ gallon}} \approx \$1500 \text{ per year}$$

An estimate of the annual fuel cost would be approximately $1500 per year.

**21.** $\dfrac{24 \text{ ounces}}{2 \text{ pounds}} = \dfrac{24 \text{ ounces}}{2(16) \text{ ounces}} = \dfrac{24}{32} = \dfrac{3}{4}$

**22.** *Verbal Model:*   $\boxed{\text{Sale price}} = \boxed{\text{List price}} - \boxed{\text{Discount}}$

Labels:

Sale price $= x$        (dollars)

List price $= \$1150$

Discount rate $= 0.20$  $\left(\begin{array}{l}\text{percent in}\\\text{decimal form}\end{array}\right)$

Discount $= 0.20(1150) = \$230$

Equation:      $x = 1150 - 230$

$x = 920$

The sale price of the camcorder is $920.

**23.** *Verbal Model:*   $\dfrac{\boxed{\text{Assessed value of larger piece}}}{\text{Area of larger piece}} = \dfrac{\boxed{\text{Assessed value of smaller piece}}}{\text{Area of smaller piece}}$

Labels:

Assessed value of larger piece $= 95,000$     (dollars)

Area of larger piece $= (100)(80) = 8000$     (square units)

Assessed value of smaller piece $= x$     (dollars)

Area of smaller piece $= (60)(80) = (4800)$     (square units)

Proportion:

$$\frac{95,000}{8000} = \frac{x}{4800}$$

$$8000x = 95,000(4800)$$

$$\frac{8000x}{8000} = \frac{95,000(4800)}{8000}$$

$$x = 57,000$$

The assessed value of the smaller piece is $57,000.

# C H A P T E R 4
# Equations and Inequalities in Two Variables

# CHAPTER 4
## Equations and Inequalities in Two Variables

### Section 4.1   Ordered Pairs and Graphs

**1.**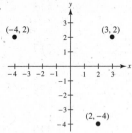

**3.**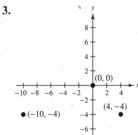

**5.**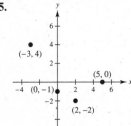

**7.**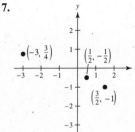

**9.**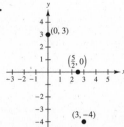

**11.** (a)  $(5, 2)$

    (b)  $(-3, 4)$

    (c)  $(2, -5)$

    (d)  $(-2, -2)$

**13.** (a)  $(-1, 3)$

    (b)  $(5, 0)$

    (c)  $(2, 1)$

    (d)  $(-1, -2)$

**15.** The point $(-3, 1)$ is located 3 units to the left of the vertical axis and 1 unit above the horizontal axis; it is in Quadrant II.

**17.** The point $\left(-\frac{1}{8}, -\frac{2}{7}\right)$ is located $\frac{1}{8}$ unit to the left of the vertical axis and $\frac{2}{7}$ unit below the horizontal axis; it is in Quadrant III.

**19.** The point $(-100, -365.6)$ is located 100 units to the left of the vertical axis and 365.6 units below the horizontal axis; it is in Quadrant III.

**21.** Quadrants II or III

This point is located 5 units to the left of the vertical axis. If $y$ is positive, the point would be above the horizontal axis and in the second quadrant. If $y$ is negative, the point would be below the horizontal axis and in the third quadrant. (If $y$ is zero, the point would be on the horizontal axis between the second and third quadrants.)

**23.** Quadrants III or IV

This point is located 2 units below the horizontal axis. If $x$ is positive, the point would be to the right of the vertical axis and in the fourth quadrant. If $x$ is negative, the point would be to the left of the vertical axis and in the third quadrant. (If $x$ is zero, the point would be on the vertical axis between the third and fourth quadrants.)

**25.** Quadrants II or IV

The two coordinates must have opposite signs if $xy < 0$. If $x$ is positive and $y$ is negative, the point would be located to the right of the vertical axis and below the horizontal axis in the fourth quadrant. If $x$ is negative and $y$ is positive, the point would be located to the left of the vertical axis and above the horizontal axis in the second quadrant.

**27.**

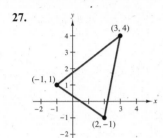

**29.**

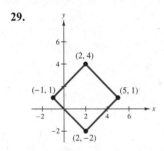

**31.**

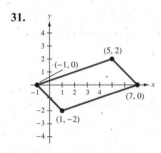

**33.**

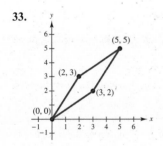

**35.**

| $x$ | −2 | 0 | 2 | 4 | 6 |
|---|---|---|---|---|---|
| $y = 3x - 4$ | −10 | −4 | 2 | 8 | 14 |

When $x = -2$, $y = 3(-2) - 4 = -6 - 4 = -10$.

When $x = 0$, $y = 3(0) - 4 = 0 - 4 = -4$.

When $x = 2$, $y = 3(2) - 4 = 6 - 4 = 2$.

When $x = 4$, $y = 3(4) - 4 = 12 - 4 = 8$.

When $x = 6$, $y = 3(6) - 4 = 18 - 4 = 14$.

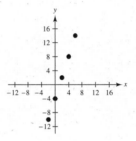

**37.**

| $x$ | −2 | 0 | 2 | 4 | 6 |
|---|---|---|---|---|---|
| $y = 3x - 4$ | −10 | −4 | 2 | 8 | 14 |

When $x = -2$, $y = -\frac{3}{2}(-2) + 5 = 3 + 5 = 8$.

When $x = 0$, $y = -\frac{3}{2}(0) + 5 = 0 + 5 = 5$.

When $x = 4$, $y = -\frac{3}{2}(4) + 5 = -6 + 5 = -1$.

When $x = 6$, $y = -\frac{3}{2}(6) + 5 = -9 + 5 = -4$.

When $x = 8$, $y = -\frac{3}{2}(8) + 5 = -12 + 5 = -7$.

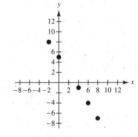

**39.**

| $x$ | $-2$ | $-1$ | $0$ | $1$ | $2$ |
|---|---|---|---|---|---|
| $y = -4x - 5$ | $3$ | $-1$ | $-5$ | $-9$ | $-13$ |

When $x = -2$, $y = -4(-2) - 5 = 8 - 5 = 3$.

When $x = -1$, $y = -4(-1) - 5 = 4 - 5 = -1$.

When $x = 0$, $y = -4(0) - 5 = 0 - 5 = -5$.

When $x = 1$, $y = -4(1) - 5 = -4 - 5 = -9$.

When $x = 2$, $y = -4(2) - 5 = -8 - 5 = -13$.

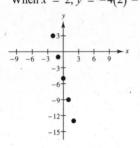

**41.**
$$7x + y = 8$$
$$7x - 7x + y = -7x + 8$$
$$y = -7x + 8$$

**43.**
$$10x - y = 2$$
$$10x - 10x - y = -10x + 2$$
$$-y = -10x + 2$$
$$-1(-y) = -1(-10x + 2)$$
$$y = 10x - 2$$

**45.**
$$6x - 3y = 3$$
$$6x - 6x - 3y = 3 - 6x$$
$$-3y = -6x + 3$$
$$\frac{-3y}{-3} = \frac{-6x + 3}{-3}$$
$$y = 2x - 1$$

**47.**
$$x + 4y = 8$$
$$x - x + 4y = 8 - x$$
$$4y = -x + 8$$
$$\frac{4y}{4} = \frac{-x + 8}{4}$$
$$y = -\frac{1}{4}x + 2$$

**49.**
$$4x - 5y = 3$$
$$4x - 4x - 5y = 3 - 4x$$
$$-5y = -4x + 3$$
$$\frac{-5y}{-5} = \frac{-4x + 3}{-5}$$
$$y = \frac{4}{5}x - \frac{3}{5}$$

**51.** (a) The ordered pair $(3, 10)$ *is* a solution because
$$10 = 2(3) + 4.$$

(b) The ordered pair $(-1, 3)$ *is not* a solution because
$$3 \neq 2(-1) + 4.$$

(c) The ordered pair $(0, 0)$ *is not* a solution because
$$0 \neq 2(0) + 4.$$

(d) The ordered pair $(-2, 0)$ *is* a solution because
$$0 = 2(-2) + 4.$$

**53.** (a) The ordered pair $(1, 1)$ *is* a solution because
$$2(1) - 3(1) + 1 = 2 - 3 + 1 = 0.$$

(b) The ordered pair $(5, 7)$ *is* a solution because
$$2(7) - 3(5) + 1 = 14 - 15 + 1 = 0.$$

(c) The ordered pair $(-3, -1)$ *is not* a solution because
$$2(-1) - 3(-3) + 1 = -2 + 9 + 1 = 8 \neq 0.$$

(d) The ordered pair $(-3, -5)$ *is* a solution because
$$2(-5) - 3(-3) + 1 = 0.$$

**55.** (a) The ordered pair $(6, 6)$ *is not* a solution because
$$6 \neq \frac{2}{3}(6).$$

(b) The ordered pair $(-9, -6)$ *is* a solution because
$$-6 = \frac{2}{3}(-9).$$

(c) The ordered pair $(0, 0)$ *is* a solution because
$$0 = \frac{2}{3}(0).$$

(d) The ordered pair $\left(-1, \frac{2}{3}\right)$ *is not* a solution because
$$\frac{2}{3} \neq \frac{2}{3}(-1).$$

**57.** (a) The ordered pair $\left(-\frac{1}{2}, 5\right)$ *is* a solution because
$$5 = 3 - 4\left(-\frac{1}{2}\right).$$

(b) The ordered pair $(1, 7)$ *is not* a solution because
$$7 \neq 3 - 4(1).$$

(c) The ordered pair $(0, 0)$ *is not* a solution because
$$0 \neq 3 - 4(0).$$

(d) The ordered pair $\left(-\frac{3}{4}, 0\right)$ *is not* a solution because
$$0 \neq 3 - 4\left(-\frac{3}{4}\right).$$

**59.** $y = 3x + 4$

(a)  When $y = 0$, $0 = 3x + 4$

$$0 - 4 = 3x + 4 - 4$$

$$-4 = 3x$$

$$\frac{-4}{3} = \frac{3x}{3}$$

$$-\frac{4}{3} = x$$

$$\left(-\frac{4}{3}, 0\right)$$

(b)  When $x = 4$, $y = 3(4) + 4$

$$y = 12 + 4$$

$$y = 16$$

$$(4, 16)$$

(c)  When $y = -2$, $-2 = 3x + 4$

$$-2 - 4 = 3x + 4 - 4$$

$$-6 = 3x$$

$$\frac{-6}{3} = \frac{3x}{3}$$

$$-2 = x$$

$$(-2, -2)$$

**61.**

| $x$ | 20 | 40 | 60 | 80 | 100 |
|---|---|---|---|---|---|
| $y = 0.066x$ | 1.32 | 2.64 | 3.96 | 5.28 | 6.60 |

When $x = 20$, $y = 0.066(20) = 1.32$.

When $x = 40$, $y = 0.066(40) = 2.64$.

When $x = 60$, $y = 0.066(60) = 3.96$.

When $x = 80$, $y = 0.066(80) = 5.28$.

When $x = 100$, $y = 0.066(100) = 6.60$.

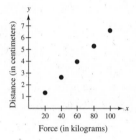

**63.**

| $x$ | 100 | 150 | 200 | 250 | 300 |
|---|---|---|---|---|---|
| $y = 25x + 5000$ | 7500 | 8750 | 10,000 | 11,250 | 12,500 |

When $x = 100$, $y = 25(100) + 5000 = 7500$.

When $x = 150$, $y = 25(150) + 5000 = 8750$.

When $x = 200$, $y = 25(200) + 5000 = 10,000$.

When $x = 250$, $y = 25(250) + 5000 = 11,250$.

When $x = 300$, $y = 25(300) + 5000 = 12,500$.

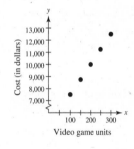

**65.** (a)

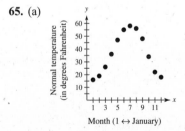

No, the scale is not the same on both axes on the graph shown above. Each mark on the horizontal axis represents 1 unit and each mark on the vertical axis represents 5 units. There are only 12 months, but the temperature ranges from 16°F to 58°F.

(b)  August is the month when the temperature changed the least from the previous month.

**67.** (a)

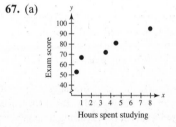

(b)  As the number of hours of study increases, the exam score increases.

**69.** The price of gasoline per gallon in 1999 was approximately $1.20.

**71.** The increase in the price of gasoline from 2003 to 2004 was approximately $0.30.

**73.** The per capita personal income in 2002 was approximately $30,750.

**75.** From the graph, it appears that the per capita personal income in 2005 was approximately $34,500 and in 2006 it was approximately $36,500.

*Verbal model:* $\boxed{\begin{array}{c}2006\\ \text{Income}\end{array}} = \boxed{\begin{array}{c}\text{Percent}\\ \text{(in decimal form)}\end{array}} \cdot \boxed{\begin{array}{c}2005\\ \text{Income}\end{array}} + \boxed{\begin{array}{c}2005\\ \text{Income}\end{array}}$

*Label:*      2006 income = 36,500            (dollars)

           Percent = $p$            (percent in decimal form)

           2005 income = 34,500            (dollars)

*Equation:*
$$36,500 = p(34,500) + 34,500$$
$$36,500 - 34,500 = p(34,500) + 34,500 - 34,500$$
$$2000 = p(34,500)$$
$$\frac{2000}{34,500} = p$$
$$0.06 \approx p$$

So, the increase in per capita income was approximately 6%.

**77.** The percent of gross domestic product spent on health care in Mexico is approximately 6.5%.

**79.** The highest recorded temperature in Chicago during January is approximately 65 degrees Fahrenheit.

**81.** (a) and (b)

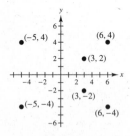

| Original Points | New Points |
|---|---|
| $(3, 2)$ | $(3, -2)$ |
| $(-5, 4)$ | $(-5, -4)$ |
| $(6, -4)$ | $(6, 4)$ |

(c) When the sign of the $y$-coordinate is changed, the location of the point is reflected about the $x$-axis.

**83.** The coordinates of the fourth vertex are $(6, 4)$.

**85.** No. The scales are determined by the magnitudes of the quantities being measured by $x$ and $y$. If $y$ is measuring revenue for a product and $x$ is measuring time in years, the scale on the $y$-axis may be in units of $100,000 and the scale on the $x$-axis may be in units of one year.

**87.**
$$-y = 10$$
$$-1(-y) = -1(10)$$
$$y = -10$$

**89.**
$$3x - 42 = 0$$
$$3x - 42 + 42 = 0 + 42$$
$$3x = 42$$
$$\frac{3x}{3} = \frac{42}{3}$$
$$x = 14$$

**91.**
$$125(r - 1) = 625$$
$$125r - 125 = 625$$
$$125r - 125 + 125 = 625 + 125$$
$$125r = 750$$
$$\frac{125r}{125} = \frac{750}{125}$$
$$r = 6$$

**93.**
$$20 - \frac{1}{9}x = 4$$
$$9\left(20 - \frac{1}{9}x\right) = 9(4)$$
$$180 - x = 36$$
$$180 - x - 180 = 36 - 180$$
$$-x = -144$$
$$\frac{-x}{-1} = \frac{-144}{-1}$$
$$x = 144$$

**95.**   $x + 3 > 2$
   $x + 3 - 3 > 2 - 3$
       $x > -1$

**97.**   $3x < 12$
   $\dfrac{3x}{3} < \dfrac{12}{3}$
       $x < 4$

# Section 4.2   Graphs of Equations in Two Variables

**1.** Graph (g)

**3.** Graph (a)

**5.** Graph (h)

**7.** Graph (d)

**9.**

| $x$ | $-2$ | $-1$ | $0$ | $1$ | $2$ |
|---|---|---|---|---|---|
| $y$ | 7 | 4 | 3 | 4 | 7 |

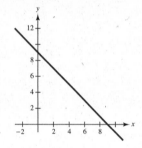

**11.**

| $x$ | $-2$ | $0$ | $2$ | $4$ | $6$ |
|---|---|---|---|---|---|
| $y$ | 3 | 2 | 1 | 0 | $-1$ |

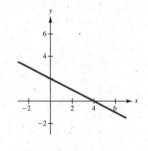

**13.**

| $x$ | $-2$ | $-1$ | $0$ | $1$ | $2$ |
|---|---|---|---|---|---|
| $y$ | 11 | 10 | 9 | 8 | 7 |

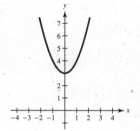

**15.**

| $x$ | $-3$ | $-2$ | $-1$ | $0$ | $1$ |
|---|---|---|---|---|---|
| $y$ | 2 | 1 | 0 | 1 | 2 |

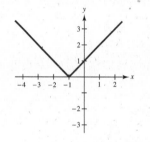

**17.** *Graphical solution:* It appears that the $x$-intercept is $(-2, 0)$ and the $y$-intercept is $(0, 4)$.

*Algebraic check:* $4x - 2y = -8$ $\qquad$ $4x - 2y = -8$

Let $y = 0.$ $\qquad\qquad$ Let $x = 0.$

$4x - 2(0) = -8$ $\qquad$ $4(0) - 2y = -8$

$\qquad 4x = -8$ $\qquad\qquad -2y = -8$

$\qquad\quad x = -2$ $\qquad\qquad\quad y = 4$

$x$-intercept: $(-2, 0)$ $\qquad$ $y$-intercept: $(0, 4)$

**19.** *Graphical solution:* It appears that the $x$-intercept is $(6, 0)$ and the $y$-intercept is $(0, 2)$.

*Algebraic check:* $x + 3y = 6$      $x + 3y = 6$

         Let $y = 0$.           Let $x = 0$.

          $x + 3(0) = 6$      $0 + 3y = 6$

            $x = 6$            $3y = 6$

         $x$-intercept: $(6, 0)$         $y = 2$

                           $y$-intercept: $(0, 2)$

**21.** *Graphical solution:* It appears that the $x$-intercepts are $(-3, 0)$ and $(3, 0)$; it appears that the $y$-intercept is $(0, -3)$.

*Algebraic check:* $y = |x| - 3$      $y = |x| - 3$

         Let $y = 0$.           Let $x = 0$.

         $0 = |x| - 3$        $y = |0| - 3$

         $3 = |x|$            $y = -3$

         $x = -3$ or $x = 3$     $y$-intercept: $(0, -3)$

         $x$-intercepts: $(-3, 0)$, $(3, 0)$

**23.** *Graphical solution:* It appears that the $x$-intercepts are $(4, 0)$ and $(-4, 0)$ and the $y$-intercept is $(0, 16)$.

*Algebraic check:* $y = 16 - x^2$      $y = 16 - x^2$

         Let $y = 0$.           Let $x = 0$.

           $0 = 16 - x^2$      $y = 16 - 0^2$

         $x^2 = 16$          $y = 16$

           $x = 4$ or $-4$     $y$-intercept: $(0, 16)$

         $x$-intercepts: $(4, 0)$ and $(-4, 0)$

**25.**    *x-intercept*        *y-intercept*

      Let $y = 0$.         Let $x = 0$.

       $0 = -2x + 7$      $y = -2(0) + 7$

     $-7 = -2x$          $y = 0 + 7$

     $\dfrac{-7}{-2} = x$          $y = 7$

      $\dfrac{7}{2} = x$           $(0, 7)$

     $\left(\dfrac{7}{2}, 0\right)$

The graph has one $x$-intercept at $\left(\dfrac{7}{2}, 0\right)$ and one $y$-intercept at $(0, 7)$.

**27.**    *x-intercept*        *y-intercept*

      Let $y = 0$.         Let $x = 0$.

       $0 = \frac{1}{2}x - 1$      $y = \frac{1}{2}(0) - 1$

     $2(0) = 2\left(\frac{1}{2}x - 1\right)$    $y = 0 - 1$

       $0 = x - 2$          $y = -1$

        $2 = 0$           $(0, -1)$

    $(2, 0)$

The graph has one $x$-intercept at $(2, 0)$ and one $y$-intercept at $(0, -1)$.

**29.**    *x-intercept*        *y-intercept*

      Let $y = 0$.         Let $x = 0$.

      $x - 0 = 1$         $0 - y = 1$

        $x = 1$         $-y = 1$

     $(1, 0)$           $y = -1$

                 $(0, -1)$

The graph has one $x$-intercept at $(1, 0)$ and one $y$-intercept at $(0, -1)$.

**31.**

| *x-intercept* | *y-intercept* |
|---|---|
| Let $y = 0$. | Let $x = 0$. |
| $2x + 0 = -2$ | $2(0) + y = -2$ |
| $2x = -2$ | $0 + y = -2$ |
| $x = -1$ | $y = -2$ |
| $(-1, 0)$ | $(0, -2)$ |

The graph has one *x*-intercept at $(-1, 0)$ and one *y*-intercept at $(0, -2)$.

**33.**

| *x-intercept* | *y-intercept* |
|---|---|
| Let $y = 0$. | Let $x = 0$. |
| $2x + 6(0) - 9 = 0$ | $2(0) + 6y - 9 = 0$ |
| $2x - 9 = 0$ | $6y - 9 = 0$ |
| $2x = 9$ | $6y = 9$ |
| $x = \frac{9}{2}$ | $y = \frac{9}{6}$ |
| | $y = \frac{3}{2}$ |
| $\left(\frac{9}{2}, 0\right)$ | $\left(0, \frac{3}{2}\right)$ |

The graph has one *x*-intercept at $\left(\frac{9}{2}, 0\right)$ and one *y*-intercept at $\left(0, \frac{3}{2}\right)$.

**35.**

| *x-intercept* | *y-intercept* |
|---|---|
| Let $y = 0$. | Let $x = 0$. |
| $\frac{3}{4}x - \frac{1}{2}(0) = 3$ | $\frac{3}{4}(0) - \frac{1}{2}y = 3$ |
| $\frac{3}{4}x = 3$ | $-\frac{1}{2}y = 3$ |
| $4\left(\frac{3}{4}x\right) = 4(3)$ | $2\left(-\frac{1}{2}y\right) = 2(3)$ |
| $3x = 12$ | $-y = 6$ |
| $x = 4$ | $y = -6$ |
| $(4, 0)$ | $(0, -6)$ |

The graph has one *x*-intercept at $(4, 0)$ and one *y*-intercept at $(0, -6)$.

**37.**

| $x$ | 0 | 2 | 3 |
|---|---|---|---|
| $y = 2 - x$ | 2 | 0 | -1 |

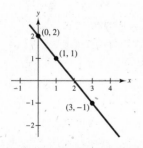

**39.**

| $x$ | -1 | 0 | 2 |
|---|---|---|---|
| $y = 3x$ | -3 | 0 | 6 |

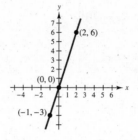

**41.** $4x + y = 6$

$\quad y = -4x + 6$

| $x$ | 0 | 1 | 2 |
|---|---|---|---|
| $y = -4x + 6$ | 6 | 2 | -2 |

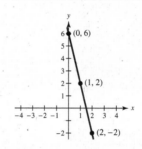

**43.** $7x + 7y = 14$

$\quad 7y = -7x + 14$

$\quad\quad y = -x + 2$

| $x$ | 0 | 2 | 4 |
|---|---|---|---|
| $y = -x + 2$ | 2 | 0 | -2 |

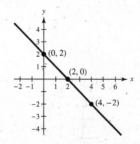

**45.**

| $x$ | $-40$ | $0$ | $8$ |
|---|---|---|---|
| $y = \frac{3}{8}x + 15$ | $0$ | $15$ | $18$ |

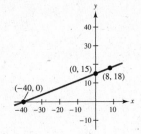

**53.**

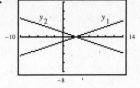

No, the graphs are not identical.

The two equations are not equivalent.

**47.**

| $x$ | $-3$ | $-2$ | $0$ | $1$ | $3$ |
|---|---|---|---|---|---|
| $y = -x^2 + 9$ | $0$ | $5$ | $9$ | $6$ | $0$ |

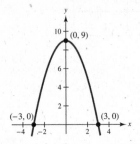

**55.**

Yes, the graphs are identical. By the Distributive Property, $2(x - 2) = 2x - 4$.

**49.**

| $x$ | $0$ | $4$ | $5$ | $6$ |
|---|---|---|---|---|
| $y = \lvert x - 5 \rvert$ | $5$ | $1$ | $0$ | $1$ |

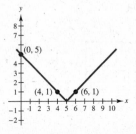

**57.**

No:

| |
|---|
| Xmin $= -10$ |
| Xmax $= 10$ |
| Xscl $= 1$ |
| Ymin $= -10$ |
| Ymax $= 30$ |
| Yscl $= 5$ |

**51.**

| $x$ | $-5$ | $0$ | $5$ |
|---|---|---|---|
| $y = 5 - \lvert x \rvert$ | $0$ | $5$ | $0$ |

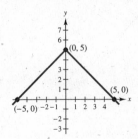

**59.**

Yes—standard viewing window

**61.**

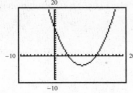

No:   

| Xmin = −10 |
| Xmax = 20 |
| Xscl = 1 |
| Ymin = −10 |
| Ymax = 20 |
| Yscl = 1 |

**63.**

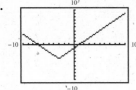

Yes—standard viewing window

**65.** $y = 35t$

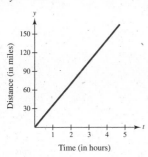

**67.** (a)  $y = 1120 - 80x$

(b)

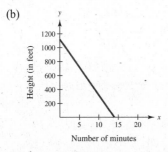

(c) The $y$-intercept $(0, 1120)$ represents the initial height of the hot air balloon; this is the height for the time $x = 0$.

**69.** (a)

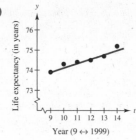

(b)  $y = 0.22(25) + 71.9$

$y = 5.5 + 71.9$

$y = 77.4$

According to the model, the life expectancy for a male child born in 2015 would be approximately 77.4 years.

**71.** Yes. For any linear equation in two variables, $x$ and $y$, there is a resulting value for $y$ when $x = 0$. The corresponding point $(0, y)$ is the $y$-intercept of the graph of the equation.

**73.**

The distance between you and the tree decreases as you move from left to right on the graph. The $x$-intercept represents the number of seconds it takes you to reach the tree.

**75.** $-4 + (-7) - 3 + 1 = -4 - 7 - 3 + 1 = -13$

**77.** $-(-3) + 5 - 4 + 9 = 3 + 5 - 4 + 9 = 13$

**79.** $\frac{3}{4}\left(-\frac{2}{9}\right) = -\frac{3(2)}{2(2)(3)(3)} = -\frac{1}{6}$

**81.** $\left(-\frac{7}{12}\right)\left(-\frac{18}{35}\right) = -\frac{7(3)(6)}{2(6)(7)(5)} = \frac{3}{10}$

**83.** $\frac{12}{5} \div \left(-\frac{1}{3}\right) = -\frac{12}{5} \cdot \frac{3}{1} = -\frac{36}{5}$

**85.** (a) The ordered pair $(0, 5)$ *is not* a solution because
$5 \neq 3(0) - 5.$

(b) The ordered pair $(-1, -2)$ *is not* a solution because
$-2 \neq 3(-1) - 5.$

(c) The ordered pair $(3, 4)$ *is* a solution because
$4 = 3(3) - 5.$

(d) The ordered pair $(-2, -11)$ *is* a solution because
$-11 = 3(-2) - 5.$

**87.** (a) The ordered pair $(1, 1)$ *is not* a solution because
$3(1) - 4(1) \neq 7.$

(b) The ordered pair $(-5, 9)$ *is not* a solution because
$3(9) - 4(-5) \neq 7.$

(c) The ordered pair $(4, -3)$ *is not* a solution because
$3(-3) - 4(4) \neq 7.$

(d) The ordered pair $(7, 7)$ *is not* a solution because
$3(7) - 4(7) \neq 7.$

## Section 4.3  Slope and Graphs of Linear Equations

**1.** $m = \frac{1}{1} = 1$

**3.** $m = 0$

**5.** $m = \frac{-1}{3} = -\frac{1}{3}$

**7.** (a) $L_2$ has slope $m = \frac{3}{2}$

(b) $L_3$ has slope $m = 0$

(c) $L_4$ has slope $m = -\frac{2}{3}$

(d) $L_1$ has slope $m = -2$

**9.** $m = \frac{5 - 0}{4 - 0} = \frac{5}{4}$

The line rises.

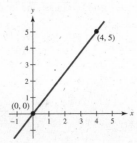

**11.** $m = \frac{-4 - 0}{8 - 0} = \frac{-4}{8} = -\frac{1}{2}$

The line falls.

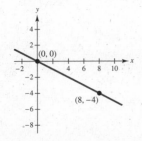

**13.** $m = \frac{4 - 0}{0 - 6} = \frac{4}{-6} = -\frac{2}{3}$

The line falls.

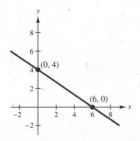

**15.** $m = \frac{6 - (-1)}{2 - (-4)} = \frac{7}{6}$

The line rises.

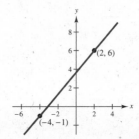

**17.** $m = \frac{4 - (-1)}{-6 - (-6)} = \frac{5}{0}$ (undefined)

$m$ is undefined. The line is vertical.

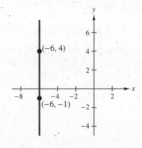

**19.** $m = \dfrac{-4 - (-4)}{8 - 3} = \dfrac{0}{5} = 0$

The line is horizontal.

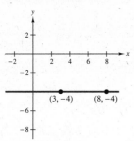

**21.** $m = \dfrac{-3 - \dfrac{3}{2}}{\dfrac{9}{2} - \dfrac{1}{4}}$

$= \dfrac{-\dfrac{6}{2} - \dfrac{3}{2}}{\dfrac{18}{4} - \dfrac{1}{4}}$

$= \dfrac{-\dfrac{9}{2}}{\dfrac{17}{4}}$

$= -\dfrac{9}{2} \cdot \dfrac{4}{17} = -\dfrac{9 \cdot 4}{2 \cdot 17} = -\dfrac{18}{17}$

The line falls.

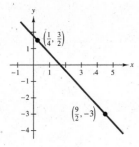

**23.** $m = \dfrac{4 - (-1)}{-3.2 - 3.2} = \dfrac{5}{-6.4} = -\dfrac{50}{64} = -\dfrac{25}{32}$

The line falls.

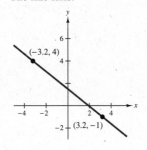

**25.** $m = \dfrac{4.25 - (-1)}{5.75 - 3.5}$

$= \dfrac{5.25}{2.25}$

$= \dfrac{525}{225} = \dfrac{7}{3}$

The line rises.

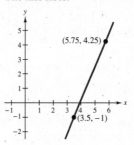

**27.** $m = \dfrac{3 - 3}{a - 4} = \dfrac{0}{a - 4} = 0$

The line is horizontal.

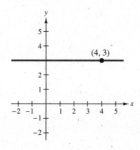

Regardless of the value of $a$, the point $(a, 3)$ lies on the

horizontal line through $(4, 3)$.

**29.**

| $x$ | $-2$ | $0$ | $2$ | $4$ |
|---|---|---|---|---|
| $y = -2x - 2$ | 2 | $-2$ | $-6$ | $-10$ |
| Solution points | $(-2, 2)$ | $(0, -2)$ | $(2, -6)$ | $(4, -10)$ |

Using $(-2, 2)$ and $(2, -6)$: $m = \dfrac{-6 - 2}{2 - (-2)} \doteq \dfrac{-8}{4} = -2$

Using $(0, -2)$ and $(4, -10)$: $m = \dfrac{-10 - (-2)}{4 - 0} = \dfrac{-8}{4} = -2$

**31.** $\dfrac{y - (-1)}{0 - 4} = -4$

$\dfrac{y + 1}{-4} = -4$

$-4\left(\dfrac{y + 1}{-4}\right) = -4(-4)$

$y + 1 = 16$

$y = 15$

**33.** $\dfrac{6 - y}{7 - (-4)} = \dfrac{5}{2}$

$\dfrac{6 - y}{11} = \dfrac{5}{2}$

$2(6 - y) = 55$  (Cross-multiply)

$12 - 2y = 55$

$-2y = 43$

$y = -\dfrac{43}{2}$

**35.**

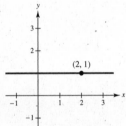

$(0, 1), (1, 1), (-2, 1)$, etc.

**37.**

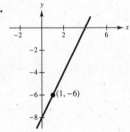

$(2, -4), (3, -2), (4, 0)$, etc.

**39.**

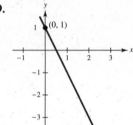

$(1, -1), (2, -3), (3, -5)$, etc.

**41.**

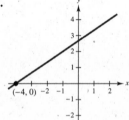

$(-1, 2), (2, 4), (5, 6)$, etc.

**43.**

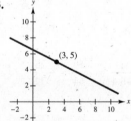

$(5, 4), (7, 3), (9, 2), (11, 1)$, etc.

**45.**

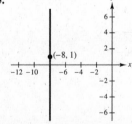

$(-8, 0), (-8, -1), (-8, -3)$, etc.

**47.**

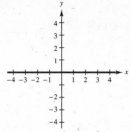

**49.**

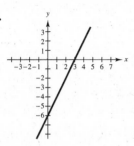

**51.**

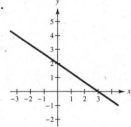

**53.** *x-intercept*          *y-intercept*

Let $y = 0$.          Let $x = 0$.

$2x + 3(0) + 6 = 0$          $2(0) + 3y + 6 = 0$

$2x + 6 = 0$          $3y + 6 = 0$

$2x = -6$          $3y = -6$

$x = -3$          $y = -2$

$(-3, 0)$          $(0, -2)$

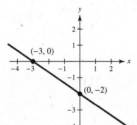

**55.** *x-intercept*          *y-intercept*

Let $y = 0$.          Let $x = 0$.

$5x - 2(0) - 10 = 0$          $5(0) - 2y - 10 = 0$

$5x - 10 = 0$          $-2y - 10 = 0$

$5x = 10$          $-2y = 10$

$x = 2$          $y = -5$

$(2, 0)$          $(0, -5)$

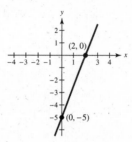

**57.** *x-intercept*          *y-intercept*

Let $y = 0$.          Let $x = 0$.

$6x - 4(0) + 12 = 0$          $6(0) - 4y + 12 = 0$

$6x + 12 = 0$          $-4y + 12 = 0$

$6x = -12$          $-4y = -12$

$x = -2$          $y = 3$

$(-2, 0)$          $(0, 3)$

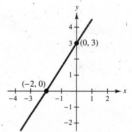

**59.** $x - y = 0$

$-y = -x$

$y = x$

$m = 1$

*y*-intercept: $(0, b) = (0, 0)$

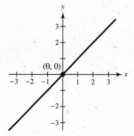

**61.** $\frac{1}{2}x + y = 0$

$$y = -\frac{1}{2}x$$

$$m = -\frac{1}{2}$$

$y$-intercept: $(0, b) = (0, 0)$

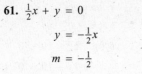

**63.** $2x - y - 3 = 0$

$$-y = -2x + 3$$

$$y = 2x - 3$$

$$m = 2$$

$y$-intercept: $(0, b) = (0, -3)$

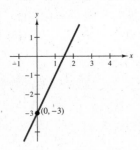

**65.** $x - 2y + 2 = 0$

$$-2y = -x - 2$$

$$\frac{-2y}{-2} = \frac{-x}{-2} + \frac{-2}{-2}$$

$$y = \frac{1}{2}x + 1$$

$$m = \frac{1}{2}$$

$y$-intercept: $(0, b) = (0, 1)$

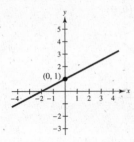

**67.** $2x - 6y - 15 = 0$

$$-6y = -2x + 15$$

$$\frac{-6y}{-6} = \frac{-2x}{-6} + \frac{15}{-6}$$

$$y = \frac{1}{3}x - \frac{5}{2}$$

$$m = \frac{1}{3}$$

$y$-intercept: $(0, b) = \left(0, -\frac{5}{2}\right)$

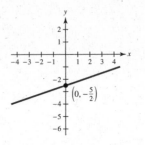

**69.** $3x - 4y + 2 = 0$

$$-4y = -3x - 2$$

$$y = \frac{3}{4}x + \frac{1}{2}$$

$$m = \frac{3}{4}$$

$y$-intercept: $(0, b) = \left(0, \frac{1}{2}\right)$

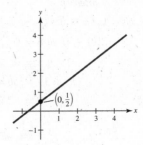

**71.** $y + 5 = 0$

$$y = -5$$

$$m = 0$$

$y$-intercept: $(0, b) = (0, -5)$

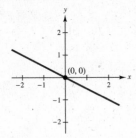

**73.** $y_1 = 4x$

    $m_1 = 4$

    *y*-intercept: $(0, 0)$

$y_2 = 4x - 3$

    $m_2 = 4$

    *y*-intercept: $(0, -3)$

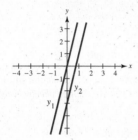

**75.** $y_1 = \frac{1}{3}x - 5$

    $m_1 = \frac{1}{3}$

    *y*-intercept: $(0, -5)$

$y_2 = -3x + 2$

    $m_2 = -3$

    *y*-intercept: $(0, 2)$

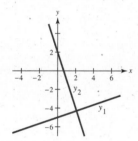

**77.** $y_1 = \frac{4}{5}x - 2$

    $m_1 = \frac{4}{5}$

    *y*-intercept: $(0, -2)$

$y_2 = -\frac{4}{5}x - 5$

    $m_2 = -\frac{4}{5}$

    *y*-intercept: $(0, -5)$

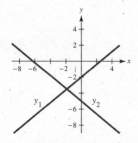

**79.** $L_1: y = 2x - 3$

    $m_1 = 2$

$L_2: y = 2x + 1$

    $m_2 = 2$

Since the two slopes are equal, the two lines are *parallel.*

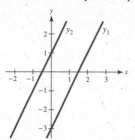

**81.** $L_1: y = 2x - 3$

    $m_1 = 2$

$L_2: y = -\frac{1}{2}x + 1$

    $m_2 = -\frac{1}{2}$

Since the two slopes are negative reciprocals of each other, the two lines are *perpendicular.*

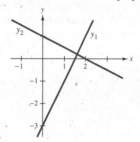

**83.** $L_1: (0, -1), (5, 9)$

$$m_1 = \frac{9 - (-1)}{5 - 0} = \frac{10}{5} = 2$$

$L_2: (0, 3), (4, 1)$

$$m_2 = \frac{1 - 3}{4 - 0} = \frac{-2}{4} = -\frac{1}{2}$$

Since the two slopes are negative reciprocals of each other, $L_1$ and $L_2$ are *perpendicular.*

**85.** $L_1: (3, 6), (-6, 0)$

$$m_1 = \frac{0 - 6}{-6 - 3} = \frac{-6}{-9} = \frac{2}{3}$$

$L_2: (0, -1), \left(5, \frac{7}{3}\right)$

$$m_2 = \frac{\frac{7}{3} - (-1)}{5 - 0} = \frac{\frac{7}{3} + \frac{3}{3}}{5} = \frac{\frac{10}{3}}{5} = \frac{10}{3} \cdot \frac{1}{5} = \frac{10}{15} = \frac{2}{3}$$

Since the two slopes are equal, $L_1$ and $L_2$ are *parallel.*

**87.** From the top of the slide to the bottom, the slide falls 8 feet. So, the vertical change is −8 feet. The horizontal change is 12 feet. The slope of the slide is the ratio of −8 feet to 12 feet.

$$m = \frac{-8}{12} = -\frac{2}{3}$$

The slope of the slide is $-\frac{2}{3}$.

**89.** The vertical change, or rise, of the path of takeoff is 4 miles. The horizontal change, or run, of the path of takeoff is 20 miles. The slope of the path is the ratio of 4 miles to 20 miles.

$$m = \frac{4}{20} = \frac{1}{5}$$

The slope of the path followed during takeoff is $\frac{1}{5}$.

**91. (a)**

200   *Not drawn to scale*

**(b)** $m = \frac{3}{200}$

**(c)** Yes. If the track rose 3 feet over a distance of 100 feet, the track would rise a total of 6 feet over the 200-foot distance. This is twice as much rise over that distance, and thus it would be a steeper slope.

$$\left|\frac{3}{100}\right| > \left|\frac{3}{200}\right|$$

**93. (a)** 2002–2003: $\frac{473 - 466}{2003 - 2002} = \frac{7}{1} = 7$

2003–2004: $\frac{528 - 473}{2004 - 2003} = \frac{55}{1} = 55$

2004–2005: $\frac{549 - 528}{2005 - 2004} = \frac{21}{1} = 21$

2005–2006: $\frac{607 - 549}{2006 - 2005} = \frac{58}{1} = 58$

**(b)** 2002–2006: $\frac{607 - 466}{2006 - 2002} = \frac{141}{4} = 35.25$

The slope of 35.25 indicates the average annual increase in the number of theatrical films released during the four years from 2002 to 2006.

**95. (a)** The slopes of the linear models represent the estimated yearly increase in profits.

**(b)** Model $P_2$ represents a faster increase in profits.

**(c)** $P_1 = 0.2(10) + 2.4$    $P_2 = 0.3(10) + 2.4$
$= 2 + 2.4$    $= 3 + 2.4$
$= 4.4$ million dollars    $= 5.4$ million dollars

**(d)**

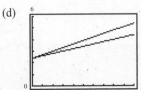

**97.** $m = \dfrac{\text{change in sales}}{\text{change in years}} = \dfrac{76}{1}$

The sales increase by 76 units.

**99.** $m = \dfrac{\text{change in sales}}{\text{change in year}} = \dfrac{0}{1}$

The sales remain the same.

**101.** $m = \dfrac{\text{change in sales}}{\text{change in years}} = \dfrac{-14}{1}$

The sales decrease by 14 units.

**103.** No. The slopes of nonvertical perpendicular lines have opposite signs. The slopes are the negative reciprocals of each other.

**105.** The rate of change is the slope of the line.

**107.** Yes. You are free to consider either of the points as $(x_1, y_1)$ and the other as $(x_2, y_2)$. However, once this is done, you must form the numerator and denominator using the same order of subtraction.

**109.** $x^2 \cdot x^3 = x \cdot x \cdot x \cdot x \cdot x = x^5$

**111.** $(-y^2)y = -y \cdot y \cdot y = -y^3$

**113.** $(25x^3)(2x^2) = 25 \cdot x \cdot x \cdot x \cdot 2 \cdot x \cdot x$
$= 25 \cdot 2 \cdot x \cdot x \cdot x \cdot x \cdot x = 50x^5$

**115.** $x^2 - 2x - x^2 + 3x + 2 = x^2 - x^2 - 2x + 3x + 2$
$= x + 2$

**117.** *x-intercept*          *y-intercept*
Let $y = 0$.          Let $x = 0$.
$0 = 6x - 3$          $y = 6(0) - 3$
$3 = 6x$          $y = 0 - 3$
$\frac{3}{6} = x$          $y = -3$
$\frac{1}{2} = x$          $(0, -3)$
$\left(\frac{1}{2}, 0\right)$

The graph has one *x*-intercept at $\left(\frac{1}{2}, 0\right)$ and one *y*-intercept at $(0, -3)$.

**119.** *x-intercept*          *y-intercept*
Let $y = 0$.          Let $x = 0$.
$2x + 0 = -3$          $2(0) + y = -3$
$2x = -3$          $0 + y = -3$
$x = -\frac{3}{2}$          $y = -3$
$\left(-\frac{3}{2}, 0\right)$          $(0, -3)$

The graph has one *x*-intercept at $\left(-\frac{3}{2}, 0\right)$ and one *y*-intercept at $(0, -3)$.

**121.**  *x-intercept*                  *y-intercept*

Let $y = 0$.                 Let $x = 0$.

$x + 6(0) - 9 = 0$         $0 + 6y - 9 = 0$

$x + 0 - 9 = 0$                 $6y = 9$

$x - 9 = 0$                     $y = \frac{9}{6}$

$x = 9$                         $y = \frac{3}{2}$

$(9, 0)$                          $\left(0, \frac{3}{2}\right)$

The graph has one *x*-intercept at $(9, 0)$ and one

*y*-intercept at $\left(0, \frac{3}{2}\right)$.

# Mid Chapter Quiz for Chapter 4

**1.**

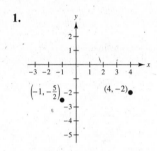

**2.** Quadrants I or IV.

The point $(3, y)$ is located 3 units to the right of the

vertical axis. If *y* is positive, the point would be above
the horizontal axis in the first quadrant. If *y* is negative,
the point would be below the horizontal axis in the fourth
quadrant.

**3.** (a)  Yes, the ordered pair $(2, 7)$ *is* a solution because

$7 = 9 - |2|$.

   (b)  No, the ordered pair $(-3, 12)$ is *not* a solution

because $12 \neq 9 - |-3|$.

   (c)  Yes, the ordered pair $(-9, 0)$ *is* a solution because

$0 = 9 - |-9|$.

   (d)  No, the ordered pair $(0, -9)$ is *not* a solution because

$-9 \neq 9 - |0|$.

**4.** Estimates from the graph may vary:

2000: Approximately $145 billion

2001: Approximately $165 billion

2002: Approximately $185 billion

2003: Approximately $205 billion

2004: Approximately $220 billion

2005: Approximately $230 billion

**5.** *x-intercept*              *y-intercept*

Let $y = 0$.             Let $x = 0$.

$x - 3(0) = 12$         $0 - 3y = 12$

$x = 12$             $-3y = 12$

$(12, 0)$                 $\dfrac{-3y}{-3} = \dfrac{12}{-3}$

$y = -4$

$(0, -4)$

The graph has one *x*-intercept at $(12, 0)$ and one

*y*-intercept at $(0, -4)$.

**6.** *x-intercept*              *y-intercept*

Let $y = 0$.             Let $x = 0$.

$0 = -7x + 2$         $y = -7(0) + 2$

$-2 = -7x$             $y = 0 + 2$

$\dfrac{-2}{-7} = x$             $y = 2$

$(0, 2)$

$\dfrac{2}{7} = x$

$\left(\dfrac{2}{7}, 0\right)$

The graph has one *x*-intercept at $\left(\dfrac{2}{7}, 0\right)$ and one

*y*-intercept at $(0, 2)$.

**7.**

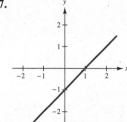

**8.**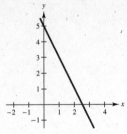

**9.**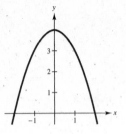

**10.**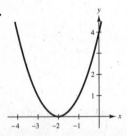

**11.**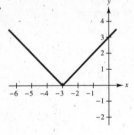

**12.**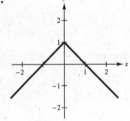

**13.** $(0, 0), (-3, 9)$

$$m = \frac{9 - 0}{-3 - 0} = \frac{9}{-3} = -3$$

The line falls.

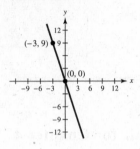

**14.** $(-4, 3), (-4, -5)$

$$m = \frac{-5 - 3}{-4 - (-4)} = \frac{-8}{0} \text{ undefined}$$

$m$ is undefined. The line is vertical.

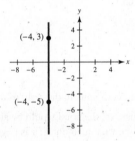

**15.** $\left(4, \dfrac{1}{2}\right), \left(-1, -\dfrac{5}{2}\right)$

$$m = \frac{-\dfrac{5}{2} - \dfrac{1}{2}}{-1 - 4} = \frac{-\dfrac{6}{2}}{-5} = \frac{-3}{-5} = \frac{3}{5}$$

The line rises.

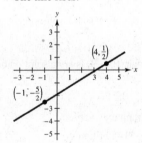

**16.** $3x - 3y + 9 = 0$

$$-3y = -3x - 9$$

$$\frac{-3y}{-3} = \frac{-3x}{-3} + \frac{-9}{-3}$$

$$y = x + 3$$

Slope: $m = 1$

$y$-intercept: $(0, b) = (0, 3)$

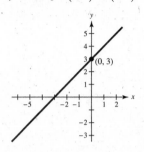

**17.** $-2x + 3y - 6 = 0$

$$3y = 2x + 6$$

$$\frac{3y}{3} = \frac{2x}{3} + \frac{6}{3}$$

$$y = \frac{2}{3}x + 2$$

Slope: $m = \frac{2}{3}$

$y$-intercept: $(0, b) = (0, 2)$

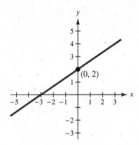

## Section 4.4   Equations of Lines

**1.**
$$m = \frac{y_2 - y_1}{x_2 - x_1}$$
$$-2 = \frac{y - 0}{x - 0}$$
$$-2 = \frac{y}{x}$$
$$-2x = y$$
$$-2x - y = 0$$
$$2x + y = 0$$

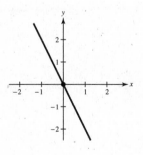

**18.**

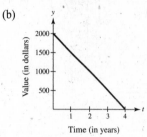

Substitute the coordinates for the respective variables in the equation and determine whether the equation is true.

**19.** (a) $y = 2000 - 500t$

(b)

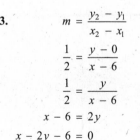

(c) $y$-intercept: Let $t = 0$

$$y = 2000 - 500(0)$$
$$y = 2000 - 0$$
$$y = 2000$$
$$(0, 2000)$$

The $y$-intercept is $(0, 2000)$, and it corresponds to the original value of the computer system.

**3.**
$$m = \frac{y_2 - y_1}{x_2 - x_1}$$
$$\frac{1}{2} = \frac{y - 0}{x - 6}$$
$$\frac{1}{2} = \frac{y}{x - 6}$$
$$x - 6 = 2y$$
$$x - 2y - 6 = 0$$
$$x - 2y = 6$$

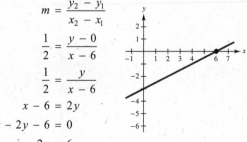

**5.** $m = \dfrac{y_2 - y_1}{x_2 - x_1}$

$$2 = \dfrac{y - 1}{x - (-2)}$$

$$2 = \dfrac{y - 1}{x + 2}$$

$$2(x + 2) = y - 1$$

$$2x + 4 = y - 1$$

$$2x - y + 4 = -1$$

$$2x - y = -5$$

**7.** $m = \dfrac{y_2 - y_1}{x_2 - x_1}$

$$-\dfrac{1}{5} = \dfrac{y - (-1)}{x - (-8)}$$

$$\dfrac{-1}{5} = \dfrac{y + 1}{x + 8}$$

$$-1(x + 8) = 5(y + 1)$$

$$-x - 8 = 5y + 5$$

$$-x = 5y + 13$$

$$-x - 5y = 13$$

$$x + 5y = -13$$

**9.** $m = \dfrac{y_2 - y_1}{x_2 - x_1}$

$$0 = \dfrac{y - (-3)}{x - \dfrac{1}{2}}$$

$$0 = \dfrac{y + 3}{x - \dfrac{1}{2}}$$

$$0 = y + 3$$

$$y = -3$$

**11.** $m = \dfrac{y_2 - y_1}{x_2 - x_1}$

$$\dfrac{2}{3} = \dfrac{y - \dfrac{3}{2}}{x - 0}$$

$$\dfrac{2}{3} = \dfrac{y - \dfrac{3}{2}}{x}$$

$$2x = 3\left(y - \dfrac{3}{2}\right)$$

$$2x = 3y - \dfrac{9}{2}$$

$$2x - 3y = -\dfrac{9}{2}$$

$$2(2x - 3y) = 2\left(-\dfrac{9}{2}\right)$$

$$4x - 6y = -9$$

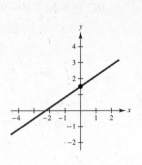

**13.** $m = \dfrac{y_2 - y_1}{x_2 - x_1}$

$$-0.8 = \dfrac{y - 4}{x - 2}$$

$$\dfrac{-4}{5} = \dfrac{y - 4}{x - 2}$$

$$-4(x - 2) = 5(y - 4)$$

$$-4x + 8 = 5y - 20$$

$$-4x - 5y + 8 = -20$$

$$-4x - 5y = -28$$

$$4x + 5y = 28$$

**15.** $y - y_1 = m(x - x_1)$

$$y - (-4) = 4(x - 0)$$

$$y + 4 = 4x$$

$$y = 4x - 4$$

**17.** $y - y_1 = m(x - x_1)$

$$y - 6 = -3[x - (-3)]$$

$$y - 6 = -3(x + 3)$$

$$y - 6 = -3x - 9$$

$$y = -3x - 3$$

**19.** $y - y_1 = m(x - x_1)$

$$y - 0 = -\tfrac{1}{3}(x - 9)$$

$$y = -\tfrac{1}{3}x + 3$$

**21.** The slope $m = 0$ indicates a horizontal line with an equation in the form $y = b$. Thus, the equation is $y = 4$.

**23.**
$$y - y_1 = m(x - x_1)$$
$$y - (-1) = -\tfrac{3}{4}(x - 8)$$
$$y + 1 = -\tfrac{3}{4}x + 6$$
$$y = -\tfrac{3}{4}x + 5$$

**25.**
$$y - y_1 = m(x - x_1)$$
$$y - 1 = \tfrac{3}{8}[x - (-2)]$$
$$y - 1 = \tfrac{3}{8}(x + 2)$$
$$y - 1 = \tfrac{3}{8}x + \tfrac{3}{4}$$
$$y = \tfrac{3}{8}x + \tfrac{7}{4}$$

**27.** $y = \tfrac{3}{8}x - 4$

$m = \tfrac{3}{8}$

The equation is in slope-intercept form, $y = mx + b$.

**29.** $y - 2 = 5(x + 3)$

$m = 5$

The equation is in point-slope form.
$y - y_1 = m(x - x_1)$.

**31.** $y + \tfrac{5}{6} = \tfrac{2}{3}(x + 4)$

$m = \tfrac{2}{3}$

The equation is in point-slope form,
$y - y_1 = m(x - x_1)$.

**33.** $y + 9 = 0$

$y = -9$

$m = 0$

This is a horizontal line.

**35.** $x - 12 = 0$

$x = 12$

The slope is undefined because this is a vertical line.

**37.** $3x - 2y + 10 = 0$
$$-2y = -3x - 10$$
$$y = \frac{-3x}{-2} - \frac{10}{-2} = \frac{3}{2}x + 5$$
$$m = \frac{3}{2}$$

The equation was rewritten in slope-intercept form, $y = mx + b$.

**39.** $y = mx + b$

$y = 3x - 1$

**41.** $y = mx + b$

$y = -\tfrac{1}{2}x + 2$

**43.**
$$y - y_1 = m(x - x_1)$$
$$y - 2 = -\tfrac{1}{3}[x - (-1)]$$
$$y - 2 = -\tfrac{1}{3}(x + 1)$$

**45.** $m = \dfrac{1 - (-1)}{4 - (-2)} = \dfrac{2}{6} = \dfrac{1}{3}$

$y - y_1 = m(x - x_1)$

$y - 1 = \tfrac{1}{3}(x - 4)$ or $y + 1 = \tfrac{1}{3}(x + 2)$

**47.** $m = \dfrac{-4 - 0}{4 - 0} = \dfrac{-4}{4} = -1$

$y = mx + b$

$y = -1x + 0$

$y = -x$

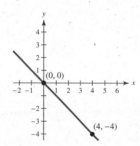

**49.** $m = \dfrac{-4 - 0}{2 - 4} = \dfrac{-4}{-2} = 2$

$$y - y_1 = m(x - x_1)$$

$$y - 0 = 2(x - 4)$$

$$y = 2x - 8$$

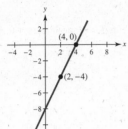

**51.** $m = \dfrac{-5 - 3}{6 - (-2)} = \dfrac{-8}{8} = -1$

$$y - y_1 = m(x - x_1)$$

$$y - 3 = -1\big[x - (-2)\big]$$

$$y - 3 = -(x + 2)$$

$$y - 3 = -x - 2$$

$$y = -x + 1$$

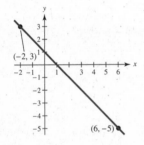

**53.** $m = \dfrac{5 - 2}{3 + 6} = \dfrac{3}{9} = \dfrac{1}{3}$

$$y - y_1 = m(x - x_1)$$

$$y - 2 = \dfrac{1}{3}(x + 6)$$

$$y - 2 = \dfrac{1}{3}x + 2$$

$$y = \dfrac{1}{3}x + 4$$

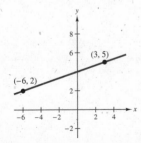

**55.** $m = \dfrac{2 - (-1)}{5 - 5} = \dfrac{3}{0}$ (undefined)

$m$ is undefined

The undefined slope indicates that this is a vertical line with an equation of the form $x = a$. The equation is $x = 5$.

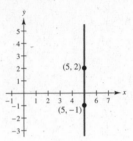

**57.** $m = \dfrac{7 - (-1)}{9/2 - 5/2} = \dfrac{8}{4/2} = \dfrac{8}{2} = 4$

$$y - y_1 = m(x - x_1)$$

$$y - 7 = 4\left(x - \dfrac{9}{2}\right)$$

$$y - 7 = 4x - 18$$

$$y = 4x - 11$$

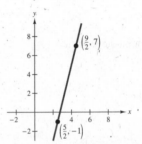

**59.** $m = \dfrac{0 - 3}{3 - 0} = \dfrac{-3}{3} = -1$

$$y - y_1 = m(x - x_1)$$

$$y - 3 = -1(x - 0)$$

$$y - 3 = -x$$

$$x + y - 3 = 0$$

**61.** $m = \dfrac{5 + 1}{-5 - 5} = \dfrac{6}{-10} = -\dfrac{3}{5}$

$$y - y_1 = m(x - x_1)$$

$$y + 1 = -\dfrac{3}{5}(x - 5)$$

$$y + 1 = -\dfrac{3}{5}x + 3$$

$$5(y + 1) = 5\left(-\dfrac{3}{5}x + 3\right)$$

$$5y + 5 = -3x + 15$$

$$3x + 5y - 10 = 0$$

**63.** $m = \dfrac{1 - 13}{-6 - 2} = \dfrac{-12}{-8} = \dfrac{3}{2}$

$$y - y_1 = m(x - x_1)$$

$$y - 13 = \frac{3}{2}(x - 2)$$

$$y - 13 = \frac{3}{2}x - 3$$

$$2(y - 13) = 2\left(\frac{3}{2}x - 3\right)$$

$$2y - 26 = 3x - 6$$

$$-3x + 2y - 20 = 0 \quad \text{or} \quad 3x - 2y + 20 = 0$$

**65.** $m = \dfrac{-4 - (-1)}{-7 - 5} = \dfrac{-3}{-12} = \dfrac{1}{4}$

$$y - y_1 = m(x - x_1)$$

$$y - (-1) = \frac{1}{4}(x - 5)$$

$$y + 1 = \frac{1}{4}x - \frac{5}{4}$$

$$4(y + 1) = 4\left(\frac{1}{4}x - \frac{5}{4}\right)$$

$$4y + 4 = x - 5$$

$$-x + 4y + 9 = 0 \quad \text{or} \quad x - 4y - 9 = 0$$

**67.** $m = \dfrac{5 - 8}{2 - (-3)} = \dfrac{-3}{5} = -\dfrac{3}{5}$

$$y - y_1 = m(x - x_1)$$

$$y - 8 = -\frac{3}{5}(x + 3)$$

$$y - 8 = -\frac{3}{5}x - \frac{9}{5}$$

$$5(y - 8) = 5\left(-\frac{3}{5}x - \frac{9}{5}\right)$$

$$5y - 40 = -3x - 9$$

$$3x + 5y - 31 = 0$$

**69.** $m = \dfrac{\frac{5}{2} - \frac{1}{2}}{\frac{1}{2} - 2} = \dfrac{2}{-\frac{3}{2}} = -\dfrac{4}{3}$

$$y - y_1 = m(x - x_1)$$

$$y - \frac{1}{2} = -\frac{4}{3}(x - 2)$$

$$y - \frac{1}{2} = -\frac{4}{3}x + \frac{8}{3}$$

$$6\left(y - \frac{1}{2}\right) = 6\left(-\frac{4}{3}x + \frac{8}{3}\right)$$

$$6y - 3 = -8x + 16$$

$$8x + 6y - 19 = 0$$

**71.** $m = \dfrac{-0.6 - 0.6}{2 - 1} = \dfrac{-1.2}{1} = -1.2$

$$y - y_1 = m(x - x_1)$$

$$y - 0.6 = -1.2(x - 1)$$

$$y - 0.6 = -1.2x + 1.2$$

$$1.2x + y - 1.8 = 0$$

Note: The equation could also be written as
$12x + 10y - 18 = 0$ or $6x + 5y - 9 = 0$.

**73.** $x - y = 3$

$$-y = -x + 3$$

$$y = x - 3$$

$$m = 1$$

(a) Parallel line: $m = 1; (2, 1)$

$$y - y_1 = m(x - x_1)$$

$$y - 1 = 1(x - 2)$$

$$y - 1 = x - 2$$

$$y = x - 1$$

$$\text{or } x - y - 1 = 0$$

(b) Perpendicular line: $m = -1; (2, 1)$

$$y - y_1 = m(x - x_1)$$

$$y - 1 = -1(x - 2)$$

$$y - 1 = -x + 2$$

$$y = -x + 3$$

$$\text{or } x + y - 3 = 0$$

**75.** $3x + 4y = 7$

$$4y = -3x + 7$$

$$y = -\frac{3}{4}x + \frac{7}{4}$$

$$m = -\frac{3}{4}$$

(a) Parallel line: $m = -\frac{3}{4}; (-12, 4)$

$$y - y_1 = m(x - x_1)$$

$$y - 4 = -\frac{3}{4}(x + 12)$$

$$y - 4 = -\frac{3}{4}x - 9$$

$$y = -\frac{3}{4}x - 5$$

$$\text{or } 3x + 4y + 20 = 0$$

(b) Perpendicular line: $m = \frac{4}{3}; (-12, 4)$

$$y - y_1 = m(x - x_1)$$

$$y - 4 = \frac{4}{3}(x + 12)$$

$$y - 4 = \frac{4}{3}x + 16$$

$$y = \frac{4}{3}x + 20$$

$$\text{or } 4x - 3y + 60 = 0$$

**77.** $2x + y = 0$

$$y = -2x$$
$$m = -2$$

(a) Parallel line: $m = -2; (1, 3)$

$$y - y_1 = m(x - x_1)$$
$$y - 3 = -2(x - 1)$$
$$y - 3 = -2x + 2$$
$$y = -2x + 5$$

or $2x + y - 5 = 0$

(b) Perpendicular line: $m = \frac{1}{2}; (1, 3)$

$$y - y_1 = m(x - x_1)$$
$$y - 3 = \frac{1}{2}(x - 1)$$
$$y - 3 = \frac{1}{2}x - \frac{1}{2}$$
$$2y - 6 = x - 1$$
$$y = \frac{1}{2}x + \frac{5}{2}$$

or $0 = x - 2y + 5$

**79.** $y + 3 = 0$

$$y = -3 \quad \text{(Horizontal line)}$$
$$m = 0.$$

(a) Parallel line: Horizontal line through $(-1, 0)$

$$y = 0$$

(b) Perpendicular line: Vertical line through $(-1, 0)$

$$x = -1$$

**81.** $3y - 2x = 7$

$$3y = 2x + 7$$
$$y = \frac{2}{3}x + \frac{7}{3}$$
$$m = \frac{2}{3}$$

(a) Parallel line: $m = \frac{2}{3}; (4, -1)$

$$y - y_1 = m(x - x_1)$$
$$y - (-1) = \frac{2}{3}(x - 4)$$
$$y + 1 = \frac{2}{3}x - \frac{8}{3}$$
$$y = \frac{2}{3}x - \frac{11}{3}$$

or $2x - 3y - 11 = 0$

(b) Perpendicular line: $m = -\frac{3}{2}; (4, -1)$

$$y - y_1 = m(x - x_1)$$
$$y + 1 = -\frac{3}{2}(x - 4)$$
$$y + 1 = -\frac{3}{2}x + 6$$
$$y = -\frac{3}{2}x + 5$$

or $3x + 2y - 10 = 0$

**83.** Because the line is vertical and passes through the point $(-2, 4)$, every point on the line has an $x$-coordinate of $-2$. So, the equation of the line is $x = -2$.

**85.** Because the line is horizontal and passes through the point $\left(\frac{1}{2}, \frac{2}{3}\right)$, every point on the line has a $y$-coordinate of $\frac{2}{3}$. So, the equation of the line is $y = \frac{2}{3}$.

**87.** Because both points have the same $x$-coordinate, the line through $(4, 1)$ and $(4, 8)$ is vertical. So, its equation is $x = 4$.

**89.** Because both points have the same $y$-coordinate, the line through $(1, -8)$ and $(7, -8)$ is horizontal. So, its equation is $y = -8$.

**91.**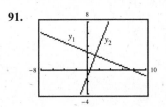

The lines are perpendicular.

**93.**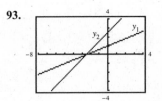

The lines are neither parallel nor perpendicular.

**95.**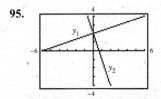

$y_1$ and $y_2$ are perpendicular.

**97.** $W = 2000 + 0.02S$

**99.** (a) $S = L - 0.2L = 0.8L$

(b)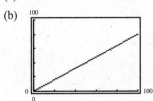

(c) The sale price is approximately \$40. You can estimate it more accurately with your graphing calculator, and you can confirm it algebraically.

$$S = 0.8(49.98) \approx 39.98$$

**101.** (a) $(0, 25,000)$ and $(1, 22,700)$

$$m = \frac{22,700 - 25,000}{1 - 0} = \frac{-2300}{1} = -2300$$

$y$-intercept: $(0, b) = (0, 25,000)$

$y = mx + b$

$y = -2300x + 25,000$ or

$V = -2300t + 25,000$

(b) $V = -2300(3) + 25,000$

$= -6900 + 25,000 = 18,100$

The value of the equipment after 3 years is estimated at $18,100.

**103.** $(5 \text{ minutes}, 45 \text{ gallons})$

$(30 \text{ minutes}, 120 \text{ gallons})$

$(5, 45)$ and $(30, 120)$

$$m = \frac{120 - 45}{30 - 5} = \frac{75}{25} = 3$$

The average rate of change of the number of gallons of water in the pool is given by the slope $m$. This rate of change is 3 gallons per minute.

$y - y_1 = m(x - x_1)$

$y - 45 = 3(x - 5)$

$y - 45 = 3x - 15$

$y = 3x + 30$

The linear model is $y = 3x + 30$, where $x$ is the time in minutes and $y$ is the number of gallons of water in the pool.

**105.** (a) When $x = 50$, $p = 580$. When $x = 47$, $p = 625$. The two points are $(50, 580)$ and $(47, 625)$.

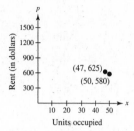

(b) $$m = \frac{625 - 580}{47 - 50} = \frac{45}{-3} = -15$$

$p - 580 = -15(x - 50)$

$p - 580 = -15x + 750$

$p = -15x + 1330$ or $15x + p - 1330 = 0$

As the rent increases, the demand decreases.

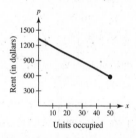

(c) $655 = -15x + 1330$

$-675 = -15x$

$\dfrac{-675}{-15} = x$

$45 = x$

Thus, 45 units would be rented if the rent were $655.

(d) $595 = -15x + 1330$

$-735 = -15x$

$\dfrac{-735}{-15} = x$

$49 = x$

Thus, 49 units would be rented if the rent were $595.

**107.** (year 2005, revenue 6.6 billion dollars)

(year 2006, revenue 10.1 billion dollars)

$(5, 6.6)$ and $(6, 10.1)$

$$m = \frac{10.1 - 6.6}{6 - 5} = \frac{3.5}{1} = 3.5$$

$$R - R_1 = m(t - t_1)$$
$$R - 6.6 = 3.5(t - 5)$$
$$R - 6.6 = 3.5t - 17.5$$
$$R = 3.5t - 10.9$$

For 2007, $t = 7$ and $R = 3.5(7) - 10.9$, so

$R = 24.5 - 10.9$ or $13.6 billion.

**109. (a)** $(7, 2540)$ and $m = 125$

$$V - V_1 = m(t - t_1)$$
$$V - 2540 = 125(t - 7)$$
$$V - 2540 = 125t - 875$$
$$V = 125t + 1665$$

**(b)** $(7, 156)$ and $m = 4.50$

$$V - V_1 = m(t - t_1)$$
$$V - 156 = 4.50(t - 7)$$
$$V - 156 = 4.50t - 31.50$$
$$V = 4.50t + 124.50$$

**(c)** $(7, 20{,}400)$ and $m = -2000$

$$V - V_1 = m(t - t_1)$$
$$V - 20{,}400 = -2000(t - 7)$$
$$V - 20{,}400 = -2000t + 14{,}000$$
$$V = -2000t + 34{,}400$$

**(d)** $(7, 45{,}000)$ and $m = -2300$

$$V - V_1 = m(t - t_1)$$
$$V - 45{,}000 = -2300(t - 7)$$
$$V - 45{,}000 = -2300t + 16{,}100$$
$$V = -2300t + 61{,}100$$

**111.** Yes. When different pairs of points are selected, the change in $y$ and the change in $x$ are the lengths of the sides of similar triangles. Corresponding sides of similar triangles are proportional, so the ratios, or slopes, are equal.

**113.** To find the $x$-intercept, set $y = 0$ and solve for $x$.

$$y = mx + b$$
$$0 = mx + b$$
$$-b = mx$$
$$-\frac{b}{m} = x$$

The $x$-intercept is $\left(-\dfrac{b}{m}, 0\right)$.

**115.** These ratios are the opposites of the slopes of the two lines. Because the ratios are equal, their opposites are equal. Therefore, the two slopes are equal, and so the lines are parallel.

**117.** $4(3 - 2x) = 12 - 8x$

**119.** $3x - 2(x - 5) = 3x - 2x + 10 = x + 10$

**121.** $3x + y = 4$

$y = -3x + 4$

**123.** $4x - 5y = -2$

$$-5y = -4x - 2$$
$$\frac{-5y}{-5} = \frac{-4x}{-5} + \frac{-2}{-5}$$
$$y = \frac{4}{5}x + \frac{2}{5}$$

**125.** $(3, 0)$ and $(4, 2)$

$$m = \frac{2 - 0}{4 - 3} = \frac{2}{1} = 2$$

**127.** $(0, -2)$ and $(-7, -1)$

$$m = \frac{-1 - (-2)}{-7 - 0} = \frac{1}{-7} = -\frac{1}{7}$$

## Section 4.5   Graphs of Linear Inequalities

**1.** (a)  $(0, 0)$

$$0 + 4(0) \overset{?}{>} 10$$

$$0 \not> 10$$

The point $(0, 0)$ *is not* a solution.

(b)  $(3, 2)$

$$3 + 4(2) \overset{?}{>} 10$$

$$11 > 10$$

The point $(3, 2)$ *is* a solution.

(c)  $(1, 2)$

$$1 + 4(2) \overset{?}{>} 10$$

$$9 \not> 10$$

The point $(1, 2)$ *is not* a solution.

(d)  $(-2, 4)$

$$-2 + 4(4) \overset{?}{>} 10$$

$$14 > 10$$

The point $(-2, 4)$ *is* a solution.

**3.** (a)  $(1, 2)$

$$-3(1) + 5(2) \overset{?}{\le} 12$$

$$-3 + 10 \overset{?}{\le} 12$$

$$7 \le 12$$

The point $(1, 2)$ *is* a solution.

(b)  $(2, -3)$

$$-3(2) + 5(-3) \overset{?}{\le} 12$$

$$-6 - 15 \overset{?}{\le} 12$$

$$-21 \le 12$$

The point $(2, -3)$ *is* a solution.

(c)  $(1, 3)$

$$-3(1) + 5(3) \overset{?}{\le} 12$$

$$-3 + 15 \overset{?}{\le} 12$$

$$12 \le 12$$

The point $(1, 3)$ *is* a solution.

(d)  $(2, 8)$

$$-3(2) + 5(8) \overset{?}{\le} 12$$

$$-6 + 40 \overset{?}{\le} 12$$

$$34 \not\le 12$$

The point $(2, 8)$ *is not* a solution.

**5.** (a)  $(1, 3)$

$$3(1) - 2(3) \overset{?}{<} 2$$

$$3 - 6 \overset{?}{<} 2$$

$$-3 < 2$$

The point $(1, 3)$ *is* a solution.

(b)  $(2, 0)$

$$3(2) - 2(0) \overset{?}{<} 2$$

$$6 - 0 \overset{?}{<} 2$$

$$6 \not< 2$$

The point $(2, 0)$ *is not* a solution.

(c)  $(0, 0)$

$$3(0) - 2(0) \overset{?}{<} 2$$

$$0 - 0 \overset{?}{<} 2$$

$$0 < 2$$

The point $(0, 0)$ *is* a solution.

(d)  $(3, -5)$

$$3(3) - 2(-5) \overset{?}{<} 2$$

$$9 + 10 \overset{?}{<} 2$$

$$19 \not< 2$$

The point $(3, -5)$ *is not* a solution

**7.** (a)  $(2, -4)$

$$6(2) + 4(-4) \overset{?}{\ge} -4$$

$$-4 \ge -4$$

The point $(2, -4)$ *is* a solution.

(b)  $(6, -9)$

$$6(6) + 4(-9) \overset{?}{\ge} -4$$

$$0 \ge -4$$

The point $(6, -9)$ *is* a solution.

(c)  $(-3, 4)$

$$6(-3) + 4(4) \overset{?}{\ge} -4$$

$$-2 \ge -4$$

The point $(-3, 4)$ *is* a solution.

(d)  $(3, -1)$

$$6(3) + 4(-1) \overset{?}{\ge} -4$$

$$14 \ge -4$$

The point $(3, -1)$ *is* a solution.

**9.** The boundary of the graph of $2x + 3y < 6$ should be *dashed.*

**11.** The boundary of the graph of $2x + 3y \geq 6$ should be *solid.*

**13.** Graph (b)

**15.** Graph (d)

**17.** Graph (c)

**19.** Graph (b)

**21.**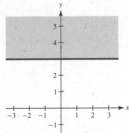

**23.**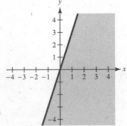

**25.**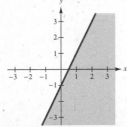

**27.**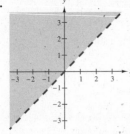

**29.**

**31.**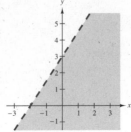

**33.**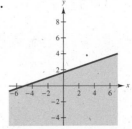

**35.**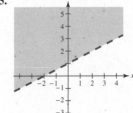

**37.**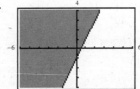

**39.**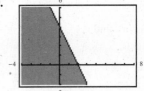

**41.**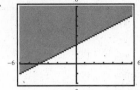

**43.**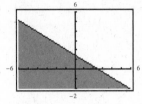

**45.** $y > 6x$

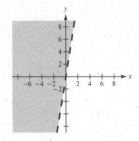

**47.** $x + y \geq 9$

or $y \geq -x + 9$

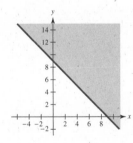

**49.** $y \leq x + 3$

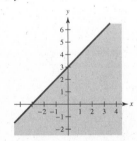

**51.** The graph shows the points on and above the horizontal line $y = 2$.

$y \geq 2$

**53.** The inequality can be written in slope-intercept form. The slope is $-2$ and the $y$-intercept is 2.

$y \leq -2x + 2$

(This could also be written as $2x + y \leq 2$ or as $2x + y - 2 \leq 0$.)

**55.** The inequality can be written in slope-intercept form. The slope is 2 and the $y$-intercept is 0.

$y < 2x$

(This could also be written as $2x - y > 0$.)

**57.** (a) *Labels:*     $x$ = total calories consumed per day

          $y$ = fat calories consumed per day

*Inequality:*     $y \leq 0.35x$

(b)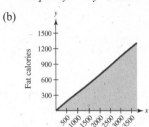

*Representative solutions:*

$(1500, 500)$

$(1800, 635)$

$(2000, 700)$

$(2500, 800)$

**59.** (a) *Labels:*

$x$ = hours worked at the grocery store

$y$ = hours worked mowing lawns

*Inequality:*     $9x + 12y \geq 210$

(b)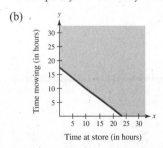

*Representative solutions:*

$(2, 16)$

$(15, 7)$

$(20, 20)$

$(24, 0)$

**61.** (a) *Labels:*     $t$ = number of tables

                  $c$ = number of chairs

      *Inequality:*  $t + \frac{3}{2}c \le 12$

(b)

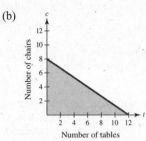

      *Representative solutions:*

      $(5, 4)$

      $(2, 6)$

      $(0, 8)$

      $(12, 0)$

**63.** The $x$-axis is the line $y = 0$. The inequality whose graph consists of all points above the $x$-axis is $y > 0$.

**65.** Yes, the two graphs are the same. If $2x < 2y$, then $x < y$, and this means the same thing as $y > x$.

**67.** (a) The graph of $x \ge 1$ on the real number line is the set of all points on the real number line that are to the right of 1 or at 1 itself. A square bracket is used to indicate that the endpoint 1 is included in the solution interval.

(b) The graph of $x \ge 1$ on a rectangular coordinate system is the set of all points $(x, y)$ that lie on or to the right of the vertical line $x = 1$. A solid vertical line is used to indicate the points on the line are included in the solution.

**69.** $\dfrac{8 \text{ dimes}}{36 \text{ nickels}} = \dfrac{80 \text{ cents}}{180 \text{ cents}} = \dfrac{4}{9}$

**71.** $\dfrac{50 \text{ centimeters}}{3 \text{ meters}} = \dfrac{50 \text{ centimeters}}{300 \text{ centimeters}} = \dfrac{1}{6}$

**73.** $\dfrac{46 \text{ inches}}{4 \text{ feet}} = \dfrac{46 \text{ inches}}{48 \text{ inches}} = \dfrac{23}{24}$

**75.** $m = \dfrac{-4 - 8}{1 - 5} = \dfrac{-12}{-4} = 3$

$y - y_1 = m(x - x_1)$

$y - 8 = 3(x - 5)$

$y - 8 = 3x - 15$

$y = 3x - 7$

**77.** $m = \dfrac{5 - 3}{-4 - 4} = \dfrac{2}{-8} = -\dfrac{1}{4}$

$y - y_1 = m(x - x_1)$

$y - 3 = -\dfrac{1}{4}(x - 4)$

$y - 3 = -\dfrac{1}{4}x + 1$

$y = -\dfrac{1}{4}x + 4$

**79.** $m = \dfrac{4 - (-2)}{\dfrac{1}{6} - \dfrac{5}{6}} = \dfrac{6}{-\dfrac{4}{6}} = \dfrac{6}{-\dfrac{2}{3}} = \dfrac{6}{1} \cdot \dfrac{-3}{2} = \dfrac{-18}{2} = -9$

$y - y_1 = m(x - x_1)$

$y - 4 = -9\left(x - \dfrac{1}{6}\right)$

$y - 4 = -9x + \dfrac{3}{2}$

$y = -9x + \dfrac{11}{2}$

# Review Exercises for Chapter 4

**1.**

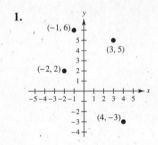

**3.**

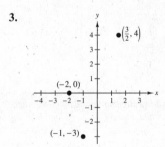

**5.** $A: (3, -2)$

$B: (0, 5)$

$C: (-1, 3)$

$D: (-5, -2)$

**7.** Quadrant II

The point $(-5, 3)$ is located 5 units to the *left* of the vertical axis and 3 units *above* the horizontal axis in Quadrant II.

**9.** The point $(4, 0)$ is located on the *x*-axis between the first and fourth quadrants.

**11.** Quadrant II

The point $(x, 5)$, $x < 0$, is located to the left of the vertical axis because *x* is negative; it is 5 units above the horizontal axis. Therefore, the point is in Quadrant II.

**13.** Quadrant II or III

The point $(-6, y)$ is located to the left of the vertical axis. If *y* is positive, the point is above the horizontal axis in the second quadrant, and if *y* is negative, the point is below the horizontal axis in the third quadrant. If $y = 0$, the point is on the *x*-axis between the second and third quadrants.

**15.**

| $x$ | $-1$ | $0$ | $1$ | $2$ |
|---|---|---|---|---|
| $y = 4x - 1$ | $-5$ | $-1$ | $3$ | $7$ |

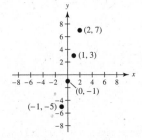

**17.**

| $x$ | $-1$ | $0$ | $1$ | $2$ |
|---|---|---|---|---|
| $y = -\frac{1}{2}x - 1$ | $-\frac{1}{2}$ | $-1$ | $-\frac{3}{2}$ | $-2$ |

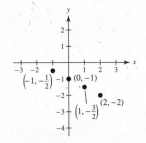

**19.** $3x + 4y = 12$

$4y = -3x + 12$

$y = -\frac{3}{4}x + 3$

**21.** $9x - 3y = 12$

$-3y = -9x + 12$

$y = 3x - 4$

**23.** $x - 2y = 8$

$-2y = -x + 8$

$y = \frac{1}{2}x - 4$

**25.** (a) $(1, -1)$ *is* a solution because $1 - 3(-1) = 4$.

(b) $(0, 0)$ *is not* a solution because $1 - 3(0) \neq 4$.

(c) $(2, 1)$ *is not* a solution because $2 - 3(1) \neq 4$.

(d) $(5, -2)$ *is not* a solution because $5 - 3(-2) \neq 4$.

**27.** (a) $(3, 5)$ *is* a solution because $5 = \frac{2}{3}(3) + 3$.

(b) $(-3, 1)$ *is* a solution because $1 = \frac{2}{3}(-3) + 3$.

(c) $(-6, 0)$ *is not* a solution because $0 \neq \frac{2}{3}(-6) + 3$.

(d) $(0, 3)$ *is* a solution because $3 = \frac{2}{3}(0) + 3$.

**29.**

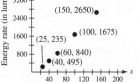

The relationship between the wattage and the energy rate is approximately linear.

**31.**

| $x$ | $-2$ | $-1$ | $0$ | $1$ | $2$ |
|---|---|---|---|---|---|
| $y = x - 5$ | $-7$ | $-6$ | $-5$ | $-4$ | $-3$ |

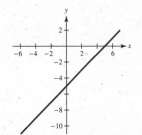

**33.**

| $x$ | $-2$ | $-1$ | $0$ | $1$ | $2$ |
|---|---|---|---|---|---|
| $y = x^2 - 1$ | 3 | 0 | $-1$ | 0 | 3 |

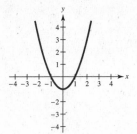

**35.**

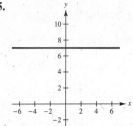

**37.**

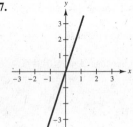

**39.**

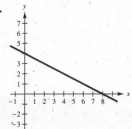

**41.**

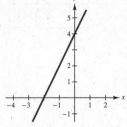

**43.**

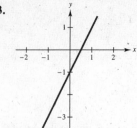

**45.**

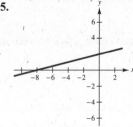

**47.**

| *x-intercept* | *y-intercept* |
|---|---|
| Let $y = 0$. | Let $x = 0$. |

$$0 = 6x + 2 \qquad y = 6(0) + 2$$
$$-2 = 6x \qquad y = 0 + 2$$
$$\frac{-2}{6} = x \qquad y = 2$$
$$-\frac{1}{3} = x \qquad (0, 2)$$
$$\left(-\frac{1}{3}, 0\right)$$

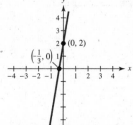

The graph has one *x*-intercept at $\left(-\frac{1}{3}, 0\right)$ and one

*y*-intercept at $(0, 2)$.

**49.** *x-intercept*   *y-intercept*

Let $y = 0$.   Let $x = 0$.

$0 = \frac{2}{5}x - 2$   $y = \frac{2}{5}(0) - 2$

$5(0) = 5\left(\frac{2}{5}x - 2\right)$   $y = 0 - 2$

$0 = 2x - 10$   $y = -2$

$10 = 2x$   $(0, -2)$

$5 = x$

$(5, 0)$

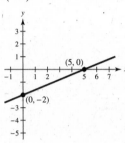

The graph has one *x*-intercept at $(5, 0)$ and one *y*-intercept at $(0, -2)$.

**51.** *x-intercept*   *y-intercept*

Let $y = 0$.   Let $x = 0$.

$2x - 0 = 4$   $2(0) - y = 4$

$2x = 4$   $-y = 4$

$x = 2$   $y = -4$

$(2, 0)$   $(0, -4)$

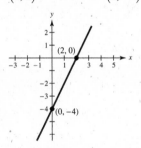

The graph has one *x*-intercept at $(2, 0)$ and one *y*-intercept at $(0, -4)$.

**53.** *x-intercept*   *y-intercept*

Let $y = 0$.   Let $x = 0$.

$4x + 2(0) = 8$   $4(0) + 2y = 8$

$4x = 8$   $2y = 8$

$x = \frac{8}{4}$   $y = 4$

$x = 2$   $(0, 4)$

$(2, 0)$

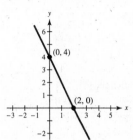

The graph has one *x*-intercept at $(2, 0)$ and one *y*-intercept at $(0, 4)$.

**55.** $C = 3x + 125$

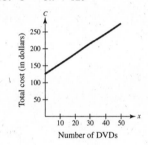

**57.** The slope $m = \frac{1}{2}$.

**59.** The slope $m = \frac{3}{1} = 3$.

**61.** $m = \frac{6 - 1}{14 - 2} = \frac{5}{12}$

The line rises.

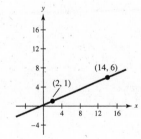

**63.** $m = \dfrac{-2 - 0}{6 - (-1)} = \dfrac{-2}{7} = -\dfrac{2}{7}$

The line falls.

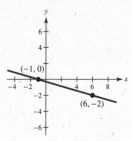

**65.** $m = \dfrac{6 - 0}{4 - 4} = \dfrac{6}{0}$ (undefined)

$m$ is undefined. The line is vertical.

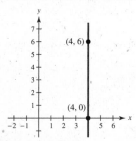

**67.** $m = \dfrac{1 - 5}{1 - (-2)} = -\dfrac{4}{3}$

The line falls.

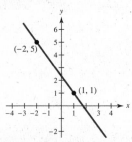

**69.** $m = \dfrac{-10 - (-4)}{-5 - (-1)} = \dfrac{-6}{-4} = \dfrac{3}{2}$

The line rises.

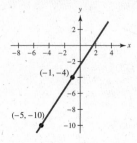

**71.** $m = \dfrac{0 - \dfrac{5}{2}}{\dfrac{5}{6} - 0} = \dfrac{-\dfrac{5}{2}}{\dfrac{5}{6}}$

$\quad = -\dfrac{5}{2} \div \dfrac{5}{6} = -\dfrac{5}{2} \cdot \dfrac{6}{5}$

$\quad = -\dfrac{30}{10} = -3$

The line falls.

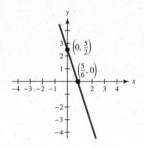

**73.** $m = \dfrac{4}{6} = \dfrac{2}{3}$

**75.** $x + y = 6$

$\quad y = -x + 6$

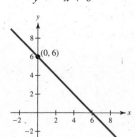

**77.** $2x - y = -1$

$\quad -y = -2x - 1$

$\quad y = 2x + 1$

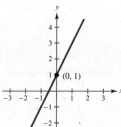

**79.** $3x + 6y = 12$

$$6y = -3x + 12$$

$$y = -\tfrac{1}{2}x + 2$$

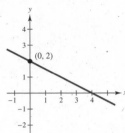

**81.** $5y - 2x = 5$

$$5y = 2x + 5$$

$$y = \tfrac{2}{5}x + 1$$

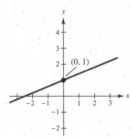

**83.** Line 1: $m = \dfrac{1 - 3}{-2 - 0} = \dfrac{-2}{-2} = 1$

Line 2: $m = \dfrac{9 - (-3)}{4 - (-8)} = \dfrac{12}{12} = 1$

The slopes are equal; the lines are parallel.

**85.** Line 1: $m = \dfrac{-5 - 6}{-1 - 3} = \dfrac{-11}{-4} = \dfrac{11}{4}$

Line 2: $m = \dfrac{7 - 3}{4 - (-2)} = \dfrac{4}{6} = \dfrac{2}{3}$

The lines are neither parallel nor perpendicular.

**87.** $y + 1 = 2(x - 4)$

$$y + 1 = 2x - 8$$

$$y = 2x - 9$$

**89.** $y - 2 = -4(x - 1)$

$$y - 2 = -4x + 4$$

$$y = -4x + 6$$

**91.** $y + 2 = \tfrac{4}{5}(x + 5)$

$$y + 2 = \tfrac{4}{5}x + 4$$

$$y = \tfrac{4}{5}x + 2$$

**93.** $y - 3 = -\tfrac{8}{3}(x + 1)$

$$y - 3 = -\tfrac{8}{3}x - \tfrac{8}{3}$$

$$y = -\tfrac{8}{3}x - \tfrac{8}{3} + \tfrac{9}{3}$$

$$y = -\tfrac{8}{3}x + \tfrac{1}{3}$$

**95.** The undefined slope indicates that this is a vertical line through $(3, 8)$. All the points have the same $x$-coordinate of 3. The equation of the line is $x = 3$, or, in general form, $x - 3 = 0$.

**97.** $m = \dfrac{-2 - 0}{0 - (-4)} = \dfrac{-2}{4} = -\dfrac{1}{2}$

$$y - 0 = -\tfrac{1}{2}(x + 4)$$

$$y = -\tfrac{1}{2}x - 2$$

$$2y = 2\left(-\tfrac{1}{2}x - 2\right)$$

$$2y = -x - 4$$

$$x + 2y + 4 = 0$$

Note: We could also obtain the solution using $m = -\dfrac{1}{2}$ and the $y$-intercept $(0, -2)$.

$$y = mx + b$$

$$y = -\tfrac{1}{2}x - 2$$

$$2y = -x - 4$$

$$x + 2y + 4 = 0$$

**99.** $m = \dfrac{8 - 8}{6 - 0} = \dfrac{0}{6} = 0$

$$y - 8 = 0(x - 0)$$

$$y - 8 = 0$$

Note: The slope of 0 indicates a horizontal line with an equation of the form $y = b$. The horizontal line through $(0, 8)$ has the equation $y = 8$ or $y - 8 = 0$.

**101.** $m = \dfrac{-7 - (-2)}{-4 - (-1)} = \dfrac{-5}{-3} = \dfrac{5}{3}$

$$y - y_1 = m(x - x_1)$$

$$y - (-2) = \frac{5}{3}\left[x - (-1)\right]$$

$$y + 2 = \frac{5}{3}(x + 1)$$

$$y + 2 = \frac{5}{3}x + \frac{5}{3}$$

$$3(y + 2) = 3\left(\frac{5}{3}x + \frac{5}{3}\right)$$

$$3y + 6 = 5x + 5$$

$$-5x + 3y + 1 = 0 \quad \text{or} \quad 5x - 3y - 1 = 0$$

**103.** $m = \dfrac{7.8 - 3.3}{6 - 2.4} = \dfrac{4.5}{3.6} = \dfrac{45}{36} = \dfrac{5}{4}$

$$y - 7.8 = \frac{5}{4}(x - 6)$$

$$y - \frac{39}{5} = \frac{5}{4}x - \frac{15}{2}$$

$$20\left(y - \frac{39}{5}\right) = 20\left(\frac{5}{4}x - \frac{15}{2}\right)$$

$$20y - 156 = 25x - 150$$

$$-25x + 20y - 6 = 0$$

$$25x - 20y + 6 = 0$$

Note: You could use decimals instead of fractions.

$$m = \frac{7.8 - 3.3}{6 - 2.4} = \frac{4.5}{3.6} = 1.25$$

$$y - 7.8 = 1.25(x - 6)$$

$$y - 7.8 = 1.25x - 7.5$$

$$-1.25x + y - 0.3 = 0$$

$$-125x + 100y - 30 = 0 \quad \text{(Multiply both sides by 100.)}$$

$$25x - 20y + 6 = 0 \quad \text{(Divide both sides by $-5$.)}$$

**105.** $x - y = -2$

$-y = -x - 2$

$y = x + 2$

$m = 1$

(a) Parallel line: $m = 1; (-6, 3)$

$y - y_1 = m(x - x_1)$

$y - 3 = 1[x - (-6)]$

$y - 3 = x + 6$

$y = x + 9$

or $x - y + 9 = 0$

(b) Perpendicular line: $m = -1; (-6, 3)$

$y - y_1 = m(x - x_1)$

$y - 3 = -1[x - (-6)]$

$y - 3 = -(x + 6)$

$y - 3 = -x - 6$

$y = -x - 3$

or $x + y + 3 = 0$

**107.** $y - 9 = 0$

$y = 9$ (Horizontal line)

$m = 0.$

(a) Parallel line: Horizontal line through $\left(\frac{3}{8}, 4\right)$

$y = 4$

(b) Perpendicular line: Vertical line through $\left(\frac{3}{8}, 4\right)$

$x = \frac{3}{8}$

**109.** Because the line is horizontal and passes through the point $(-4, 5)$, every point on the line has a $y$-coordinate of 5. So, the equation of the line is $y = 5$.

**111.** Because the line is vertical and passes through the point $(5, -1)$, every point on the line has an $x$-coordinate of 5. So, the equation of the line is $x = 5$.

**113.** The line is a horizontal line, so it has an equation of the form $y = b$. The line is located 3 units above the origin, so the equation is $y = 3$.

**115.** $W = 0.07S + 5500$

**117.** (a) $(-1, -5)$ *is not* a solution because

$-1 - (-5) = -1 + 5 \not> 4.$

(b) $(0, 0)$ *is not* a solution because $0 - 0 \not> 4.$

(c) $(3, -2)$ *is* a solution because

$3 - (-2) = 3 + 2 > 4.$

(d) $(8, 1)$ *is* a solution because $8 - 1 > 4.$

**119.** (a) $(3, 4)$

$3(3) - 2(4) \overset{?}{<} -1$

$1 \not< -1$

The point $(3, 4)$ is *not* a solution.

(b) $(-1, 2)$

$3(-1) - 2(2) \overset{?}{<} -1$

$-7 < -1$

The point $(-1, 2)$ *is* a solution.

(c) $(1, 8)$

$3(1) - 2(8) \overset{?}{<} -1$

$-13 < -1$

The point $(1, 8)$ *is* a solution.

(d) $(0, 0)$

$3(0) - 2(0) \overset{?}{<} -1$

$0 \not< -1$

The point $(0, 0)$ is *not* a solution.

**121.** $x - 2 \geq 0$

$x \geq 2$

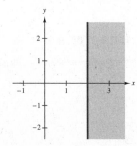

**123.** $2x + y < 1$

$\quad\quad y < -2x + 1$

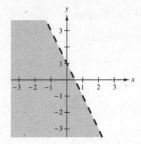

**125.** $\quad x \le 4y - 2$

$\quad\quad x + 2 \le 4y$

$\quad\quad \frac{1}{4}x + \frac{1}{2} \le y$

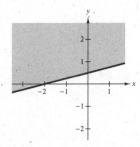

**127.** $3x - y + 4 < -3$

$\quad\quad -y < -3x - 7$

$\quad\quad y > 3x + 7 \quad$ (Reverse inequality.)

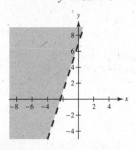

# Chapter Test for Chapter 4

**1.**

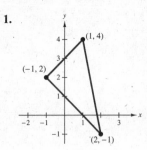

**129.** The graph shows the points below the horizontal line $y = 2$. The inequality is $y < 2$.

**131.** The inequality can be written in slope-intercept form. The slope is 1 and the $y$-intercept is $(0, 1)$. The inequality is $y \le x + 1$.

**133.** (a) *Labels:* $\quad x = $ number of DVD players

$\quad\quad\quad\quad\quad y = $ number of camcorders

$\quad\quad$ *Inequality:* $\quad 2x + 3y \le 120$

(b)

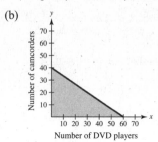

*Representative solutions:*

$(10, 15)$

$(25, 20)$

$(30, 20)$

$(40, 10)$

**2.** (a) No, $(0, -2)$ is not a solution because

$\quad -2 \ne |0| + |0 - 2|$ or $0 + 2$.

(b) Yes, $(0, 2)$ is a solution because $2 = |0| + |0 - 2|$ or $0 + 2$.

(c) Yes, $(-4, 10)$ is a solution because

$\quad 10 = |-4| + |-4 - 2|$ or $4 + 6$.

(d) No, $(-2, -2)$ is not a solution because

$\quad -2 \ne |-2| + |-2 - 2|$ or $2 + 4$.

**3.** The $y$-coordinate of any point on the $x$-axis is 0.

**4.** *x-intercept*

Let $y = 0$.

$8x - 2(0) = -16$

$8x = -16$

$x = -2$

$(-2, 0)$

*y-intercept*

Let $x = 0$.

$8(0) - 2y = -16$

$-2y = -16$

$y = 8$

$(0, 8)$

The graph has one *x*-intercept at $(-2, 0)$ and one *y*-intercept at $(0, 8)$.

**5.**

| $x$ | $-2$ | $-1$ | $0$ | $1$ | $2$ |
|---|---|---|---|---|---|
| $y = -3x - 4$ | $2$ | $-1$ | $-4$ | $-7$ | $-10$ |

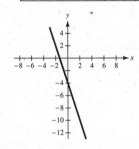

**6.**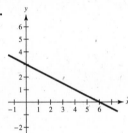

**7.**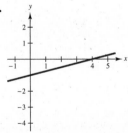

**8.**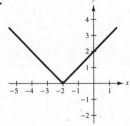

**9.**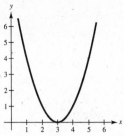

**10.** $m = \dfrac{\dfrac{3}{2} - 0}{2 - (-5)} = \dfrac{\dfrac{3}{2}}{7} = \dfrac{3}{2} \cdot \dfrac{1}{7} = \dfrac{3}{14}$

$y - y_1 = m(x - x_1)$

$y - 0 = \dfrac{3}{14}\big[x - (-5)\big]$

$y = \dfrac{3}{14}(x + 5)$

$y = \dfrac{3}{14}x + \dfrac{15}{14}$

**11.**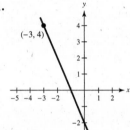

Additional points: $(-2, 2), (-1, 0), (-3, -2)$

Answers may vary.

**12.** $7x - 8y + 5 = 0$

$-8y = -7x - 5$

$\dfrac{-8y}{-8} = \dfrac{-7x}{-8} + \dfrac{-5}{-8}$

$y = \dfrac{7}{8}x + \dfrac{5}{8}$

Slope: $m = \dfrac{7}{8}$

Perpendicular slope: $m = -\dfrac{8}{7}$

**13.** $y = mx + b$

$m = -\tfrac{3}{8}$ and $(0, b) = (0, 6)$

$y = -\tfrac{3}{8}x + 6$ or

$3x + 8y - 48 = 0$

**14.** Because the line is vertical and passes through the point $(3, -7)$, every point on the line has an *x*-coordinate of 3. So, the equation of the line is $x = 3$.

**15.** (a)  $(2, 2)$

$$3 \cdot 2 + 5 \cdot 2 \overset{?}{\leq} 16$$

$$6 + 10 \overset{?}{\leq} 16$$

$$16 \leq 16$$

The point $(2, 2)$ *is* a solution.

(b)  $(6, -1)$

$$3 \cdot 6 + 5(-1) \overset{?}{\leq} 16$$

$$18 + (-5) \overset{?}{\leq} 16$$

$$18 - 5 \overset{?}{\leq} 16$$

$$13 \leq 16$$

The point $(6, -1)$ *is* a solution.

(c)  $(-2, 4)$

$$3(-2) + 5 \cdot 4 \overset{?}{\leq} 16$$

$$-6 + 20 \overset{?}{\leq} 16$$

$$14 \leq 16$$

The point $(-2, 4)$ *is* a solution.

(d)  $(7, -1)$

$$3 \cdot 7 + 5(-1) \overset{?}{\leq} 16$$

$$21 - 5 \overset{?}{\leq} 16$$

$$16 \leq 16$$

The point $(7, -1)$ *is* a solution.

**16.**

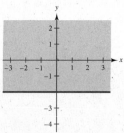

**17.**

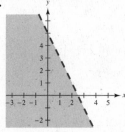

**18.**

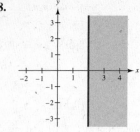

**19.**  $-y + 4x > 3$

$$-y > -4x + 3$$

$$y < 4x - 3 \qquad \text{(Reverse inequality)}$$

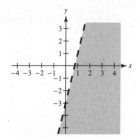

**20.** The slope of 230 indicates that sales are increasing at the rate of 230 units per year.

# C H A P T E R 5
# Exponents and Polynomials

# CHAPTER 5
## Exponents and Polynomials

## Section 5.1   Negative Exponents and Scientific Notation

**1.** $u^2 \cdot u^4 = u^{2+4} = u^6$

**3.** $3x^3 \cdot x^4 = 3x^{3+4} = 3x^7$

**5.** $5x\left(x^6\right) = 5x^{1+6} = 5x^7$

**7.** $\left(-5z^3\right)\left(3z^2\right) = (-5 \cdot 3)\left(z^3 \cdot z^2\right)$
$$= -15z^{3+2} = -15z^5$$

**9.** $(-xz)\left(-2y^2z\right) = 2xy^2z^{1+1} = 2xy^2z^2$

**11.** $2b^4(-ab)\left(3b^2\right) = (2)(-1)(3)(a)\left(b^4 \cdot b \cdot b^2\right)$
$$= -6ab^{4+1+2} = -6ab^7$$

**13.** $\left(t^2\right)^4 = t^{2(4)} = t^8$

**15.** $5(uv)^5 = 5u^5v^5$

**17.** $(-2s)^3 = (-2)^3s^3 = -8s^3$

**19.** $\left(a^2b\right)^3\left(ab^2\right)^4 = \left(a^2\right)^3 \cdot b^3 \cdot a^4\left(b^2\right)^4$
$$= a^{2\cdot3} \cdot a^4 \cdot b^3 \cdot b^{2\cdot4}$$
$$= a^6 \cdot a^4 \cdot b^3 \cdot b^8$$
$$= a^{6+4}b^{3+8}$$
$$= a^{10}b^{11}$$

**21.** $\left(uv^3\right)\left(-2uv^2\right)^5 = u \cdot v^3 \cdot (-2)^5 \cdot u^5 \cdot \left(v^2\right)^5$
$$= (-2)^5 \cdot u \cdot v^3 \cdot u^5 \cdot v^{10}$$
$$= -32u^{1+5}v^{3+10}$$
$$= -32u^6v^{13}$$

**23.** $\left[(x-3)^4\right]^2 = (x-3)^{4\cdot2} = (x-3)^8$

**25.** $(x-2y)^3(x-2y)^3 = (x-2y)^{3+3} = (x-2y)^6$

**27.** By Dividing Out
$$\frac{x^5}{x^2} = \frac{x \cdot x \cdot x \cdot \cancel{x} \cdot \cancel{x}}{\cancel{x} \cdot \cancel{x}} = x^3$$

By Subtracting Exponents
$$\frac{x^5}{x^2} = x^{5-2} = x^3$$

**29.** By Dividing Out
$$\frac{x^2}{x} = \frac{x \cdot \cancel{x}}{\cancel{x}} = x$$

By Subtracting Exponents
$$\frac{x^2}{x} = x^{2-1} = x$$

**31.** By Dividing Out
$$\frac{z^9}{z^3} = \frac{z \cdot z \cdot z \cdot z \cdot z \cdot z \cdot \cancel{z} \cdot \cancel{z} \cdot \cancel{z}}{\cancel{z} \cdot \cancel{z} \cdot \cancel{z}} = z^6$$

By Subtracting Exponents
$$\frac{z^9}{z^3} = z^{9-3} = z^6$$

**33.** By Dividing Out $\dfrac{3u^4}{u^3} = \dfrac{3 \cdot u \cdot \cancel{u} \cdot \cancel{u} \cdot \cancel{u}}{\cancel{u} \cdot \cancel{u} \cdot \cancel{u}} = 3u$

By Subtracting Exponents $\dfrac{3u^4}{u^3} = 3u^{4-3} = 3u$

**35.** By Dividing Out
$$\frac{2^3y^4}{2^2y^2} = \frac{2 \cdot \cancel{2} \cdot \cancel{2} \cdot y \cdot y \cdot \cancel{y} \cdot \cancel{y}}{\cancel{2} \cdot \cancel{2} \cdot \cancel{y} \cdot \cancel{y}} = 2y^2$$

By Subtracting Exponents
$$\frac{2^3y^4}{2^2y^2} = 2^{3-2}y^{4-2} = 2y^2$$

**37.** By Dividing Out
$$\frac{4^5x^5}{4x^3} = \frac{4 \cdot 4 \cdot 4 \cdot 4 \cdot \cancel{4} \cdot \cancel{x} \cdot \cancel{x} \cdot \cancel{x} \cdot x \cdot x}{\cancel{4} \cdot \cancel{x} \cdot \cancel{x} \cdot \cancel{x}} = 256x^2$$

By Subtracting Exponents
$$\frac{4^5x^5}{4x^3} = 4^{5-1}x^{5-3} = 4^4x^2 = 256x^2$$

**39.** By Dividing Out
$$\frac{3^4(ab)^3}{3(ab)^2} = \frac{3 \cdot 3 \cdot 3 \cdot \cancel{3}\left(\cancel{ab}\right)\left(\cancel{ab}\right)(ab)}{\cancel{3}\left(\cancel{ab}\right)\left(\cancel{ab}\right)} = 27ab$$

By Subtracting Exponents
$$\frac{3^4(ab)^3}{3(ab)^2} = 3^{4-1}(ab)^{3-2} = 3^3(ab) = 27ab$$

**41.** $\dfrac{-3x^2}{x} = -3x^{2-1} = -3x$

**43.** $\dfrac{4x^6}{x^3} = 4x^{6-3} = 4x^3$

**45.** $\dfrac{-12z^3}{-3z} = \left(\dfrac{-12}{-3}\right)(z^{3-1}) = 4z^2$

**47.** $\dfrac{32b^4}{12b^3} = \left(\dfrac{32}{12}\right)(b^{4-3}) = \dfrac{8}{3}b$ or $\dfrac{8b}{3}$

**49.** $\dfrac{-22y^2}{4y} = \left(\dfrac{-22}{4}\right)y^{2-1} = -\dfrac{11}{2}y$ or $-\dfrac{11y}{2}$

**51.** $\dfrac{-18s^4}{-12r^2s} = \left(\dfrac{-18}{-12}\right)\cdot\dfrac{s^{4-1}}{r^2} = \dfrac{3}{2}\cdot\dfrac{s^3}{r^2} = \dfrac{3s^3}{2r^2}$

**53.** $\dfrac{24u^2v^6}{18u^2v^4} = \dfrac{24}{18}u^{2-2}v^{6-4} = \dfrac{4}{3}u^0v^2$
$= \dfrac{4}{3}(1)v^2 = \dfrac{4}{3}v^2$ or $\dfrac{4v^2}{3}$

**55.** $\dfrac{(-3z)^3}{18z^2} = \dfrac{-27z^3}{18z^2} = \dfrac{-27}{18}z^{3-2}$
$= -\dfrac{3}{2}z$ or $-\dfrac{3z}{2}$

**57.** $\dfrac{(2x^2y)^3}{(4y)^2x^4} = \dfrac{2^3x^6y^3}{4^2y^2x^4} = \dfrac{8}{16}x^{6-4}y^{3-2}$
$= \dfrac{1}{2}x^2y$ or $\dfrac{x^2y}{2}$

**59.** $2^{-3} = \dfrac{1}{2^3}$

**61.** $y^{-5} = \dfrac{1}{y^5}$

**63.** $8x^{-7} = \dfrac{8}{x^7}$

**65.** $7x^{-4}y^{-1} = \dfrac{7}{x^4y}$

**67.** $\dfrac{1}{2z^{-4}} = \dfrac{1z^4}{2} = \dfrac{z^4}{2}$

**69.** $\dfrac{2x}{3y^{-2}} = \dfrac{2xy^2}{3}$

**71.** $\dfrac{1}{4} = 4^{-1}$

**73.** $\dfrac{1}{5^2} = 5^{-2}$

**75.** $\dfrac{10}{t^5} = 10t^{-5}$

**77.** $\dfrac{5}{x^n} = 5x^{-n}$

**79.** $\dfrac{2x^2}{y^4} = 2x^2y^{-4} = \dfrac{2y^{-4}}{x^{-2}}$

**81.** $6^{-2} = \dfrac{1}{6^2} = \dfrac{1}{36}$

**83.** $(-2)^{-5} = \dfrac{1}{(-2)^5} = \dfrac{1}{-32} = -\dfrac{1}{32}$

**85.** $\dfrac{1}{4^{-4}} = 4^4 = 256$

**87.** $\dfrac{2}{5^{-2}} = 2(5)^2 = 2(25) = 50$

**89.** $\dfrac{2^{-4}}{3} = \dfrac{1}{3(2)^4} = \dfrac{1}{48}$

**91.** $\dfrac{4^{-2}}{3^{-4}} = \dfrac{3^4}{4^2} = \dfrac{81}{16}$

**93.** $\left(\dfrac{2}{3}\right)^{-2} = \dfrac{2^{-2}}{3^{-2}} = \dfrac{3^2}{2^2} = \dfrac{9}{4}$

**95.** The exponent is $-2$.
$5^{-2} = \dfrac{1}{5^2} = \dfrac{1}{25}$

**97.** The exponent is $-3$.
$\dfrac{1}{3^{-3}} = 3^3 = 27$

**99.** The exponent is $-2$.
$(x^{-2}y^3)^{-2} = x^{-2(-2)}y^{3(-2)} = x^4y^{-6} = \dfrac{x^4}{y^6}$

**101.** The exponent is $-3$.
$(x^4y^{-3})^{-3} = x^{4(-3)}y^{-3(-3)} = x^{-12}y^9 = \dfrac{y^9}{x^{12}}$

**103.** $3.8^{-4} = 0.0048$

**105.** $100(1.355)^{-15} \approx 100(-0.010493) \approx 1.0493$

**107.** $4^{-2}\cdot4^3 = 4^{-2+3} = 4^1 = 4$

**109.** $x^{-4}\cdot x^6 = x^{-4+6} = x^2$

**111.** $u^{-6} \cdot u^3 = u^{-6+3} = u^{-3} = \dfrac{1}{u^3}$

**113.** $xy^{-3} \cdot y^2 = xy^{-3+2} = xy^{-1} = \dfrac{x}{y}$

**115.** $\dfrac{x^2}{x^{-3}} = x^2 \cdot x^3 = x^{2+3} = x^5$

Note: We could also use the rule $\dfrac{a^m}{a^n} = a^{m-n}$.

$$\dfrac{x^2}{x^{-3}} = x^{2-(-3)} = x^{2+3} = x^5$$

**119.** $\dfrac{x^{-4}}{x^{-2}} = \dfrac{x^2}{x^4} = \dfrac{1}{x^2}$

Note: There are several ways to work this problem.

$$\dfrac{x^{-4}}{x^{-2}} = x^{-4-(-2)} \qquad \text{or} \qquad \dfrac{x^{-4}}{x^{-2}} = \dfrac{1}{x^{-2} \cdot x^4} \qquad \text{or} \qquad \dfrac{x^{-4}}{x^{-2}} = x^{-4}x^2$$

$$= x^{-4+2} \qquad\qquad\qquad\quad = \dfrac{1}{x^{-2+4}} \qquad\qquad\qquad = x^{-4+2}$$

$$= x^{-2} \qquad\qquad\qquad\qquad = \dfrac{1}{x^2} \qquad\qquad\qquad\qquad = x^{-2}$$

$$= \dfrac{1}{x^2} \qquad\qquad\qquad\qquad\qquad\qquad\qquad\qquad\qquad = \dfrac{1}{x^2}$$

**121.** $\left(y^{-3}\right)^2 = y^{-6} = \dfrac{1}{y^6}$

**123.** $\left(s^2\right)^{-1} = s^{-2} = \dfrac{1}{s^2}$

**125.** $\left(2x^{-2}\right)^0 = 1$

**127.** $\dfrac{b^2 b^{-3}}{b^4} = \dfrac{b^{2+(-3)}}{b^4} = \dfrac{b^{-1}}{b^4} = b^{-1-4} = b^{-5} = \dfrac{1}{b^5}$

Note: Here is another way to work this problem.

$$\dfrac{b^2 b^{-3}}{b^4} = \dfrac{b^2}{b^3 b^4} = \dfrac{b^2}{b^{3+4}} = \dfrac{b^2}{b^7} = b^{2-7} = b^{-5} = \dfrac{1}{b^5}$$

**129.** $\left(3x^2 y\right)^{-2} = 3^{-2}\left(x^2\right)^{-2} y^{-2}$

$$= 3^{-2} x^{-4} y^{-2}$$

$$= \dfrac{1}{3^2 x^4 y^2}$$

$$= \dfrac{1}{9x^4 y^2}$$

**117.** $\dfrac{y^{-5}}{y} = \dfrac{1}{y \cdot y^5} = \dfrac{1}{y^{1+5}} = \dfrac{1}{y^6}$

Note: We could also use the rule $\dfrac{a^m}{a^n} = a^{m-n}$.

$$\dfrac{y^{-5}}{y} = y^{-5-1} = y^{-6} = \dfrac{1}{y^6}$$

**131.** $\left(-5a^{-2}b^3\right)^{-3} = (-5)^{-3} a^{(-2)(-3)} b^{3(-3)}$

$$= (-5)^{-3} a^6 b^{-9}$$

$$= \dfrac{a^6}{(-5)^3 b^9}$$

$$= \dfrac{a^6}{-125 b^9}$$

$$= -\dfrac{a^6}{125 b^9}$$

**133.** $\left(-2x^2\right)\left(4x^{-3}\right) = -8x^{2-3} = -8x^{-1} = -\dfrac{8}{x}$

**135.** $\left(\dfrac{x}{10}\right)^{-1} = \dfrac{x^{-1}}{10^{-1}} = \dfrac{10}{x}$

**137.** $\left(\dfrac{3z^2}{x}\right)^{-5} = \dfrac{3^{-5} z^{2(-5)}}{x^{-5}} = \dfrac{3^{-5} z^{-10}}{x^{-5}} = \dfrac{x^5}{3^5 z^{10}} = \dfrac{x^5}{243 z^{10}}$

**139.** $\dfrac{(2y)^{-4}}{(2y)^{-4}} = (2y)^{-4-(-4)} = (2y)^0 = 1$

**141.** $\dfrac{3}{2}\cdot\left(\dfrac{-2}{3}\right)^{-6} = \dfrac{3}{2}\cdot\dfrac{(-2)^{-6}}{3^{-6}}$

$= \dfrac{3}{2}\cdot\dfrac{3^6}{(-2)^6}$

$= \dfrac{3}{2}\cdot\dfrac{729}{64}$

$= \dfrac{3\cdot 729}{2\cdot 64}$

$= \dfrac{2187}{128}$

**143.** $\dfrac{(-2x)^{-3}}{-4x^{-2}} = \dfrac{(-2)^{-3}x^{-3}}{-4x^{-2}} = \dfrac{x^{-3-(-2)}}{(-4)(-2)^3} = \dfrac{x^{-1}}{-4(-8)} = \dfrac{1}{32x}$

**145.** $\left(5x^2y^4z^6\right)^3\left(5x^2y^4z^6\right)^{-3} = \left(5x^2y^4z^6\right)^{3+(-3)}$

$= \left(5x^2y^4z^6\right)^0$

$= 1$

**147.** $(x+y)^{-8}(x+y)^8 = (x+y)^{-8+8} = (x+y)^0 = 1$

**149.** $93,000,000 = 9.3\times 10^7$

**151.** $1,637,000,000 = 1.637\times 10^9$

**153.** $0.0000212 = 2.12\times 10^{-5}$

**155.** $0.004392 = 4.392\times 10^{-3}$

**157.** $0.0678 = 6.78\times 10^{-2}$

**159.** $1.09\times 10^6 = 1,090,000$

**161.** $8.67\times 10^2 = 867$

**163.** $8.52\times 10^{-3} = 0.00852$

**165.** $6.21\times 10^{-6} = 0.00000621$

**167.** $\left(8\times 10^3\right)+\left(3\times 10^0\right)+\left(5\times 10^{-2}\right) = 8000 + 3 + 0.05$

$= 8003.05$

**169.** $5,000,000\times 654,000 = \left(5\times 10^6\right)\left(6.54\times 10^5\right)$

$= 32.7\times 10^{11}$

$= 3.27\times 10^{12}$

**171.** $0.000495\times 7,840,000,000 = \left(4.95\times 10^{-4}\right)\left(7.84\times 10^9\right)$

$= 38.808\times 10^5$

$= 3.8808\times 10^6$

**173.** $\left(3.28\times 10^{-6}\right)^4 \approx 115.74\times 10^{-24} \approx 1.1574\times 10^{-22}$

**175.** $3,200,000^5 = \left(3.2\times 10^6\right)^5$

$\approx 335.54\times 10^{30}$

$\approx 3.3554\times 10^{32}$

**177.** $\dfrac{8.6\times 10^4}{3.9\times 10^7} = \left(\dfrac{8.6}{3.9}\right)\times 10^{-3} \approx 2.2051\times 10^{-3}$

**179.** $\dfrac{848,000,000}{1,620,000} = \dfrac{8.48\times 10^8}{1.62\times 10^6}$

$= \left(\dfrac{8.48}{1.62}\right)\times 10^2$

$\approx 5.2346\times 10^2$

**181.** $\left(4.85\times 10^5\right)\left(2.04\times 10^8\right) = (4.85)(2.04)\times 10^{13}$

$= 9.8940\times 10^{13}$

**183.** $25\times\left(5.8657\times 10^{12}\right) = 25\times 5.8657\times 10^{12}$

$\approx 146.64\times 10^{12}$

$\approx 1.4664\times 10^{14}$

The estimated distance from the Earth to the star is $1.4664\times 10^{14}$ miles.

**185.** $\dfrac{3.537\times 10^6}{2.97\times 10^8} = \dfrac{3.537}{2.97}\times 10^{-2} \approx 1.1909\times 10^{-2}$

The amount of land in the United States is $1.1909\times 10^{-2}$ square miles per person.

**187.** (a)

| $t$ | 0 | 2 | 4 | 6 | 8 |
|---|---|---|---|---|---|
| $22{,}000(1.2)^{-t}$ | \$22,000 | \$15,278 | \$10,610 | \$7368 | \$5116 |

(b)

Value (in dollars) vs. Year

(0, 22,000)
(2, 15,278)
(4, 10,610)
(6, 7368)
(8, 5116)

(c) The car will be valued at less than \$1000 when it is 17 years old.

**189.** $\boxed{\begin{array}{c}\text{Decreased}\\\text{volume}\end{array}} = \boxed{\begin{array}{c}\text{Original}\\\text{volume}\end{array}} - \boxed{\begin{array}{c}\text{Amount of}\\\text{decrease}\end{array}}$

$$v = 2 - 2(150)(20 \times 10^{-6})$$
$$= 2 - 6000 \times 10^{-6}$$
$$= 2 - 6 \times 10^{-3}$$
$$= 2 - 0.006$$
$$= 1.994$$

The original volume of 2 gallons is decreased by 0.006 gallon, so the decreased volume is 1.994 gallons.

**191.** False.

If $x = 5$ and $y = 2$,

$x^3 y^3 = 5^3 2^3 = 125(8) = 1000$, but

$xy^3 = 5(2^3) = 5(8) = 40$.

**193.** False. If $x = 5$ and $y = 2$.

$$x^{-1} + y^{-1} = 5^{-1} + 2^{-1} = \frac{1}{5} + \frac{1}{2} = \frac{7}{10}.$$

but $\dfrac{1}{x + y} = \dfrac{1}{5 + 2} = \dfrac{1}{7}.$

**195.** True. $\left(x \times 10^3\right)^4 = x^4 \times 10^{12}$

**197.** Yes

$$a^{-1}b^{-1} = a^{-1} \cdot b^{-1} = \frac{1}{a} \times \frac{1}{b} = \frac{1}{ab}$$

**199.** $4 \times 10^{-3} = \dfrac{4}{10^3}$

The reciprocal of $\dfrac{4}{10^3}$ is $\dfrac{10^3}{4}$, or 250.

**201.** $\left(3 \times 10^5\right)\left(4 \times 10^6\right) = \left(3 \times 10^5\right)\left(10^6 \times 4\right)$    Commutative Property of Multiplication

$= 3\left(10^5 \times 10^6\right)(4)$    Associative Property of Multiplication

$= 3\left(10^{5+6}\right)(4)$    Product Rule of Exponents

$= (3 \times 4)10^{11}$    Commutative Property of Multiplication

$= 12 \times 10^{11}$    Multiplication

$= 1.2 \times 10^{12}$    Scientific notation

**203.** $\dfrac{3xy^2}{4} \cdot \dfrac{8}{9} = \dfrac{\cancel{3}(\cancel{4})(2)xy^2}{\cancel{4}(\cancel{3})(3)} = \dfrac{2xy^2}{3}$

**205.** $-\left(\dfrac{2x}{y}\right)\left(\dfrac{2x}{y}\right) = -\dfrac{2 \cdot 2 \cdot x \cdot x}{y \cdot y} = -\dfrac{4x^2}{y^2}$

**207.**

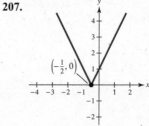

$\left(-\frac{1}{2}, 0\right)$

The graph appears to have an $x$-intercept at $\left(-\frac{1}{2}, 0\right)$ and a $y$-intercept at $(0, 1)$.

**209.**

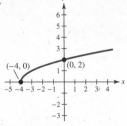

The graph appears to have an $x$-intercept at $(-4, 0)$ and a $y$-intercept at $(0, 2)$.

**211.**

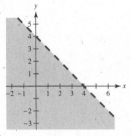

**213.**

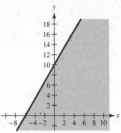

## Section 5.2   Adding and Subtracting Polynomials

**1.** Yes, the expression $9 - z$ is a polynomial.

**3.** The expression $p^{3/4} - 16$ is not a polynomial because the exponent in the first term is not an integer.

**5.** No, the expression $6x^{-1}$ is not a polynomial because the exponent is negative.

**7.** Yes, the expression $z^2 - 3z + \frac{1}{4}$ is a polynomial.

**9.** Polynomial: $12x + 9$

Standard Form: $12x + 9$

Degree: 1

Leading Coefficient: 12

**11.** Polynomial: $7x - 5x^2 + 10$

Standard Form: $-5x^2 + 7x + 10$

Degree: 2

Leading Coefficient: $-5$

**13.** Polynomial: $6m - 3m^5 - m^2 + 12$

Standard form: $-3m^5 - m^2 + 6m + 12$

Degree: 5

Leading Coefficient: $-3$

**15.** Polynomial: 10

Standard Form: 10

Degree: 0

Leading Coefficient: 10

**17.** Polynomial: $v_0 t - 16t^2$

Standard Form: $-16t^2 + v_0 t$

Degree: 2

Leading coefficient: $-16$

**19.** The polynomial $14y - 2$ has two terms; it is a binomial.

**21.** The polynomial $-32$ has one term; it is a monomial.

**23.** The polynomial $4x + 18x^2 - 5$ has three terms; it is a trinomial.

**25.** $-8x^3 + 5x$ or $5x^3 - 10$

**27.** $10x^2$ or $3y^2$

**29.** $x^6 - 4x^3 - 2$ or $10x^6 - x^5 + 3x^4$

**31.** $(4w + 5) + (16w - 9) = (4w + 16w) + (5 - 9)$
$$= 20w - 4$$

**33.** $(3z^2 - z + 2) + (z^2 - 4) = (3z^2 + z^2) + (-z) + (2 - 4)$
$$= 4z^2 - z - 2$$

**35.** $\left(\frac{2}{3}y^2 - \frac{3}{4}\right) + \left(\frac{5}{6}y^2 + 2\right) = \left(\frac{2}{3}y^2 + \frac{5}{6}y^2\right) + \left(-\frac{3}{4} + 2\right)$
$$= \left(\frac{4}{6}y^2 + \frac{5}{6}y^2\right) + \left(-\frac{3}{4} + \frac{8}{4}\right)$$
$$= \left(\frac{9}{6}y^2\right) + \left(\frac{5}{4}\right)$$
$$= \frac{3}{2}y^2 + \frac{5}{4}$$

**37.** $\left(0.1t^3 - 3.4t^2\right) + \left(1.5t^3 - 7.3\right) = \left(0.1t^3 + 1.5t^3\right) - 3.4t^2 - 7.3$

$$= 1.6t^3 - 3.4t^2 - 7.3$$

**39.** $b^2 + \left(b^3 - 2b^2 + 3\right) + \left(b^3 - 3\right) = \left(b^3 + b^3\right) + \left(b^2 - 2b^2\right) + (3 - 3)$

$$= 2b^3 - b^2$$

**41.** $(2ab - 3) + \left(a^2 - 2ab\right) + \left(4b^2 - a^2\right) = \left(a^2 - a^2\right) + (2ab - 2ab) + 4b^2 - 3 = 4b^2 - 3$

**43.**  $2x + 5$
$\phantom{xx}3x + 8$
$\overline{\phantom{xx}5x + 13}$

**45.**  $-2x + 10$
$\phantom{xxx}x - 38$
$\overline{\phantom{xxx}-x - 28}$

**47.**  $x^2 - 4$
$\phantom{xx}2x^2 + 6$
$\overline{\phantom{xx}3x^2 + 2}$

**49.**  $-x^3 \phantom{xxxxx} + 3$
$\phantom{xx}3x^3 + 2x^2 + 5$
$\overline{\phantom{xx}2x^3 + 2x^2 + 8}$

**51.** $3x^4 - 2x^3 - 4x^2 + 2x - 5$
$\phantom{xxxxxxxx}x^2 - 7x + 5$
$\overline{3x^4 - 2x^3 - 3x^2 - 5x}$

**53.**  $x^2 - 2x + 2$
$\phantom{xx}x^2 + 4x$
$\phantom{xx}2x^2$
$\overline{\phantom{xx}4x^2 + 2x + 2}$

**55.**  $-3y^3 + 5$
$\phantom{xx}8y^3 + 7$
$\overline{\phantom{xx}5y^3 + 12}$

**57.** $(11x - 8) - (2x + 3) = 11x - 8 - 2x - 3$

$$= (11x - 2x) + (-8 - 3)$$

$$= 9x - 11$$

**59.** $\left(x^2 - x\right) - (x - 2) = x^2 - x - x + 2$

$$= \left(x^2\right) + (-x - x) + 2$$

$$= x^2 - 2x + 2$$

**61.** $\left(4 - 2x - x^3\right) - \left(3 - 2x + 2x^3\right) = 4 - 2x - x^3 - 3 + 2x - 2x^3 = \left(-x^3 - 2x^3\right) + (-2x + 2x) + (4 - 3) = -3x^3 + 1$

**63.** $8 - \left(w^3 - w\right) = 8 - w^3 + w = -w^3 + w + 8$

**65.** $\left(x^5 - 3x^4 + x^3 - 5x + 1\right) - \left(4x^5 - x^3 + x - 5\right) = x^5 - 3x^4 + x^3 - 5x + 1 - 4x^5 + x^3 - x + 5$

$$= \left(x^5 - 4x^5\right) - 3x^4 + \left(x^3 + x^3\right) + (-5x - x) + (1 + 5)$$

$$= -3x^5 - 3x^4 + 2x^3 - 6x + 6$$

**67.**  $2x - 2 \Rightarrow 2x - 2$
$\phantom{xx}-(x - 1) \Rightarrow \underline{-x + 1}$
$\phantom{xxxxxxxxxx}x - 1$

**69.**  $2x^2 - x + 2 \Rightarrow 2x^2 - x + 2$
$\phantom{xx}-\left(3x^2 + x - 1\right) \Rightarrow \underline{-3x^2 - x + 1}$
$\phantom{xxxxxxxxxxxxx}-x^2 - 2x + 3$

**71.** $\phantom{xxx}-3x^3 - 4x^2 + 2x - 5 \Rightarrow \phantom{xxxx}-3x^3 - 4x^2 + 2x - 5$
$\underline{-\left(2x^4 + 2x^3 \phantom{xxxx}- 4x + 5\right)} \Rightarrow \underline{-2x^4 - 2x^3 \phantom{xxxx}+ 4x - 5}$
$\phantom{xxxxxxxxxxxxxxxxxxx}-2x^4 - 5x^3 - 4x^2 + 6x - 10$

**73.**  $-x^3 + 2 \Rightarrow -x^3 + 2$
$\phantom{xx}-\left(x^3 + 2\right) \Rightarrow \underline{-x^3 - 2}$
$\phantom{xxxxxxxxxxx}-2x^3$

**75.**  $4t^3 - 3t + 5 \Rightarrow 4t^3 \phantom{xxx}- 3t + 5$
$\phantom{xx}-\left(3t^2 - 3t - 10\right) \Rightarrow \underline{\phantom{xx}-3t^2 + 3t + 10}$
$\phantom{xxxxxxxxxxxxx}4t^3 - 3t^2 \phantom{xxx}+ 15$

**77.**
$$
\begin{array}{rcl}
7x^2 - x & \Rightarrow & 7x^2 - x \\
-(x^3 - 2x^2 + 10) & \Rightarrow & \underline{-x^3 + 2x^2 \qquad - 10} \\
& & -x^3 + 9x^2 - x - 10
\end{array}
$$

**79.**
$$
\begin{array}{rcl}
10x^3 \qquad + 15 & \Rightarrow & 10x^3 \qquad + 15 \\
-(7x^3 - 4x + 5) & \Rightarrow & \underline{-7x^3 + 4x - 5} \\
& & 3x^3 + 4x + 10
\end{array}
$$

**81.** $(6x - 5) - (8x + 15) = 6x - 5 - 8x - 15$
$$= (6x - 8x) + (-5 - 15)$$
$$= -2x - 20$$

**83.** $-(x^3 - 2) + (4x^3 - 2x) = -x^3 + 2 + 4x^3 - 2x$
$$= (-x^3 + 4x^3) + (-2x) + 2$$
$$= 3x^3 - 2x + 2$$

**85.** $2(x^4 + 2x) + (5x + 2) = 2x^4 + 4x + 5x + 2$
$$= 2x^4 + (4x + 5x) + 2$$
$$= 2x^4 + 9x + 2$$

**87.** $(15x^2 - 6) - (-8x^3 - 14x^2 - 17) = 15x^2 - 6 + 8x^3 + 14x^2 + 17$
$$= (8x^3) + (15x^2 + 14x^2) + (-6 + 17) = 8x^3 + 29x^2 + 11$$

**89.** $5z - \big[3z - (10z + 8)\big] = 5z - [3z - 10z - 8]$
$$= 5z - [-7z - 8]$$
$$= 5z + 7z + 8$$
$$= 12z + 8$$

**91.** $(y^3 + 1) - \big[(y^2 + 1) + (3y - 7)\big] = (y^3 + 1) - \big[y^2 + 1 + 3y - 7\big]$
$$= (y^3 + 1) - \big[y^2 + 3y + (1 - 7)\big]$$
$$= (y^3 + 1) - \big[y^2 + 3y - 6\big]$$
$$= y^3 + 1 - y^2 - 3y + 6$$
$$= y^3 - y^2 - 3y + (1 + 6)$$
$$= y^3 - y^2 - 3y + 7$$

**93.** $2(t^2 + 5) - 3(t^2 + 5) + 5(t^2 + 5) = 2t^2 + 10 - 3t^2 - 15 + 5t^2 + 25$
$$= (2t^2 - 3t^2 + 5t^2) + (10 - 15 + 25)$$
$$= 4t^2 + 20$$

**95.** $8v - 6(3v - v^2) + 10(10v + 3) = 8v - 18v + 6v^2 + 100v + 30$
$$= (6v^2) + (8v - 18v + 100v) + 30 = 6v^2 + 90v + 30$$

**97.** Perimeter $= 2z + 4z + 2x + z + 1 + 2 + 1 + z = 10z + 4$

**99.** *Verbal model:*   | Area of shaded region | = | Area of larger rectangle | − | Area of smaller rectangle |

*Labels:*   Area of shaded region = $A$

Length of larger rectangle = $2x$

Width of larger rectangle = $x$

Area of larger rectangle = $2x(x) = 2x^2$

Length of smaller rectangle = $4$

Width of smaller rectangle = $\dfrac{x}{2}$

Area of smaller rectangle = $4\left(\dfrac{x}{2}\right) = 2x$

*Solution:*   $A = 2x^2 - 2x$

The area of the shaded region is $2x^2 - 2x$.

Note: To find the area of each rectangle, we use the formula $A = lw$ or Area = (length)(width).

**101.** *Verbal model:*   | Area of shaded region | = | Area of larger rectangle | − | Area of smaller rectangle |

*Labels:*   Area of shaded region = $A$

Length of larger rectangle = $6x$

Width of larger rectangle = $\frac{7}{2}x$

Area of larger rectangle = $6x\left(\frac{7}{2}x\right)$

Length of smaller rectangle = $10$

Width of smaller rectangle = $\frac{4}{5}x$

Area of smaller rectangle = $10\left(\frac{4}{5}x\right)$

*Solution:*   $A = \left(6x\right)\left(\frac{7}{2}x\right) - \left(10\right)\left(\frac{4}{5}x\right) = 21x^2 - 8x$

The area of the shaded region is $21x^2 - 8x$.

Note: To find the area of each rectangle, we use the formula $A = lw$ or Area = (length)(width).

**103.** (a)  $T = P + R$

$\qquad = \left(0.53t + 58.1\right) + \left(0.007t^2 - 0.06t + 11.0\right),\ 10 \le t \le 15$

$\qquad = 0.53t + 58.1 + 0.007t^2 - 0.06t + 11.0,\ 10 \le t \le 15$

$\qquad = 0.007t^2 + 0.47t + 69.1,\ 10 \le t \le 15$

(b)  Enrollment in public schools $P$

Enrollment in private schools $R$

Total school enrollment $T$

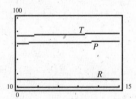

(c)  Public school enrollment: Increasing

Private school enrollment: Increasing

Total school enrollment: Increasing

**105.** The terms are like terms if they are both constants or if they have the same variable factor or factors. Like terms can differ only by their numerical coefficients.

**107.** Yes, a polynomial is an algebraic expression whose terms are all of the form $ax^k$, where $a$ is any real number and $k$ is a nonnegative integer.

**109.**
$$\frac{4x}{27} = \frac{8}{9}$$
$$4x(9) = 27(8)$$
$$36x = 216$$
$$x = \frac{216}{36}$$
$$x = \frac{\cancel{9}(\cancel{4})(3)(2)}{\cancel{9}(\cancel{4})}$$
$$x = 6$$

**111.**
$$\frac{x+3}{6} = \frac{2}{5}$$
$$5(x+3) = 6(2)$$
$$5x + 15 = 12$$
$$5x = -3$$
$$x = -\frac{3}{5}$$

**113.**

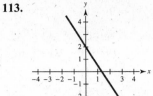

**115.** The exponent is $-5$.
$$2^{-5} = \frac{2}{2^5} = \frac{1}{32}$$

**117.**
$$\left(4.15 \times 10^3\right)^{-4} = (4.15)^{-4} \times 10^{-12}$$
$$\approx 0.0033714 \times 10^{-12}$$
$$\approx 3.3714 \times 10^{-15}$$

# Mid-Chapter Quiz for Chapter 5

**1.** $\left(4m^3n^2\right)^4 = 4^4 m^{3\cdot4} n^{2\cdot4} = 256m^{12}n^8$

**2.** $(-3xy)^2\left(2x^2y\right)^3 = (-3)^2 x^2 y^2 \cdot 2^3 x^{2\cdot3} y^3$
$$= 9x^2 y^2 \cdot 8x^6 y^3$$
$$= 72x^{2+6} y^{2+3} = 72x^8 y^5$$

**3.** $\dfrac{-12x^3 y}{9x^5 y^2} = \left(-\dfrac{12}{9}\right)x^{3-5} y^{1-2} = -\dfrac{4}{3} x^{-2} y^{-1} = -\dfrac{4}{3x^2 y}$

**4.** $\dfrac{3t^3}{(-6t)^2} = \dfrac{3t^3}{36t^2} = \left(\dfrac{3}{36}\right)t^{3-2} = \dfrac{1}{12} t^1 = \dfrac{t}{12}$

**5.** $5x^{-2} y^{-3} = \dfrac{5}{x^2 y^3}$

**6.** $\dfrac{3x^{-2} y}{5z^{-1}} = \dfrac{3yz}{5x^2}$

**7.** $\left(3a^{-3}b^2\right)^{-2} = 3^{-2} a^{-3(-2)} b^{2(-2)} = 3^{-2} a^6 b^{-4} = \dfrac{a^6}{3^2 b^4} = \dfrac{a^6}{9b^4}$

**8.** $\left(4t^{-3}\right)^0 = 1$

**9.** $8{,}168{,}000{,}000{,}000 = 8.168 \times 10^{12}$

**10.** $5.021 \times 10^{-8} = 0.00000005021$

**11.** The expression $x^2 + 2x - 3x^{-1}$ is not a polynomial because the exponent of the third term, $-3x^{-1}$, is negative.

**12.** Polynomial: $10 + x^2 - 4x^3$
Degree: 3
Leading coefficient: $-4$

**13.** $6x^5 + x^2 - 10$ or $3x^5 - 3x + 1$
Note: There are many correct answers.

**14.** $\left(y^2 + 3y - 1\right) + (4 + 3y) = y^2 + (3y + 3y) + (-1 + 4)$
$$= y^2 + 6y + 3$$

**15.** $\left(3v^2 - 5\right) - \left(v^3 + 2v^2 - 6v\right) = 3v^2 - 5 - v^3 - 2v^2 + 6v$
$$= -v^3 + v^2 + 6v - 5$$

**16.** $9s - [6 - (s - 5) + 7s] = 9s - [6 - s + 5 + 7s]$
$= 9s - 6 + s - 5 - 7s$
$= (9s + s - 7s) + (-6 - 5)$
$= 3s - 11$

**17.** $-3(4 - x) + 4(x^2 + 2) - (x^2 - 2x) = -12 + 3x + 4x^2 + 8 - x^2 + 2x$
$= (4x^2 - x^2) + (3x + 2x) + (-12 + 8) = 3x^2 + 5x - 4$

**18.**
$\begin{array}{r} 5x^4 \quad\quad + 2x^2 + x - 3 \\ + \quad 3x^3 - 2x^2 - 3x + 5 \\ \hline 5x^4 + 3x^3 \quad\quad - 2x + 2 \end{array}$

**22.** Perimeter $= 5x + (18 - 2x) + 3x + 2x + 2x + 18$
$= 5x + 18 - 2x + 3x + 2x + 2x + 18$
$= (5x - 2x + 3x + 2x + 2x) + (18 + 18)$
$= 10x + 36$

**19.**
$\begin{array}{r} x^3 - 3x^2 \quad\quad - 15 \\ 2x^2 + 5x - 4 \\ \hline x^3 - x^2 + 5x - 19 \end{array}$

**20.** $x^2 - x + 2 \Rightarrow x^2 - x + 2$
$\underline{-(x - 4)} \Rightarrow \underline{\quad -x + 4}$
$x^2 - 2x + 6$

**21.** $6x^4 + 3x^3 + 8 \Rightarrow 6x^4 + 3x^3 \quad\quad + 8$
$\underline{-(x^4 + 4x^2 + 2)} \Rightarrow \underline{-x^4 \quad\quad - 4x^2 - 2}$
$5x^4 + 3x^3 - 4x^2 + 6$

## Section 5.3   Multiplying Polynomials: Special Products

**1.** $x(-2x) = -2x^2$

**3.** $t^2(4t) = 4t^3$

**5.** $\left(\dfrac{x}{4}\right)(10x) = \dfrac{5}{2}x^2$ or $\dfrac{5x^2}{2}$

**7.** $(-2b^2)(-3b) = 6b^3$

**9.** $y(3 - y) = (y)(3) - (y)(y)$
$= 3y - y^2$ or $-y^2 + 3y$

**11.** $-x(x^2 - 4) = (-x)(x^2) - (-x)(4)$
$= -x^3 + 4x$

**13.** $-3x(2x^2 + 5) = -3x(2x^2) + (-3x)(5)$
$= -6x^3 - 15x$

**15.** $-4x(3 + 3x^2 - 6x^3) = (-4x)(3) + (-4x)(3x^2) - (-4x)(6x^3)$
$= -12x - 12x^3 + 24x^4$ or $24x^4 - 12x^3 - 12x$

**17.** $3x(x^2 - 2x + 1) = (3x)(x^2) - (3x)(2x) + (3x)(1)$
$= 3x^3 - 6x^2 + 3x$

**21.** $4t^3(t - 3) = (4t^3)(t) - (4t^3)(3)$
$= 4t^4 - 12t^3$

**19.** $2x(x^2 - 2x + 8) = (2x)(x^2) - (2x)(2x) + (2x)(8)$
$= 2x^3 - 4x^2 + 16x$

**23.** $x^2(4x^2 - 3x + 1) = (x^2)(4x^2) - (x^2)(3x) + (x^2)(1)$
$= 4x^4 - 3x^3 + x^2$

**25.** $-3x^3(4x^2 - 6x + 2) = -3x^3(4x^2) - (-3x^3)(6x) + (-3x^3)(2) = -12x^5 + 18x^4 - 6x^3$

**27.** $-2x(-3x)(5x + 2) = [(-2x)(-3x)](5x + 2) = (6x^2)(5x + 2) = (6x^2)(5x) + (6x^2)(2) = 30x^3 + 12x^2$

**29.** $-2x(-6x^4) - 3x^2(2x^2) = 12x^5 - 6x^4$

**31.**

$$\quad\quad\quad\quad\text{F}\quad\text{O}\quad\text{I}\quad\text{L}$$
$$(x + 3)(x + 4) = x^2 + 4x + 3x + 12$$
$$= x^2 + 7x + 12$$

**33.**

$$\quad\quad\quad\quad\text{F}\quad\text{O}\quad\text{I}\quad\text{L}$$
$$(3x - 5)(x + 1) = 3x^2 + 3x - 5x - 5$$
$$= 3x^2 - 2x - 5$$

**35.**

$$\quad\quad\quad\quad\text{F}\quad\text{O}\quad\text{I}\quad\text{L}$$
$$(2x - y)(x - 2y) = 2x^2 - 4xy - xy + 2y^2$$
$$= 2x^2 - 5xy + 2y^2$$

**37.**

$$\quad\quad\quad\text{F}\quad\text{O}\quad\text{I}\quad\text{L}$$
$$(5x + 6)(3x + 1) = 15x^2 + 5x + 18x + 6$$
$$= 15x^2 + 23x + 6$$

**39.**

$$\quad\quad\quad\text{F}\quad\text{O}\quad\text{I}\quad\text{L}$$
$$(6 - 2x)(4x + 3) = 24x + 18 - 8x^2 - 6x$$
$$= -8x^2 + 18x + 18$$

**41.**

$$\quad\quad\quad\quad\text{F}\quad\text{O}\quad\text{I}\quad\text{L}$$
$$(3x - 2y)(x - y) = 3x^2 - 3xy - 2xy + 2y^2$$
$$= 3x^2 - 5xy + 2y^2$$

**43.**

$$\quad\quad\quad\quad\text{F}\quad\text{O}\quad\text{I}\quad\text{L}$$
$$(3x^2 - 4)(x + 2) = 3x^3 + 6x^2 - 4x - 8$$

**45.**

$$\quad\quad\quad\quad\quad\quad\text{F}\quad\text{O}\quad\text{I}\quad\text{L}$$
$$(2x + 4)^2 = (2x + 4)(2x + 4) = 4x^2 + 8x + 8x + 16$$
$$= 4x^2 + 16x + 16$$

**47.**

$$\quad\quad\quad\quad\quad\quad\text{F}\quad\text{O}\quad\text{I}\quad\text{L}$$
$$(8x + 2)^2 = (8x + 2)(8x + 2) = 64x^2 + 16x + 16x + 4$$
$$= 64x^2 + 32x + 4$$

**49.** $(3s + 1)(3s + 4) - (3s)^2 = 9s^2 + 12s + 3s + 4 - 9s^2$
$$= 15s + 4$$

**51.** $(4x^2 - 1)(2x + 8) + (-x^2)^3 = 8x^3 + 32x^2 - 2x - 8 - x^6 = -x^6 + 8x^3 + 32x^2 - 2x - 8$

**53.** $(x + 10)(x + 2) = x(x + 2) + 10(x + 2)$
$$= x^2 + 2x + 10x + 20$$
$$= x^2 + 12x + 20$$

**55.** $(2x - 5)(7x + 2) = 2x(7x + 2) - 5(7x + 2)$
$$= 14x^2 + 4x - 35x - 10$$
$$= 14x^2 - 31x - 10$$

**57.** $(x + 1)(x^2 + 2x - 1) = x(x^2 + 2x - 1) + 1(x^2 + 2x - 1)$
$$= x^3 + 2x^2 - x + x^2 + 2x - 1$$
$$= x^3 + 3x^2 + x - 1$$

**59.** $(x^3 - 2x + 1)(x - 5) = x^3(x - 5) - 2x(x - 5) + 1(x - 5)$
$$= x^4 - 5x^3 - 2x^2 + 10x + x - 5$$
$$= x^4 - 5x^3 - 2x^2 + 11x - 5$$

**61.** $(x - 2)(5x^2 + 2x + 4) = x(5x^2 + 2x + 4) - 2(5x^2 + 2x + 4)$
$$= 5x^3 + 2x^2 + 4x - 10x^2 - 4x - 8$$
$$= 5x^3 - 8x^2 - 8$$

**63.** $(x^2 + 3)(x^2 - 6x + 2) = x^2(x^2 - 6x + 2) + 3(x^2 - 6x + 2)$
$$= x^4 - 6x^3 + 2x^2 + 3x^2 - 18x + 6$$
$$= x^4 - 6x^3 + 5x^2 - 18x + 6$$

**65.** $(3x^2 - 4x - 2)(3x^2 + 1) = 3x^2(3x^2 + 1) - 4x(3x^2 + 1) - 2(3x^2 + 1)$
$$= 9x^4 + 3x^2 - 12x^3 - 4x - 6x^2 - 2$$
$$= 9x^4 - 12x^3 - 3x^2 - 4x - 2$$

**67.**
$$
\begin{array}{r}
x + 3 \\
\times \quad x - 2 \\
\hline
-2x - 6 \\
x^2 + 3x \quad\quad \\
\hline
x^2 + x - 6
\end{array}
$$

**69.**
$$
\begin{array}{r}
4x^4 \quad - 6x^2 \quad + 9 \\
\times \quad\quad\quad\quad 2x + 3 \\
\hline
12x^4 \quad - 18x^2 \quad + 27 \\
8x^5 \quad - 12x^3 \quad + 18x \quad\quad \\
\hline
8x^5 + 12x^4 - 12x^3 - 18x^2 + 18x + 27
\end{array}
$$

**75.** $\left(x + 3 - 2x^2\right)\left(5x + x^2 - 4\right) = \left(-2x^2 + x + 3\right)\left(x^2 + 5x - 4\right)$

$$
\begin{array}{r}
-2x^2 + x + 3 \\
\times \quad\quad x^2 + 5x - 4 \\
\hline
8x^2 - 4x - 12 \\
-10x^3 + 5x^2 + 15x \quad\quad \\
-2x^4 + x^3 + 3x^2 \quad\quad\quad\quad \\
\hline
-2x^4 - 9x^3 + 16x^2 + 11x - 12
\end{array}
$$

**77.**
$$
\begin{array}{r}
x - 2 \\
\times \quad x - 2 \\
\hline
-2x + 4 \\
x^2 - 2x \quad\quad \\
\hline
x^2 - 4x + 4
\end{array}
$$

$(x - 2)^3 = (x - 2)^2(x - 2) = \left(x^2 - 4x + 4\right)(x - 2)$

$$
\begin{array}{r}
x^2 - 4x + 4 \\
\times \quad\quad x - 2 \\
\hline
-2x^2 + 8x - 8 \\
x^3 - 4x^2 + 4x \quad\quad \\
\hline
x^3 - 6x^2 + 12x - 8
\end{array}
$$

**79.**
$$
\begin{array}{r}
x - 1 \\
\times \quad x - 1 \\
\hline
-x + 1 \\
x^2 - x \quad\quad \\
\hline
x^2 - 2x + 1
\end{array}
$$

$(x - 1)^2(x - 1)^2 = \left(x^2 - 2x + 1\right)\left(x^2 - 2x + 1\right)$

$$
\begin{array}{r}
x^2 - 2x + 1 \\
\times \quad x^2 - 2x + 1 \\
\hline
x^2 - 2x + 1 \\
-2x^3 + 4x^2 - 2x \quad\quad \\
x^4 - 2x^3 + x^2 \quad\quad\quad\quad \\
\hline
x^4 - 4x^3 + 6x^2 - 4x + 1
\end{array}
$$

**71.**
$$
\begin{array}{r}
3x^3 \quad\quad + x + 7 \\
\times \quad\quad\quad x^2 \quad + 1 \\
\hline
3x^3 \quad\quad + x + 7 \\
3x^5 + x^3 + 7x^2 \quad\quad\quad \\
\hline
3x^5 + 4x^3 + 7x^2 + x + 7
\end{array}
$$

**73.**
$$
\begin{array}{r}
x^2 + x - 2 \\
\times \quad x^2 - x + 2 \\
\hline
2x^2 + 2x - 4 \\
-x^3 - x^2 + 2x \quad\quad \\
x^4 + x^3 - 2x^2 \quad\quad\quad\quad \\
\hline
x^4 \quad - x^2 + 4x - 4
\end{array}
$$

**81.**
$$
\begin{array}{r}
x + 2 \\
\times \quad x + 2 \\
\hline
2x + 4 \\
x^2 + 2x \quad\quad \\
\hline
x^2 + 4x + 4
\end{array}
$$

$(x + 2)^2(x - 4) = \left(x^2 + 4x + 4\right)(x - 4)$

$$
\begin{array}{r}
x^2 + 4x + 4 \\
\times \quad\quad x - 4 \\
\hline
-4x^2 - 16x - 16 \\
x^3 + 4x^2 + 4x \quad\quad \\
\hline
x^3 \quad\quad - 12x - 16
\end{array}
$$

**83.**
$$
\begin{array}{r}
u - 1 \\
\times \quad 2u + 3 \\
\hline
3u - 3 \\
2u^2 - 2u \quad\quad \\
\hline
2u^2 + u - 3
\end{array}
$$

$(u - 1)(2u + 3)(2u + 1) = \left(2u^2 + u - 3\right)(2u + 1)$

$$
\begin{array}{r}
2u^2 + u - 3 \\
\times \quad\quad 2u + 1 \\
\hline
2u^2 + u - 3 \\
4u^3 + 2u^2 - 6u \quad\quad \\
\hline
4u^3 + 4u^2 - 5u - 3
\end{array}
$$

**85.** $(x + 3)(x - 3) = x^2 - 3^2 = x^2 - 9$

**87.** $(x + 20)(x - 20) = x^2 - 20^2 = x^2 - 400$

**89.** $(2u + 3)(2u - 3) = (2u)^2 - (3)^2 = 4u^2 - 9$

**91.** $(4t - 6)(4t + 6) = (4t)^2 - 6^2 = 16t^2 - 36$

**93.** $\left(2x^2 + 5\right)\left(2x^2 - 5\right) = \left(2x^2\right)^2 - 5^2 = 4x^4 - 25$

**95.** $\left(4x + y\right)\left(4x - y\right) = \left(4x\right)^2 - \left(y\right)^2 = 16x^2 - y^2$

**97.** $\left(9u + 7v\right)\left(9u - 7v\right) = \left(9u\right)^2 - \left(7v\right)^2 = 81u^2 - 49v^2$

**99.** $\left(x + 6\right)^2 = \left(x\right)^2 + 2\left(x\right)\left(6\right) + \left(6\right)^2 = x^2 + 12x + 36$

**101.** $\left(t - 3\right)^2 = t^2 - 2\left(t\right)\left(3\right) + \left(3\right)^2 = t^2 - 6t + 9$

**103.** $\left(3x + 2\right)^2 = \left(3x\right)^2 + 2\left(3x\right)\left(2\right) + 2^2 = 9x^2 + 12x + 4$

**105.** $\left(8 - 3x\right)^2 = \left(8\right)^2 - 2\left(8\right)\left(3z\right) + \left(3z\right)^2$

$$= 64 - 48z + 9z^2 \text{ or } 9z^2 - 48z + 64$$

**107.** $\left(4 + 7s^2\right)^2 = \left(4\right)^2 + 2\left(4\right)\left(7s^2\right) + \left(7s^2\right)^2$

$$= 16 + 56s^2 + 49s^4$$

**109.** $\left(2x - 5y\right)^2 = \left(2x\right)^2 - 2\left(2x\right)\left(5y\right) + \left(5y\right)^2$

$$= 4x^2 - 20xy + 25y^2$$

**111.** $\left[\left(x + 1\right) + y\right]^2 = \left(x + 1\right)^2 + 2\left(x + 1\right)\left(y\right) + \left(y\right)^2$

$$= \left(x + 1\right)^2 + 2y\left(x + 1\right) + y^2$$

$$= x^2 + 2x + 1 + 2y\left(x + 1\right) + y^2$$

$$= x^2 + 2x + 1 + 2xy + 2y + y^2$$

$$\text{or } x^2 + y^2 + 2xy + 2x + 2y + 1$$

**113.** $\left[u - \left(3 + v\right)\right]^2 = \left(u\right)^2 - 2\left(u\right)\left(3 + v\right) + \left(3 + v\right)^2$

$$= u^2 - 2u\left(3 + v\right) + \left(3 + v\right)^2$$

$$= u^2 - 6u - 2uv + 3^2 + 2\left(3\right)\left(v\right) + v^2$$

$$= u^2 + v^2 - 2uv - 6u + 6v + 9$$

**115.** $\left(x + 2\right)^2 - \left(x - 2\right)^2 = \left[\left(x\right)^2 + 2\left(x\right)\left(2\right) + \left(2\right)^2\right] - \left[\left(x\right)^2 - 2\left(x\right)\left(2\right) + \left(2\right)^2\right]$

$$= \left[x^2 + 4x + 4\right] - \left[x^2 - 4x + 4\right] = x^2 + 4x + 4 - x^2 + 4x - 4 = 8x$$

**117.** Yes, this is an identity.

$$\left(x + y\right)^3 = \left(x + y\right)\left(x + y\right)\left(x + y\right)$$

$$= \left(x^2 + 2xy + y^2\right)\left(x + y\right)$$

$$= x^3 + x^2y + 2x^2y + 2xy^2 + xy^2 + y^3$$

$$= x^3 + 3x^2y + 3xy^2 + y^3$$

**119.** $\left(x + 2\right)^3 = \left(x\right)^3 + 3\left(x\right)^2\left(2\right) + 3\left(x\right)\left(2\right)^2 + \left(2\right)^3$

$$= x^3 + 6x^2 + 12x + 8$$

*Pattern:* $\left(a + b\right)^3 = a^3 + 3a^2b + 3ab^2 + b^3$

**121.** Area $A = \frac{1}{2}\left(\text{Base}\right)\left(\text{Height}\right)$

$$= \frac{1}{2}\left(2x\right)\left(x + 10\right) = x\left(x + 10\right) = x^2 + 10x$$

**123.** *Verbal model:* $\boxed{\text{Area of new sandbox}} = \boxed{\text{Area of original sandbox}} + \boxed{26}$

*Labels:*    Length of side of original sandbox $= x$ (feet)

Area of original sandbox $= x^2$ (square feet)

Area of new sandbox $= \left(x + 2\right)\left(x + 3\right)$ (square feet)

*Equation:*    $\left(x + 2\right)\left(x + 3\right) = x^2 + 26$

$$x^2 + 3x + 2x + 6 = x^2 + 26$$

$$x^2 + 5x + 6 - x^2 = x^2 + 26 - x^2$$

$$5x + 6 = 26$$

$$5x = 20$$

$$x = 4$$

The dimensions of the original sandbox are 4 feet by 4 feet.

**125.** $x^2 + 5x + 4 = \left(x + 4\right)\left(x + 1\right)$

**127.** Area = Area 1 + Area 2 + Area 3 + Area 4

$$(x + 2)^2 = (x)(x) + x(2) + x(2) + (2)(2)$$

$$(x + 2)^2 = x^2 + 2x + 2x + 4$$

$$(x + 2)^2 = x^2 + 4x + 4$$

The geometric model represents the special product.

$$(a + b)^2 = a^2 + 2ab + b^2.$$

**129.** Area = (Width)(Length)

The width is $x + 5$, and the length is $x + 4$.

$$(x + 5)(x + 4) = x^2 + 4x + 5x + 20 = x^2 + 9x + 20$$

**131. (a)** $T = MP$

$$T = (-0.31t + 25.6)(-0.034t^2 + 3.72t + 248.6)$$

$$T = -0.31t(-0.034t^2 + 3.72t + 248.6) + 25.6(-0.034t^2 + 3.72t + 248.6)$$

$$T = 0.01054t^3 - 1.1532t^2 - 77.066t - 0.8704t^2 + 95.232t + 6364.16$$

$$T = 0.01054t^3 - 2.0236t^2 + 18.166t + 6364.16, \quad 6 \le t \le 15$$

**(b)**

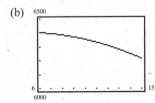

**(c)** The total consumption of milk in 2004 was approximately 6251 million gallons.

**133.** $500(1 + r)^2 = 500\left[(1)^2 + 2(1)(r) + (r)^2\right] = 500\left[1 + 2r + r^2\right] = 500 + 1000r + 500r^2$ or $500r^2 + 1000r + 500$

**135.** Multiplying a polynomial by a monomial is a direct application of the Distributive Property. Multiplying two polynomials requires repeated applications of the Distributive Property.

$$3x(x + 4) = 3x^2 + 12x$$

$$(5x + 2)(x - 6) = 5x(x - 6) + 2(x - 6)$$

$$= 5x^2 - 30x + 2x - 12$$

$$= 5x^2 - 28x - 12$$

**137.** The product of the terms of highest degree in each polynomial will be of the form $(ax^m)(bx^n) = abx^{m+n}$.

This will be the term of highest degree in the product, and therefore the degree of the product is $m + n$.

**139.** False, the product of two binomials is not always a binomial. For example, $(x - 2)(x + 3) = x^2 + x - 6$.

**141.** $(12x - 3) + (3x - 4) = 12x - 3 + 3x - 4 = 15x - 7$

**143.** $(-8x + 11) - (-4x - 6) = -8x + 11 + 4x + 6$

$$= -4x + 17$$

**145.** *Verbal model:* $\boxed{\text{What number}} = \boxed{25\% \text{ of } 45}$

*Label:*  Unknown number = $a$

*Equation:*  $a = 0.25(45)$

$a = 11.25$

The number that is 25% of 45 is 11.25.

**147.** *Verbal model:* $\boxed{20} = \boxed{\text{what percent of } 60}$

*Label:*  Unknown percent (in decimal form) = $p$

*Equation:*  $20 = p(60)$

$\frac{20}{60} = p$

$0.333\ldots = p$

20 is $33\frac{1}{3}\%$ of 60.

**149.** $\frac{2}{5} = \frac{x}{10}$

$20 = 5x$

$4 = x$

**151.** $\dfrac{z}{6} = \dfrac{5}{8}$

$8z = 30$

$z = \dfrac{30}{8}$

$z = \dfrac{15}{4}$

# Section 5.4   Dividing Polynomials

**1.** $\dfrac{3z+3}{3} = \dfrac{3z}{3} + \dfrac{3}{3} = z+1$

**3.** $\dfrac{4z-8}{4} = \dfrac{4z}{4} - \dfrac{8}{4} = z-2$

**5.** $\dfrac{5x+5}{-5} = \dfrac{5x}{-5} + \dfrac{5}{-5} = -x-1$

**7.** $\dfrac{9x-5}{3} = \dfrac{9x}{3} - \dfrac{5}{3} = 3x - \dfrac{5}{3}$

**9.** $\dfrac{8a+5}{-4} = \dfrac{8a}{-4} + \dfrac{5}{-4} = -2a - \dfrac{5}{4}$

**11.** $\dfrac{b^2-2b}{b} = \dfrac{b^2}{b} - \dfrac{2b}{b} = b-2$

**13.** $(5x^2 - 2x) \div x = \dfrac{5x^2}{x} - \dfrac{2x}{x} = 5x-2$

**15.** $\dfrac{25z^3+10z^2}{-5z} = \dfrac{25z^3}{-5z} + \dfrac{10z^2}{-5z} = -5z^2 - 2z$

**17.** $\dfrac{8z^3+3z^2-2z}{2z} = \dfrac{8z^3}{2z} + \dfrac{3z^2}{2z} - \dfrac{2z}{2z} = 4z^2 + \dfrac{3z}{2} - 1$

**19.** $\dfrac{4x^2-12x}{4x^2} = \dfrac{4x^2}{4x^2} - \dfrac{12x}{4x^2} = 1 - \dfrac{3}{x}$

**21.** $\dfrac{8x^2+12x^3}{-4x^3} = \dfrac{8x^2}{-4x^3} + \dfrac{12x^3}{-4x^3} = -\dfrac{2}{x} - 3$

**23.** $\dfrac{6x^4-2x^3+3x^2-x+4}{2x^3} = \dfrac{6x^4}{2x^3} - \dfrac{2x^3}{2x^3} + \dfrac{3x^2}{2x^3} - \dfrac{x}{2x^3} + \dfrac{4}{2x^3} = 3x - 1 + \dfrac{3}{2x} - \dfrac{1}{2x^2} + \dfrac{2}{x^3}$

**25.** $\dfrac{15x^5+10x^4-8x^2-3x}{-5x^2} = \dfrac{15x^5}{-5x^2} + \dfrac{10x^4}{-5x^2} + \dfrac{-8x^2}{-5x^2} + \dfrac{-3x}{-5x^2} = -3x^3 - 2x^2 + \dfrac{8}{5} + \dfrac{3}{5x}$

**27.**
$$\begin{array}{r} x-2 \\ x+1\overline{)x^2-x-2} \\ \underline{x^2+x} \\ -2x-2 \\ \underline{-2x-2} \\ 0 \end{array}$$

So, $\dfrac{x^2-x-2}{x+1} = x-2$.

**29.**
$$\begin{array}{r} x+5 \\ x+4\overline{)x^2+9x+20} \\ \underline{x^2+4x} \\ 5x+20 \\ \underline{5x+20} \\ 0 \end{array}$$

So, $(x^2+9x+20) \div (x+4) = x+5$.

**31.**
$$\begin{array}{r} x+4 \\ x+4\overline{)x^2+8x+7} \\ \underline{x^2+4x} \\ 4x+7 \\ \underline{4x+16} \\ -9 \end{array}$$

So, $\dfrac{x^2+8x+7}{x+4} = x+4 - \dfrac{9}{x+4}$.

**33.**
$$\begin{array}{r} z-3 \\ 2z+1\overline{)2z^2-5z-3} \\ \underline{2z^2+z} \\ -6z-3 \\ \underline{-6z-3} \\ 0 \end{array}$$

So, $\dfrac{2z^2-5z-3}{2z+1} = z-3$.

**35.**
$$\begin{array}{r} y + 1 \\ 3y+1\overline{\smash)3y^2 + 4y - 4} \\ \underline{3y^2 + \phantom{0}y} \\ 3y - 4 \\ \underline{3y + 1} \\ -5 \end{array}$$

So, $\dfrac{3y^2 + 4y - 4}{3y + 1} = y + 1 - \dfrac{5}{3y + 1}$.

**37.**
$$\begin{array}{r} 6t + 1 \\ 3t-4\overline{\smash)18t^2 - 21t - 4} \\ \underline{18t^2 - 24t} \\ 3t - 4 \\ \underline{3t - 4} \\ 0 \end{array}$$

So, $\dfrac{18t^2 - 21t - 4}{3t - 4} = 6t + 1$.

**39.**
$$\begin{array}{r} x^2 - 2x + 5 + \dfrac{3}{x-2} \\ x-2\overline{\smash)x^3 - 4x^2 + 9x - 7} \\ \underline{x^3 - 2x^2} \\ -2x^2 + 9x \\ \underline{-2x^2 + 4x} \\ 5x - 7 \\ \underline{5x - 10} \\ 3 \end{array}$$

So, $\dfrac{x^3 - 4x^2 + 9x - 7}{x - 2} = x^2 - 2x + 5 + \dfrac{3}{x - 2}$.

**41.** $\left(9x^4 + 7x^2 - 12x^3 + 9 + 10x\right) \div (-3x + 5) = \left(9x^4 - 12x^3 + 7x^2 + 10x + 9\right) \div (-3x + 5)$

$$\begin{array}{r} -3x^3 - x^2 - 4x - 10 \\ -3x+5\overline{\smash)9x^4 - 12x^3 + 7x^2 + 10x + 9} \\ \underline{9x^4 - 15x^3} \\ 3x^3 + 7x^2 \\ \underline{3x^3 - 5x^2} \\ 12x^2 + 10x \\ \underline{12x^2 - 20x} \\ 30x + \phantom{0}9 \\ \underline{30x - 50} \\ 59 \end{array}$$

So, $\dfrac{9x^4 - 12x^3 + 7x^2 + 10x + 9}{-3x + 5} = -3x^3 - x^2 - 4x - 10 + \dfrac{59}{-3x + 5}$.

**43.**
$$\begin{array}{r} 7 - \dfrac{11}{x+2} \\ x+2\overline{\smash)7x + 3} \\ \underline{7x + 14} \\ -11 \end{array}$$

So, $\dfrac{7x + 3}{x + 2} = 7 - \dfrac{11}{x + 2}$.

**45.**
$$\begin{array}{r} x^2 + 2x + 4 \\ x-2\overline{\smash)x^3 + 0x^2 + 0x - 8} \\ \underline{x^3 - 2x^2} \\ 2x^2 \\ \underline{2x^2 - 4x} \\ 4x - 8 \\ \underline{4x - 8} \\ 0 \end{array}$$

So, $\dfrac{x^3 - 8}{x - 2} = x^2 + 2x + 4$.

**47.**

$$
x + 4 \overline{) 3x^4 \qquad - 40x^2 + 28x - 18}
$$

quotient: $3x^3 - 12x^2 + 8x - 4$

$$
\begin{array}{r}
\underline{3x^4 + 12x^3} \\
-12x^3 - 40x^2 \\
\underline{-12x^3 - 48x^2} \\
8x^2 + 28x \\
\underline{8x^2 + 32x} \\
-4x - 18 \\
\underline{-4x - 16} \\
-2
\end{array}
$$

So, $\dfrac{3x^4 - 40x^2 + 28x - 18}{x + 4} = 3x^3 - 12x^2 + 8x - 4 - \dfrac{2}{x + 4}$.

**49.**

$$
3x + 1 \overline{) 9x^2 + 0x - 1}
$$

quotient: $3x - 1$

$$
\begin{array}{r}
\underline{9x^2 + 3x} \\
-3x - 1 \\
\underline{-3x - 1} \\
0
\end{array}
$$

So, $\dfrac{9x^2 - 1}{3x + 1} = 3x - 1$.

**51.**

$$
x - 1 \overline{) x^4 + 0x^3 + 0x^2 + 0x - 1}
$$

quotient: $x^3 + x^2 + x + 1$

$$
\begin{array}{r}
\underline{x^4 - x^3} \\
x^3 \\
\underline{x^3 - x^2} \\
x^2 \\
\underline{x^2 - x} \\
x - 1 \\
\underline{x - 1} \\
0
\end{array}
$$

So, $\dfrac{x^4 - 1}{x - 1} = x^3 + x^2 + x + 1$.

**53.** $\dfrac{4x^3}{x^2} - \dfrac{8x}{4} = 4x - 2x = 2x$

**55.** $\dfrac{25x^2}{10x} + \dfrac{3x}{2} = \dfrac{5x}{2} + \dfrac{3x}{2} = \dfrac{8x}{2} = 4x$

**57.** $\dfrac{8u^2v}{2u} + \dfrac{(uv)^2}{uv} = \dfrac{8u^2v}{2u} + \dfrac{u^2v^2}{uv} = 4uv + uv = 5uv$

**59.** $\dfrac{9x^5y}{3x^4} - \dfrac{(x^2y)^3}{x^5y^2} = \dfrac{9x^5y}{3x^4} - \dfrac{x^6y^3}{x^5y^2} = 3xy - xy = 2xy$

**61.**

$$
x + 1 \overline{) x^2 + 2x + 1}
$$

quotient: $x + 1$

$$
\begin{array}{r}
\underline{x^2 + x} \\
x + 1 \\
\underline{x + 1} \\
0
\end{array}
$$

$\dfrac{x^2 + 2x + 1}{x + 1} - (3x - 4) = x + 1 - (3x - 4)$

$\qquad\qquad\qquad = x + 1 - 3x + 4 = -2x + 5$

**63.**

$$
x - 1 \overline{) x^2 - 3x + 2}
$$

quotient: $x - 2$

$$
\begin{array}{r}
\underline{x^2 - x} \\
-2x + 2 \\
\underline{-2x + 2} \\
0
\end{array}
$$

$\dfrac{x^2 - 3x + 2}{x - 1} + (4x - 3) = x - 2 + 4x - 3 = 5x - 5$

**65.** $(\text{Length})(\text{Width}) = \text{Area}; \ \text{Length} = \dfrac{\text{Area}}{\text{Width}}$

$$
x - 1 \overline{) x^2 + 5x - 6}
$$

quotient: $x + 6$

$$
\begin{array}{r}
\underline{x^2 - x} \\
6x - 6 \\
\underline{6x - 6} \\
0
\end{array}
$$

$\text{Length} = \dfrac{x^2 + 5x - 6}{x - 1} = x + 6$

**67.** (a)  Yes, the graphs are the same.

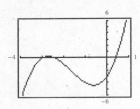

(b)  $(x + 3)(x^2 + 2x - 1) = x(x^2 + 2x - 1) + 3(x^2 + 2x - 1) = x^3 + 2x^2 - x + 3x^2 + 6x - 3 = x^3 + 5x^2 + 5x - 3$

(c)
$$
\begin{array}{r}
x^2 + 2x - 1 \\
x + 3 {\overline{\smash{\big)}\,x^3 + 5x^2 + 5x - 3}} \\
\underline{x^3 + 3x^2\phantom{{}+5x-3}} \\
2x^2 + 5x\phantom{{}-3} \\
\underline{2x^2 + 6x\phantom{{}-3}} \\
-x - 3 \\
\underline{-x - 3} \\
\end{array}
$$

$$\frac{x^3 + 5x^2 + 5x - 3}{x + 3} = x^2 + 2x - 1$$

**69.** (a)
$$
\begin{array}{r}
1\phantom{2} \\
t + 8 {\overline{\smash{\big)}\,t + 12}} \\
\underline{t + 8\phantom{2}} \\
4 \\
\end{array}
$$

So, $\dfrac{t + 12}{t + 8} = 1 + \dfrac{4}{t + 8}$.

(b)

| $t$ | 0 | 10 | 20 | 30 | 40 | 50 | 60 |
|---|---|---|---|---|---|---|---|
| $\dfrac{t + 12}{t + 8}$ | $\dfrac{12}{8} = 1.5$ | $\dfrac{22}{18} \approx 1.22$ | $\dfrac{32}{28} \approx 1.14$ | $\dfrac{42}{38} \approx 1.11$ | $\dfrac{52}{48} \approx 1.08$ | $\dfrac{62}{58} \approx 1.07$ | $\dfrac{72}{68} \approx 1.06$ |

(c)  As the value of $t$ increases, the value of the ratio approaches 1.

From part (a), the ratio $\dfrac{t + 12}{t + 8} = 1 + \dfrac{4}{t + 8}$. As $t$ increases, the value of the expression $\dfrac{4}{t + 8}$ approaches 0, so the sum

$1 + \dfrac{4}{t + 8}$ approaches $1 + 0$, or 1.

**71.**  The degree of the quotient is 2.

**73.**  Use the reverse order of the rule for adding fractions, $\dfrac{a + b}{c} = \dfrac{a}{c} + \dfrac{b}{c}$. This allows you to separate the original division

problem into multiple division problems, each involving the division of one term of the polynomial by the monomial.
Then reduce the resulting expressions to simplest form.

**75.**  $\dfrac{-8}{12} = -\dfrac{\cancel{4}(2)}{\cancel{4}(3)} = -\dfrac{2}{3}$

**77.**  $\dfrac{-60}{-150} = \dfrac{\cancel{30}(2)}{\cancel{30}(5)} = \dfrac{2}{5}$

**79.**  $m = \dfrac{9 - 5}{8 - 4} = \dfrac{4}{4} = 1$

$y - y_1 = m(x - x_1)$

$y - 5 = 1(x - 4)$

$y - 5 = x - 4$

$y = x + 1$

**81.** $m = \dfrac{5 - (-8)}{1 - (-5)} = \dfrac{13}{6}$

$$y - y_1 = m(x - x_1)$$

$$y - 5 = \dfrac{13}{6}(x - 1)$$

$$y - 5 = \dfrac{13}{6}x - \dfrac{13}{6}$$

$$y = \dfrac{13}{6}x - \dfrac{13}{6} + \dfrac{30}{6}$$

$$y = \dfrac{13}{6}x + \dfrac{17}{6}$$

**83.** $-2x^2(5x^3) = -2(5)x^2 \cdot x^3 = -10x^5$

**85.** $(x + 7)^2 = x^2 + 2(x)(7) + 7^2 = x^2 + 14x + 49$

# Review Exercises for Chapter 5

**1.** $x^2 \cdot x^2 \cdot x^4 = x^{2+2+4} = x^8$

**3.** $(x^9)^2 = x^{9 \cdot 2} = x^{18}$

**5.** $t^5(-2t^7) = -2t^{5+7} = -2t^{12}$

**7.** $(xy)^4(-5x^2y^3) = x^4y^4(-5x^2y^3)$

$$= -5x^{4+2}y^{4+3} = -5x^6y^7$$

**9.** $\dfrac{24x^3}{12x^2} = \dfrac{24}{12}x^{3-2} = 2x$

**11.** $\dfrac{64u^3v^2}{32uv} = \dfrac{64}{32}u^{3-1}v^{2-1} = 2u^2v$

**13.** $2^{-2} = \dfrac{1}{2^2}$

**15.** $6x^{-3} = \dfrac{6}{x^3}$

**17.** $\dfrac{4}{t^{-8}} = 4t^8$

**19.** $t^{-2} \cdot t = t^{-2+1} = t^{-1} = \dfrac{1}{t}$

**21.** $\dfrac{a^3}{a^{-4}} = a^{3-(-4)} = a^7$

**23.** $4x^{-6}y^2 \cdot x^6 = 4x^{-6+6}y^2 = 4x^0y^2 = 4(1)y^2 = 4y^2$

**25.** $(-3a^2)^{-2}(a^2)^0 = (-3)^{-2}a^{2(-2)}(1)$

$$= (-3)^{-2}a^{-4} = \dfrac{1}{(-3)^2 a^4} = \dfrac{1}{9a^4}$$

**27.** $(2x^2y^{-3})^2 = 2^2 x^{2 \cdot 2} y^{-3 \cdot 2} = 4x^4 y^{-6} = \dfrac{4x^4}{y^6}$

**29.** $\left(\dfrac{y}{5}\right)^{-2} = \dfrac{y^{-2}}{5^{-2}} = \dfrac{5^2}{y^2} = \dfrac{25}{y^2}$

**31.** $\dfrac{(-3y)^{-4}}{-9y^{-6}} = \dfrac{(-3)^{-4}y^{-4}}{-9y^{-6}}$

$$= \dfrac{1}{(-3)^4(-9)}y^{-4-(-6)}$$

$$= \dfrac{1}{81(-9)}y^2 = -\dfrac{1}{729}y^2 \text{ or } -\dfrac{y^2}{729}$$

**33.** $(2u^{-2}v)^3(4u^{-5}v^4)^{-1} = 2^3u^{-6}v^3(4^{-1})u^5v^{-4}$

$$= \dfrac{8}{4}u^{-6+5}v^{3-4}$$

$$= 2u^{-1}v^{-1} = \dfrac{2}{uv}$$

**35.** $0.000728 = 7.28 \times 10^{-4}$

**37.** $1.809 \times 10^8 = 180{,}900{,}000$

**39.** Polynomial: $10x - 4 - 5x^3$

Standard form: $-5x^3 + 10x - 4$

Degree: 3

Leading coefficient: $-5$

**41.** Polynomial: $4x^3 - 2x + 5x^4 - 7x^2$

Standard form: $5x^4 + 4x^3 - 7x^2 - 2x$

Degree: 4

Leading coefficient: 5

**43.** Polynomial: $7x^4 - 1 + 11x^2$

Standard form: $7x^4 + 11x^2 - 1$

Degree: 4

Leading coefficient: 7

**45.** $(8x + 4) + (x - 4) = 8x + 4 + x - 4 = 9x$

**47.** $(3y^3 + 5y^2 - 9y) + (2y^3 - 3y + 10) = 3y^3 + 5y^2 - 9y + 2y^3 - 3y + 10 = 5y^3 + 5y^2 - 12y + 10$

**49.** $(3u + 4u^2) + 5(u + 1) + 3u^2 = 3u + 4u^2 + 5u + 5 + 3u^2 = 7u^2 + 8u + 5$

**51.**
$$\begin{array}{r} -x^4 - 2x^2 + 3 \\ + 3x^4 - 5x^2 \phantom{+ 3} \\ \hline 2x^4 - 7x^2 + 3 \end{array}$$

**53.** *Common formula:*  $P = 2l + 2w$

  *Labels:*    $P$ = perimeter of wall

       $l = x$

       $w = x - 3$

  *Equation:*    $P = 2x + 2(x - 3)$

        $= 2x + 2x - 6$

        $= 4x - 6$

  The perimeter of the wall is $(4x - 6)$ units.

**55.** $(3t - 5) - (3t - 9) = 3t - 5 - 3t + 9 = 4$

**57.** $(6x^2 - 9x - 5) - 3(4x^2 - 6x + 1) = 6x^2 - 9x - 5 - 12x^2 + 18x - 3 = -6x^2 + 9x - 8$

**59.** $4y^2 - \left[y - 3(y^2 + 2)\right] = 4y^2 - \left[y - 3y^2 - 6\right] = 4y^2 - y + 3y^2 + 6 = 7y^2 - y + 6$

**61.**
$$\begin{array}{l} 5x^2 + 2x - 27 \Rightarrow \phantom{-}5x^2 + 2x - 27 \\ \underline{-(2x^2 - 2x - 13)} \Rightarrow \underline{-2x^2 + 2x + 13} \\ \phantom{-(2x^2 - 2x - 13) \Rightarrow} 3x^2 + 4x - 14 \end{array}$$

**63.** (a)  $P = R - C$

  $P = 40x - \frac{1}{2}x^2 - (15 + 26x)$

   $= 40x - \frac{1}{2}x^2 - 15 - 26x$

   $= -\frac{1}{2}x^2 + 14x - 15$

(b)

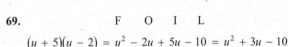

(c)  $P = -\frac{1}{2}(14)^2 + 14(14) - 15$

   $= -\frac{1}{2}(196) + 196 - 15$

   $= -98 + 196 - 15$

   $= 83$

So, when 14 units are sold, the profit is $83. As the number of units produced, $x$, increases to 14, the profit increases. At 14 units sold, the maximum profit is reached. As the number of units produced increases beyond 14, the profit decreases.

**65.** $3y(y - 1) = 3y^2 - 3y$

**67.** $(5 - 7y)(-6y^2) = 5(-6y^2) - 7y(-6y^2)$

   $= -30y^2 + 42y^3$ or $42y^3 - 30y^2$

**69.**
$$\phantom{xxxxxx}\text{F}\phantom{xx}\text{O}\phantom{xx}\text{I}\phantom{xx}\text{L}$$
$(u + 5)(u - 2) = u^2 - 2u + 5u - 10 = u^2 + 3u - 10$

**71.**
$$\phantom{xxxxxx}\text{F}\phantom{xxx}\text{O}\phantom{xxx}\text{I}\phantom{xxx}\text{L}$$
$(6 - 2x)(7x - 10) = 42x - 60 - 14x^2 + 20x$

        $= -14x^2 + 62x - 60$

**73.** $(s^2 + 4s - 3)(s - 3) = s^2(s - 3) + 4s(s - 3) - 3(s - 3)$

      $= s^3 - 3s^2 + 4s^2 - 12s - 3s + 9$

      $= s^3 + s^2 - 15s + 9$

**75.** $(4x + 2)(x^2 + 6x - 5) = 4x(x^2 + 6x - 5) + 2(x^2 + 6x - 5)$

$\qquad\qquad\qquad\qquad = 4x^3 + 24x^2 - 20x + 2x^2 + 12x - 10$

$\qquad\qquad\qquad\qquad = 4x^3 + 26x^2 - 8x - 10$

**77.**
$$
\begin{array}{r}
5y^2 - 2y + 9 \\
\times \qquad\quad 3y + 4 \\
\hline
20y^2 - 8y + 36 \\
15y^3 - 6y^2 + 27y \qquad\quad \\
\hline
15y^3 + 14y^2 + 19y + 36
\end{array}
$$

**79.**
$$
\begin{array}{r}
x^2 + 8x - 12 \\
\times \qquad x^2 - 9x + 2 \\
\hline
2x^2 + 16x - 24 \\
-9x^3 - 72x^2 + 108x \qquad\quad \\
x^4 + 8x^3 - 12x^2 \qquad\qquad\qquad \\
\hline
x^4 - x^3 - 82x^2 + 124x - 24
\end{array}
$$

**81.** $(3y - 2)^3 = (3y - 2)(3y - 2)(3y - 2)$

$\qquad\qquad\quad = (9y^2 - 6y - 6y + 4)(3y - 2)$

$\qquad\qquad\quad = (9y^2 - 12y + 4)(3y - 2)$

$\qquad\qquad\quad = 27y^3 - 18y^2 - 36y^2 + 24y + 12y - 8$

$\qquad\qquad\quad = 27y^3 - 54y^2 + 36y - 8$

**83.** Area $= (\text{Length})(\text{Width})$

$\qquad\quad = (x + 30)(x + 25)$

$\qquad\quad = x^2 + 25x + 30x + 750$

$\qquad\quad = x^2 + 55x + 750$

**85.** $(x - 5)^2 = (x)^2 - 2(x)(5) + (5)^2 = x^2 - 10x + 25$

**87.** $(4 + 3b)^2 = (4)^2 + 2(4)(3b) + (3b)^2 = 16 + 24b + 9b^2$

**89.** $(r + 7)(r - 7) = r^2 - 7^2 = r^2 - 49$

**91.** $(3a + b)^2 = (3a)^2 + 2(3a)(b) + b^2 = 9a^2 + 6ab + b^2$

**93.** $(4u + 5v)(4u - 5v) = (4u)^2 - (5v)^2 = 16u^2 - 25v^2$

**95.** $\dfrac{8u^3 + 4u^2}{2u} = \dfrac{8u^3}{2u} + \dfrac{4u^2}{2u} = 4u^2 + 2u$

**97.** $(5x^2 + 15x - 25) \div (-5x) = \dfrac{5x^2}{-5x} + \dfrac{15x}{-5x} + \dfrac{-25}{-5x}$

$\qquad\qquad\qquad\qquad\qquad = -x - 3 + \dfrac{5}{x}$

**99.**
$$
\begin{array}{r}
x - 4 \\
x + 5 \overline{)\, x^2 + x - 20} \\
\underline{x^2 + 5x \qquad} \\
-4x - 20 \\
\underline{-4x - 20} \\
0
\end{array}
$$

So, $\dfrac{x^2 + x - 20}{x + 5} = x - 4.$

**101.**
$$
\begin{array}{r}
8x - 4 \\
2x + 1 \overline{)\, 16x^2 \qquad - 5} \\
\underline{16x^2 + 8x \qquad} \\
-8x - 5 \\
\underline{-8x - 4} \\
-1
\end{array}
$$

So, $\dfrac{16x^2 - 5}{2x + 1} = 8x - 4 - \dfrac{1}{2x + 1}.$

**103.**
$$
\begin{array}{r}
2x^3 \qquad\quad - 9x \\
3x - 2 \overline{)\, 6x^4 - 4x^3 - 27x^2 + 18x + 0} \\
\underline{6x^4 - 4x^3 \qquad\qquad\qquad\qquad} \\
-27x^2 + 18x \\
\underline{-27x^2 + 18x} \\
0
\end{array}
$$

So, $\dfrac{6x^4 - 4x^3 - 27x^2 + 18x}{3x - 2} = 2x^3 - 9x$

**105.** $(\text{Length})(\text{Width}) = \text{Area}$

$\text{Length} = \dfrac{\text{Area}}{\text{Width}}$

$$
\begin{array}{r}
3x - 4 \\
x + 3 \overline{)\, 3x^2 + 5x - 3} \\
\underline{3x^2 + 9x \qquad} \\
-4x - 3 \\
\underline{-4x - 12} \\
9
\end{array}
$$

$\text{Length} = \dfrac{3x^2 + 5x - 3}{x + 3} = 3x - 4 + \dfrac{9}{x + 3}$

# Chapter Test for Chapter 5

**1.** $\left(3x^2y\right)\left(-xy\right)^2 = 3x^2y \cdot x^2y^2$

$\qquad = 3x^{2+2}y^{1+2}$

$\qquad = 3x^4y^3$

**2.** $\left(5x^2y^3\right)^2\left(-2xy\right)^3 = 5^2x^{2\cdot2}y^{3\cdot2}(-2)^3x^3y^3$

$\qquad = 25x^4y^6(-8)x^3y^3$

$\qquad = -200x^{4+3}y^{6+3}$

$\qquad = -200x^7y^9$

**3.** $\dfrac{-6a^2b}{-9ab} = \dfrac{-6}{-9}a^1b^0 = \dfrac{2}{3}a \text{ or } \dfrac{2a}{3}$

**4.** $\left(3x^{-2}y^3\right)^{-2} = 3^{-2}\left(x^{-2}\right)^{-2}\left(y^3\right)^{-2}$

$\qquad = 3^{-2}x^4y^{-6}$

$\qquad = \dfrac{x^4}{3^2y^6} = \dfrac{x^4}{9y^6}$

**5.** $\left(4u^{-3}v^2\right)^{-2}\left(8u^{-1}v^{-2}\right)^0 = 4^{-2}u^6v^{-4}(1)$

$\qquad = \dfrac{u^6}{4^2v^4}$

$\qquad = \dfrac{u^6}{16v^4}$

**6.** $\dfrac{12x^{-3}y^5}{4x^{-2}y^{-1}} = \dfrac{12}{4}x^{-3-(-2)}y^{5-(-1)} = 3x^{-1}y^6 = \dfrac{3y^6}{x}$

**7. (a)** $\dfrac{2^{-3}}{3^{-1}} = \dfrac{3^1}{2^3} = \dfrac{3}{8}$

**(b)** $\left(1.5 \times 10^5\right)^2 = (1.5)^2 \times \left(10^5\right)^2$

$\qquad = 2.25 \times 10^{10} = 22,500,000,000$

**(c)** $\dfrac{6.3 \times 10^{-3}}{2.1 \times 10^2} = \dfrac{6.3}{2.1} \times 10^{-3-2} = 3 \times 10^{-5} = 0.00003$

**8. (a)** $0.00015 = 1.5 \times 10^{-4}$

**(b)** $8 \times 10^7 = 80,000,000$

**9.** Degree: 4

Leading Coefficient: $-3$

The degree of a polynomial is the degree of the term with the highest power, and the coefficient of this term is the leading coefficient of the polynomial.

**10.** There are many correct answers.

Examples:

$z^5 - 3z^3 + 5$

$4x^5 + x^4 - 7x^2$

$-2y^5 + y + 1$

**11.** $\left(3z^2 - 3z + 7\right) + \left(8 - z^2\right) = 3z^2 - 3z + 7 + 8 - z^2$

$\qquad\qquad = 2z^2 - 3z + 15$

**12.** $\left(8u^3 + 3u^2 - 2u - 1\right) - \left(u^3 + 3u^2 - 2u\right) = 8u^3 + 3u^2 - 2u - 1 - u^3 - 3u^2 + 2u = 7u^3 - 1$

**13.** $6y - \left[2y - \left(3 + 4y - y^2\right)\right] = 6y - \left[2y - 3 - 4y + y^2\right]$

$\qquad\qquad = 6y - \left[y^2 - 2y - 3\right]$

$\qquad\qquad = 6y - y^2 + 2y + 3$

$\qquad\qquad = -y^2 + 8y + 3$

**14.** $-5\left(x^2 - 1\right) + 3(4x + 7) - \left(x^2 + 26\right) = -5x^2 + 5 + 12x + 21 - x^2 - 26 = -6x^2 + 12x$

**15.** $(x - 7)^2 = x^2 - 2(x)(7) + 7^2 = x^2 - 14x + 49$

**16.** $(2x - 3)(2x + 3) = (2x)^2 - (3)^2 = 4x^2 - 9$

**17.** $(z + 2)\left(2z^2 - 3z + 5\right) = 2z^3 - 3z^2 + 5z + 4z^2 - 6z + 10$

$\qquad\qquad = 2z^3 + z^2 - z + 10$

**18.** $(y + 3)\left(y^4 + y^2 - 4\right) = y\left(y^4 + y^2 - 4\right) + 3\left(y^4 + y^2 - 4\right)$

$\qquad\qquad = y^5 + y^3 - 4y + 3y^4 + 3y^2 - 12$

$\qquad\qquad = y^5 + 3y^4 + y^3 + 3y^2 - 4y - 12$

**19.** $\dfrac{4z^3 + z}{2z} = \dfrac{4z^3}{2z} + \dfrac{z}{2z} = 2z^2 + \dfrac{1}{2}$

**20.** $\dfrac{16x^2 - 12}{-8} = \dfrac{16x^2}{-8} + \dfrac{-12}{-8} = -2x^2 + \dfrac{3}{2}$

**21.**

$$
\begin{array}{r}
x^2 + 2x + 3 \\
x - 2 \overline{\big)\, x^3 + 0x^2 - x - 6} \\
\underline{x^3 - 2x^2} \\
2x^2 - x \\
\underline{2x^2 - 4x} \\
3x - 6 \\
\underline{3x - 6} \\
0
\end{array}
$$

So, $\dfrac{x^2 - x - 6}{x - 2} = x^2 + 2x + 3.$

**22.**

$$
\begin{array}{r}
2x^2 + 9x + 17 \\
2x - 4 \overline{\big)\, 4x^3 + 10x^2 - 2x - 5} \\
\underline{4x^3 - 8x^2} \\
18x^2 - 2x \\
\underline{18x^2 - 36x} \\
34x - 5 \\
\underline{34x - 68} \\
63
\end{array}
$$

So, $\dfrac{4x^3 + 10x^2 - 2x - 5}{2x - 4} = 2x^2 + 9x + 17 + \dfrac{63}{2x - 4}.$

**23.** $\boxed{\begin{array}{c}\text{Area of} \\ \text{shaded region}\end{array}} = \boxed{\begin{array}{c}\text{Area of} \\ \text{large rectangle}\end{array}} - \boxed{\begin{array}{c}\text{Area of} \\ \text{small rectangle}\end{array}}$

$$
\begin{aligned}
\text{Area of shaded region} &= (3x + 2)(2x) - (2x + 5)(x) \\
&= 6x^2 + 4x - 2x^2 - 5x \\
&= 4x^2 - x
\end{aligned}
$$

**24.** $\begin{aligned}
\text{Area of triangle} &= \tfrac{1}{2}(\text{Base})(\text{Height}) \\
&= \tfrac{1}{2}(4x - 2)(x + 6) \\
&= (2x - 1)(x + 6) \\
&= 2x^2 + 12x - x - 6 \\
&= 2x^2 + 11x - 6
\end{aligned}$

**25.** $\begin{aligned}
1500(1 + r)^2 &= 1500\left[1^2 + 2(1)(r) + r^2\right] \\
&= 1500(1 + 2r + r^2) \\
&= 1500 + 3000r + 1500r^2
\end{aligned}$

**26.** $(\text{Length})(\text{Width}) = \text{Area};\ \text{Width} = \dfrac{\text{Area}}{\text{Length}}$

$$
\begin{array}{r}
x - 3 \\
x + 1 \overline{\big)\, x^2 - 2x - 3} \\
\underline{x^2 + x} \\
-3x - 3 \\
\underline{-3x - 3} \\
\end{array}
$$

$\text{Width} = \dfrac{x^2 - 2x - 3}{x + 1} = x - 3$

# CHAPTER 6
## Factoring and Solving Equations

# C H A P T E R  6
# Factoring and Solving Equations

## Section 6.1   Factoring Polynomials with Common Factors

**1.** $z^2 = z \cdot z = z^2(1)$

$-z^6 = -(z \cdot z \cdot z \cdot z \cdot z \cdot z) = z^2(-z^4)$

The greatest common factor is $z^2$.

**3.** $2x^2 = 2 \cdot x \cdot x = 2x(x)$

$12x = 2 \cdot 2 \cdot 3 \cdot x = 2x(6)$

The greatest common factor is $2x$.

**7.** $9y^8z^4 = 3 \cdot 3 \cdot y \cdot y \cdot y \cdot y \cdot y \cdot y \cdot y \cdot y \cdot z \cdot z \cdot z \cdot z = 3y^5z^4(3y^3)$

$-12y^5z^4 = -(2 \cdot 2 \cdot 3 \cdot y \cdot y \cdot y \cdot y \cdot y \cdot z \cdot z \cdot z \cdot z) = 3y^5z^4(-4)$

The greatest common factor is $3y^5z^4$.

**9.** $14x^2 = 2 \cdot 7 \cdot x \cdot x = 1(14x^2)$

$1 = 1(1)$

$7x^4 = 7 \cdot x \cdot x \cdot x \cdot x = 1(7x^4)$

The greatest common factor is 1.

**11.** $28a^4b^2 = 2 \cdot 2 \cdot 7 \cdot a \cdot a \cdot a \cdot a \cdot b \cdot b = 14a^2(2a^2b^2)$

$14a^3 = 2 \cdot 7 \cdot a \cdot a \cdot a = 14a^2(a)$

$42a^2b^5 = 2 \cdot 3 \cdot 7 \cdot a \cdot a \cdot b \cdot b \cdot b \cdot b \cdot b = 14a^2(3b^5)$

The greatest common factor is $14a^2$.

**13.** $2(x + 3)$

$3(x + 3)$

The greatest common factor is $x + 3$.

**15.** $x(7x + 5)$

$7x + 5$ or $1(7x + 5)$

The greatest common factor is $7x + 5$.

**17.** $3x + 3 = 3(x + 1)$

**19.** $6z + 36 = 6(z + 6)$

**21.** $8t - 16 = 8(t - 2)$

**23.** $-25x - 10 = -5(5x + 2)$ or $5(-5x - 2)$

**25.** $24y^2 - 18 = 6(4y^2 - 3)$

**5.** $u^2v = u \cdot u \cdot v = u^2v(1)$

$u^3v^2 = u \cdot u \cdot u \cdot v \cdot v = u^2v(uv)$

The greatest common factor is $u^2v$.

**27.** $x^2 + x = x(x + 1)$

**29.** $25u^2 - 14u = u(25u - 14)$

**31.** $2x^4 + 6x^3 = 2x^3(x + 3)$

**33.** $7s^2 + 9t^2$ (No common factor)

**35.** $12x^2 - 2x = 2x(6x - 1)$

**37.** $-10r^3 - 35r = -5r(2r^2 + 7)$

**39.** $12x^2 + 16x - 8 = 4(3x^2 + 4x - 2)$

**41.** $100 + 75z - 50z^2 = 25(4 + 3z - 2z^2)$

**43.** $9x^4 + 6x^3 + 18x^2 = 3x^2(3x^2 + 2x + 6)$

**45.** $5u^2 + 5u^2 + 5u = 10u^2 + 5u = 5u(2u + 1)$

**47.** $16a^3b^3 + 24a^4b^3 = 8a^3b^3(2 + 3a)$

**49.** $10ab + 10a^2b = 10ab(1 + a)$

**51.** $4xy + 8x^2y - 24x^4y^5 = 4xy(1 + 2x - 6x^3y^4)$

**53.** $5 - 10x = -5(-1 + 2x) = -5(2x - 1)$

**55.** $-10x - 3000 = -10(x + 300)$

**57.** $-x^2 + 5x + 10 = -1(x^2 - 5x - 10)$

$\qquad\qquad$ or $-(x^2 - 5x - 10)$

**59.** $4 + 12x - 2x^2 = -2(-2 - 6x + x^2)$

$\qquad\qquad = -2(x^2 - 6x - 2)$

**61.** $x(x - 3) + 5(x - 3) = (x - 3)(x + 5)$

**63.** $y(q - 5) - 10(q - 5) = (q - 5)(y - 10)$

**65.** $x^3(y + 4) + y(y + 4) = (y + 4)(x^3 + y)$

**67.** $(a + b)(a - b) + a(a + b) = (a + b)[(a - b) + a]$

$\qquad\qquad\qquad\qquad\quad = (a + b)(a - b + a)$

$\qquad\qquad\qquad\qquad\quad = (a + b)(2a - b)$

**69.** $x^2 + 10x + x + 10 = x(x + 10) + 1(x + 10)$

$\qquad\qquad\qquad\quad = (x + 10)(x + 1)$

**71.** $x^2 + 3x + 4x + 12 = x(x + 3) + 4(x + 3)$

$\qquad\qquad\qquad\quad = (x + 3)(x + 4)$

**73.** $x^2 + 3x - 5x - 15 = x(x + 3) - 5(x + 3)$

$\qquad\qquad\qquad\quad = (x + 3)(x - 5)$

**75.** $4x^2 - 14x + 14x - 49 = 2x(2x - 7) + 7(2x - 7)$

$\qquad\qquad\qquad\qquad\quad = (2x - 7)(2x + 7)$

**77.** $6x^2 + 3x - 2x - 1 = 3x(2x + 1) - (2x + 1)$

$\qquad\qquad\qquad\quad = (2x + 1)(3x - 1)$

**79.** $8x^2 + 32x + x + 4 = 8x(x + 4) + (x + 4)$

$\qquad\qquad\qquad\quad = (x + 4)(8x + 1)$

**81.** $9x^2 - 21x + 6x - 14 = 3x(3x - 7) + 2(3x - 7)$

$\qquad\qquad\qquad\qquad = (3x - 7)(3x + 2)$

**83.** $2x^2 - 4x - 3x + 6 = 2x(x - 2) - 3(x - 2)$

$\qquad\qquad\qquad\quad = (x - 2)(2x - 3)$

**85.** $t^3 - 3t^2 + 2t - 6 = t^2(t - 3) + 2(t - 3)$

$\qquad\qquad\qquad\quad = (t - 3)(t^2 + 2)$

**87.** $16z^3 + 8z^2 + 2z + 1 = 8z^2(2z + 1) + (2z + 1)$

$\qquad\qquad\qquad\qquad = (2z + 1)(8z^2 + 1)$

**89.** $x^3 - 3x - x^2 + 3 = x(x^2 - 3) - 1(x^2 - 3)$

$\qquad\qquad\qquad\quad = (x^2 - 3)(x - 1)$

**91.** $3x^2 + x^3 - 18 - 6x = x^2(3 + x) - 6(3 + x)$

$\qquad\qquad\qquad\qquad = (3 + x)(x^2 - 6)$

$\qquad\qquad\qquad\qquad$ or $(x + 3)(x^2 - 6)$

**93.** $ky^2 - 4ky + 2y - 8 = ky(y - 4) + 2(y - 4)$

$\qquad\qquad\qquad\qquad = (y - 4)(ky + 2)$

**95.** $\frac{1}{4}x + \frac{3}{4} = \frac{1}{4}(x + 3)$

The missing factor is $x + 3$.

**97.** $2y - \frac{1}{5} = \frac{10}{5}y - \frac{1}{5} = \frac{1}{5}(10y - 1)$

The missing factor is $10y - 1$.

**99.** $\frac{7}{8}x + \frac{5}{16}y = \frac{14}{16}x + \frac{5}{16}y = \frac{1}{16}(14x + 5y)$

The missing factor is $14x + 5y$.

**101.** Area $= (\text{Length})(\text{Width})$

Area $= 2x^2 + 2x$ and Width $= 2x$

$2x^2 + 2x = 2x(x + 1)$

Therefore, the length of the rectangle must be $x + 1$.

**103.** Area of large rectangle $= 4x(2x) = 8x^2$

Area of small rectangle $= 2x(x) = 2x^2$

Shaded area

$= $ Area of large rectangle $-$ Area of small rectangle

$= 8x^2 - 2x^2$

$= 6x^2$

**105.** $S = 2\pi r^2 + 2\pi rh = 2\pi r(r + h)$

**107.** $kQx - kx^2 = kx(Q - x)$

**109.** The expression is in factored form because it is written as a product and there are no common monomial factors that can be factored out of the binomial.

**111.** $\qquad\qquad 2x + 4 = 2(x + 2)$

$x(x^2 + 1) - 3(x^2 + 1) = (x^2 + 1)(x - 3)$

There are many correct answers.

**113.** $x + 2 = 3$

    (a)  $5 + 2 \overset{?}{=} 3$

         $7 \neq 3$

       0 *is not* a solution.

    (b)  $1 + 2 \overset{?}{=} 3$

         $3 = 3$

       1 *is* a solution.

**115.** $2x - 4 = 0$

    (a)  $2(2) - 4 \overset{?}{=} 0$

         $0 = 0$

       2 *is* a solution.

    (b)  $2(-2) - 4 \overset{?}{=} 0$

         $-8 \neq 0$

       $-2$ *is not* a solution.

**117.** $\left(\dfrac{3y}{2x^3}\right)^2 = \dfrac{9y^2}{4x^6}$

**119.** $z^2 \cdot z^{-6} = z^{2+(-6)} = z^{-4} = \dfrac{1}{z^4}$

**121.** $\dfrac{3m^2n}{m} = 3mn$

**123.**

$$
\begin{array}{r}
x + 1 \phantom{000} \\
x + 8 \overline{\smash{)}\, x^2 + 9x + 8} \\
\underline{x^2 + 8x} \phantom{00000} \\
x + 8 \\
\underline{x + 8} \\
0
\end{array}
$$

So, $\dfrac{x^2 + 9x + 8}{x + 8} = x + 1.$

# Section 6.2   Factoring Trinomials

**1.** $x^2 + 8x + 7 = (x + 7)(x + 1)$

The missing factor is $x + 1$.

**Check:**       F   O   I   L

$(x + 7)(x + 1) = x^2 + x + 7x + 7 = x^2 + 8x + 7$

**3.** $y^2 + 11y - 12 = (y - 1)(y + 12)$

The missing factor is $y + 12$.

**Check:**       F   O   I   L

$(y - 1)(y + 12) = y^2 + 12y - y - 12 = y^2 + 11y - 12$

**5.** $z^2 - 7z + 12 = (z - 4)(z - 3)$

The missing factor is $z - 3$.

**Check:**       F   O   I   L

$(z - 4)(z - 3) = z^2 - 3z - 4z + 12 = z^2 - 7z + 12$

**7.** $(x + 11)(x + 1)$

$(x - 11)(x - 1)$

**9.** $(x + 14)(x + 1)$

$(x - 14)(x - 1)$

$(x + 7)(x + 2)$

$(x - 7)(x - 2)$

**11.** $(x + 12)(x - 1)$

$(x - 12)(x + 1)$

$(x + 6)(x - 2)$

$(x - 6)(x + 2)$

$(x + 4)(x - 3)$

$(x - 4)(x + 3)$

**13.** $x^2 + 6x + 8 = (x + 4)(x + 2)$

**15.** $x^2 + 2x - 15 = (x + 5)(x - 3)$

**17.** $x^2 - 9x - 22 = (x - 11)(x + 2)$

**19.** $x^2 - 9x + 14 = (x - 7)(x - 2)$

**21.** $2x + 15 - x^2 = -x^2 + 2x + 15$

$\phantom{2x + 15 - x^2} = -(x^2 - 2x - 15)$

$\phantom{2x + 15 - x^2} = -(x - 5)(x + 3)$

**23.** $u^2 - 22u - 48 = (u - 24)(u + 2)$

**25.** $x^2 + 3x - 70 = (x + 10)(x - 7)$

**27.** $x^2 + 19x + 60 = (x + 15)(x + 4)$

**29.** $x^2 - 17x + 72 = (x - 8)(x - 9)$

**31.** $y^2 + 5y + 11$

The trinomial is prime.

**33.** $x^2 - 7xz - 18z^2 = (x - 9z)(x + 2z)$

**35.** $x^2 - 5xy + 6y^2 = (x - 3y)(x - 2y)$

**37.** $x^2 + 8xy + 15y^2 = (x + 3y)(x + 5y)$

**39.** $a^2 + 2ab - 15b^2 = (a + 5b)(a - 3b)$

**41.** $4x^2 - 32x + 60 = 4(x^2 - 8x + 15)$

$= 4(x - 5)(x - 3)$

**43.** $3z^2 + 5z + 6$

This trinomial is prime.

**45.** $9x^2 + 18x - 18 = 9(x^2 + 2x - 2)$

**47.** $x^3 - 13x^2 + 30x = x(x^2 - 13x + 30)$

$= x(x - 3)(x - 10)$

**49.** $3x^3 + 18x^2 + 24x = 3x(x^2 + 6x + 8)$

$= 3x(x + 4)(x + 2)$

**51.** $x^4 - 5x^3 + 6x^2 = x^2(x^2 - 5x + 6)$

$= x^2(x - 3)(x - 2)$

**53.** $2x^4 - 20x^3 + 42x^2 = 2x^2(x^2 - 10x + 21)$

$= 2x^2(x - 7)(x - 3)$

**55.** $x^3 + 5x^2y + 6xy^2 = x(x^2 + 5xy + 6y^2)$

$= x(x + 3y)(x + 2y)$

**57.** $-3y^2x - 9xy + 54x = -3x(y^2 + 3y - 18)$

$= -3x(y + 6)(y - 3)$

**59.** $2x^3y + 4x^2y^2 - 6xy^3 = 2xy(x^2 + 2xy - 3y^2)$

$= 2xy(x + 3y)(x - y)$

**61.** $b = 9$     $(x + 6)(x + 3)$

$b = -9$     $(x - 6)(x - 3)$

$b = 11$     $(x + 9)(x + 2)$

$b = -11$     $(x - 9)(x - 2)$

$b = 19$     $(x + 18)(x + 1)$

$b = -19$     $(x - 18)(x - 1)$

**63.** $b = 2$     $(x + 7)(x - 5)$

$b = -2$     $(x - 7)(x + 5)$

$b = 34$     $(x + 35)(x - 1)$

$b = -34$     $(x - 35)(x + 1)$

**65.** $b = 12$     $(x + 6)(x + 6)$

$b = -12$     $(x - 6)(x - 6)$

$b = 13$     $(x + 9)(x + 4)$

$b = -13$     $(x - 9)(x - 4)$

$b = 15$     $(x + 12)(x + 3)$

$b = -15$     $(x - 12)(x - 3)$

$b = 20$     $(x + 18)(x + 2)$

$b = -20$     $(x - 18)(x - 2)$

$b = 37$     $(x + 36)(x + 1)$

$b = -37$     $(x - 36)(x - 1)$

**67.** $c = 2$     $(x + 2)(x + 1)$

$c = -4$     $(x + 4)(x - 1)$

$c = -10$     $(x + 5)(x - 2)$

$c = -18$     $(x + 6)(x - 3)$

$c = -28$     $(x + 7)(x - 4)$

There are many correct answers.

**69.** $c = 3$     $(x - 3)(x - 1)$

$c = 4$     $(x - 2)(x - 2)$

$c = -12$     $(x - 6)(x + 2)$

$c = -5$     $(x - 5)(x + 1)$

$c = -21$     $(x - 7)(x + 3)$

$c = -32$     $(x - 8)(x + 4)$

There are many correct answers.

**71.** $c = 20$     $(x - 4)(x - 5)$

$c = 18$     $(x - 6)(x - 3)$

$c = 14$     $(x - 7)(x - 2)$

$c = 8$     $(x - 8)(x - 1)$

$c = -10$     $(x - 10)(x + 1)$

$c = -22$     $(x - 11)(x + 2)$

$c = -36$     $(x - 12)(x + 3)$

There are many correct answers.

**73.** (a) $V = 4x^3 - 20x^2 + 24x$, $0 < x < 2$

$$= 4x(x^2 - 5x + 6)$$

$$= 4x(x - 3)(x - 2) \text{ or } x(2)(x - 3)(2)(x - 2) \quad \text{Note: } 4 = 2(2)$$

$$\text{or } x(2x - 6)(2x - 4)$$

$$\text{or } x(2x - 6)(-1)(-1)(2x - 4) \quad \text{Note: } (-1)(-1) = 1$$

$$\text{or } x\big[(2x - 6)(-1)\big]\big[(-1)(2x - 4)\big]$$

$$\text{or } x(-2x + 6)(-2x + 4).$$

$$\text{or } x(6 - 2x)(4 - 2x)$$

The volume is the product of the length, width, and height. The length is $6 - 2x$ feet, the width is $4 - 2x$ feet, and the height is $x$ feet.

Note: Here is an alternate factoring approach which also points out the dimensions of the box.

$$V = 4x^3 - 20x^2 + 24x$$

$$= 4x(x^2 - 5x + 6)$$

$$= 4x(6 - 5x + x^2)$$

$$= 4x(3 - x)(2 - x)$$

$$= x(2)(3 - x)(2)(2 - x) = x\big[2(3 - x)\big]\big[2(2 - x)\big] = x(6 - 2x)(4 - 2x)$$

(b)

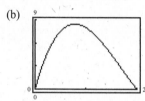

The volume is greatest when $x$ is approximately 0.785 feet.

**75.** $x^2 + 30x + 200 = (x + 20)(x + 10)$

Width of rectangle: $x + 10$

Length of rectangle: $x + 20$

Width of shaded region: 10

Length of shaded region: $x + 20 - x = 20$

Area of shaded region: $10(20) = 200$ square units

**77.** (a) and (d) are factorizations of the polynomial.

(a) $(2x - 4)(x + 5) = 2x^2 + 6x - 20$, but this is not a complete factorization because $2x - 4$ can be factored as $2(x - 2)$.

(b) This is not a factorization of $2x^2 + 6x - 20$.

$(2x - 4)(2x + 10) = 4x^2 + 12x - 40$

(c) This is not a factorization of $2x^2 + 6x - 20$.

$(x - 2)(x + 5) = x^2 + 3x - 10$

(d) Yes, $2(x - 2)(x + 5) = 2x^2 + 6x - 20$, and this is a complete factorization.

**79.** The process of factoring $x^2 + bx + c$ is easier if $c$ is a prime number because there are fewer factorizations to examine.

**81.**
$$3x - 5 = 16$$
$$3x - 5 + 5 = 16 + 5$$
$$3x = 21$$
$$\frac{3x}{3} = \frac{21}{3}$$
$$x = 7$$

**83.**
$$7 - 2x = 9$$
$$7 - 2x - 7 = 9 - 7$$
$$-2x = 2$$
$$\frac{-2x}{-2} = \frac{2}{-2}$$
$$x = -1$$

**85.**
$$10x + 24 = 2x$$
$$10x + 24 - 10x = 2x - 10x$$
$$24 = -8x$$
$$\frac{24}{-8} = \frac{-8x}{-8}$$
$$-3 = x$$

**87.**
$$3x - 8 = 9x + 4$$
$$3x - 8 - 9x = 9x + 4 - 9x$$
$$-6x - 8 = 4$$
$$-6x - 8 + 8 = 4 + 8$$
$$-6x = 12$$
$$\frac{-6x}{-6} = \frac{12}{-6}$$
$$x = -2$$

**93.** $12xy^3 = 2 \cdot 2 \cdot 3 \cdot x \cdot y \cdot y \cdot y = 4xy^3(3)$

$28x^3y^4 = 2 \cdot 2 \cdot 7 \cdot x \cdot x \cdot x \cdot y \cdot y \cdot y \cdot y = 4xy^3\left(7x^2y\right)$

Greatest common factor is $4xy^3$.

**95.** $21xy^4 = 3 \cdot 7 \cdot x \cdot y \cdot y \cdot y \cdot y = 3xy\left(7y^3\right)$

$42x^2y^2 = 2 \cdot 3 \cdot 7 \cdot x \cdot x \cdot y \cdot y = 3xy(14xy)$

$9x^4y = 3 \cdot 3 \cdot x \cdot x \cdot x \cdot x \cdot y = 3xy\left(3x^3\right)$

Greatest common factor is $3xy$.

**89.** $35t^2 = 5 \cdot 7 \cdot t \cdot t = 7t^2(5)$

$7t^8 = 7 \cdot t \cdot t \cdot t \cdot t \cdot t \cdot t \cdot t \cdot t = 7t^2\left(t^6\right)$

Greatest common factor is $7t^2$.

**91.** $xy^3 = x \cdot y \cdot y \cdot y = xy^2(y)$

$x^2y^2 = x \cdot x \cdot y \cdot y = xy^2(x)$

Greatest common factor is $xy^2$.

# Section 6.3   More About Factoring Trinomials

**1.** $2x^2 + 7x - 4 = (2x - 1)(x + 4)$

The missing factor is $x + 4$.

**3.** $3t^2 + 4t - 15 = (3t - 5)(t + 3)$

The missing factor is $t + 3$.

**5.** $7x^2 + 15x + 2 = (7x + 1)(x + 2)$

The missing factor is $x + 2$.

**7.** $5x^2 + 18x + 9 = (5x + 3)(x + 3)$

The missing factor is $x + 3$.

**9.** $5a^2 + 12a - 9 = (a + 3)(5a - 3)$

The missing factor is $5a - 3$.

**11.** $4z^2 - 13z + 3 = (z - 3)(4z - 1)$

The missing factor is $4z - 1$.

**13.** $6x^2 - 23x + 7 = (3x - 1)(2x - 7)$

The missing factor is $2x - 7$.

**15.** $9a^2 - 6a - 8 = (3a + 2)(3a - 4)$

The missing factor is $3a - 4$.

**17.** $18t^2 + 3t - 10 = (6t + 5)(3t - 2)$

The missing factor is $3t - 2$.

**19.** $(5x + 3)(x + 1)$

$(5x - 3)(x - 1)$

$(5x + 1)(x + 3)$

$(5x - 1)(x - 3)$

**21.** $(5x + 4)(x + 7)$

$(5x - 4)(x - 7)$

$(5x + 7)(x + 4)$

$(5x - 7)(x - 4)$

$(5x + 14)(x + 2)$

$(5x - 14)(x - 2)$

$(5x + 2)(x + 14)$

$(5x - 2)(x - 14)$

$(5x + 28)(x + 1)$

$(5x - 28)(x - 1)$

$(5x + 1)(x + 28)$

$(5x - 1)(x - 28)$

**23.** $2x^2 + 5x + 3 = (2x + 3)(x + 1)$

**25.** $4y^2 + 5y + 1 = (4y + 1)(y + 1)$

**27.** $6y^2 - 7y + 1 = (6y - 1)(y - 1)$

**29.** $12x^2 + 7x - 5 = (12x - 5)(x + 1)$

**31.** $5x^2 - 2x + 1$

This trinomial is prime.

**33.** $2x^2 + x + 3$

This trinomial is prime.

**35.** $8s^2 - 14s + 3 = (4s - 1)(2s - 3)$

**37.** $4x^2 + 13x - 12 = (4x - 3)(x + 4)$

**39.** $9x^2 - 18x + 8 = (3x - 2)(3x - 4)$

**41.** $18u^2 - 3u - 28 = (3u - 4)(6u + 7)$

**43.** $15a^2 + 14a - 8 = (5a - 2)(3a + 4)$

**45.** $10t^2 - 3t - 18 = (5t + 6)(2t - 3)$

**47.** $15m^2 + 13m - 20 = (5m - 4)(3m + 5)$

**49.** $16z^2 - 34z + 15 = (8z - 5)(2z - 3)$

**51.** $6x^2 - 3x = 3x(2x - 1)$

**53.** $15y^2 - 40y = 5y(3y - 8)$

**55.** $u(u - 3) + 9(u - 3) = (u - 3)(u + 9)$

**57.** $2v^2 + 8v - 42 = 2(v^2 + 4v - 21)$

$= 2(v + 7)(v - 3)$

**59.** $-3x^2 - 3x - 60 = -3(x^2 + x + 20)$

**61.** $9z^2 - 24z + 15 = 3(3z^2 - 8z + 5)$

$= 3(3z - 5)(z - 1)$

**63.** $4x^2 - 4x - 8 = 4(x^2 - x - 2) = 4(x - 2)(x + 1)$

**65.** $-15x^4 - 2x^3 + 8x^2 = -x^2(15x^2 + 2x - 8)$

$= -x^2(3x - 2)(5x + 4)$

**67.** $3x^3 + 4x^2 + 2x = x(3x^2 + 4x + 2)$

**69.** $6x^3 + 24x^2 - 192x = 6x(x^2 + 4x - 32)$

$= 6x(x + 8)(x - 4)$

**71.** $18u^4 + 18u^3 - 27u^2 = 9u^2(2u^2 + 2u - 3)$

**73.** $-2x^2 + 7x + 9 = -(2x^2 - 7x - 9)$

$= -(2x - 9)(x + 1)$

or $(-2x + 9)(x + 1)$

or $(2x - 9)(-x - 1)$

**75.** $4 - 4x - 3x^2 = (2 - 3x)(2 + x)$ or

$4 - 4x - 3x^2 = -1(-4 + 4x + 3x^2)$

$= -1(3x^2 + 4x - 4)$

$= -(3x - 2)(x + 2)$

or $(-3x + 2)(x + 2)$

or $(3x - 2)(-x - 2)$

**77.** $-6x^2 + 7x + 10 = (-1)(6x^2 - 7x - 10)$

$= -(6x + 5)(x - 2)$

or $(-6x - 5)(x - 2)$

or $(6x + 5)(-x + 2)$

**79.** $1 - 4x - 60x^2 = (1 - 10x)(1 + 6x)$ or

$1 - 4x - 60x^2 = (-1)(60x^2 + 4x - 1)$

$= -(10x - 1)(6x + 1)$

or $(-10x + 1)(6x + 1)$

or $(10x - 1)(-6x - 1)$

**81.** $16 - 8x - 15x^2 = (4 + 3x)(4 - 5x)$ or

$\quad 16 - 8x - 15x^2 = -15x^2 - 8x + 16$

$\qquad\qquad = -1(15x^2 + 8x - 16)$

$\qquad\qquad = -(5x - 4)(3x + 4)$

$\qquad\quad$ or $(-5x + 4)(3x + 4)$

$\qquad\quad$ or $(5x - 4)(-3x - 4)$

**83.** $b = 13 \qquad (3x + 10)(x + 1)$

$\quad b = -13 \qquad (3x - 10)(x - 1)$

$\quad b = 31 \qquad (3x + 1)(x + 10)$

$\quad b = -31 \qquad (3x - 1)(x - 10)$

$\quad b = 11 \qquad (3x + 5)(x + 2)$

$\quad b = -11 \qquad (3x - 5)(x - 2)$

$\quad b = 17 \qquad (3x + 2)(x + 5)$

$\quad b = -17 \qquad (3x - 2)(x - 5)$

**85.** $b = 4 \qquad 2(x + 3)(x - 1)$ or $(2x + 6)(x - 1)$ or $(2x - 2)(x + 3)$

$\quad b = -4 \qquad 2(x - 3)(x + 1)$ or $(2x - 6)(x + 1)$ or $(2x + 2)(x - 3)$

$\quad b = -11 \qquad (2x + 1)(x - 6)$

$\quad b = 11 \qquad (2x - 1)(x + 6)$

$\quad b = -1 \qquad (2x + 3)(x - 2)$

$\quad b = 1 \qquad (2x - 3)(x + 2)$

**87.** $b = 26 \qquad (6x + 20)(x + 1)$ or $(3x + 10)(2x + 2)$

$\quad b = -26 \qquad (6x - 20)(x - 1)$ or $(3x - 10)(2x - 2)$

$\quad b = 22 \qquad (6x + 10)(x + 2)$ or $(3x + 5)(2x + 4)$

$\quad b = -22 \qquad (6x - 10)(x - 2)$ or $(3x - 5)(2x - 4)$

$\quad b = 34 \qquad (6x + 4)(x + 5)$ or $(3x + 2)(2x + 10)$

$\quad b = -34 \qquad (6x - 4)(x - 5)$ or $(3x - 2)(2x - 10)$

$\quad b = 62 \qquad (6x + 2)(x + 10)$ or $(3x + 1)(2x + 20)$

$\quad b = -62 \qquad (6x - 2)(x - 10)$ or $(3x - 1)(2x - 20)$

$\quad b = 121 \qquad (6x + 1)(x + 20)$

$\quad b = -121 \qquad (6x - 1)(x - 20)$

$\quad b = 29 \qquad (6x + 5)(x + 4)$

$\quad b = -29 \qquad (6x - 5)(x - 4)$

$\quad b = 43 \qquad (3x + 20)(2x + 1)$

$\quad b = -43 \qquad (3x - 20)(2x - 1)$

$\quad b = 23 \qquad (3x + 4)(2x + 5)$

$\quad b = -23 \qquad (3x - 4)(2x - 5)$

Note: The common factor, 2, could be factored out from the first 16 of the 24 factorizations listed above.

**89.** $c = -1 \qquad (4x - 1)(x + 1)$

$\quad c = -10 \qquad (4x - 5)(x + 2)$

$\quad c = -27 \qquad (4x - 9)(x + 3)$

$\quad c = -7 \qquad (4x + 7)(x - 1)$

$\quad c = -22 \qquad (4x + 11)(x - 2)$

$\quad c = -45 \qquad (4x + 15)(x - 3)$

There are many correct answers.

**91.** $c = 8 \qquad (3x - 4)(x - 2)$

$\quad c = 7 \qquad (3x - 7)(x - 1)$

$\quad c = 3 \qquad (3x - 1)(x - 3)$

$\quad c = -8 \qquad (3x + 2)(x - 4)$

$\quad c = -25 \qquad (3x + 5)(x - 5)$

$\quad c = -13 \qquad (3x - 13)(x + 1)$

There are many correct answers.

**93.** $c = -1 \qquad (6x + 1)(x - 1)$

$\quad c = -14 \qquad (6x + 7)(x - 2)$

$\quad c = -11 \qquad (6x - 11)(x + 1)$

$\quad c = 1 \qquad (3x - 1)(2x - 1)$

$\quad c = -6 \qquad (3x + 2)(2x - 3)$

$\quad c = -4 \qquad (3x - 4)(2x + 1)$

There are many correct answers.

**95.** $ac = 3(1) = 3$

$b = 4$

The two numbers with a product of 3 and a sum of 4 are 3 and 1.

$$3x^2 + 4x + 1 = 3x^2 + 3x + x + 1$$
$$= (3x^2 + 3x) + (x + 1)$$
$$= 3x(x + 1) + (x + 1)$$
$$= (x + 1)(3x + 1)$$

**97.** $ac = 7(-3) = -21$

$b = 20$

The two numbers with a product of $-21$ and a sum of 20 are 21 and $-1$.

$$7x^2 + 20x - 3 = 7x^2 + 21x - x - 3$$
$$= (7x^2 + 21x) + (-x - 3)$$
$$= 7x(x + 3) - (x + 3)$$
$$= (x + 3)(7x - 1)$$

**99.** $ac = 6(-4) = -24$

$b = 5$

The two numbers with a product of $-24$ and a sum of 5 are $-3$ and 8.

$$6x^2 + 5x - 4 = 6x^2 - 3x + 8x - 4$$
$$= (6x^2 - 3x) + (8x - 4)$$
$$= 3x(2x - 1) + 4(2x - 1)$$
$$= (2x - 1)(3x + 4)$$

**101.** $ac = 15(2) = 30$

$b = -11$

The two numbers with a product of 30 and a sum of $-11$ are $-6$ and $-5$.

$$15x^2 - 11x + 2 = 15x^2 - 6x - 5x + 2$$
$$= (15x^2 - 6x) + (-5x + 2)$$
$$= 3x(5x - 2) - 1(5x - 2)$$
$$= (5x - 2)(3x - 1)$$

**103.** $ac = 3(10) = 30$

$b = 11$

The two numbers with a product of 30 and a sum of 11 are 6 and 5.

$$3a^2 + 11a + 10 = 3a^2 + 6a + 5a + 10$$
$$= 3a(a + 2) + 5(a + 2)$$
$$= (a + 2)(3a + 5)$$

**105.** $ac = 16(-3) = -48$

$b = 2$

The two numbers with a product of $-48$ and a sum of 2 are 8 and $-6$.

$$16x^2 + 2x - 3 = 16x^2 + 8x - 6x - 3$$
$$= (16x^2 + 8x) - (6x + 3)$$
$$= 8x(2x + 1) - 3(2x + 1)$$
$$= (2x + 1)(8x - 3)$$

**107.** $ac = 12(6) = 72$

$b = -17$

The two numbers with a product of 72 and a sum of $-17$ are $-9$ and $-8$.

$$12x^2 - 17x + 6 = 12x^2 - 9x - 8x + 6$$
$$= (12x^2 - 9x) + (-8x + 6)$$
$$= 3x(4x - 3) - 2(4x - 3)$$
$$= (4x - 3)(3x - 2)$$

**109.** $ac = 6(-14) = -84$

$b = -5$

The two numbers with a product of $-84$ and a sum of $-5$ are $-12$ and 7.

$$6u^2 - 5u - 14 = 6u^2 - 12u + 7u - 14$$
$$= (6u^2 - 12u) + (7u - 14)$$
$$= 6u(u - 2) + 7(u - 2)$$
$$= (u - 2)(6u + 7)$$

**111.** Volume = (Length)(Width)(Height)

$$2x^3 + 7x^2 + 6x = x(2x^2 + 7x + 6)$$
$$= x(2x + 3)(x + 2)$$
$$= (2x + 3)(x + 2)x$$

So, $2x + 3$ is the length of the box.

**113.** $2x^2 + 9x + 10 = (2x + 5)(x + 2)$

Length of largest rectangle: $2x + 5$

Width of largest rectangle: $x + 2$

Length of shaded region: $(2x + 5) - x = x + 5$

Width of shaded region: 2

Area of shaded region: $(x + 5)2 = 2x + 10$

**115.** (a) $y_1 = 2x^3 + 3x^2 - 5x$

$= x(2x^2 + 3x - 5)$

$= x(2x + 5)(x - 1)$

$y_1 = y_2$

(b)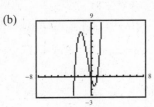

(c) The $x$-intercepts are $(0, 0)$, $(1, 0)$, and $\left(-\frac{5}{2}, 0\right)$.

The $y$-intercept is $(0, 0)$.

**117.** The constant of the trinomial is $-15$, but the product of the last terms of the binomials is 15.

**119.** Four

Here are the four factorizations that need to be tested:

$(ax + 1)(x + c)$      $(ax + c)(x + 1)$

$(ax - 1)(x - c)$      $(ax - c)(x - 1)$

**121.** Examples of third-degree trinomials that have a common factor of $2x$:

$2x(x^2 + x + 1) = 2x^3 + 2x^2 + 2x$

$2x(x^2 - 7x + 3) = 2x^3 - 14x^2 + 6x$

$2x(x + 5)(x - 1) = 2x(x^2 + 4x - 5)$

$= 2x^3 + 8x^2 - 10x$

$2x(x + 3)(x + 3) = 2x(x^2 + 6x + 9)$

$= 2x^3 + 12x^2 + 18x$

$2x(3x - 7)(x - 2) = 2x(3x^2 - 13x + 14)$

$= 6x^3 - 26x^2 + 28x$

There are many correct answers.

**123.** Results of Guess, Check and Revise strategy:

$6x^2 - 13x + 6 = (3x - 2)(2x - 3)$

$2x^2 + 5x - 12 = (2x - 3)(x + 4)$

$3x^2 + 11x - 4 = (3x - 1)(x + 4)$

Factoring by grouping:

$6x^2 - 13x + 6 = 6x^2 - (4x + 9x) + 6$

$= (6x^2 - 4x) - (9x - 6)$

$= 2x(3x - 2) - 3(3x - 2)$

$= (3x - 2)(2x - 3)$

$2x^2 + 5x - 12 = 2x^2 + (8x - 3x) - 12$

$= (2x^2 + 8x) - (3x + 12)$

$= 2x(x + 4) - 3(x + 4)$

$= (x + 4)(2x - 3)$

$3x^2 + 11x - 4 = 3x^2 + (12x - x) - 4$

$= (3x^2 + 12x) - (x + 4)$

$= 3x(x + 4) - (x + 4)$

$= (x + 4)(3x - 1)$

Answers about preferences will vary. Explanations of advantages and disadvantages will vary.

**125.** $315 = 3 \cdot 3 \cdot 5 \cdot 7 = 3^2 \cdot 5 \cdot 7$

**127.** $2275 = 5 \cdot 5 \cdot 7 \cdot 13 = 5^2 \cdot 7 \cdot 13$

**129.** $(3x - 2)^2 = (3x)^2 - 2(3x)(2) + (2)^2$

$= 9x^2 - 12x + 4$

**131.** $(3y + 8)(9y + 3) = 27y^2 + 9y + 72y + 24$

$= 27y^2 + 81y + 24$

**133.** $y^2 + 2y - 15 = (y + 5)(y - 3)$

**135.** $x^2 - 12x + 20 = (x - 10)(x - 2)$

# Mid-Chapter Quiz for Chapter 6

**1.** $\frac{2}{3}x - 1 = \frac{2}{3}x - \frac{3}{3}$

$\qquad = \frac{1}{3}(2x - 3)$

The missing factor is $2x - 3$.

**2.** $x^2y - xy^2 = xy(x - y)$

The missing factor is $x - y$.

**3.** $y^2 + y - 42 = (y + 7)(y - 6)$

The missing factor is $y - 6$.

**4.** $3y^2 - y - 30 = (3y - 10)(y + 3)$

The missing factor is $y + 3$.

**5.** $9x^2 + 21 = 3(3x^2 + 7)$

**6.** $5a^3b - 25a^2b^2 = 5a^2b(a - 5b)$

**7.** $x(x + 7) - 6(x + 7) = (x + 7)(x - 6)$

**8.** $t^3 - 3t^2 + t - 3 = t^2(t - 3) + (t - 3)$

$\qquad = t^2(t - 3) + 1(t - 3)$

$\qquad = (t - 3)(t^2 + 1)$

**9.** $y^2 + 11y + 30 = (y + 6)(y + 5)$

**10.** $u^2 + u - 56 = (u + 8)(u - 7)$

**11.** $x^3 - x^2 - 30x = x(x^2 - x - 30)$

$\qquad = x(x - 6)(x + 5)$

**12.** $2x^2y + 8xy - 64y = 2y(x^2 + 4x - 32)$

$\qquad = 2y(x + 8)(x - 4)$

**13.** $2y^2 - 3y - 27 = (2y - 9)(y + 3)$

**14.** $6 - 13z - 5z^2 = (3 + z)(2 - 5z)$ or

$6 - 13z - 5z^2 = -1(5z^2 + 13z - 6)$

$\qquad = -(5z - 2)(z + 3)$

$\qquad$ or $(-5z + 2)(z + 3)$

$\qquad$ or $(5z - 2)(-z - 3)$

**15.** $12x^2 - 5x - 2 = (4x + 1)(3x - 2)$

**16.** $10s^4 - 14s^3 + 2s^2 = 2s^2(5s^2 - 7s + 1)$

**17.** $b = 7 \qquad (x + 3)(x + 4)$

$b = -7 \qquad (x - 3)(x - 4)$

$b = 8 \qquad (x + 6)(x + 2)$

$b = -8 \qquad (x - 6)(x - 2)$

$b = 13 \qquad (x + 12)(x + 1)$

$b = -13 \qquad (x - 12)(x - 1)$

These integers are the sums of the factors of 12.

**18.** $c = 21 \qquad (x - 7)(x - 3)$

$c = 24 \qquad (x - 6)(x - 4)$

$c = 16 \qquad (x - 2)(x - 8)$

$c = -24 \qquad (x + 2)(x - 12)$

$c = -11 \qquad (x - 11)(x + 1)$

$c = 25 \qquad (x - 5)(x - 5)$

These are some of the possible values of $c$. There are many correct answers. The factors of $c$ have a sum of $-10$.

**19.** $(3x + 2)(x + 3) \qquad (3x - 2)(x - 3)$

$(3x + 3)(x + 2) \qquad (3x - 3)(x - 2)$

$(3x + 6)(x + 1) \qquad (3x - 6)(x - 1)$

$(3x + 1)(x + 6) \qquad (3x - 1)(x - 6)$

$m$ and $n$ are factors of 6.

**20.** $3x^2 + 38x + 80 = (3x + 8)(x + 10)$

Width of the large rectangle: $x + 10$

Length of the large rectangle: $3x + 8$

Width of shaded region: 10

Length of shaded region: $3x + 8 - x = 2x + 8$

Area of shaded region: $10(2x + 8) = 20x + 80$

**21.** $y_1 = y_2$

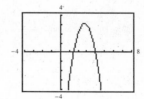

The graphs are identical, and the two expressions are equivalent.

## Section 6.4   Factoring Polynomials with Special Forms

1. $x^2 - 9 = x^2 - 3^2 = (x + 3)(x - 3)$

3. $u^2 - 64 = u^2 - 8^2 = (u + 8)(u - 8)$

5. $144 - x^2 = 12^2 - x^2 = (12 + x)(12 - x)$

7. $u^2 - \frac{1}{4} = u^2 - \left(\frac{1}{2}\right)^2 = \left(u + \frac{1}{2}\right)\left(u - \frac{1}{2}\right)$

9. $v^2 - \frac{4}{9} = v^2 - \left(\frac{2}{3}\right)^2 = \left(v - \frac{2}{3}\right)\left(v + \frac{2}{3}\right)$

11. $16y^2 - 9 = (4y)^2 - 3^2 = (4y + 3)(4y - 3)$

13. $100 - 49x^2 = 10^2 - (7x)^2 = (10 - 7x)(10 + 7x)$

15. $(x - 1)^2 - 4 = (x - 1)^2 - 2^2$
$= \left[(x - 1) - 2\right]\left[(x - 1) + 2\right]$
$= (x - 3)(x + 1)$

17. $25 - (z + 5)^2 = 5^2 - (z + 5)^2$
$= \left[5 + (z + 5)\right]\left[5 - (z + 5)\right]$
$= [5 + z + 5][5 - z - 5]$
$= (z + 10)(-z)$
$= -z(z + 10)$

19. $16 - (a + 2)^2 = 4^2 - (a + 2)^2$
$= \left[4 + (a + 2)\right]\left[4 - (a + 2)\right]$
$= [4 + a + 2][4 - a - 2]$
$= (6 + a)(2 - a)$

21. $9y^2 - 25z^2 = (3y)^2 - (5z)^2 = (3y + 5z)(3y - 5z)$

23. $2x^2 - 72 = 2(x^2 - 36)$
$= 2(x^2 - 6^2)$
$= 2(x + 6)(x - 6)$

25. $4x - 25x^3 = x(4 - 25x^2)$
$= x\left[(2)^2 - (5x)^2\right]$
$= x(2 + 5x)(2 - 5x)$

27. $8y^3 - 50y = 2y(4y^2 - 25)$
$= 2y\left[(2y)^2 - 5^2\right]$
$= 2y(2y + 5)(2y - 5)$

29. $y^4 - 81 = \left(y^2\right)^2 - 9^2$
$= \left(y^2 + 9\right)\left(y^2 - 9\right)$
$= \left(y^2 + 9\right)\left(y^2 - 3^2\right)$
$= \left(y^2 + 9\right)(y + 3)(y - 3)$

31. $1 - x^4 = 1^2 - \left(x^2\right)^2$
$= \left(1 + x^2\right)\left(1 - x^2\right)$
$= \left(1 + x^2\right)\left(1^2 - x^2\right)$
$= \left(1 + x^2\right)(1 + x)(1 - x)$

33. $2x^4 - 162 = 2\left(x^4 - 81\right)$
$= 2\left[\left(x^2\right)^2 - 9^2\right]$
$= 2\left(x^2 + 9\right)\left(x^2 - 9\right)$
$= 2\left(x^2 + 9\right)\left(x^2 - 3^2\right)$
$= 2\left(x^2 + 9\right)(x + 3)(x - 3)$

35. $81x^4 - 16y^4 = \left(9x^2\right)^2 - \left(4y^2\right)^2$
$= \left(9x^2 + 4y^2\right)\left(9x^2 - 4y^2\right)$
$= \left(9x^2 + 4y^2\right)\left[(3x)^2 - (2y)^2\right]$
$= \left(9x^2 + 4y^2\right)(3x + 2y)(3x - 2y)$

37. $9b^2 + 24b + 16$

Yes. This trinomial *is* a perfect square trinomial. It is the square of the sum of $3b$ and 4 because the first term is the square of $3b$, the third term is the square of 4, and the middle term is twice the product of $3b$ and 4.

$24b = 2(3b)(4)$

$9b^2 + 24b + 16 = (3b + 4)^2$

39. $m^2 - 2m - 1$

No. This trinomial *is not* a perfect square trinomial because the third term is not positive.

41. $4k^2 - 20k + 25$

Yes. This trinomial *is* a perfect square trinomial. It is the square of the difference of $2k$ and 5 because the first term is the square of $2k$, the third term is the square of 5, and the middle term is the opposite of twice the product of $2k$ and 5.

$-20k = -\left[2(2k)(5)\right]$

$4k^2 - 20k + 25 = (2k - 5)^2$

**43.** $x^2 - 8x + 16 = x^2 - 2(x)(4) + 4^2 = (x - 4)^2$

**45.** $x^2 + 14x + 49 = x^2 + 2(x)(7) + 7^2 = (x + 7)^2$

**47.** $b^2 + b + \frac{1}{4} = b^2 + 2(b)\left(\frac{1}{2}\right) + \left(\frac{1}{2}\right)^2 = \left(b + \frac{1}{2}\right)^2$

**49.** $4t^2 + 4t + 1 = (2t)^2 + 2(2t)(1) + 1^2 = (2t + 1)^2$

**51.** $25y^2 - 10y + 1 = (5y)^2 - 2(5y)(1) + 1^2 = (5y - 1)^2$

**53.** $4x^2 - x + \frac{1}{16} = (2x)^2 - (2)(2x)\left(\frac{1}{4}\right) + \left(\frac{1}{4}\right)^2$

$\qquad = \left(2x - \frac{1}{4}\right)^2$

**55.** $x^2 - 6xy + 9y^2 = x^2 - 2(x)(3y) + (3y)^2 = (x - 3y)^2$

**57.** $4y^2 + 20yz + 25z^2 = (2y)^2 + 2(10yz) + (5z)^2$

$\qquad = (2y)^2 + 2(2y)(5z) + (5z)^2$

$\qquad = (2y + 5z)^2$

**59.** $9a^2 - 12ab + 4b^2 = (3a)^2 - 2(3a)(2b) + (2b)^2$

$\qquad = (3a - 2b)^2$

**61.** $x^2 + bx + 1 = x^2 + bx + 1^2$

$(x + 1)^2 = x^2 + 2(x)(1) + 1^2 = x^2 + 2x + 1$

$(x - 1)^2 = x^2 - 2(x)(1) + 1^2 = x^2 - 2x + 1$

So, $b = 2$ or $b = -2$.

**63.** $x^2 + bx + \frac{16}{25} = x^2 + bx + \left(\frac{4}{5}\right)^2$

$\left(x + \frac{4}{5}\right)^2 = x^2 + 2(x)\left(\frac{4}{5}\right) + \left(\frac{4}{5}\right)^2 = x^2 + \frac{8}{5}x + \frac{16}{25}$

$\left(x - \frac{4}{5}\right)^2 = x^2 - 2(x)\left(\frac{4}{5}\right) + \left(\frac{4}{5}\right)^2 = x^2 - \frac{8}{5}x + \frac{16}{25}$

So, $b = \frac{8}{5}$ or $b = -\frac{8}{5}$.

**65.** $4x^2 + bx + 81 = (2x)^2 + bx + 9^2$

$(2x + 9)^2 = (2x)^2 + 2(2x)(9) + 9^2$

$\qquad = 4x^2 + 36x + 81$

$(2x - 9)^2 = (2x)^2 - 2(2x)(9) + 9^2$

$\qquad = 4x^2 - 36x + 81$

So, $b = 36$ or $b = -36$.

**67.** $x^2 + 6x + c = x^2 + 2(x)(3) + c$

$(x + 3)^2 = x^2 + 2(x)(3) + 3^2 = x^2 + 6x + 9$

So, $c = 9$.

**69.** $y^2 - 4y + c = (y)^2 - 2(y)(2) + c$

$(y - 2)^2 = y^2 - 2(y)(2) + 2^2 = y^2 - 4y + 4$

So, $c = 4$.

**71.** $x^3 - 8 = x^3 - 2^3$

$\qquad = (x - 2)(x^2 + (x)(2) + 2^2)$

$\qquad = (x - 2)(x^2 + 2x + 4)$

**73.** $y^3 + 64 = y^3 + 4^3$

$\qquad = (y + 4)(y^2 - (y)(4) + 4^2)$

$\qquad = (y + 4)(y^2 - 4y + 16)$

**75.** $1 + 8t^3 = 1^3 + (2t)^3$

$\qquad = (1 + 2t)\left[1^2 - (1)(2t) + (2t)^2\right]$

$\qquad = (1 + 2t)(1 - 2t + 4t^2)$

**77.** $27u^3 - 8 = (3u)^2 - 2^3$

$\qquad = (3u - 2)\left[(3u)^2 + (3u)(2) + 2^2\right]$

$\qquad = (3u - 2)(9u^2 + 6u + 4)$

**79.** $4x^3 - 32s^3 = 4(x^3 - 8s^3)$

$\qquad = 4\left[x^3 - (2s)^3\right]$

$\qquad = 4(x - 2s)(x^2 + 2sx + 4s^2)$

**81.** $27x^3 + 64y^3 = (3x)^3 + (4y)^3$

$\qquad = (3x + 4y)\left[(3x)^2 - (3x)(4y) + (4y)^2\right]$

$\qquad = (3x + 4y)(9y^2 - 12xy + 16y^2)$

**83.** $4x - 28 = 4(x - 7)$

**85.** $u^2 + 3u = u(u + 3)$

**87.** $5y^2 - 25y = 5y(y - 5)$

**89.** $5y^2 - 125 = 5(y^2 - 25)$

$\qquad = 5(y^2 - 5^2)$

$\qquad = 5(y + 5)(y - 5)$

**91.** $y^4 - 25y^2 = y^2(y^2 - 25)$

$\qquad = y^2(y^2 - 5^2)$

$\qquad = y^2(y + 5)(y - 5)$

**93.** $x^2 - 4xy + 4y^2 = x^2 - 2(x)(2y) + (2y)^2$
$$= (x - 2y)^2$$

**95.** $x^2 - 2x + 1 = x^2 - 2(x)(1) + 1^2 = (x - 1)^2$

**97.** $9x^2 + 10x + 1 = (9x + 1)(x + 1)$

**99.** $2x^3 - 2x^2y - 4xy^2 = 2x(x^2 - xy - 2y^2)$
$$= 2x(x - 2y)(x + y)$$

**101.** $9t^2 - 16 = (3t)^2 - 4^2 = (3t + 4)(3t - 4)$

**103.** $36 - (z + 6)^2 = 6^2 - (z + 6)^2$
$$= [6 - (z + 6)][6 + (z + 6)]$$
$$= -z(z + 12)$$

**105.** $(t - 1)^2 - 121 = (t - 1)^2 - 11^2$
$$= [(t - 1) + 11][(t - 1) - 11]$$
$$= (t + 10)(t - 12)$$

**107.** $u^3 + 2u^2 + 3u = u(u^2 + 2u + 3)$

**109.** $x^2 + 81$

This polynomial is prime.

**111.** $2t^3 - 16 = 2(t^3 - 8)$
$$= 2(t^3 - 2^3)$$
$$= 2(t - 2)(t^2 + (t)(2) + 2^2)$$
$$= 2(t - 2)(t^2 + 2t + 4)$$

**113.** $3a^3 + 24b^3 = 3(a^3 + 8b^3)$
$$= 3[a^3 + (2b)^3]$$
$$= 3(a + 2b)[a^2 - a(2b) + (2b)^2]$$
$$= 3(a + 2b)(a^2 - 2ab + 4b^2)$$

**115.** $x^4 - 81 = (x^2)^2 - 9^2$
$$= (x^2 + 9)(x^2 - 9)$$
$$= (x^2 + 9)(x^2 - 3^2)$$
$$= (x^2 + 9)(x + 3)(x - 3)$$

**117.** $x^4 - y^4 = (x^2)^2 - (y^2)^2$
$$= (x^2 + y^2)(x^2 - y^2)$$
$$= (x^2 + y^2)(x + y)(x - y)$$

**119.** $x^3 - 4x^2 - x + 4 = x^2(x - 4) - 1(x - 4)$
$$= (x - 4)(x^2 - 1)$$
$$= (x - 4)(x^2 - 1^2)$$
$$= (x - 4)(x + 1)(x - 1)$$

**121.** $x^4 + 3x^3 - 16x^2 - 48x = x(x^3 + 3x^2 - 16x - 48)$
$$= x[x^2(x + 3) - 16(x + 3)]$$
$$= x[(x + 3)(x^2 - 16)]$$
$$= x(x + 3)(x^2 - 4^2)$$
$$= x(x + 3)(x + 4)(x - 4)$$

**123.** $64 - y^6 = 8^2 - (y^3)^2$
$$= (8 - y^3)(8 + y^3)$$
$$= (2^3 - y^3)(2^3 + y^3)$$
$$= (2 - y)(2^2 + 2y + y^2)(2 + y)(2^2 - 2y + y^2)$$
$$= (2 - y)(4 + 2y + y^2)(2 + y)(4 - 2y + y^2)$$
$$\text{or } (2 - y)(2 + y)(y^2 + 2y + 4)(y^2 - 2y + 4)$$

**125.** $21^2 = (20 + 1)^2$
$$= 20^2 + 2(20)(1) + 1^2$$
$$= 400 + 40 + 1$$
$$= 441$$

**127.** $59 \cdot 61 = (60 - 1)(60 + 1)$
$$= 60^2 - 1^2$$
$$= 3600 - 1$$
$$= 3599$$

**129.** $-16t^2 + 841 = -1(16t^2 - 841)$
$$= -1[(4t)^2 - 29^2]$$
$$= -1(4t + 29)(4t - 29)$$

**131.** $x^2 + 6x + 8 = (x^2 + 6x + 9) - 1$
$$= (x + 3)^2 - 1^2$$
$$= [(x + 3) + 1][(x + 3) - 1]$$
$$= (x + 4)(x + 2)$$

**133.**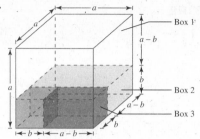

Volume of larger cube $= a^3$

Volume of smaller cube $= b^3$

Volume of Box 1: $a \cdot a(a - b) = a^2(a - b)$

Volume of Box 2: $a \cdot b(a - b)$

Volume of Box 3: $b \cdot b(a - b) = b^2(a - b)$

Special product pattern: $a^3 - b^3 = (a - b)(a^2 + ab + b^2)$

The first expression represents the total volume as a difference of volumes—a difference of two cubes:

$a^3 - b^3 =$ the volume of the large cube $-$ the volume of the smaller cube

The second expression represents the total volume as a sum of 3 volumes:

$(a - b)(a^2 + ab + b^2) = a^2(a - b) + ab(a - b) + b^2(a - b) =$ Box 1 + Box 2 + Box 3

**135.** No, $x(x + 2) - 2(x + 2)$ is not in factored form.

$x(x + 2) - 2(x + 2) = (x + 2)(x - 2)$

**137.** No. $x^3 - 27 = x^3 - 3^3 = (x - 3)(x^2 + 3x + 9)$

Also, $(x - 3)^3 = (x - 3)(x - 3)(x - 3)$

$= (x - 3)(x^2 - 6x + 9)$

$= x^3 - 6x^2 + 9x - 3x^2 + 18x - 27$

$= x^3 - 9x^2 + 27x - 27.$

So, $x^3 - 27 \neq (x - 3)^3.$

**139.**
$$7 + 5x = 7x - 1$$
$$7 + 5x - 7x = 7x - 1 - 7x$$
$$7 - 2x = -1$$
$$7 - 2x - 7 = -1 - 7$$
$$-2x = -8$$
$$\frac{-2x}{-2} = \frac{-8}{-2}$$
$$x = 4$$

**141.**
$$2(x + 1) = 0$$
$$2x + 2 = 0$$
$$2x + 2 - 2 = 0 - 2$$
$$2x = -2$$
$$\frac{2x}{2} = \frac{-2}{2}$$
$$x = -1$$

**143.**
$$\frac{x}{5} + \frac{1}{5} = \frac{7}{10}$$
$$10\left(\frac{x}{5} + \frac{1}{5}\right) = 10\left(\frac{7}{10}\right)$$
$$10\left(\frac{x}{5}\right) + 10\left(\frac{1}{5}\right) = 10\left(\frac{7}{10}\right)$$
$$2x + 2 = 7$$
$$2x + 2 - 2 = 7 - 2$$
$$2x = 5$$
$$\frac{2x}{2} = \frac{5}{2}$$
$$x = \frac{5}{2}$$

**145.** $2x^2 + 7x + 3 = (2x + 1)(x + 3)$

**147.** $6m^2 + 7m - 20 = (3m - 4)(2m + 5)$

# Section 6.5  Solving Quadratic Equations by Factoring

**1.** $x(x - 5) = 0$

$x = 0$

$x - 5 = 0 \Rightarrow x = 5$

Note: There are *two* solutions, 0 and 5.

**3.** $(y - 2)(y - 3) = 0$

$y - 2 = 0 \Rightarrow y = 2$

$y - 3 = 0 \Rightarrow y = 3$

**5.** $(2t - 5)(3t + 1) = 0$

$2t - 5 = 0 \Rightarrow 2t = 5 \Rightarrow t = \frac{5}{2}$

$3t + 1 = 0 \Rightarrow 3t = -1 \Rightarrow t = -\frac{1}{3}$

**7.** $x^2 - 16 = 0$

$(x + 4)(x - 4) = 0$

$x + 4 = 0 \Rightarrow x = -4$

$x - 4 = 0 \Rightarrow x = 4$

**9.** $3y^2 - 27 = 0$

$3(y^2 - 9) = 0$

$3(y + 3)(y - 3) = 0$

$3 \neq 0$

$y + 3 = 0 \Rightarrow y = -3$

$y - 3 = 0 \Rightarrow y = 3$

**11.** $2x^2 + 4x = 0$

$2x(x + 2) = 0$

$2x = 0 \Rightarrow x = 0$

$x + 2 = 0 \Rightarrow x = -2$

**13.** $x^2 - 2x - 8 = 0$

$(x - 4)(x + 2) = 0$

$x - 4 = 0 \Rightarrow x = 4$

$x + 2 = 0 \Rightarrow x = -2$

**15.** $t^2 - 7t - 18 = 0$

$(t - 9)(t + 2) = 0$

$t - 9 = 0 \Rightarrow t = 9$

$t + 2 = 0 \Rightarrow t = -2$

**17.** $3 + 5x - 2x^2 = 0$

$(3 - x)(1 + 2x) = 0$

$3 - x = 0 \Rightarrow -x = -3 \Rightarrow x = 3$

$1 + 2x = 0 \Rightarrow 2x = -1 \Rightarrow x = -\frac{1}{2}$

**19.** $6x^2 + 4x - 10 = 0$

$2(3x^2 + 2x - 5) = 0$

$2(3x + 5)(x - 1) = 0$

$2 \neq 0$

$3x + 5 = 0 \Rightarrow 3x = -5 \Rightarrow x = -\frac{5}{3}$

$x - 1 = 0 \Rightarrow x = 1$

**21.** $t^2 + 6t = 16$

$t^2 + 6t - 16 = 0$

$(t + 8)(t - 2) = 0$

$t + 8 = 0 \Rightarrow t = -8$

$t - 2 = 0 \Rightarrow t = 2$

**23.** $2y^2 - y = 3$

$2y^2 - y - 3 = 0$

$(2y - 3)(y + 1) = 0$

$2y - 3 = 0 \Rightarrow 2y = 3 \Rightarrow y = \frac{3}{2}$

$y + 1 = 0 \Rightarrow y = -1$

**25.** $5x^2 - 24 = 37x$

$5x^2 - 37x - 24 = 0$

$(5x + 3)(x - 8) = 0$

$5x + 3 = 0 \Rightarrow 5x = -3 \Rightarrow x = -\frac{3}{5}$

$x - 8 = 0 \Rightarrow x = 8$

**27.** $1 + x^2 = 2x$

$x^2 - 2x + 1 = 0$

$x^2 - 2(x)(1) + 1^2 = 0$

$(x - 1)^2 = 0$

$x - 1 = 0 \Rightarrow x = 1$

**29.** $x^2 + 18x + 49 = 4x$

$x^2 + 14x + 49 = 0$

$(x + 7)^2 = 0$

$x + 7 = 0 \Rightarrow x = -7$

**31.** $4t^2 + 2t + 9 = 14t$

$4t^2 - 12t + 9 = 0$

$(2t - 3)^2 = 0$

$2t - 3 = 0 \Rightarrow 2t = 3 \Rightarrow t = \frac{3}{2}$

**33.**
$$3x^2 - 2x = 9 - 8x$$
$$3x^2 + 6x - 9 = 0$$
$$3(x^2 + 2x - 3) = 0$$
$$3(x + 3)(x - 1) = 0$$
$$3 \neq 0$$
$$x + 3 = 0 \Rightarrow x = -3$$
$$x - 1 = 0 \Rightarrow x = 1$$

**35.**
$$x(x + 10) = 24$$
$$x^2 + 10x = 24$$
$$x^2 + 10x - 24 = 0$$
$$(x + 12)(x - 2) = 0$$
$$x + 12 = 0 \Rightarrow x = -12$$
$$x - 2 = 0 \Rightarrow x = 2$$

**37.**
$$x(3x + 1) = 2$$
$$3x^2 + x = 2$$
$$3x^2 + x - 2 = 0$$
$$(3x - 2)(x + 1) = 0$$
$$3x - 2 = 0 \Rightarrow 3x = 2 \Rightarrow x = \tfrac{2}{3}$$
$$x + 1 = 0 \Rightarrow x = -1$$

**39.** $(x + 1)(x - 3) = 12$
$$x^2 - 2x - 3 = 12$$
$$x^2 - 2x - 15 = 0$$
$$(x - 5)(x + 3) = 0$$
$$x - 5 = 0 \Rightarrow x = 5$$
$$x + 3 = 0 \Rightarrow x = -3$$

**41.** $(x - 6)(x - 3) = -2$
$$x^2 - 9x + 18 = -2$$
$$x^2 - 9x + 20 = 0$$
$$(x - 5)(x - 4) = 0$$
$$x - 5 = 0 \Rightarrow x = 5$$
$$x - 4 = 0 \Rightarrow x = 4$$

**43.** $(x - 7)(x + 9) = 17$
$$x^2 + 2x - 63 = 17$$
$$x^2 + 2x - 80 = 0$$
$$(x + 10)(x - 8) = 0$$
$$x + 10 = 0 \Rightarrow x = -10$$
$$x - 8 = 0 \Rightarrow x = 8$$

**45.** The $x$-intercepts are $(1, 0)$ and $(-3, 0)$.
$$y = x^2 + 2x - 3$$
$$0 = x^2 + 2x - 3$$
$$0 = (x + 3)(x - 1)$$
$$x + 3 = 0 \Rightarrow x = -3$$
$$x - 1 = 0 \Rightarrow x = 1$$
$x$-intercepts: $(-3, 0), (1, 0)$

The number of solutions is equal to the number of $x$-intercepts, and the solutions correspond to the first coordinates of the $x$-intercepts.

**47.** The $x$-intercepts are $(-3, 0)$ and $(4, 0)$.
$$y = 12 + x - x^2$$
$$0 = 12 + x - x^2$$
$$0 = (4 - x)(3 + x)$$
$$4 - x = 0 \Rightarrow -x = -4 \Rightarrow x = 4$$
$$3 + x = 0 \Rightarrow x = -3$$
$x$-intercepts: $(4, 0)$ and $(-3, 0)$

The number of solutions is equal to the number of $x$-intercepts, and the solutions correspond to the first coordinates of the $x$-intercepts.

**49.**

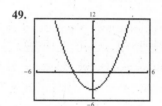

The $x$-intercepts are $(2, 0)$ and $(-2, 0)$.

**Check:**
$$y = x^2 - 4 \qquad\qquad y = x^2 - 4$$
$$y = 2^2 - 4 \qquad\qquad y = (-2)^2 - 4$$
$$y = 4 - 4 \qquad\qquad y = 4 - 4$$
$$y = 0 \qquad\qquad\qquad y = 0$$

**51.**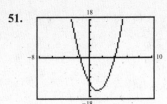

The $x$-intercepts are $\left(-\frac{3}{2}, 0\right)$ and $(4, 0)$.

**Check:** $y = 2x^2 - 5x - 12$

$y = 2\left(-\frac{3}{2}\right)^2 - 5\left(-\frac{3}{2}\right) - 12$

$y = 2\left(\frac{9}{4}\right) + \frac{15}{2} - 12$

$y = \frac{9}{2} + \frac{15}{2} - 12$

$y = \frac{24}{2} - 12$

$y = 12 - 12$

$y = 0$

$y = 2x^2 - 5x - 12$

$y = 2(4)^2 - 5(4) - 12$

$y = 2(16) - 20 - 12$

$y = 32 - 20 - 12$

$y = 0$

**53.** *Verbal Model:* $\boxed{\begin{array}{l}\text{First consecutive}\\\text{positive integer}\end{array}} \cdot \boxed{\begin{array}{l}\text{Second consecutive}\\\text{positive integer}\end{array}} = \boxed{72}$

*Labels:*     First integer $= n$

Second integer $= n + 1$

*Equation:*     $n(n + 1) = 72$

$n^2 + n = 72$

$n^2 + n - 72 = 0$

$(n + 9)(n - 8) = 0$

$n + 9 = 0 \Rightarrow n = -9$

$n - 8 = 0 \Rightarrow n = 8$ and $n + 1 = 9$

Because the problem states that the integers are positive, we discard $-9$ as a solution. So, $n = 8$ and $n + 1 = 9$; the two consecutive integers are 8 and 9.

**55.** *Verbal Model:* $\boxed{\begin{array}{l}\text{First consecutive}\\\text{even integer}\end{array}} \cdot \boxed{\begin{array}{l}\text{Second consecutive}\\\text{even integer}\end{array}} = \boxed{440}$

*Labels:*     First consecutive even integer $= 2n$

Second consecutive even integer $= 2n + 2$

*Equation:*     $2n(2n + 2) = 440$

$4n^2 + 4n = 440$

$4n^2 + 4n - 440 = 0$

$4(n^2 + n - 110) = 0$

$4(n + 11)(n - 10) = 0$

$4 \neq 0$

$n + 11 = 0 \Rightarrow n = -11$

$n - 10 = 0 \Rightarrow n = 10$ so $2(n) = 20$ and $2n + 2 = 22$

Because the problem states that the integers are positive, discard the negative solution. The two consecutive positive even integers are $2(10) = 20$ and $2(10) + 2 = 22$.

**57.**
$$\text{Height} = -16t^2 + 1600$$
$$0 = -16t^2 + 1600 \quad \text{(Set height equal to 0.)}$$
$$16t^2 - 1600 = 0$$
$$16\left(t^2 - 100\right) = 0$$
$$16(t + 10)(t - 10) = 0$$
$$16 \neq 0$$
$$t + 10 = 0 \Rightarrow t = -10$$
$$t - 10 = 0 \Rightarrow t = 10$$

Because the time could not be $-10$ seconds, we discard this negative answer. So, $t = 10$; the object reaches the ground in 10 seconds.

**59.**
$$\text{Height} = -16t^2 + 16t + 32$$
$$0 = -16t^2 + 16t + 32$$
$$16t^2 - 16t - 32 = 0$$
$$16\left(t^2 - t - 2\right) = 0$$
$$16(t - 2)(t + 1) = 0$$
$$16 \neq 0$$
$$t - 2 = 0 \Rightarrow t = 2$$
$$t + 1 = 0 \Rightarrow t = -1$$

Because the time cannot be $t = -1$ seconds, discard the negative answer. The diver reaches the water after $t = 2$ seconds.

**61.** *Verbal Model:*   $\boxed{\begin{array}{c}\text{Length of}\\\text{frame}\end{array}} \cdot \boxed{\begin{array}{c}\text{Width of}\\\text{frame}\end{array}} = \boxed{\begin{array}{c}\text{Area of}\\\text{frame}\end{array}}$

*Labels:*   $\text{Width} = x$    (inches)

   $\text{Length} = x + 3$    (inches)

   $\text{Area} = 108$    (square inches)

*Equation:*
$$(x + 3)x = 108$$
$$x^2 + 3x = 108$$
$$x^2 + 3x - 108 = 0$$
$$(x + 12)(x - 9) = 0$$
$$x + 12 = 0 \Rightarrow x = -12$$
$$x - 9 = 0 \Rightarrow x = 9 \text{ and } x + 3 = 12$$

Because the width of the frame could not be $-12$, we discard this answer. So, $x = 9$ and $x + 3 = 12$; the width of the frame is 9 inches and the length is 12 inches.

**63.** *Verbal Model:* $\boxed{\text{Length}} \cdot \boxed{\text{Width}} = \boxed{\text{Area}}$

*Labels:* Width $= w$ (inches)

Length $= 2w$ (inches)

Area $= 648$ (square inches)

*Equation:*
$$2w(w) = 648$$
$$2w^2 = 648$$
$$w^2 = 324$$
$$w^2 - 324 = 0$$
$$(w + 18)(w - 18) = 0$$
$$w + 18 = 0 \Rightarrow w = -18$$
$$w - 18 = 0 \Rightarrow w = 18 \text{ and } 2w = 36$$

Discard the negative solution. The length of the poster is 36 inches and the width is 18 inches.

**65.** (a) Volume $= (\text{length})(\text{width})(\text{height})$

Length $= x$

Width $= x$

Height $= 2$

Volume $= (x)(x)(2) = 2x^2$

(b)

| $x$ | 2 | 4 | 6 | 8 |
|---|---|---|---|---|
| $V$ | 8 | 32 | 72 | 128 |

$x = 2, V = 2(2)^2 = 2 \cdot 4 = 8$

$x = 4, V = 2(4)^2 = 2 \cdot 16 = 32$

$x = 6, V = 2(6)^2 = 2 \cdot 36 = 72$

$x = 8, V = 2(8)^2 = 2 \cdot 64 = 128$

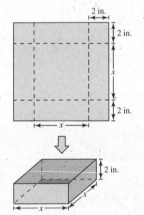

(c)
$$V = 200 \Rightarrow 2x^2 = 200$$
$$2x^2 = 200$$
$$2x^2 - 200 = 0$$
$$2(x^2 - 100) = 0$$
$$2(x + 10)(x - 10) = 0$$
$$2 \neq 0$$
$$x + 10 = 0 \Rightarrow x = -10$$
$$x - 10 = 0 \Rightarrow x = 10$$

Because the length of the box could not be $-10$, we discard this negative solution. So, $x = 10$. The original piece of cardboard is a square with each side $2 + x + 2$ or $x + 4$ inches long. We know $x = 10$, so $x + 4 = 14$. The original piece of cardboard is a 14 inch $\times$ 14 inch square.

**67.** (a)

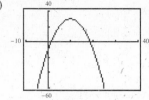

(b) The profit is $20 for two $x$-values, $x = 5$ and $x = 15$.

(c)
$$P = -0.4x^2 + 8x - 10$$
$$20 = -0.4x^2 + 8x - 10$$
$$0 = -0.4x^2 + 8x - 30$$
$$10(0) = 10(-0.4x^2 + 8x - 30)$$
$$0 = -4x^2 + 80x - 300$$
$$0 = -4(x^2 - 20x + 75)$$
$$0 = -4(x - 5)(x - 15)$$
$$-4 \neq 0$$
$$x - 5 = 0 \Rightarrow x = 5$$
$$x - 15 = 0 \Rightarrow x = 15$$

**69.** (a)  $1 + 2 + 3 + 4 + 5 = 15$

$1 + 2 + 3 + 4 + 5 + 6 + 7 + 8 = 36$

$1 + 2 + 3 + 4 + 5 + 6 + 7 + 8 + 9 + 10 = 55$

(b) For $n = 5$: $\frac{1}{2}n(n + 1) = \frac{1}{2}(5)(5 + 1)$
$$= \frac{1}{2}(5)(6) = 15$$

For $n = 8$: $\frac{1}{2}n(n + 1) = \frac{1}{2}(8)(8 + 1)$
$$= \frac{1}{2}(8)(9) = 36$$

For $n = 10$: $\frac{1}{2}n(n + 1) = \frac{1}{2}(10)(10 + 1)$
$$= \frac{1}{2}(10)(11) = 55$$

(c)
$$\frac{1}{2}n(n + 1) = 210$$
$$n(n + 1) = 420$$
$$n^2 + n = 420$$
$$n^2 + n - 420 = 0$$
$$(n + 21)(n - 20) = 0$$
$$n + 21 = 0 \Rightarrow n = -21$$
$$n - 20 = 0 \Rightarrow n = 20$$

Discard the negative solution. If the sum of the first $n$ natural numbers is 210, then $n = 20$.

**71.** False. The equation has two solutions, $x = 0$ and $x = 4$.
$$x^2 - 4x = 0$$
$$x(x - 4) = 0$$
$$x = 0$$
$$x - 4 = 0 \Rightarrow x = 4$$

**73.** True.
$$ax^2 + bx = 0$$
$$x(ax + b) = 0$$
$$x = 0$$
$$ax + b = 0 \Rightarrow ax = -b \Rightarrow x = -\frac{b}{a}$$

**75.** The student should have written the equation in general form before factoring. Then set each factor equal to 0 instead of setting the factors equal to 12.
$$x^2 + 4x = 12$$
$$x^2 + 4x - 12 = 0$$
$$(x + 6)(x - 2) = 0$$
$$x + 6 = 0 \Rightarrow x = -6$$
$$x - 2 = 0 \Rightarrow x = 2$$

**Check:** $(-6)^2 + 4(-6) \overset{?}{=} 12$ $\qquad$ $2^2 + 4(2) \overset{?}{=} 12$

$\qquad\qquad\quad 36 - 24 \overset{?}{=} 12$ $\qquad\qquad\quad 4 + 8 \overset{?}{=} 12$

$\qquad\qquad\quad 12 = 12$ $\qquad\qquad\qquad\quad 12 = 12$

**77.** $7x - 2y > -5$

   (a) $7(-2) - 2(-6) \overset{?}{>} -5$

       $-14 + 12 \overset{?}{>} -5$

          $-2 > -5$

     $(-2, -6)$ *is* a solution.

   (b) $7(1) - 2(8) \overset{?}{>} -5$

       $7 - 16 \overset{?}{>} -5$

        $-9 \not> -5$

     $(1, 8)$ *is not* a solution.

   (c) $7(0) - 2(0) \overset{?}{>} -5$

        $0 - 0 \overset{?}{>} -5$

          $0 > -5$

     $(0, 0)$ *is* a solution.

   (d) $7(3) - 2(12) \overset{?}{>} -5$

       $21 - 24 \overset{?}{>} -5$

        $-3 > -5$

     $(3, 12)$ *is* a solution.

**79.** $\frac{9}{16} - x^2 = \left(-x^2 + \frac{9}{16}\right)$

$= -\left(x^2 - \frac{9}{16}\right)$

$= -\left[x^2 - \left(\frac{3}{4}\right)^2\right]$

$= -\left(x + \frac{3}{4}\right)\left(x - \frac{3}{4}\right)$

**81.** $9x^2 - 324 = 9\left(x^2 - 36\right)$

$= 9\left(x^2 - 6^2\right)$

$= 9(x + 6)(x - 6)$

**83.** $x^2 - 22x + 121 = x^2 - 2(x)(11) + 11^2 = (x - 11)^2$

**85.** $9x^2 + 42x + 49 = (3x)^2 + 2(3x)(7) + 7^2 = (3x + 7)^2$

# Review Exercises for Chapter 6

**1.** $t^2 = t \cdot t = t^2(1)$

$t^5 = t \cdot t \cdot t \cdot t \cdot t = t^2(t^3)$

The greatest common factor is $t^2$.

**3.** $3x^4 = 3 \cdot x \cdot x \cdot x \cdot x = 3x^2(x^2)$

$21x^2 = 3 \cdot 7 \cdot x \cdot x = 3x^2(7)$

The greatest common factor is $3x^2$.

**5.** $14x^2y^3 = 2 \cdot 7 \cdot x \cdot x \cdot y \cdot y \cdot y = 7x^2y^3(2)$

$-21x^3y^5 = -1 \cdot 3 \cdot 7 \cdot x \cdot x \cdot x \cdot y \cdot y \cdot y \cdot y \cdot y = 7x^2y^3(-3xy^2)$

The greatest common factor is $7x^2y^3$.

**7.** $8x^2y = 2 \cdot 2 \cdot 2 \cdot x \cdot x \cdot y = 4xy(2x)$

$24xy^2 = 2 \cdot 2 \cdot 2 \cdot 3 \cdot x \cdot y \cdot y = 4xy(6y)$

$4xy = 2 \cdot 2 \cdot x \cdot y = 4xy(1)$

The greatest common factor is $4xy$.

**9.** $3x - 6 = 3(x - 2)$

**11.** $3t - t^2 = t(3 - t)$

**13.** $5x^2 + 10x^3 = 5x^2(1 + 2x)$

**15.** $8a^2 - 12a^3 = 4a^2(2 - 3a)$

**17.** $5x^3 + 5x^2 - 5x = 5x(x^2 + x - 1)$

**19.** $8y^2 + 4y + 12 = 4(2y^2 + y + 3)$

**21.** $p(p - 4) - 2(p - 4) = (p - 4)(p - 2)$

**23.** Area of larger rectangle $= 2x(2x + 5) = 4x^2 + 10x$

Area of smaller rectangle $= x(x + 6) = x^2 + 6x$

Area of shaded region $= \left(4x^2 + 10x\right) - \left(x^2 + 6x\right)$

$= 4x^2 + 10x - x^2 - 6x$

$= 3x^2 + 4x$

$= x(3x + 4)$

**25.** $x(x + 1) - 3(x + 1) = (x + 1)(x - 3)$

**27.** $2u(u - 2) + 5(u - 2) = (u - 2)(2u + 5)$

**29.** $y^3 + 3y^2 + 2y + 6 = y^2(y + 3) + 2(y + 3)$
$$= (y + 3)(y^2 + 2)$$

**31.** $x^3 + 2x^2 + x + 2 = (x^3 + 2x^2) + (x + 2)$
$$= x^2(x + 2) + 1(x + 2)$$
$$= (x + 2)(x^2 + 1)$$

**33.** $x^2 - 4x + 3x - 12 = x(x - 4) + 3(x - 4)$
$$= (x - 4)(x + 3)$$

**35.** $x^2 - 3x - 28 = (x - 7)(x + 4)$

**37.** $u^2 + 5u - 36 = (u + 9)(u - 4)$

**39.** $x^2 - 2x - 24 = (x - 6)(x + 4)$

**41.** $y^2 + 10y + 21 = (y + 3)(y + 7)$

**43.** $b^2 + 13b - 30 = (b + 15)(b - 2)$

**45.** $w^2 + 3w - 40 = (w + 8)(w - 5)$

**47.** $b = 6 \qquad (x + 3)(x + 3)$
$b = -6 \qquad (x - 3)(x - 3)$
$b = 10 \qquad (x + 9)(x + 1)$
$b = -10 \qquad (x - 9)(x - 1)$

**49.** $b = 12 \qquad (z + 11)(z + 1)$
$b = -12 \qquad (z - 11)(z - 1)$

**51.** $x^2 + 9xy - 10y^2 = (x + 10y)(x - y)$

**53.** $y^2 - 6xy - 27x^2 = (y - 9x)(y + 3x)$

**55.** $x^2 - 2xy - 8y^2 = (x - 4y)(x + 2y)$

**57.** $4x^2 - 24x + 32 = 4(x^2 - 6x + 8) = 4(x - 4)(x - 2)$

**59.** $x^3 + 9x^2 + 18x = x(x^2 + 9x + 18)$
$$= x(x + 6)(x + 3)$$

**61.** $4x^3 + 36x^2 + 56x = 4x(x^2 + 9x + 14)$
$$= 4x(x + 7)(x + 2)$$

**63.** $3x^2 + 18x - 81 = 3(x^2 + 6x - 27) = 3(x + 9)(x - 3)$

**65.** $5 - 2x - 3x^2 = (5 + 3x)(1 - x)$
or $-(3x + 5)(x - 1)$

**67.** $50 - 5x - x^2 = (10 + x)(5 - x)$ or
$$50 - 5x - x^2 = -1(-50 + 5x + x^2)$$
$$= -1(x^2 + 5x - 50)$$
$$= -1(x + 10)(x - 5)$$

**69.** $6x^2 + 7x + 2 = (3x + 2)(2x + 1)$

**71.** $4y^2 - 3y - 1 = (4y + 1)(y - 1)$

**73.** $3x^2 + 7x - 6 = (3x - 2)(x + 3)$

**75.** $3x^2 + 5x - 2 = (3x - 1)(x + 2)$

**77.** $2x^2 - 3x + 1 = (2x - 1)(x - 1)$

**79.** $b = 23 \qquad (b + 24)(b - 1)$
$b = -23 \qquad (b - 24)(b + 1)$
$b = 10 \qquad (b + 12)(b - 2)$
$b = -10 \qquad (b - 12)(b + 2)$
$b = 5 \qquad (b + 8)(b - 3)$
$b = -5 \qquad (b - 8)(b + 3)$
$b = 2 \qquad (b + 6)(b - 4)$
$b = -2 \qquad (b - 6)(b + 4)$

**81.** $c = 2 \qquad 2(x - 1)^2$
$c = -6 \qquad 2(x + 1)(x - 3)$
$c = -16 \qquad 2(x + 2)(x - 4)$
$c = -30 \qquad 2(x + 3)(x - 5)$
$c = -48 \qquad 2(x + 4)(x - 6)$

There are many other correct answers.

**83.** $3x^2 + 33x + 90 = 3(x^2 + 11x + 30)$
$$= 3(x + 6)(x + 5)$$

**85.** $6y^2 + 39y - 21 = 3(2y^2 + 13y - 7)$
$$= 3(2y - 1)(y + 7)$$

**87.** $6u^3 + 3u^2 - 30u = 3u(2u^2 + u - 10)$
$$= 3u(2u + 5)(u - 2)$$

**89.** $8y^3 - 20y^2 + 12y = 4y(2y^2 - 5y + 3)$
$$= 4y(2y - 3)(y - 1)$$

**91.** $6x^3 + 14x^2 - 12x = 2x(3x^2 + 7x - 6)$
$$= 2x(3x - 2)(x + 3)$$

**93.** $3x^3 + 4x^2 + x = x(3x^2 + 4x + 1)$
$$= x(3x + 1)(x + 1)$$

Volume = (Length)(Width)(Height)

The height of the cake box is $x$ and the width is $x + 1$.
The remaining factor, $3x + 1$, is the length of the box.

**95.** $2x^2 - 13x + 21 = 2x^2 - 6x - 7x + 21$
$$= (2x^2 - 6x) - (7x - 21)$$
$$= 2x(x - 3) - 7(x - 3)$$
$$= (x - 3)(2x - 7)$$

**97.** $4y^2 + y - 3 = 4y^2 + 4y - 3y - 3$
$$= 4y(y + 1) - 3(y + 1)$$
$$= (y + 1)(4y - 3)$$

**99.** $6x^2 + 11x - 10 = 6x^2 + 15x - 4x - 10$
$$= 3x(2x + 5) - 2(2x + 5)$$
$$= (2x + 5)(3x - 2)$$

**101.** $ac = 14(5) = 70$

$b = 17$

The two numbers with the product of 70 and a sum of 17 are 7 and 10.

$14x^2 + 17x + 5 = 14x^2 + 7x + 10x + 5$
$$= (14x^2 + 7x) + (10x + 5)$$
$$= 7x(2x + 1) + 5(2x + 1)$$
$$= (2x + 1)(7x + 5)$$

**103.** $a^2 - 100 = a^2 - 10^2 = (a + 10)(a - 10)$

**105.** $25 - 4y^2 = 5^2 - (2y)^2 = (5 + 2y)(5 - 2y)$

**107.** $12x^2 - 27 = 3(4x^2 - 9)$
$$= 3\left[(2x)^2 - 3^2\right]$$
$$= 3(2x + 3)(2x - 3)$$

**109.** $(u + 1)^2 - 4 = (u + 1)^2 - 2^2$
$$= \left[(u + 1) + 2\right]\left[(u + 1) - 2\right]$$
$$= (u + 3)(u - 1)$$

**111.** $16 - (z - 5)^2 = 4^2 - (z - 5)^2$
$$= \left[4 + (z - 5)\right]\left[4 - (z - 5)\right]$$
$$= (4 + z - 5)(4 - z + 5)$$
$$= (z - 1)(-z + 9) \text{ or } -(z - 1)(z - 9)$$

**113.** $x^3 - x = x(x^2 - 1) = x(x^2 - 1^2) = x(x - 1)(x + 1)$

The two missing factors are $(x - 1)(x + 1)$.

**115.** $3y^3 - 75y = 3y(y^2 - 25)$
$$= 3y(y^2 - 5^2)$$
$$= 3y(y + 5)(y - 5)$$

**117.** $s^3t - st^3 = st(s^2 - t^2) = st(s + t)(s - t)$

**119.** $x^4 - 81 = (x^2)^2 - 9^2$
$$= (x^2 + 9)(x^2 - 9)$$
$$= (x^2 + 9)(x^2 - 3^2)$$
$$= (x^2 + 9)(x + 3)(x - 3)$$

**121.** $x^3 - 2x^2 + 4x - 8 = (x^3 - 2x^2) + (4x - 8)$
$$= x^2(x - 2) + 4(x - 2)$$
$$= (x - 2)(x^2 + 4)$$

**123.** $x^2 - 8x + 16 = x^2 - 2(x)(4) + 4^2 = (x - 4)^2$

**125.** $9s^2 + 12s + 4 = (3s)^2 + 2(3s)(2) + 2^2 = (3s + 2)^2$

**127.** $y^2 + 4yz + 4z^2 = y^2 + 2(y)(2z) + (2z)^2$
$$= (y + 2z)^2$$

**129.** $x^2 + \frac{2}{3}x + \frac{1}{9} = x^2 + 2(x)\left(\frac{1}{3}\right) + \left(\frac{1}{3}\right)^2 = \left(x + \frac{1}{3}\right)^2$

**131.** $a^3 + 1 = a^3 + 1^3$
$$= (a + 1)\left[a^2 - a(1) + 1^2\right]$$
$$= (a + 1)(a^2 - a + 1)$$

**133.** $27 - 8t^3 = 3^3 - (2t)^3$

$$= (3 - 2t)\left[3^2 + (3)(2t) + (2t)^2\right]$$

$$= (3 - 2t)(9 + 6t + 4t^2)$$

**135.** $8x^3 + y^3 = (2x)^3 + y^3$

$$= (2x + y)\left[(2x)^2 + (2x)(y) + y^2\right]$$

$$= (2x + y)(4x^2 - 2xy + y^2)$$

**137.** $x(2x - 3) = 0$

$$x = 0$$

$$2x - 3 = 0 \Rightarrow 2x = 3 \Rightarrow x = \tfrac{3}{2}$$

**139.** $(x + 3)(x - 2) = 0$

$$x + 3 = 0 \Rightarrow x = -3$$

$$x - 2 = 0 \Rightarrow x = 2$$

**141.** $(3x - 7)(2x + 1) = 0$

$$3x - 7 = 0 \Rightarrow 3x = 7 \Rightarrow x = \tfrac{7}{3}$$

$$2x + 1 = 0 \Rightarrow 2x = -1 \Rightarrow x = -\tfrac{1}{2}$$

**143.** $x(x + 8)(3x - 4) = 0$

$$x = 0$$

$$x + 8 = 0 \Rightarrow x = -8$$

$$3x - 4 = 0 \Rightarrow 3x = 4 \Rightarrow x = \tfrac{4}{3}$$

**145.** $x^2 - 81 = 0$

$$(x + 9)(x - 9) = 0$$

$$x + 9 = 0 \Rightarrow x = -9$$

$$x - 9 = 0 \Rightarrow x = 9$$

**147.** $x^2 - 12x + 36 = 0$

$$(x - 6)^2 = 0$$

$$x - 6 = 0 \Rightarrow x = 6$$

**149.** $2t^2 - 3t - 2 = 0$

$$(2t + 1)(t - 2) = 0$$

$$2t + 1 = 0 \Rightarrow 2t = -1 \Rightarrow t = -\tfrac{1}{2}$$

$$t - 2 = 0 \Rightarrow t = 2$$

**151.** $(z - 2)^2 - 4 = 0$

$$(z - 2)^2 - 2^2 = 0$$

$$\left[(z - 2) + 2\right]\left[(z - 2) - 2\right] = 0$$

$$z(z - 4) = 0$$

$$z = 0$$

$$z - 4 = 0 \Rightarrow z = 4$$

**153.** $x(7 - x) = 12$

$$7x - x^2 = 12$$

$$-x^2 + 7x - 12 = 0$$

$$x^2 - 7x + 12 = 0 \quad \left(\begin{array}{l}\text{Multiply both sides}\\ \text{of equation by } -1.\end{array}\right)$$

$$(x - 3)(x - 4) = 0$$

$$x - 3 = 0 \Rightarrow x = 3$$

$$x - 4 = 0 \Rightarrow x = 4$$

**155.** $(x - 1)(x + 2) = 10$

$$x^2 + x - 2 = 10$$

$$x^2 + x - 12 = 0$$

$$(x + 4)(x - 3) = 0$$

$$x + 4 = 0 \Rightarrow x = -4$$

$$x - 3 = 0 \Rightarrow x = 3$$

**157.** *Verbal Model:* $\boxed{\begin{array}{l}\text{First positive consecutive}\\ \text{even integer}\end{array}} \cdot \boxed{\begin{array}{l}\text{Second positive consecutive}\\ \text{even integer}\end{array}} = 168$

*Labels:* First positive consecutive even integer $= 2n$

Second positive consecutive even integer $= 2n + 2$

*Equation:* $2n(2n + 2) = 168$

$$4n^2 + 4n = 168$$

$$4n^2 + 4n - 168 = 0$$

$$4(n^2 + n - 42) = 0$$

$$4(n + 7)(n - 6) = 0$$

$$4 \neq 0$$

$$n + 7 = 0 \Rightarrow n = -7$$

$$n - 6 = 0 \Rightarrow n = 6$$

Because the problem states that the integers are positive, discard the negative answer. The two positive even integers are $2(6) = 12$ and $2(6) + 2 = 12 + 2 = 14$.

**159.** *Common formula:*    $lw = A$

*Labels:*      $A = 2400 \,(\text{square inches})$

$w = \text{width (inches)}$

$l = height = \frac{3}{2}w$

*Equation:*

$$\left(\tfrac{3}{2}w\right)(w) = 2400$$

$$\tfrac{3}{2}w^2 = 2400$$

$$2\left(\tfrac{3}{2}w^2\right) = 2(2400)$$

$$3w^2 = 4800$$

$$3w^2 - 4800 = 0$$

$$3\left(w^2 - 1600\right) = 0$$

$$3(w + 40)(w - 40) = 0$$

$$3 \neq 0$$

$$w + 40 = 0 \Rightarrow w = -40$$

$$w - 40 = 0 \Rightarrow w = 40 \text{ and } \tfrac{3}{2}w = 60$$

Because the width of the window could not be a negative number, discard the negative answer. So, $w = 40$ and $\frac{3}{2}w = 60$. The window is 60 inches by 40 inches.

**161.**

$$\text{Height} = -16t^2 + 32t + 48$$

$$0 = -16t^2 + 32t + 48$$

$$16t^2 - 32t - 48 = 0$$

$$16\left(t^2 - 2t - 3\right) = 0$$

$$16(t - 3)(t + 1) = 0$$

$$16 \neq 0$$

$$t - 3 = 0 \Rightarrow t = 3$$

$$t + 1 = 0 \Rightarrow t = -1$$

Because the time cannot be $-1$, discard the negative answer. It takes $t = 3$ seconds for the rock to reach the ground.

## Chapter Test for Chapter 6

**1.** $9x^2 - 63x^5 = 9x^2\left(1 - 7x^3\right)$

**2.** $z(z + 17) - 10(z + 17) = (z + 17)(z - 10)$

**3.** $t^2 - 2t - 80 = (t - 10)(t + 8)$

**4.** $6x^2 - 11x + 4 = (3x - 4)(2x - 1)$

**5.** $3y^3 + 72y^2 - 75y = 3y\left(y^2 + 24y - 25\right)$
$$= 3y(y + 25)(y - 1)$$

**6.** $4 - 25v^2 = 2^2 - (5v)^2 = (2 + 5v)(2 - 5v)$

**7.** $x^3 + 8 = x^3 + 2^3$
$$= (x + 2)\left(x^2 - x \cdot 2 + 2^2\right)$$
$$= (x + 2)\left(x^2 - 2x + 4\right)$$

**8.** $100 - (z + 11)^2 = 10^2 - (z + 11)^2$
$$= \left[10 - (z + 11)\right]\left[10 + (z + 11)\right]$$
$$= (10 - z - 11)(10 + z + 11)$$
$$= (-z - 1)(z + 21)$$
$$= -(z + 1)(z + 21)$$

**9.** $9x^3 + 2x^2 - 9x - 18 = \left(x^3 + 2x^2\right) - \left(9x + 18\right)$

$$= x^2(x + 2) - 9(x + 2)$$
$$= (x + 2)\left(x^2 - 9\right)$$
$$= (x + 2)(x + 3)(x - 3)$$

**10.** $16 - z^4 = \left[4^2 - \left(z^2\right)^2\right]$

$$= \left(4 + z^2\right)\left(4 - z^2\right)$$
$$= \left(4 + z^2\right)\left(2^2 - z^2\right)$$
$$= \left(4 + z^2\right)(2 + z)(2 - z)$$

or $16 - z^4 = -1\left(z^2 + 4\right)(z + 2)(z - 2)$

**11.** $\frac{2}{5}x - \frac{3}{5} = \frac{1}{5}(2x - 3)$

The missing factor is $2x - 3$.

**12.** $b = 6 \qquad (x + 5)(x + 1)$

$\quad\; b = -6 \qquad (x - 5)(x - 1)$

**13.** $x^2 + 12x + c = x^2 + 2(6x) + c = x^2 + 2(x)(6) + c$

$$(x + 6)^2 = x^2 + 2(x)(6) + 6^2$$
$$= x^2 + 12x + 36$$

So, if $c = 36$, $x^2 + 12x + c$ is a perfect square trinomial.

**14.** This factorization is not complete because the second factor, $3x - 6$, has a common factor of 3. The complete factorization of the polynomial is $3(x + 1)(x - 2)$.

**15.** $(x + 4)(2x - 3) = 0$

$x + 4 = 0 \Rightarrow x = -4$

$2x - 3 = 0 \Rightarrow 2x = 3 \Rightarrow x = \frac{3}{2}$

**16.** $3x^2 + 7x - 6 = 0$

$(3x - 2)(x + 3) = 0$

$3x - 2 = 0 \Rightarrow 3x = 2 \Rightarrow x = \frac{2}{3}$

$x + 3 = 0 \Rightarrow x = -3$

**17.** $\quad y(2y - 1) = 6$

$2y^2 - y - 6 = 0$

$(2y + 3)(y - 2) = 0$

$2y + 3 = 0 \Rightarrow 2y = -3 \Rightarrow y = -\frac{3}{2}$

$y - 2 = 0 \Rightarrow y = 2$

**18.** $\quad 2x^2 - 3x = 8 + 3x$

$2x^2 - 6x - 8 = 0$

$2\left(x^2 - 3x - 4\right) = 0$

$2(x - 4)(x + 1) = 0$

$2 \neq 0$

$x - 4 = 0 \Rightarrow x = 4$

$x + 1 = 0 \Rightarrow x = -1$

**19.** $\quad x^3 + 6x^2 + 8x = 0$

$x\left(x^2 + 6x + 8\right) = 0$

$x(x + 4)(x + 2) = 0$

The volume is the product of the length, width, and height. The width is $x + 2$ and the height is $x$, so the length is the remaining factor of $x + 4$.

**20.** *Common formula:* $\quad lw = A$

*Labels:* $\qquad l$ = length $\qquad$ (inches)

$\qquad\qquad\quad w$ = width = $l - 5$ (inches)

$\qquad\qquad\quad A$ = area = 84 $\qquad$ (square inches)

*Equation:* $\qquad\quad l(l - 5) = 84$

$\qquad\qquad\qquad\quad l^2 - 5l = 84$

$\qquad\qquad\quad l^2 - 5l - 84 = 0$

$\qquad\quad (l - 12)(l + 7) = 0$

$\qquad\qquad\qquad l - 12 = 0 \Rightarrow l = 12$ and $l - 5 = 7$

$\qquad\qquad\qquad l + 7 = 0 \Rightarrow l = -7$

We discard the negative answer for the length of the rectangle. So, $l = 12$ and $l - 5 = 7$. The dimensions of the rectangle are 12 inches by 7 inches.

**21.**
$$h = -16t^2 + 14t + 1136$$
$$0 = -16t^2 + 14t + 1136$$
$$0 = -2(8t^2 - 7t - 568)$$
$$0 = -2(8t - 71)(t + 8)$$
$$-2 \neq 0$$
$$8t - 71 = 0 \Rightarrow 8t = 71 \Rightarrow t = \frac{71}{8} \text{ or } 8.875$$
$$t + 8 = 0 \Rightarrow t = -8$$

Discard negative solution. It will take 8.875 seconds for the object to reach the ground.

$$h = -16t^2 + 14t + 1136$$
$$806 = -16t^2 + 14t + 1136$$
$$0 = -16t^2 + 14t + 330$$
$$0 = -2(8t^2 - 7t - 165)$$
$$0 = -2(8t + 33)(t - 5)$$
$$-2 \neq 0$$
$$8t + 33 = 0 \Rightarrow 8t = -33 \Rightarrow t = -\frac{33}{8} \text{ or } -4.125$$
$$t - 5 = 0 \Rightarrow t = 5$$

Discard the negative solution. It will take 5 seconds for the object to fall to a height of 806 feet.

**22.** *Verbal model:* $\boxed{\text{First positive consecutive even integer}} \cdot \boxed{\text{Second positive consecutive even integer}} = 624$

*Labels:*   First positive consecutive even integer $= 2n$

Second positive consecutive even integer $= 2n + 2$

*Equation:*
$$2n(2n + 2) = 624$$
$$4n^2 + 4n = 624$$
$$4n^2 + 4n - 624 = 0$$
$$4(n^2 + n - 156) = 0$$
$$4(n + 13)(n - 12) = 0$$
$$4 \neq 0$$
$$n + 13 = 0 \Rightarrow n = -13$$
$$n - 12 = 0 \Rightarrow n = 12 \text{ so } 2n = 24 \text{ and } 2n + 2 = 26.$$

Because the problem states that the integers are positive, discard the negative answer. So, the two positive even integers are $2(12) = 24$ and $2(12) + 2 = 24 + 2 = 26$.

# Cumulative Test for Chapters 4–6

**1.** $(-2, y)$, $y$ is a real number.

Quadrant II or III.

This point is located 2 units to the *left* of the vertical axis. If $y$ is positive, the point would be *above* the horizontal axis and in the second quadrant. If $y$ is negative, the point would be *below* the horizontal axis and in the third quadrant. If $y$ is zero, the point would be on the horizontal axis *between* the second and third quadrants.

**2.** (a) The ordered pair $(-1, -1)$ *is not* a solution because
$$9(-1) - 4(-1) + 36 = -9 + 4 + 36 \neq 0.$$

(b) The ordered pair $(8, 27)$ *is* a solution because
$$9(8) - 4(27) + 36 = 72 - 108 + 36 = 0.$$

(c) The ordered pair $(-4, 0)$ *is* a solution because
$$9(-4) - 4(0) + 36 = -36 - 0 + 36 = 0.$$

(d) The ordered pair $(3, -2)$ *is not* a solution because
$$9(3) - 4(-2) + 36 = 27 + 8 + 36 \neq 0.$$

**3.**

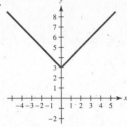

The $y$-intercept is $(0, 3)$; there is no $x$-intercept.

**4.**

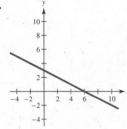

The $x$-intercept is $(6, 0)$ and the $y$-intercept is $(0, 3)$.

**5.** Slope: $m = -\dfrac{1}{4}$

$$\frac{\text{rise}}{\text{run}} = \frac{-1}{4}$$

Here are some other points on the line with $(2, 1)$:

$(6, 0)$  $(10, -1)$  $(14, -2)$

$(-2, 2)$  $(-6, 3)$  $(-10, 4)$

There are many correct answers because there are infinitely many points on the line.

**6.** Point: $\left(0, -\frac{3}{2}\right)$  Note: This point is the $y$-intercept.

Slope: $m = \frac{2}{5}$

$y = mx + b$

$y = \frac{2}{5}x - \frac{3}{2}$

**7.** $y = \frac{2}{3}x - 3 \Rightarrow m_1 = \frac{2}{3}$

$y = -\frac{3}{2}x + 1 \Rightarrow m_2 = -\frac{3}{2}$

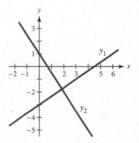

The slopes are negative reciprocals, so the lines are perpendicular.

**8.** $y = 2 - 0.4x$ or $y = -0.4x + 2 \Rightarrow m_1 = -0.4$

$y = -\frac{2}{5}x \Rightarrow m_2 = -\frac{2}{5}$

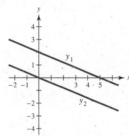

The slopes, $-0.4$ and $-\frac{2}{5}$, are equal, so the lines are parallel.

**9.** $\left(x^3 - 3x^2\right) - \left(x^3 + 2x^2 - 5\right) = x^3 - 3x^2 - x^3 - 2x^2 + 5 = \left(x^3 - x^3\right) + \left(-3x^2 - 2x^2\right) + 5 = -5x^2 + 5$

**10.** $(6z)(-7z)\left(z^2\right) = (6)(-7)z \cdot z \cdot z^2 = -42z^{1+1+2} = -42z^4$

**11.**         **F   O   I   L**

$(3x + 5)(x - 4) = 3x^2 - 12x + 5x - 20 = 3x^2 - 7x - 20$

**12.** $(5x - 3)(5x + 3) = (5x)^2 - 3^2$     Pattern: $(a - b)(a + b) = a^2 - b^2$

$= 25x^2 - 9$

**13.** $(5x + 6)^2 = (5x)^2 + 2(5x)(6) + 6^2$     Pattern: $(a + b)^2 = a^2 + 2ab + b^2$

$= 25x^2 + 60x + 36$

**14.** $\left(6x^2 + 72x\right) \div 6x = \dfrac{6x^2}{6x} + \dfrac{72x}{6x} = x + 12$

**15.**

$$\begin{array}{r} x + 1 + \dfrac{2}{x-4} \\ x-4\overline{)x^2 - 3x - 2} \\ \underline{x^2 - 4x} \\ x - 2 \\ \underline{x - 4} \\ 2 \end{array}$$

So, $\dfrac{x^2 - 3x - 2}{x - 4} = x + 1 + \dfrac{2}{x-4}$.

**16.** $\dfrac{\left(3xy^2\right)^{-2}}{6x^{-3}} = \dfrac{3^{-2}x^{-2}y^{-4}}{6x^{-3}}$

$= \dfrac{1}{3^2(6)}x^{-2-(-3)}y^{-4} = \dfrac{1}{54}xy^{-4} = \dfrac{x}{54y^4}$

**17.** $2u^2 - 6u = 2u(u - 3)$

**18.** $(x - 4)^2 - 36 = (x - 4)^2 - 6^2$

$= \left[(x - 4) + 6\right]\left[(x - 4) - 6\right]$

$= (x - 4 + 6)(x - 4 - 6)$

$= (x + 2)(x - 10)$

**19.** $x^3 + 8x^2 + 16x = x\left(x^2 + 8x + 16\right)$

$= x\left(x^2 + 2(x)(4) + 4^2\right)$

$= x(x + 4)^2$

**20.** $x^3 + 2x^2 - 4x - 8 = \left(x^3 + 2x^2\right) - (4x + 8)$

$= x^2(x + 2) - 4(x + 2)$

$= (x + 2)\left(x^2 - 4\right)$

$= (x + 2)\left(x^2 - 2^2\right)$

$= (x + 2)(x + 2)(x - 2)$

$= (x + 2)^2(x - 2)$

**21.** $u(u - 12) = 0$

$u = 0$

$u - 12 = 0 \Rightarrow u = 12$

**22.** $5x^2 - 12x - 9 = 0$

$(5x + 3)(x - 3) = 0$

$5x + 3 = 0 \Rightarrow 5x = -3 \Rightarrow x = -\frac{3}{5}$

$x - 3 = 0 \Rightarrow x = 3$

The solutions are $-\frac{3}{5}$ and 3.

**23.** $\left(\dfrac{x}{2}\right)^{-2} = \dfrac{x^{-2}}{2^{-2}} = \dfrac{2^2}{x^2} = \dfrac{4}{x^2}$

**24.** $C = 150 + 0.45x$

Explanations may vary.

$x = 70 \Rightarrow C = 150 + 0.45(70)$

$C = 150 + 31.50$

$C = 181.50$

The cost for a day when the representative drives 70 miles is $181.50.

**25.** $0.70x + 9000 = C$

$0.70x + 9000 \le 36,400$

$0.70x \le 27,400$

$x \le \dfrac{27,400}{0.70}$

$x \le 39,142.86$

The cost will stay at or below $36,400 if the mileage on the car is not more than 39,142 miles.

# CHAPTER 7
# Rational Expressions and Equations

# CHAPTER 7
# Rational Expressions and Equations

## Section 7.1 Simplifying Rational Expressions

**1.** Yes, the expression is a rational expression. It is one polynomial divided by another.

**3.** No, the expression is not a rational expression because the numerator is not a polynomial. (A polynomial has only positive integral exponents).

**5.** The denominator is zero when $x - 4 = 0$ or $x = 4$. So, the domain is all real values of $x$ such that $x \neq 4$.

**7.** The denominator is zero when $x + 2 = 0$ or $x = -2$. So, the domain is all real values of $x$ such that $x \neq -2$.

**9.** The denominator is not zero for any value of $x$. So, the domain is the set of *all* real numbers $x$.

**11.** The denominator is zero when $x^2 + 4 = 0$, but $x^2 + 4$ is never zero for any real values of $x$. So, the domain is the set of *all* real numbers $x$.

**13.** The denominator $t^2 - 25 = (t + 5)(t - 5)$ is zero when $t = -5$ or $t = 5$. So, the domain is all real values of $t$ such that $t \neq -5$ and $t \neq 5$.

**15.** The denominator $y^2 - 3y - 28 = (y - 7)(y + 4)$ is zero when $y = 7$ or $y = -4$. So, the domain is all real values of $y$ such that $y \neq 7$ and $y \neq -4$.

**17.** The denominator $x^2 - x - 2 = (x - 2)(x + 1)$ is zero when $x = 2$ or $x = -1$. So, the domain is all real values of $x$ such that $x \neq 2$ and $x \neq -1$.

**19.** The denominator $3z^2 - z - 2 = (3z + 2)(z - 1)$ is zero when $z = -\frac{2}{3}$ or $z = 1$. So, the domain is all real values of $z$ such that $z \neq -\frac{2}{3}$ and $z \neq 1$.

**21.** $\dfrac{x}{x - 3}$

(a) $x = 0 \Rightarrow \dfrac{x}{x - 3} = \dfrac{0}{0 - 3} = \dfrac{0}{-3} = 0$

(b) $x = 3 \Rightarrow \dfrac{x}{x - 3} = \dfrac{3}{3 - 3} = \dfrac{3}{0}$, undefined

Division by 0 is undefined.

(c) $x = 10 \Rightarrow \dfrac{x}{x - 3} = \dfrac{10}{10 - 3} = \dfrac{10}{7}$

(d) $x = -3 \Rightarrow \dfrac{x}{x - 3} = \dfrac{-3}{-3 - 3} = \dfrac{-3}{-6} = \dfrac{1}{2}$

**23.** $\dfrac{x + 1}{x^2 - 4}$

(a) $x = 2 \Rightarrow \dfrac{x + 1}{x^2 - 4} = \dfrac{2 + 1}{2^2 - 4} = \dfrac{3}{0}$, undefined

Division by 0 is undefined.

(b) $x = 1 \Rightarrow \dfrac{x + 1}{x^2 - 4} = \dfrac{1 + 1}{1^2 - 4} = \dfrac{2}{-3} = -\dfrac{2}{3}$

(c) $x = -5 \Rightarrow \dfrac{x + 1}{x^2 - 4} = \dfrac{-5 + 1}{(-5)^2 - 4} = \dfrac{-4}{21} = -\dfrac{4}{21}$

(d) $x = -2 \Rightarrow \dfrac{x + 1}{x^2 - 4} = \dfrac{-2 + 1}{(-2)^2 - 4} = \dfrac{-1}{0}$, undefined

Division by 0 is undefined.

**25.** The answer is any two of these expressions:

$\dfrac{-x}{-12}, -\dfrac{x}{-12}, -\dfrac{-x}{12}$

**27.** The answer is any two of these expressions:

$\dfrac{-t - 2}{t^2 - 1}, \dfrac{t + 2}{-t^2 + 1}, -\dfrac{-t - 2}{-t^2 + 1}, \dfrac{t + 2}{1 - t^2}, -\dfrac{-t - 2}{1 - t^2}$

**29.** $\dfrac{5}{2x} = \dfrac{5(3x)}{2x(3x)} = \dfrac{5(3x)}{6x^2}$

The missing factor is $3x$.

**31.** $\dfrac{3}{4} = \dfrac{3(x + 1)}{4(x + 1)}$

The missing factor is $(x + 1)$.

**33.** $\dfrac{-7}{3x} = \dfrac{-7(x^2)}{3x(x^2)} = \dfrac{7(-x^2)}{3x^3}$

The missing factor is $(-x^2)$.

**35.** $\dfrac{x}{2} = \dfrac{x(x + 3)}{2(x + 3)}$

The missing factor is $(x + 3)$.

**37.** $\dfrac{x + 1}{x} = \dfrac{(x + 1)[x(2x + 5)]}{x[x(2x + 5)]} = \dfrac{(x + 1)[x(2x + 5)]}{x^2(2x + 5)}$

The missing factor is $[x(2x + 5)]$.

**39.** $\dfrac{3x}{x-3} = \dfrac{3x(x+2)}{(x-3)(x+2)} = \dfrac{3x(x+2)}{x^2-x-6}$

The missing factor is $(x+2)$.

**41.** $\dfrac{9x}{27} = \dfrac{\cancel{9}(x)}{\cancel{9}(3)} = \dfrac{x}{3}$

**43.** $\dfrac{8y^4}{y^2} = \dfrac{8y^2\left(\cancel{y^2}\right)}{1\left(\cancel{y^2}\right)} = \dfrac{8y^2}{1} = 8y^2, \quad y \neq 0$

**45.** $\dfrac{16x^2}{36x} = \dfrac{4x\,(\cancel{4x})}{9\,(\cancel{4x})} = \dfrac{4x}{9}, \quad x \neq 0$

**47.** $\dfrac{44a^7b}{11ab^2} = \dfrac{4a^6\,(\cancel{11ab})}{b\,(\cancel{11ab})} = \dfrac{4a^6}{b}, \quad a \neq 0$

**49.** $\dfrac{x^2(x+1)}{x(x+1)} = \dfrac{x\,(\cancel{x})\,(\cancel{x+1})}{1\,(\cancel{x})\,(\cancel{x+1})} = \dfrac{x}{1} = x, \ x \neq 0, x \neq -1$

**51.** $\dfrac{75x(x-1)^2}{15x(x-1)} = \dfrac{5\,(\cancel{15})\,(\cancel{x})\,(\cancel{x-1})(x-1)}{\cancel{15}\,(\cancel{x})\,(\cancel{x-1})}$

$= 5(x-1), \quad x \neq 0, x \neq 1$

**53.** $\dfrac{14(x^2+3x)}{2x^2} = \dfrac{7\,(\cancel{2})\,(\cancel{x})(x+3)}{\cancel{2}\,(\cancel{x})(x)} = \dfrac{7(x+3)}{x}$

**55.** $\dfrac{x-5}{2x-10} = \dfrac{x-5}{2(x-5)}$

$= \dfrac{1\,(\cancel{x-5})}{2\,(\cancel{x-5})}$

$= \dfrac{1}{2}, \quad x \neq 5$

**57.** $\dfrac{5x^2-20x}{(x-4)^2} = \dfrac{5x\,(\cancel{x-4})}{(x-4)\,(\cancel{x-4})} = \dfrac{5x}{x-4}$

**59.** $\dfrac{x^2+10x+16}{4x+8} = \dfrac{(x+8)\,(\cancel{x+2})}{4\,(\cancel{x+2})} = \dfrac{x+8}{4}, \ x \neq -2$

**61.** $\dfrac{x^2-25}{5-x} = \dfrac{(x-5)(x+5)}{5-x}$

$= \dfrac{-1\,(\cancel{5-x})(x+5)}{\cancel{5-x}}$

$= -(x+5), \quad x \neq 5$

**63.** $\dfrac{x^2-36z^2}{x-6z} = \dfrac{(x+6z)\,(\cancel{x-6z})}{1\,(\cancel{x-6z})}$

$= \dfrac{x+6z}{1} = x+6z, \quad x \neq 6z$

**65.** $\dfrac{7s^2-28t^2}{28t^2-7s^2} = \dfrac{7s^2-28t^2}{-7s^2+28t^2}$

$= \dfrac{1\left(\cancel{7s^2-28t^2}\right)}{-1\left(\cancel{7s^2-28t^2}\right)}$

$= -1, \quad s^2 \neq 4t^2$

Note: The numerator and denominator could be factored completely. When the common factors are divided out, the simplified form remains $-1$.

**67.** $\dfrac{3y+3}{xy+x} = \dfrac{3\,(\cancel{y+1})}{x\,(\cancel{y+1})} = \dfrac{3}{x}, \quad y \neq -1$

**69.** $\dfrac{x^2-2x+1}{1-x^2} = \dfrac{(x-1)(x-1)}{(1-x)(1+x)}$

$= \dfrac{-1\,(\cancel{1-x})(x-1)}{(\cancel{1-x})(1+x)}$

$= -\dfrac{x-1}{1+x}, \quad x \neq 1 \text{ or } \dfrac{1-x}{1+x}, \quad x \neq 1$

**71.** $\dfrac{a+2}{a^2+4a+4} = \dfrac{a+2}{(a+2)(a+2)}$

$= \dfrac{1\,(\cancel{a+2})}{(\cancel{a+2})(a+2)}$

$= \dfrac{1}{a+2}$

**73.** $\dfrac{x^2-5x}{x^2-10x+25} = \dfrac{x(x-5)}{(x-5)(x-5)}$

$= \dfrac{x\,(\cancel{x-5})}{(\cancel{x-5})(x-5)}$

$= \dfrac{x}{x-5}$

**75.** $\dfrac{x^2-4x+3}{x^2-5x+6} = \dfrac{(\cancel{x-3})(x-1)}{(\cancel{x-3})(x-2)}$

$= \dfrac{x-1}{x-2}, \quad x \neq 3$

**77.** $\dfrac{x^2 + 8x - 20}{x^2 + 11x + 10} = \dfrac{(x+10)(x-2)}{(x+10)(x+1)}$

$\qquad = \dfrac{\cancel{(x+10)}(x-2)}{\cancel{(x+10)}(x+1)}$

$\qquad = \dfrac{x-2}{x+1}, \quad x \neq -10$

**79.** $\dfrac{5r + 5s}{r^2 + 7rs + 6s^2} = \dfrac{5(r+s)}{(r+6s)(r+s)}$

$\qquad = \dfrac{5}{r+6s}, \quad r \neq -s$

**81.** $\dfrac{x^2 + 4xy - 21y^2}{x^2 + 2xy - 15y^2} = \dfrac{(x-3y)(x+7y)}{(x-3y)(x+5y)}$

$\qquad = \dfrac{x+7y}{x+5y}, \quad x \neq 3y$

**83.** $\dfrac{x^3 + 5x^2 + 6x}{x^2 - 4} = \dfrac{x(x^2+5x+6)}{(x+2)(x-2)}$

$\qquad = \dfrac{x(x+3)(x+2)}{(x+2)(x-2)}$

$\qquad = \dfrac{x(x+3)\cancel{(x+2)}}{\cancel{(x+2)}(x-2)}$

$\qquad = \dfrac{x(x+3)}{x-2}, \quad x \neq -2$

**85.** $\dfrac{x^3 - 2x^2 + x - 2}{x-2} = \dfrac{(x^3 - 2x^2) + (x - 2)}{x-2}$

$\qquad = \dfrac{x^2(x-2) + 1(x-2)}{x-2}$

$\qquad = \dfrac{(x-2)(x^2+1)}{x-2}$

$\qquad = \dfrac{\cancel{(x-2)}(x^2+1)}{1\cancel{(x-2)}}$

$\qquad = x^2 + 1, \quad x \neq 2$

**87.** $\dfrac{a^3 - 8}{a^2 - 4} = \dfrac{(a-2)(a^2 + 2a + 4)}{(a+2)(a-2)}$

$\qquad = \dfrac{a^2 + 2a + 4}{a+2}, \quad a \neq 2$

**89.** $\dfrac{x^4 - y^4}{(y-x)^4} = \dfrac{(x^2 + y^2)(x^2 - y^2)}{(y-x)(y-x)^3}$

$\qquad = \dfrac{(x^2 + y^2)(x+y)\cancel{(x-y)}}{-1\cancel{(x-y)}(y-x)^3}$

$\qquad = -\dfrac{(x^2 + y^2)(x+y)}{(y-x)^3}$

**91.** $\dfrac{2x^2 + 4x}{2x} = \dfrac{2x(x+2)}{2x} = x + 2, \quad x \neq 0$

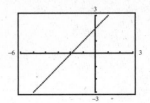

**93.** $\dfrac{3(x-3)^2}{x-3} = \dfrac{3(x-3)(x-3)}{x-3} = 3(x-3), \quad x \neq 3$

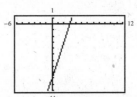

**95.** $\dfrac{x^3 - 2x^2}{x^3 + 2x} = \dfrac{x(x)(x-2)}{x(x^2+2)} = \dfrac{x(x-2)}{x^2+2}, \quad x \neq 0$

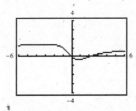

**97.**

| $x$ | 2 | 2.5 | 3 | 3.5 | 4 |
|---|---|---|---|---|---|
| $\dfrac{x^3 - 3x^2}{x-3}$ | 4 | 6.25 | Undefined | 12.25 | 16 |
| $x^2$ | 4 | 6.25 | 9 | 12.25 | 16 |

$\dfrac{x^3 - 3x^2}{x-3} = \dfrac{x^2(x-3)}{x-3} = x^2, \quad x \neq 3$

The two expressions are equal for all values of $x$ except 3. When $x = 3$, the first expression is undefined.

**99.** (a) Cost $= 3000 + 8.50x$, where $x$ is the number of units.

$$\text{Average cost per unit} = \frac{3000 + 8.50x}{x}$$

(b) The variable $x$ represents the number of units produced.

The domain is $\{1, 2, 3, 4, \dots\}$.

(c) $\dfrac{3000 + 8.50(100)}{100} = \dfrac{3000 + 850}{100}$

$$= \frac{3850}{100}$$

$$= 38.50$$

When 100 units are produced, the average cost per unit is \$38.50.

**101.** (a) $B = \dfrac{156.89x + 7.34x^2}{x + 0.017x^2}$

$$= \frac{\cancel{(x)}(156.89 + 7.34x)}{\cancel{(x)}(1 + 0.017x)}$$

$$= \frac{156.89 + 7.34x}{1 + 0.017x}, \quad 10 \le x \le 100$$

(b) $B = \dfrac{156.89 + 7.34(14.7)}{1 + 0.017(14.7)}$

$$= \frac{264.788}{1.2499}$$

$$\doteq 211.847°\text{F}$$

**103.** (a) The denominator of the expression is zero when $p = 100$. So, $p = 100$ is not in the domain. Since $p$ is a percentage of air pollutants, it must be a non-negative number. So, the domain is $0 \le p < 100$.

(b)

| P | 20 | 40 | 60 | 80 |
|---|----|----|----|----|
| C | \$20,000 | \$53,333 | \$120,000 | \$320,000 |

$C = \dfrac{80,000(20)}{100 - 20}$    $C = \dfrac{80,000(40)}{100 - 40}$    $C = \dfrac{80,000(60)}{100 - 60}$    $C = \dfrac{80,000(80)}{100 - 80}$

$= \dfrac{1,600,000}{80}$    $= \dfrac{3,200,000}{60}$    $= \dfrac{4,800,000}{40}$    $= \dfrac{6,400,000}{20}$

$= 20,000$    $= 53,333$    $= 120,000$    $= 320,000$

(c) As $P$ increases, the cost increases. According to the model, you cannot remove 100% of the pollutants because when $p = 100$, the expression $80,000p/(100 - p)$ is undefined.

**105.** Probability $= \dfrac{\text{Area of shaded area}}{\text{Total area}}$

$$= \frac{\cancel{x}(x - 3)}{\cancel{x}(x + 5)}$$

$$= \frac{x - 3}{x + 5}$$

**107.** Area of shaded region $= (x + 1)(x + 2)$

Total area of figure $= (3x + 6)(2x + 2)$

$$\frac{(x + 1)(x + 2)}{(3x + 6)(2x + 2)} = \frac{\cancel{(x + 1)}\cancel{(x + 2)}}{3\cancel{(x + 2)}(2)\cancel{(x + 1)}} = \frac{1}{6}$$

The ratio of the shaded region to the total area of the figure is $\dfrac{1}{6}$.

**109.** True; $\dfrac{ac}{bc} = \dfrac{a}{b}$ when $a$, $b$, and $c$ are real numbers, variables, or algebraic expressions, $b \ne 0, c \ne 0$.

**111.** The two expressions are equivalent for $x \ne 0$.

$$\frac{4x}{2x} = \frac{2\cancel{(2x)}}{\cancel{2x}} = 2, \quad x \ne 0$$

**113.** There are many correct answers. Here is an example:

$$\frac{x}{x^2 + x}$$

$$\frac{x}{x^2 + x} = \frac{\cancel{x}}{\cancel{x}(x + 1)} = \frac{1}{x + 1}, \quad x \ne 0$$

**115.** $14 - 2x = x + 2$

$14 - 2x - x = x - x + 2$

$14 - 3x = 2$

$14 - 14 - 3x = 2 - 14$

$-3x = -12$

$\dfrac{-3x}{-3} = \dfrac{-12}{-3}$

$x = 4$

**117.** $\dfrac{x}{3} + 5 = 8$

$3\left(\dfrac{x}{3} + 5\right) = 3(8)$

$x + 15 = 24$

$x + 15 - 15 = 24 - 15$

$x = 9$

**119.** $m = \dfrac{0 - 4}{10 - 0} = \dfrac{-4}{10} = -\dfrac{2}{5}$

**121.** $m = \dfrac{8 - 3}{4 - (-1)} = \dfrac{5}{5} = 1$

**123.** $x(x - 8) = 0$

$x = 0$

$x - 8 = 0 \Rightarrow x = 8$

**125.** $(5z + 6)(z + 14) = 0$

$5z + 6 = 0 \Rightarrow 5z = -6 \Rightarrow z = -\dfrac{6}{5}$

$z + 14 = 0 \Rightarrow z = -14$

## Section 7.2 Multiplying and Dividing Rational Expressions

**1.** $\dfrac{8x^2}{3} \cdot \dfrac{9}{16x} = \dfrac{8(9)x^2}{3(16)x} = \dfrac{\cancel{8}\,(\cancel{3})(3)(\cancel{x})(x)}{\cancel{3}\,(\cancel{8})(2)(\cancel{x})} = \dfrac{3x}{2}, \ x \neq 0$

**3.** $\dfrac{12x^2}{6x} \cdot \dfrac{12x}{8x^2} = \dfrac{144x^3}{48x^3} = \dfrac{(3)(\cancel{48})\,\cancel{x^3}}{\cancel{48}\,\cancel{x^3}} = 3, \ x \neq 0$

**5.** $\left(-\dfrac{5a^4}{6a}\right)\left(-\dfrac{2}{a}\right) = \dfrac{10a^4}{6a^2} = \dfrac{5(\cancel{2})\,(\cancel{a^2})(a^2)}{3(\cancel{2})\,(\cancel{a^2})} = \dfrac{5a^2}{3}, \ a \neq 0$

**7.** $\dfrac{20y^3}{6xy} \cdot \dfrac{4x^2y}{5y^2} = \dfrac{80x^2y^4}{30xy^3}$

$= \dfrac{8(\cancel{2})\,(\cancel{5})\,(\cancel{x})(x)(\cancel{y^3})(y)}{3(\cancel{2})\,(\cancel{5})\,(\cancel{x})\,(\cancel{y^3})}$

$= \dfrac{8xy}{3}, \ x \neq 0, y \neq 0$

**9.** $\dfrac{y - 1}{5} \cdot \dfrac{5}{y - 1} = \dfrac{(\cancel{y - 1})\,(\cancel{5})}{\cancel{5}\,(\cancel{y - 1})} = 1, \ y \neq 1$

**11.** $\dfrac{x + 1}{2} \cdot \dfrac{4x}{x + 1} = \dfrac{(\cancel{x + 1})(\cancel{2})(2)(x)}{\cancel{2}\,(\cancel{x + 1})} = 2x, \ x \neq -1$

**13.** $\dfrac{1 - r}{3} \cdot \dfrac{3}{r - 1} = \dfrac{3(1 - r)}{3(r - 1)} = \dfrac{\cancel{3}(-1)(\cancel{r - 1})}{\cancel{3}\,(\cancel{r - 1})} = -1, \ r \neq 1$

**15.** $\dfrac{x - 3}{x^2 - 16} \cdot \dfrac{x + 4}{2x^2 - 6x} = \dfrac{(x - 3)(x + 4)}{(x^2 - 16)(2x^2 - 6x)}$

$= \dfrac{(\cancel{x - 3})\,(\cancel{x + 4})}{(\cancel{x + 4})(x - 4)(2x)(\cancel{x - 3})}$

$= \dfrac{1}{2x(x - 4)}, \ x \neq -4, x \neq 3$

**17.** $\dfrac{y + 5}{y - 2} \cdot (2y) = \dfrac{y + 5}{y - 2} \cdot \dfrac{2y}{1} = \dfrac{2y(y + 5)}{y - 2}$

**19.** $\dfrac{3x}{5x - 15} \cdot (3 - x) = \dfrac{3x}{5(x - 3)} \cdot \dfrac{3 - x}{1}$

$= \dfrac{3x(-1)(\cancel{x - 3})}{5(\cancel{x - 3})}$

$= -\dfrac{3x}{5}, \ x \neq 3$

**21.** $\dfrac{(x - 5)^2}{x + 5} \cdot \dfrac{x + 5}{x - 5} = \dfrac{(x - 5)(\cancel{x - 5})\,(\cancel{x + 5})}{(\cancel{x + 5})\,(\cancel{x - 5})}$

$= x - 5, \ x \neq -5, x \neq 5$

**23.** $\dfrac{5}{x - 1} \cdot \dfrac{x - 1}{25(x - 3)} = \dfrac{5(x - 1)}{25(x - 1)(x - 3)}$

$= \dfrac{1(\cancel{5})\,(\cancel{x - 1})}{5(\cancel{5})\,(\cancel{x - 1})(x - 3)}$

$= \dfrac{1}{5(x - 3)}, \ x \neq 1$

**25.** $\dfrac{2 - t}{2 + t} \cdot \dfrac{t + 2}{t - 2} = \dfrac{(2 - t)(t + 2)}{(2 + t)(t - 2)}$

$= \dfrac{-1(\cancel{t - 2})\,(\cancel{t + 2})}{(\cancel{t + 2})\,(\cancel{t - 2})}$

$= -1, \ t \neq -2, t \neq 2$

**27.** $\dfrac{9y - 15z}{7y + 14z} \cdot \dfrac{2y + 4z}{3y - 5z} = \dfrac{(9y - 15z)(2y + 4z)}{(7y + 14z)(3y - 5z)}$

$= \dfrac{3(\cancel{3y - 5z})(2)(\cancel{y + 2z})}{7(\cancel{y + 2z})\,(\cancel{3y - 5z})}$

$= \dfrac{6}{7}, \ y \neq -2z, 3y \neq 5z$

**29.** $\dfrac{r}{r-t} \cdot \dfrac{r^2 - t^2}{r^2} = \dfrac{r(r+t)(r-t)}{r^2(r-t)}$

$\qquad = \dfrac{\cancel{r}(r+t)\cancel{(r-t)}}{r\cancel{(r)}\cancel{(r-t)}}$

$\qquad = \dfrac{r+t}{r}, \quad r \neq t$

**31.** $\left(x^2 - 4\right) \cdot \dfrac{x}{(x-2)^2} = \dfrac{(x+2)(x-2)}{1} \cdot \dfrac{x}{(x-2)^2}$

$\qquad = \dfrac{(x+2)\cancel{(x-2)}(x)}{1\cancel{(x-2)}(x-2)}$

$\qquad = \dfrac{x(x+2)}{(x-2)}$

**33.** $\dfrac{t^2 - t - 6}{t^2 + 6t + 9} \cdot \dfrac{t+3}{t^2 - 4} = \dfrac{(t-3)\cancel{(t+2)}\cancel{(t+3)}}{\cancel{(t+3)}(t+3)(t-2)\cancel{(t+2)}}$

$\qquad = \dfrac{t-3}{(t+3)(t-2)}, \quad t \neq -2$

**35.** $\dfrac{x^2 + x - 2}{x^3 + x^2} \cdot \dfrac{x}{x^2 + 3x + 2} = \dfrac{(x+2)(x-1)x}{x^2(x+1)(x+2)(x+1)}$

$\qquad = \dfrac{\cancel{x}\cancel{(x+2)}(x-1)}{\cancel{x} \cdot x(x+1)\cancel{(x+2)}(x+1)}$

$\qquad = \dfrac{(x-1)}{x(x+1)^2}, \quad x \neq -2$

**37.** $\dfrac{t^2 + 2t - 3}{t^2 + 4t - 5} \cdot \dfrac{t^2 - 3t - 10}{t^2 + 5t + 6} = \dfrac{(t+3)(t-1)(t-5)(t+2)}{(t+5)(t-1)(t+3)(t+2)}$

$\qquad = \dfrac{\cancel{(t+3)}\cancel{(t-1)}(t-5)\cancel{(t+2)}}{(t+5)\cancel{(t-1)}\cancel{(t+3)}\cancel{(t+2)}}$

$\qquad = \dfrac{t-5}{t+5}, \quad t \neq 1, t \neq -3, t \neq -2$

**39.** $\dfrac{4}{x} \cdot \dfrac{x+2}{x+6} \cdot \dfrac{x}{x+2} = \dfrac{4\cancel{(x+2)}\cancel{(x)}}{\cancel{x}(x+6)\cancel{(x+2)}} = \dfrac{4}{x+6}, \quad x \neq 0, x \neq -2$

**41.** $\dfrac{a+1}{a-1} \cdot \dfrac{a^2 - 2a + 1}{a} \cdot \left(3a^2 + 3a\right) = \dfrac{a+1}{a-1} \cdot \dfrac{a^2 - 2a + 1}{a} \cdot \dfrac{3a^2 + 3a}{1}$

$\qquad = \dfrac{(a+1)(a-1)(a-1)(3a)(a+1)}{(a-1)(a)}$

$\qquad = \dfrac{(a+1)\cancel{(a-1)}(a-1)(3)\cancel{(a)}(a+1)}{\cancel{(a-1)}\cancel{(a)}}$

$\qquad = 3(a+1)^2(a-1), \quad a \neq 0, a \neq 1$

**43.** $\dfrac{2}{z+3} \cdot \dfrac{z^2 + 6z + 9}{z-3} \cdot \dfrac{4}{z^2 - 9} = \dfrac{2\cancel{(z+3)}\cancel{(z+3)}(4)}{\cancel{(z+3)}(z-3)\cancel{(z+3)}(z-3)} = \dfrac{8}{(z-3)^2}, \quad z \neq -3$

**45.** $\dfrac{x-7}{x+10} \cdot \dfrac{x-3}{x-7} = \dfrac{\cancel{(x-7)}(x-3)}{(x+10)\cancel{(x-7)}} = \dfrac{x-3}{x+10}, \quad x \neq 7$

The missing polynomial is $x - 3$.

**47.** $\dfrac{x^2 - 2x - 8}{x-5} \cdot \dfrac{x-5}{x-4} = \dfrac{\cancel{(x-4)}(x+2)\cancel{(x-5)}}{\cancel{(x-5)}\cancel{(x-4)}} = x + 2, \quad x \neq 5, x \neq 4$

The missing polynomial is $x - 4$.

**49.** $\dfrac{x+3}{x-4} \cdot \dfrac{(x-4)(x+2)}{x+3} = \dfrac{\cancel{(x+3)}\,\cancel{(x-4)}(x+2)}{\cancel{(x-4)}\,\cancel{(x+3)}} = x+2, \quad x \neq 4, x \neq -3$

The missing polynomial is $(x-4)(x+2)$ or $x^2 - 2x - 8$.

**51.** The reciprocal of $\dfrac{3-x}{x^2+4}$ is $\dfrac{x^2+4}{3-x}$.

**53.** The reciprocal of $\dfrac{x^2+2x-5}{x^2-4x+7}$ is $\dfrac{x^2-4x+7}{x^2+2x-5}$.

**55.** The reciprocal of $a^3 - 8a$ or $\dfrac{a^3-8a}{1}$ is $\dfrac{1}{a^3-8a}$.

**57.** $\dfrac{x^3}{6} \div \dfrac{x^2}{3} = \dfrac{x^3}{6} \cdot \dfrac{3}{x^2} = \dfrac{x\cancel{(x^2)}\cancel{(3)}}{\cancel{3}(2)\cancel{(x^2)}} = \dfrac{x}{2}, \quad x \neq 0$

**59.** $\dfrac{7x^2}{10} \div \dfrac{14x^3}{15} = \dfrac{7x^2}{10} \cdot \dfrac{15}{14x^3}$

$\quad = \dfrac{7(15)\left(x^2\right)}{10(14)\left(x^3\right)}$

$\quad = \dfrac{\cancel{7}\,\cancel{(5)}(3)\cancel{\left(x^2\right)}}{\cancel{5}(2)\cancel{(7)}(2)\cancel{\left(x^2\right)}(x)}$

$\quad = \dfrac{3}{4x}$

**61.** $\dfrac{a}{a+1} \div \dfrac{9}{(a+1)^2} = \dfrac{a}{a+1} \cdot \dfrac{(a+1)^2}{9}$

$\quad = \dfrac{a(a+1)(a+1)}{9(a+1)}$

$\quad = \dfrac{a(a+1)\cancel{(a+1)}}{9\cancel{(a+1)}}$

$\quad = \dfrac{a(a+1)}{9}, \quad a \neq -1$

**67.** $\dfrac{(2x)^2}{(x+2)^2} \div \dfrac{4x}{(x+2)^3} = \dfrac{(2x)^2}{(x+2)^2} \cdot \dfrac{(x+2)^3}{4x}$

$\quad = \dfrac{(2x)^2(x+2)^3}{4x(x+2)^2}$

$\quad = \dfrac{\cancel{(2)}\,\cancel{(2)}\,\cancel{(x)}(x)\cancel{(x+2)}\,\cancel{(x+2)}(x+2)}{\cancel{(2)}\,\cancel{(2)}\,\cancel{(x)}\,\cancel{(x+2)}\,\cancel{(x+2)}(1)}$

$\quad = \dfrac{x(x+2)}{1} = x(x+2), \quad x \neq 0, x \neq -2$

**69.** $\dfrac{y^2-4}{y^2} \div \dfrac{y-2}{3y} = \dfrac{y^2-4}{y^2} \cdot \dfrac{3y}{y-2} = \dfrac{\cancel{(y-2)}(y+2)(3)\cancel{(y)}}{y\cancel{(y)}\cancel{(y-2)}} = \dfrac{3(y+2)}{y}, \quad y \neq 2$

**63.** $\dfrac{y}{y+8} \div \dfrac{y-6}{y+8} = \dfrac{y}{y+8} \cdot \dfrac{y+8}{y-6}$

$\quad = \dfrac{y(y+8)}{(y+8)(y-6)}$

$\quad = \dfrac{y\cancel{(y+8)}}{\cancel{(y+8)}(y-6)}$

$\quad = \dfrac{y}{y-6}, \quad y \neq -8$

**65.** $\dfrac{3(x+4)}{8} \div \dfrac{x+4}{2} = \dfrac{3(x+4)}{8} \cdot \dfrac{2}{x+4}$

$\quad = \dfrac{3(2)(x+4)}{8(x+4)}$

$\quad = \dfrac{3\cancel{(2)}\,\cancel{(x+4)}}{4\cancel{(2)}\,\cancel{(x+4)}}$

$\quad = \dfrac{3}{4}, \quad x \neq -4$

**71.** $\dfrac{y^2 - 3y}{y^2 + 4y - 21} \div \dfrac{3y}{4y + 28} = \dfrac{y^2 - 3y}{y^2 + 4y - 21} \cdot \dfrac{4y + 28}{3y}$

$$= \dfrac{(y)(y - 3)(4)(y + 7)}{(y + 7)(y - 3)(3)(y)}$$

$$= \dfrac{\cancel{y}\,\cancel{(y-3)}(4)\cancel{(y+7)}}{\cancel{(y+7)}\,\cancel{(y-3)}(3)\cancel{(y)}}$$

$$= \dfrac{4}{3}, \quad y \neq 0, y \neq 3, y \neq -7$$

**73.** $\dfrac{x^2 - 7x + 12}{x + 4} \div (x - 3)^2 = \dfrac{x^2 - 7x + 12}{x + 4} \cdot \dfrac{1}{(x - 3)^2}$

$$= \dfrac{(x - 3)(x - 4)(1)}{(x + 4)(x - 3)(x - 3)}$$

$$= \dfrac{\cancel{(x-3)}(x - 4)(1)}{(x + 4)\cancel{(x-3)}(x - 3)}$$

$$= \dfrac{x - 4}{(x + 4)(x - 3)}$$

**75.** $(x - 3) \div \dfrac{x^2 + 3x - 18}{x} = \dfrac{x - 3}{1} \cdot \dfrac{x}{x^2 + 3x - 18}$

$$= \dfrac{\cancel{(x-3)}(x)}{(x + 6)\cancel{(x-3)}}$$

$$= \dfrac{x}{x + 6}, \quad x \neq 3, x \neq 0$$

**77.** $\dfrac{x + 2}{7 - x} \div \dfrac{x^2 - 5x + 6}{x^2 - 9x + 14} = \dfrac{x + 2}{7 - x} \cdot \dfrac{x^2 - 9x + 14}{x^2 - 5x + 6}$

$$= \dfrac{(x + 2)\cancel{(x-7)}\cancel{(x-2)}}{(-1)\cancel{(x-7)}\cancel{(x-2)}(x - 3)}$$

$$= -\dfrac{x + 2}{x - 3}, \quad x \neq 7, x \neq 2$$

**79.** $\dfrac{25 - y^2}{4y + 20} \div \dfrac{2y^2 - 7y - 15}{24y + 4} = \dfrac{25 - y^2}{4y + 20} \cdot \dfrac{24y + 4}{2y^2 - 7y - 15}$

$$= \dfrac{(5 + y)(5 - y)(4)(6y + 1)}{(4)(y + 5)(2y + 3)(y - 5)}$$

$$= \dfrac{(y + 5)(-1)(y - 5)(4)(6y + 1)}{(4)(y + 5)(2y + 3)(y - 5)}$$

$$= \dfrac{(-1)\cancel{(y+5)}\cancel{(y-5)}\cancel{(4)}(6y + 1)}{\cancel{4}\cancel{(y+5)}(2y + 3)\cancel{(y-5)}}$$

$$= -\dfrac{6y + 1}{2y + 3}, \quad y \neq -5, y \neq 5, y \neq -\dfrac{1}{6}$$

**81.** $\dfrac{a^2 - 10a + 25}{a^2 + 7a + 12} \div \dfrac{a^2 - a - 20}{a^2 + 6a + 9} = \dfrac{a^2 - 10a + 25}{a^2 + 7a + 12} \cdot \dfrac{a^2 + 6a + 9}{a^2 - a - 20}$

$$= \dfrac{(a - 5)\cancel{(a-5)}\cancel{(a+3)}(a + 3)}{\cancel{(a+3)}(a + 4)\cancel{(a-5)}(a + 4)}$$

$$= \dfrac{(a - 5)(a + 3)}{(a + 4)^2}, \quad a \neq 5, a \neq -3$$

**83.** $\left(\dfrac{x^2}{5} \cdot \dfrac{x + a}{2}\right) \div \dfrac{x}{30} = \left(\dfrac{x^2}{5} \cdot \dfrac{x + a}{2}\right) \cdot \dfrac{30}{x} = \dfrac{x\cancel{(x)}(x + a)\cancel{(5)}\cancel{(2)}(3)}{\cancel{(5)}\cancel{(2)}\cancel{(x)}} = 3x(x + a), \quad x \neq 0$

**85.** $\left[\left(\dfrac{x+2}{3}\right)^2 \cdot \left(\dfrac{x+1}{2}\right)^2\right] \div \dfrac{(x+1)(x+2)}{36} = \left[\dfrac{(x+2)^2}{3^2} \cdot \dfrac{(x+1)^2}{2^2}\right] \div \dfrac{(x+1)(x+2)}{36}$

$$= \dfrac{(x+2)^2(x+1)^2}{9(4)} \cdot \dfrac{36}{(x+1)(x+2)}$$

$$= \dfrac{36(x+2)^2(x+1)^2}{36(x+1)(x+2)}$$

$$= \dfrac{36(x+2)(x+2)(x+1)(x+1)}{36(x+1)(x+2)}$$

$$= (x+2)(x+1), \quad x \neq -1, x \neq -2$$

**87.** $\left[\dfrac{(t+2)^3}{(t+1)^3} \div \dfrac{t^2+4t+4}{t^2+2t+1}\right] \cdot \dfrac{t+1}{t+2} = \dfrac{(t+2)^3}{(t+1)^3} \cdot \dfrac{t^2+2t+1}{t^2+4t+4} \cdot \dfrac{t+1}{t+2}$

$$= \dfrac{(t+2)^3(t+1)^2(t+1)}{(t+1)^3(t+2)^2(t+2)}$$

$$= \dfrac{(t+2)^3(t+1)^3}{(t+1)^3(t+2)^3}$$

$$= 1, \quad t \neq -1, t \neq -2$$

**89.** (a) The time required to pump 1 gallon is $\dfrac{1}{5}$ minute.

(b) The time required to pump $x$ gallons is $x\left(\dfrac{1}{5}\right) = \dfrac{x}{5}$ minutes.

(c) The time required to pump 120 gallons is $120\left(\dfrac{1}{5}\right) = \dfrac{120}{5} = 24$ minutes.

**91.** (a) Area $= \dfrac{1}{2} \cdot$ Base $\cdot$ Height $= \dfrac{1}{2}\left(\dfrac{8}{x^2+5x}\right) \cdot \left(\dfrac{x+5}{2}\right) = \dfrac{(8)(x+5)}{2(x^2+5x)(2)} = \dfrac{(2)(2)(2)(x+5)}{(2)(x)(x+5)(2)} = \dfrac{2}{x}$

(b) Base: As $x$ increases, the base decreases.

Height: As $x$ increases, the height increases.

Area: As $x$ increases, the area decreases.

(c)

| $x$ | 2 | 4 | 6 | 8 | 10 | 12 |
|---|---|---|---|---|---|---|
| Area $= \dfrac{2}{x}$ | 1 | $\dfrac{1}{2}$ | $\dfrac{1}{3}$ | $\dfrac{1}{4}$ | $\dfrac{1}{5}$ | $\dfrac{1}{6}$ |

**93.** There are many correct answers. Here is one example of three rational expressions whose product is 1.

$$\dfrac{x+2}{x} \cdot \dfrac{x^2}{(x+2)^3} \cdot \dfrac{(x+2)^2}{x} = \dfrac{(x+2)(x^2)(x+2)^2}{x(x+2)^3(x)} = \dfrac{x^2(x+2)^3}{x^2(x+2)^3} = 1, \quad x \neq 0, x \neq -2$$

**95.** The first fraction, the dividend, was mistakenly inverted. Instead, the divisor should have been inverted.

$$\dfrac{x^2-4}{5x} \div \dfrac{x+2}{x-2} = \dfrac{x^2-4}{5x} \cdot \dfrac{x-2}{x+2} = \dfrac{(x+2)(x-2)(x-2)}{5x(x+2)} = \dfrac{(x+2)(x-2)(x-2)}{5x(x+2)} = \dfrac{(x-2)^2}{5x}, \quad x \neq -2, x \neq 2$$

**97.** $3x^2 + 7x = x(3x + 7)$

**99.** $x^2 + 7x - 18 = (x + 9)(x - 2)$

**101.** The denominator 48 is never 0. So, the domain is the set of all real numbers $x$.

**103.** The denominator $4(x + 4)^2$ is zero when $x = -4$. So, the domain is all real values of $x$ such that $x \neq -4$.

**105.** The denominator $8 - 2x$ is zero when $x = 4$. So, the domain is all real values of $x$ such that $x \neq 4$.

**107.** The denominator $x^2 - 4 = (x + 2)(x - 2)$ is zero when $x = -2$ or $x = 2$. So, the domain is all real values of $x$ such that $x \neq -2$ and $x \neq 2$.

# Section 7.3   Adding and Subtracting Rational Expressions

**1.** $\dfrac{y}{4} + \dfrac{3y}{4} = \dfrac{y + 3y}{4} = \dfrac{4y}{4} = \dfrac{4y}{4} = y$

**3.** $\dfrac{5}{3a} + \dfrac{9}{3a} = \dfrac{5 + 9}{3a} = \dfrac{14}{3a}$

**5.** $\dfrac{x}{3} + \dfrac{1 - x}{3} = \dfrac{x + (1 - x)}{3} = \dfrac{x + 1 - x}{3} = \dfrac{1}{3}$

**7.** $\dfrac{3}{x + 2} - \dfrac{7x}{x + 2} = \dfrac{3 - 7x}{x + 2}$

**9.** $\dfrac{x}{x^2 - 1} + \dfrac{1}{x^2 - 1} = \dfrac{x + 1}{x^2 - 1}$

$= \dfrac{(x + 1) \cdot 1}{(x + 1)(x - 1)}$

$= \dfrac{1}{x - 1}, \quad x \neq -1$

**11.** $\dfrac{3x - 2}{(x + 1)^2} + \dfrac{4x + 5}{(x + 1)^2} = \dfrac{3x - 2 + 4x + 5}{(x + 1)^2}$

$= \dfrac{7x + 3}{(x + 1)^2}$

**13.** $\dfrac{6x}{x^2 - 7x} - \dfrac{2x - x^2}{x^2 - 7x} = \dfrac{6x - (2x - x^2)}{x^2 - 7x}$

$= \dfrac{6x - 2x + x^2}{x(x - 7)}$

$= \dfrac{x^2 + 4x}{x(x - 7)}$

$= \dfrac{(x)(x + 4)}{(x)(x - 7)}$

$= \dfrac{x + 4}{x - 7}, \quad x \neq 0$

**15.** $\dfrac{3}{x - 5} + \dfrac{9}{5 - x} = \dfrac{3}{x - 5} + \dfrac{(-1)(9)}{(-1)(5 - x)}$

$= \dfrac{3}{x - 5} + \dfrac{-9}{-5 + x}$

$= \dfrac{3}{x - 5} + \dfrac{-9}{x - 5}$

$= \dfrac{3 - 9}{x - 5}$

$= \dfrac{-6}{x - 5}$ or $-\dfrac{6}{x - 5}$ or $\dfrac{6}{5 - x}$

**17.** $\dfrac{5x}{2x - 3} - \dfrac{9}{3 - 2x} = \dfrac{5x}{2x - 3} - \dfrac{(-1)(9)}{(-1)(3 - 2x)}$

$= \dfrac{5x}{2x - 3} - \dfrac{-9}{2x - 3}$

$= \dfrac{5x - (-9)}{2x - 3}$

$= \dfrac{5x + 9}{2x - 3}$

**19.** $\dfrac{4x - 1}{x - 11} + \dfrac{2x}{11 - x} = \dfrac{4x - 1}{x - 11} + \dfrac{(-1)(2x)}{(-1)(11 - x)}$

$= \dfrac{4x - 1}{x - 11} + \dfrac{-2x}{x - 11}$

$= \dfrac{2x - 1}{x - 11}$

**21.** $\dfrac{3x + 1}{x - 15} + \dfrac{x - 2}{15 - x} = \dfrac{3x + 1}{x - 15} + \dfrac{-1(x - 2)}{-1(15 - x)}$

$= \dfrac{3x + 1}{x - 15} + \dfrac{-x + 2}{x - 15}$

$= \dfrac{3x + 1 - x + 2}{x - 15}$

$= \dfrac{2x + 3}{x - 15}$

**23.** $9y^2 = 3 \cdot 3 \cdot y \cdot y = 3^2 \cdot y^2$

$12y = 2 \cdot 2 \cdot 3 \cdot y = 2^2 \cdot 3 \cdot y$

The different factors are 2, 3, and $y$. Using the highest powers of these factors, we conclude that the least common multiple is $2^2 \cdot 3^2 \cdot y^2$, or $36y^2$.

**25.** $2(y - 3) = 2 \cdot (y - 3)$

$6(y - 3) = 2(3)(y - 3)$

The different factors are 2, 3, and $(y - 3)$. Using the highest powers of the factors, we conclude that the least common multiple is $2(3)(y - 3)$ or $6(y - 3)$.

**27.** $16x = 2 \cdot 2 \cdot 2 \cdot 2 \cdot x = 2^4 \cdot x$

$12x(x + 2) = 2 \cdot 2 \cdot 3 \cdot x \cdot (x + 2)$

$\qquad\qquad = 2^2 \cdot 3 \cdot x \cdot (x + 2)$

The different factors are 2, 3, $x$, and $(x + 2)$. Using the highest powers of these factors, we conclude that the least common multiple is $2^4 \cdot 3 \cdot x \cdot (x + 2)$, or $48x(x + 2)$.

**29.** $x - 7$

$x^2 = x \cdot x$

$x(x + 7) = x \cdot (x + 7)$

The different factors are $x$, $(x - 7)$, and $(x + 7)$. Using the highest powers of these factors, we conclude that the least common multiple is $x^2(x - 7)(x + 7)$.

**31.** $x + 2$

$x^2 - 4 = (x + 2)(x - 2)$

$x$

The different factors are $(x + 2)$, $(x - 2)$, and $x$. Using the highest powers of these factors, we conclude that the least common multiple is $x(x + 2)(x - 2)$.

**33.** $t^3 + 4t^2 + 4t = t(t^2 + 4t + 4) = t \cdot (t + 2)^2$

$t^2 - 4t = t \cdot (t - 4)$

The different factors are $t$, $(t + 2)$, and $(t - 4)$. Using the highest powers of these factors, we conclude that the least common multiple is $t(t + 2)^2(t - 4)$.

**35.** The denominators are $7x$ and $14x^3 = 2 \cdot 7 \cdot x^3$. The least common denominator is $2 \cdot 7 \cdot x^3$ or $14x^3$.

$\dfrac{3}{7x} = \dfrac{3(2x^2)}{7x(2x^2)} = \dfrac{6x^2}{14x^3}$

$\dfrac{5x}{14x^3}$

**37.** The denominators are $3x - 6 = 3(x - 2)$ and $(x - 2)$. The least common denominator is $3(x - 2)$.

$\dfrac{x + 5}{3x - 6} = \dfrac{x + 5}{3(x - 2)}$

$\dfrac{10}{x - 2} = \dfrac{10(3)}{(x - 2)(3)} = \dfrac{30}{3(x - 2)}$

**39.** The denominators are $(x + 3)^2$ and $x(x + 3)$. The least common denominator is $x(x + 3)^2$.

$\dfrac{2}{(x + 3)^2} = \dfrac{2(x)}{(x + 3)^2(x)} = \dfrac{2x}{x(x + 3)^2}$

$\dfrac{5}{x(x + 3)} = \dfrac{5(x + 3)}{x(x + 3)(x + 3)} = \dfrac{5(x + 3)}{x(x + 3)^2}$

**41.** The denominators are $x^2 - 16 = (x - 4)(x + 4)$ and $x^2 - 8x + 16 = (x - 4)^2$. The least common denominator is $(x - 4)^2(x + 4)$.

$\dfrac{x - 8}{x^2 - 16} = \dfrac{x - 8}{(x - 4)(x + 4)} = \dfrac{(x - 8)(x - 4)}{(x - 4)(x + 4)(x - 4)} = \dfrac{(x - 8)(x - 4)}{(x - 4)^2(x + 4)}$

$\dfrac{9x}{x^2 - 8x + 16} = \dfrac{9x}{(x - 4)^2} = \dfrac{9x(x + 4)}{(x - 4)^2(x + 4)}$

**43.** The denominators are $2s$ and $5s$. The least common denominator is $2 \cdot 5 \cdot s$ or $10s$.

$\dfrac{3}{2s} + \dfrac{1}{5s} = \dfrac{3(5)}{2s(5)} + \dfrac{1(2)}{5s(2)} = \dfrac{15}{10s} + \dfrac{2}{10s} = \dfrac{15 + 2}{10s} = \dfrac{17}{10s}$

**45.** The denominators are $5x = 5 \cdot x$ and 5. The least common denominator is $5x$.

$$\frac{1}{5x} - \frac{3}{5} = \frac{1}{5x} - \frac{3(x)}{5(x)} = \frac{1}{5x} - \frac{3x}{5x} = \frac{1 - 3x}{5x}$$

**47.** The denominators are $u$ and $u^2$. The least common denominator is $u^2$.

$$\frac{5}{u} + \frac{2}{u^2} = \frac{5(u)}{u(u)} + \frac{2}{u^2} = \frac{5u + 2}{u^2}$$

**49.** The denominators are $2b$ and $2b^2$. The least common denominator is $2b^2$.

$$\frac{3}{2b} + \frac{5}{2b^2} = \frac{3(b)}{2b(b)} + \frac{5}{2b^2} = \frac{3b}{2b^2} + \frac{5}{2b^2} = \frac{3b + 5}{2b^2}$$

**51.** The least common denominator is $x + 3$.

$$6 - \frac{5}{x + 3} = \frac{6}{1} - \frac{5}{x + 3}$$

$$= \frac{6(x + 3)}{1(x + 3)} - \frac{5}{x + 3}$$

$$= \frac{6(x + 3) - 5}{x + 3}$$

$$= \frac{6x + 18 - 5}{x + 3}$$

$$= \frac{6x + 13}{x + 3}$$

**53.** The denominators are 1 and $2x - 3$. The least common denominator is $2x - 3$.

$$7 + \frac{2}{2x - 3} = \frac{7}{1} + \frac{2}{2x - 3}$$

$$= \frac{7(2x - 3)}{2x - 3} + \frac{2}{2x - 3}$$

$$= \frac{14x - 21 + 2}{2x - 3}$$

$$= \frac{14x - 19}{2x - 3}$$

**55.** The denominators are $x - 5$ and $x + 3$. The least common denominator is $(x - 5)(x + 3)$.

$$\frac{3}{x - 5} + \frac{2}{x + 3} = \frac{3(x + 3)}{(x - 5)(x + 3)} + \frac{2(x - 5)}{(x + 3)(x - 5)}$$

$$= \frac{3x + 9 + 2x - 10}{(x - 5)(x + 3)}$$

$$= \frac{5x - 1}{(x - 5)(x + 3)}$$

**57.** The denominators are $(x - 1)$ and $(x + 2)$. The least common denominator is $(x - 1)(x + 2)$.

$$\frac{1}{x - 1} - \frac{1}{x + 2} = \frac{1(x + 2)}{(x - 1)(x + 2)} - \frac{1(x - 1)}{(x + 2)(x - 1)}$$

$$= \frac{1(x + 2) - 1(x - 1)}{(x - 1)(x + 2)}$$

$$= \frac{x + 2 - x + 1}{(x - 1)(x + 2)}$$

$$= \frac{3}{(x - 1)(x + 2)}$$

**59.** The denominators are $x - 3$ and $2x + 5$. The least common denominator is $(x - 3)(2x + 5)$.

$$\frac{x}{x - 3} + \frac{3}{2x + 5} = \frac{x(2x + 5)}{(x - 3)(2x + 5)} + \frac{3(x - 3)}{(x - 3)(2x + 5)}$$

$$= \frac{x(2x + 5) + 3(x - 3)}{(x - 3)(2x + 5)}$$

$$= \frac{2x^2 + 5x + 3x - 9}{(x - 3)(2x + 5)}$$

$$= \frac{2x^2 + 8x - 9}{(x - 3)(2x + 5)}$$

**61.** The denominators are $2(x - 4)$ and $2x$. The least common denominator is $2x(x - 4)$.

$$\frac{3}{2(x - 4)} - \frac{1}{2x} = \frac{3(x)}{2(x - 4)(x)} - \frac{1(x - 4)}{2x(x - 4)}$$

$$= \frac{3x - (x - 4)}{2x(x - 4)}$$

$$= \frac{3x - x + 4}{2x(x - 4)}$$

$$= \frac{2x + 4}{2x(x - 4)}$$

$$= \frac{\cancel{2}(x + 2)}{\cancel{2}(x)(x - 4)}$$

$$= \frac{x + 2}{x(x - 4)}$$

**63.** The denominators are $x^2 - 9 = (x + 3)(x - 3)$ and $(x + 3)$. The least common denominator is $(x + 3)(x - 3)$.

$$\frac{x}{x^2 - 9} + \frac{3}{x + 3} = \frac{x}{(x + 3)(x - 3)} + \frac{3(x - 3)}{(x + 3)(x - 3)}$$

$$= \frac{x + 3(x - 3)}{(x + 3)(x - 3)}$$

$$= \frac{x + 3x - 9}{(x + 3)(x - 3)}$$

$$= \frac{4x - 9}{(x + 3)(x - 3)}$$

**65.** The denominators are $v(v + 4)$ and $v^2$. The least common denominator is $v^2(v + 4)$.

$$\frac{5v}{v(v + 4)} + \frac{2v}{v^2} = \frac{5v(v)}{v(v + 4)(v)} + \frac{2v(v + 4)}{v^2(v + 4)}$$

$$= \frac{5v^2 + 2v(v + 4)}{v^2(v + 4)}$$

$$= \frac{5v^2 + 2v^2 + 8v}{v^2(v + 4)}$$

$$= \frac{7v^2 + 8v}{v^2(v + 4)}$$

$$= \frac{\cancel{v}(7v + 8)}{\cancel{v}(v)(v + 4)}$$

$$= \frac{7v + 8}{v(v + 4)}$$

Note: The fractions in the original problem could have been simplified.

$$\frac{5v}{v(v + 4)} + \frac{2v}{v^2} = \frac{5\cancel{v}}{\cancel{v}(v + 4)} + \frac{2\cancel{v}}{\cancel{v}(v)}$$

$$= \frac{5}{v + 4} + \frac{2}{v} \qquad \text{The common denominator is now } v(v + 4).$$

$$= \frac{5(v)}{(v + 4)(v)} + \frac{2(v + 4)}{v(v + 4)}$$

$$= \frac{5v + 2(v + 4)}{v(v + 4)}$$

$$= \frac{5v + 2v + 8}{v(v + 4)}$$

$$= \frac{7v + 8}{v(v + 4)}$$

**67.** The denominators are $x^2 - 5x + 6 = (x - 3)(x - 2)$ and $(2 - x) = -1(x - 2)$. The least common denominator is $(x - 3)(x - 2)$.

$$
\begin{aligned}
\frac{x + 2}{x^2 - 5x + 6} + \frac{3}{2 - x} &= \frac{x + 2}{(x - 3)(x - 2)} + \frac{(-1)(3)}{(-1)(2 - x)} \\[2mm]
&= \frac{x + 2}{(x - 3)(x - 2)} - \frac{3}{x - 2} \\[2mm]
&= \frac{x + 2}{(x - 3)(x - 2)} - \frac{3(x - 3)}{(x - 3)(x - 2)} \\[2mm]
&= \frac{x + 2 - 3(x - 3)}{(x - 3)(x - 2)} \\[2mm]
&= \frac{x + 2 - 3x + 9}{(x - 3)(x - 2)} \\[2mm]
&= \frac{-2x + 11}{(x - 3)(x - 2)}
\end{aligned}
$$

**69.** The denominators are $6x^2 + 7x + 2 = (3x + 2)(2x + 1)$ and $2x^2 - 3x - 2 = (2x + 1)(x - 2)$. The least common denominator is $(3x + 2)(2x + 1)(x - 2)$.

$$
\begin{aligned}
\frac{x}{6x^2 + 7x + 2} - \frac{5}{2x^2 - 3x - 2} &= \frac{x}{(3x + 2)(2x + 1)} - \frac{5}{(2x + 1)(x - 2)} \\[2mm]
&= \frac{x(x - 2)}{(3x + 2)(2x + 1)(x - 2)} - \frac{5(3x + 2)}{(3x + 2)(2x + 1)(x - 2)} \\[2mm]
&= \frac{x(x - 2) - 5(3x + 2)}{(3x + 2)(2x + 1)(x - 2)} \\[2mm]
&= \frac{x^2 - 2x - 15x - 10}{(3x + 2)(2x + 1)(x - 2)} \\[2mm]
&= \frac{x^2 - 17x - 10}{(3x + 2)(2x + 1)(x - 2)}
\end{aligned}
$$

**71.** The denominators are $x^2 - 16 = (x - 4)(x + 4)$ and $4 - x$. The least common denominator is $(x - 4)(x + 4)$.

$$
\begin{aligned}
\frac{2x + 1}{x^2 - 16} + \frac{4x}{4 - x} &= \frac{2x + 1}{(x - 4)(x + 4)} + \frac{4x(-1)}{(4 - x)(-1)} \\[2mm]
&= \frac{2x + 1}{(x - 4)(x + 4)} + \frac{-4x}{x - 4} \\[2mm]
&= \frac{2x + 1}{(x - 4)(x + 4)} - \frac{4x}{x - 4} \\[2mm]
&= \frac{2x + 1}{(x - 4)(x + 4)} - \frac{4x(x + 4)}{(x - 4)(x + 4)} \\[2mm]
&= \frac{2x + 1 - 4x(x + 4)}{(x - 4)(x + 4)} \\[2mm]
&= \frac{2x + 1 - 4x^2 - 16x}{(x - 4)(x + 4)} = \frac{-4x^2 - 14x + 1}{(x - 4)(x + 4)} \text{ or } \frac{4x^2 + 14x - 1}{(4 - x)(4 + x)}
\end{aligned}
$$

**73.** The denominators are $x, x^2$, and $x + 1$. The least common denominator is $x^2(x + 1)$.

$$\frac{3}{x} - \frac{1}{x^2} + \frac{1}{x + 1} = \frac{3(x)(x + 1)}{x(x)(x + 1)} - \frac{1(x + 1)}{x^2(x + 1)} + \frac{1(x^2)}{(x + 1)(x^2)}$$

$$= \frac{3x(x + 1) - (x + 1) + x^2}{x^2(x + 1)}$$

$$= \frac{3x^2 + 3x - x - 1 + x^2}{x^2(x + 1)}$$

$$= \frac{4x^2 + 2x - 1}{x^2(x + 1)}$$

**75.** The denominators are $3(x - 2)^2, 3(x - 2)$, and $2x$. The least common denominator is $3 \cdot 2 \cdot x \cdot (x - 2)^2 = 6x(x - 2)^2$.

$$\frac{x + 2}{3(x - 2)^2} + \frac{4}{3(x - 2)} + \frac{1}{2x} = \frac{(x + 2)(2x)}{3(x - 2)^2(2x)} + \frac{4(2x)(x - 2)}{3(x - 2)(2x)(x - 2)} + \frac{1(3)(x - 2)^2}{2x(3)(x - 2)^2}$$

$$= \frac{2x(x + 2) + 8x(x - 2) + 3(x - 2)^2}{6x(x - 2)^2}$$

$$= \frac{2x^2 + 4x + 8x^2 - 16x + 3(x^2 - 4x + 4)}{6x(x - 2)^2}$$

$$= \frac{2x^2 + 4x + 8x^2 - 16x + 3x^2 - 12x + 12}{6x(x - 2)^2}$$

$$= \frac{13x^2 - 24x + 12}{6x(x - 2)^2}$$

**77.** The denominators are $x^2 - 4 = (x + 2)(x - 2), x + 2$, and $4 - x^2 = -1(x^2 - 4) = -(x + 2)(x - 2)$. The least common denominator is $(x + 2)(x - 2)$.

$$\frac{x}{x^2 - 4} + \frac{3x}{x + 2} + \frac{3x^2 - 5x}{4 - x^2} = \frac{x}{(x^2 - 4)} + \frac{3x}{x + 2} + \frac{-1(3x^2 - 5x)}{-1(4 - x^2)}$$

$$= \frac{x}{x^2 - 4} + \frac{3x}{x + 2} + \frac{-3x^2 + 5x}{x^2 - 4}$$

$$= \frac{x}{(x + 2)(x - 2)} + \frac{3x(x - 2)}{(x + 2)(x - 2)} + \frac{-3x^2 + 5x}{(x - 2)(x + 2)}$$

$$= \frac{x + 3x^2 - 6x - 3x^2 + 5x}{(x + 2)(x - 2)}$$

$$= \frac{0}{(x + 2)(x - 2)}$$

$$= 0, \ x \neq -2, x \neq 2$$

**79.** The denominators are $x^2 + x - 6 = (x + 3)(x - 2)$ and $x + 3$. The least common denominator is $(x + 3)(x - 2)$.

$$\frac{6x}{x^2 + x - 6} + \frac{x}{x + 3} - 4 = \frac{6x}{(x + 3)(x - 2)} + \frac{x}{x + 3} - \frac{4}{1}$$

$$= \frac{6x}{(x + 3)(x - 2)} + \frac{x(x - 2)}{(x + 3)(x - 2)} - \frac{4(x + 3)(x - 2)}{1(x + 3)(x - 2)}$$

$$= \frac{6x + x(x - 2) - 4(x + 3)(x - 2)}{(x + 3)(x - 2)}$$

$$= \frac{6x + x(x - 2) - 4(x^2 + x - 6)}{(x + 3)(x - 2)}$$

$$= \frac{6x + x^2 - 2x - 4x^2 - 4x + 24}{(x + 3)(x - 2)}$$

$$= \frac{-3x^2 + 24}{(x + 3)(x - 2)} \text{ or } \frac{24 - 3x^2}{(x + 3)(x - 2)}$$

**81.** $\dfrac{8}{r - 2} + \dfrac{8}{r + 2} = \dfrac{8(r + 2)}{(r - 2)(r + 2)} + \dfrac{8(r - 2)}{(r - 2)(r + 2)}$

$$= \frac{8(r + 2) + 8(r - 2)}{(r - 2)(r + 2)}$$

$$= \frac{8r + 16 + 8r - 16}{(r - 4)(r + 4)}$$

$$= \frac{16r}{(r - 2)(r + 2)}$$

The total travel time of the kayaker is $\dfrac{16r}{(r - 2)(r + 2)}$ hours.

**83.** $90 - \dfrac{40 - x}{x} = \dfrac{90x}{x} - \dfrac{40 - x}{x}$

$$= \frac{90x - (40 - x)}{x}$$

$$= \frac{90x - 40 + x}{x}$$

$$= \frac{91x - 40}{x}$$

The measure of the complement is $\left(\dfrac{91x - 40}{x}\right)^{\circ}$.

**85.** Perimeter $= 2 \cdot$ Length $+ 2 \cdot$ Width

$$P = 2\left(\frac{6}{y - 2}\right) + 2\left(\frac{3}{y}\right)$$

$$= \frac{12}{y - 2} + \frac{6}{y}$$

$$= \frac{12y}{(y - 2)y} + \frac{6(y - 2)}{y(y - 2)}$$

$$= \frac{12y + 6(y - 2)}{y(y - 2)}$$

$$= \frac{12y + 6y - 12}{y(y - 2)}$$

$$= \frac{18y - 12}{y(y - 2)} \text{ or } \frac{6(3y - 2)}{y(y - 2)}$$

The perimeter $P = \dfrac{6(3y - 2)}{y(y - 2)}$ feet.

**87.** Yes, the least common multiple of two or more polynomials can be the same as one of the polynomials. Example: The least common multiple of $2(x + 2)$ and $x + 2$ is $2(x + 2)$.

**89.** The binomial is one of the factors of the trinomial. For example, the least common multiple of $x + 3$ and $x^2 + 5x + 6$ is $x^2 + 5x + 6$ because $x + 3$ is a factor of $x^2 + 5x + 6$; $x^2 + 5x + 6 = (x + 3)(x + 2)$.

**91. (a)** $x = 0$

$$3(0) - 18 \overset{?}{>} 0$$

$$0 - 18 \overset{?}{>} 0$$

$$-18 \not> 0$$

0 *is not* a solution.

**(b)** $x = 6$

$$3(6) - 18 \overset{?}{>} 0$$

$$18 - 18 \overset{?}{>} 0$$

$$0 \not> 0$$

6 *is not* a solution.

**(c)** $x = 4$

$$3(4) - 18 \overset{?}{>} 0$$

$$12 - 18 \overset{?}{>} 0$$

$$-6 \not> 0$$

4 *is not* a solution.

**(d)** $x = 11$

$$3(11) - 18 \overset{?}{>} 0$$

$$33 - 18 \overset{?}{>} 0$$

$$15 > 0$$

11 *is* a solution.

**93. (a)** $x = -1$

$$2 - \tfrac{3}{4}(-1) \overset{?}{<} -3$$

$$2 + \tfrac{3}{4} \overset{?}{<} -3$$

$$2\tfrac{3}{4} \not< -3$$

$-1$ *is not* a solution.

**(b)** $x = 4$

$$2 - \tfrac{3}{4}(4) \overset{?}{<} -3$$

$$2 - 3 \overset{?}{<} -3$$

$$-1 \not< -3$$

4 *is not* a solution.

**(c)** $x = \tfrac{1}{8}$

$$2 - \tfrac{3}{4}\left(\tfrac{1}{8}\right) \overset{?}{<} -3$$

$$2 - \tfrac{3}{32} \overset{?}{<} -3$$

$$1\tfrac{29}{32} \not< -3$$

$\tfrac{1}{8}$ *is not* a solution.

**(d)** $x = 9$

$$2 - \tfrac{3}{4}(9) \overset{?}{<} -3$$

$$\tfrac{8}{4} - \tfrac{27}{4} \overset{?}{<} -3$$

$$-\tfrac{19}{4} < -3$$

9 *is* a solution.

**95.** $\dfrac{15}{7x^2} \cdot \dfrac{28x}{3} = \dfrac{15 \cdot 28x}{7x^2 \cdot 3} = \dfrac{(5)(\cancel{3})(\cancel{7})(4)(\cancel{x})}{(\cancel{7})(\cancel{3})(\cancel{x})(x)} = \dfrac{20}{x}$

**97.** $\dfrac{x^5}{4} \div \dfrac{x^2}{12} = \dfrac{x^5}{4} \cdot \dfrac{12}{x^2} = \dfrac{12x^5}{4x^2} = \dfrac{(\cancel{4})(3)(\cancel{x^2})(x^3)}{(\cancel{4})(\cancel{x^2})} = 3x^3, \quad x \neq 0$

**99.** $\dfrac{12x}{7x - 35} \cdot (5 - x) = \dfrac{12x}{7(x - 5)} \cdot (-x + 5)$

$$= \dfrac{12x}{7(x - 5)} \cdot (-1)(x - 5) = \dfrac{12x}{7(x - 5)} \cdot \dfrac{-1(x - 5)}{1} = \dfrac{-(12x)\cancel{(x - 5)}}{7\cancel{(x - 5)}} = -\dfrac{12x}{7}, \; x \neq 5$$

**101.** $\dfrac{x^2 - 9}{x^3} \div \dfrac{x - 3}{9x} = \dfrac{(x + 3)(x - 3)}{x^3} \cdot \dfrac{9x}{x - 3} = \dfrac{9(x)(x + 3)(x - 3)}{x(x)(x)(x - 3)} = \dfrac{9\cancel{(x)}(x + 3)\cancel{(x - 3)}}{x(x)\cancel{(x)}\cancel{(x - 3)}} = \dfrac{9(x + 3)}{x^2}, \; x \neq 3$

# Mid-Chapter Quiz for Chapter 7

**1.** The domain of a rational expression is the set of all real numbers for which the denominator of the expression is not equal to zero.

   (a) For this rational expression, the denominator is not zero for any real values of $x$, so the domain is the set of *all* real numbers $x$.

   (b) For this rational expression, the denominator,
$x^2 - 4 = (x + 2)(x - 2)$, is zero when $x = -2$

   or 2. So, the domain is all real values of $x$ such that $x \neq -2$ and $x \neq 2$.

**2.** (a) $\dfrac{10 - 3}{10 + 2} = \dfrac{7}{12}$

   (b) $\dfrac{3 - 3}{3 + 2} = \dfrac{0}{5} = 0$

   (c) $\dfrac{-2 - 3}{-2 + 2} = \dfrac{-5}{0}$    Undefined

      Division by 0 is undefined.

**3.** $\dfrac{14z^4}{35z} = \dfrac{7(2)(z)(z^3)}{5(7)(z)} = \dfrac{2z^3}{5}, \quad z \neq 0$

**4.** $\dfrac{60a^2b}{45ab^3} = \dfrac{4(15)(a)(a)(b)}{3(15)(a)(b)(b)(b)}$

   $= \dfrac{4(15)(a)(a)(b)}{3(15)(a)(b)(b)(b)}$

   $= \dfrac{4a}{3b^2}, \quad a \neq 0$

**5.** $\dfrac{y^2 - 4}{8 - 4y} = \dfrac{(y + 2)(y - 2)}{4(2 - y)}$

   $= \dfrac{(y + 2)(y - 2)}{4(-1)(y - 2)}$

   $= -\dfrac{y + 2}{4}, \quad y \neq 2$

**6.** $\dfrac{15u(u - 3)^2}{25u^2(u - 3)} = \dfrac{3(5)(u)(u - 3)(u - 3)}{5(5)(u)(u)(u - 3)}$

   $= \dfrac{3(u - 3)}{5u}, \quad u \neq 3$

**7.** $\dfrac{b^2 + 3b}{b^3 + 2b^2 - 3b} = \dfrac{b(b + 3)}{b(b^2 + 2b - 3)}$

   $= \dfrac{(b)(b + 3)}{b(b + 3)(b - 1)}$

   $= \dfrac{1}{b - 1}, \quad b \neq 0, b \neq -3$

**8.** $\dfrac{4x^2 - 12x + 9}{2x^2 - x - 3} = \dfrac{(2x - 3)(2x - 3)}{(2x - 3)(x + 1)}$

   $= \dfrac{2x - 3}{x + 1}, \quad x \neq \dfrac{3}{2}$

**9.** $\dfrac{3y^3}{5} \cdot \dfrac{25}{9y} = \dfrac{(3y)(y^2)(5)(5)}{5(3)(3y)} = \dfrac{5y^2}{3}, \quad y \neq 0$

**10.** $\dfrac{s - 5}{15} \cdot \dfrac{12s}{25 - s^2} = \dfrac{(s - 5)(3)(4s)}{5(3)(5 + s)(5 - s)}$

   $= \dfrac{(s - 5)(3)(4s)}{5(3)(5 + s)(-1)(s - 5)}$

   $= -\dfrac{4s}{5(s + 5)}, \quad s \neq 5$

**11.** $(x^3 + 4x^2) \cdot \dfrac{5x}{x^2 + 2x - 8} = \dfrac{x^3 + 4x^2}{1} \cdot \dfrac{5x}{x^2 + 2x - 8}$

   $= \dfrac{x^2(x + 4)(5x)}{(x + 4)(x - 2)}$

   $= \dfrac{5x^3}{x - 2}, \quad x \neq -4$

**12.** $\dfrac{r^2 - 16}{r} \div \dfrac{r + 4}{r^2} = \dfrac{r^2 - 16}{r} \cdot \dfrac{r^2}{r + 4}$

   $= \dfrac{(r + 4)(r - 4)(r)(r)}{r(r + 4)}$

   $= \dfrac{(r + 4)(r - 4)(r)(r)}{(r)(r + 4)(1)}$

   $= \dfrac{r(r - 4)}{1}$

   $= r(r - 4), \quad r \neq 0, r \neq -4$

**13.** $\dfrac{x}{25} \div \dfrac{x^2 + 2x}{10} \cdot \dfrac{1}{x + 2} = \dfrac{x}{25} \cdot \dfrac{10}{x^2 + 2x} \cdot \dfrac{1}{x + 2}$

$$= \dfrac{\cancel{(x)}\,\cancel{(5)}\,(2)(1)}{5\,\cancel{(5)}\,\cancel{(x)}(x + 2)(x + 2)}$$

$$= \dfrac{2}{5(x + 2)^2}, \quad x \neq 0$$

**14.** $\dfrac{10x^2}{3y} \div \left( \dfrac{y}{x} \cdot \dfrac{x^3 y}{6} \right) = \dfrac{10x^2}{3y} \div \left( \dfrac{x^3 y^2}{6x} \right)$

$$= \dfrac{10x^2}{3y} \cdot \dfrac{6x}{x^3 y^2}$$

$$= \dfrac{10\,\cancel{(x^2)}\,\cancel{(3)}\,(2)\,\cancel{(x)}}{\cancel{3}\,(y)\,\cancel{(x^2)}\,\cancel{(x)}(y^2)}$$

$$= \dfrac{20}{y^3}, \quad x \neq 0$$

**17.** $\dfrac{x - 5}{x^2 + 7x + 12} + \dfrac{x - 3}{x + 4} = \dfrac{x - 5}{(x + 3)(x + 4)} + \dfrac{(x - 3)(x + 3)}{(x + 4)(x + 3)}$

$$= \dfrac{(x - 5) + (x - 3)(x + 3)}{(x + 3)(x + 4)}$$

$$= \dfrac{x - 5 + x^2 - 9}{(x + 3)(x + 4)}$$

$$= \dfrac{x^2 + x - 14}{(x + 3)(x + 4)}$$

**18.** $\dfrac{5x + 10}{x^2 - 5x - 14} - \dfrac{4}{x - 7} = \dfrac{5x + 10}{(x - 7)(x + 2)} - \dfrac{4(x + 2)}{(x - 7)(x + 2)}$

$$= \dfrac{5x + 10 - 4(x + 2)}{(x - 7)(x + 2)}$$

$$= \dfrac{5x + 10 - 4x - 8}{(x - 7)(x + 2)}$$

$$= \dfrac{\cancel{x + 2}}{(x - 7)\cancel{(x + 2)}}$$

$$= \dfrac{1}{x - 7}, \quad x \neq -2$$

**15.** $\dfrac{x}{x^2 - 36} + \dfrac{4}{x - 6} = \dfrac{x}{(x + 6)(x - 6)} + \dfrac{4(x + 6)}{(x - 6)(x + 6)}$

$$= \dfrac{x + 4(x + 6)}{(x + 6)(x - 6)}$$

$$= \dfrac{x + 4x + 24}{(x + 6)(x - 6)}$$

$$= \dfrac{5x + 24}{(x + 6)(x - 6)}$$

**16.** $\dfrac{x + 5}{x - 5} - \dfrac{x - 5}{x + 5} = \dfrac{(x + 5)(x + 5)}{(x - 5)(x + 5)} - \dfrac{(x - 5)(x - 5)}{(x + 5)(x - 5)}$

$$= \dfrac{(x + 5)^2 - (x - 5)^2}{(x - 5)(x + 5)}$$

$$= \dfrac{x^2 + 10x + 25 - \left( x^2 - 10x + 25 \right)}{(x - 5)(x + 5)}$$

$$= \dfrac{x^2 + 10x + 25 - x^2 + 10x - 25}{(x - 5)(x + 5)}$$

$$= \dfrac{20x}{(x - 5)(x + 5)}$$

**19.** (a)  Cost:  $C = 10,000 + 225x$

Average cost per unit:  $\dfrac{C}{x} = \dfrac{10,000 + 225x}{x}$

(b)

| $x$ | 2000 | 3000 | 4000 | 5000 |
|---|---|---|---|---|
| Average cost | $230.00 | $228.33 | $227.50 | $227.00 |

For $x = 2000$:  $\dfrac{C}{x} = \dfrac{10,000 + 225(2000)}{2000} \doteq \$230.00$

For $x = 3000$:  $\dfrac{C}{x} = \dfrac{10,000 + 225(3000)}{3000} \approx \$228.33$

For $x = 4000$:  $\dfrac{C}{x} = \dfrac{10,000 + 225(4000)}{4000} \doteq \$227.50$

For $x = 5000$:  $\dfrac{C}{x} = \dfrac{10,000 + 225(5000)}{5000} = \$227.00$

Trend: As the number of units increases, the cost per unit decreases.

## Section 7.4   Complex Fractions

**1.**  $\dfrac{\left(\dfrac{3}{2}\right)}{\left(\dfrac{9}{10}\right)} = \dfrac{3}{2} \cdot \dfrac{10}{9} = \dfrac{3 \cdot 5 \cdot 2}{2 \cdot 3 \cdot 3} = \dfrac{\cancel{(3)}(5)\cancel{(2)}}{\cancel{(2)}\cancel{(3)}(3)} = \dfrac{5}{3}$

**3.**  $\dfrac{\left(-\dfrac{5}{8}\right)}{\left(\dfrac{4}{15}\right)} = -\dfrac{5}{8} \cdot \dfrac{15}{4} = -\dfrac{75}{32}$

**5.**  $\dfrac{\left(\dfrac{3}{x}\right)}{\left(\dfrac{6}{x^2}\right)} = \dfrac{3}{x} \div \dfrac{6}{x^2}$

$= \dfrac{3}{x} \cdot \dfrac{x^2}{6}$

$= \dfrac{3x^2}{6x}$

$= \dfrac{\cancel{3}(x)\cancel{(x)}}{\cancel{3}(2)\cancel{(x)}} = \dfrac{x}{2}, \quad x \neq 0$

**7.**  $\dfrac{\left(\dfrac{x^3}{4}\right)}{\left(\dfrac{x}{8}\right)} = \dfrac{x^3}{4} \div \dfrac{x}{8}$

$= \dfrac{x^3}{4} \cdot \dfrac{8}{x}$

$= \dfrac{8x^3}{4x}$

$= \dfrac{\cancel{4}(2)\cancel{(x)}(x^2)}{\cancel{4}\cancel{(x)}}$

$= 2x^2, \quad x \neq 0$

**9.**  $\dfrac{\left(\dfrac{8x^2y}{3z^2}\right)}{\left(\dfrac{4xy}{9z^5}\right)} = \dfrac{8x^2y}{3z^2} \cdot \dfrac{9z^5}{4xy}$

$= \dfrac{\cancel{4}(2)\cancel{(x)}(x)\cancel{(y)}\cancel{(3)}(3)\cancel{(z^2)}(z^3)}{\cancel{3}\cancel{(z^2)}\cancel{(4)}\cancel{(x)}\cancel{(y)}}$

$= 6xz^3, \quad x \neq 0, y \neq 0, z \neq 0$

**11.**
$$\dfrac{\left[\dfrac{6x^3}{(5y)^2}\right]}{\left[\dfrac{(3x)^2}{15y^4}\right]} = \dfrac{6x^3}{25y^2}\cdot\dfrac{15y^4}{9x^2}$$

$$= \dfrac{6(15)x^3y^4}{25(9)x^2y^2}$$

$$= \dfrac{(2)\cancel{(3)}\cancel{(3)}\cancel{(5)}\left(\cancel{x^2}\right)(x)\left(\cancel{y^2}\right)(y^2)}{5\cancel{(5)}\cancel{(3)}\cancel{(3)}\left(\cancel{x^2}\right)\left(\cancel{y^2}\right)}$$

$$= \dfrac{2xy^2}{5},\quad x\neq 0,\ y\neq 0$$

**13.**
$$\dfrac{\left(\dfrac{y}{3-y}\right)}{\left(\dfrac{y^2}{y-3}\right)} = \dfrac{y}{3-y}\cdot\dfrac{y-3}{y^2}$$

$$= \dfrac{y(y-3)}{(3-y)y^2}$$

$$= \dfrac{\cancel{y}\,\cancel{(y-3)}}{-1\cancel{(y-3)}\left(\cancel{y}\right)(y)}$$

$$= -\dfrac{1}{y},\quad y\neq 3$$

**15.**
$$\dfrac{\left(\dfrac{x+4}{x+7}\right)}{\left(\dfrac{x+4}{x+14}\right)} = \dfrac{x+4}{x+7}\cdot\dfrac{x+14}{x+4}$$

$$= \dfrac{(x+4)(x+14)}{(x+7)(x+4)}$$

$$= \dfrac{\cancel{(x+4)}(x+14)}{(x+7)\cancel{(x+4)}}$$

$$= \dfrac{x+14}{x+7},\quad x\neq -4,\ x\neq -14$$

**23.**
$$\dfrac{\left(5+\dfrac{3}{4}\right)}{\left(1+\dfrac{1}{4}\right)} = \dfrac{\left(\dfrac{20}{4}+\dfrac{3}{4}\right)}{\left(\dfrac{4}{4}+\dfrac{1}{4}\right)} = \dfrac{\left(\dfrac{23}{4}\right)}{\left(\dfrac{5}{4}\right)} = \dfrac{23}{4}\cdot\dfrac{4}{5} = \dfrac{(23)\cancel{(4)}}{\cancel{(4)}(5)} = \dfrac{23}{5}$$

**Alternate method:**
$$\dfrac{\left(5+\dfrac{3}{4}\right)}{\left(1+\dfrac{1}{4}\right)} = \dfrac{\left(5+\dfrac{3}{4}\right)}{\left(1+\dfrac{1}{4}\right)}\cdot\dfrac{4}{4} = \dfrac{5(4)+\dfrac{3}{4}(4)}{1(4)+\dfrac{1}{4}(4)} = \dfrac{20+3}{4+1} = \dfrac{23}{5}$$

**17.**
$$\dfrac{\left(\dfrac{2x-10}{x+1}\right)}{\left(\dfrac{x-5}{x+1}\right)} = \dfrac{2x-10}{x+1}\div\dfrac{x-5}{x+1}$$

$$= \dfrac{2x-10}{x+1}\cdot\dfrac{x+1}{x-5}$$

$$= \dfrac{2\cancel{(x-5)}\cancel{(x+1)}}{\cancel{(x+1)}\cancel{(x-5)}}$$

$$= 2,\quad x\neq -1,\ x\neq 5$$

**19.**
$$\dfrac{\left(\dfrac{x^2+3x-10}{x+4}\right)}{3x+15} = \dfrac{\left(\dfrac{x^2+3x-10}{x+4}\right)}{\left(\dfrac{3x+15}{1}\right)}$$

$$= \dfrac{x^2+3x-10}{x+4}\cdot\dfrac{1}{3x+15}$$

$$= \dfrac{(x+5)(x-2)(1)}{(x+4)(3)(x+5)}$$

$$= \dfrac{\cancel{(x+5)}(x-2)}{(x+4)(3)\cancel{(x+5)}}$$

$$= \dfrac{x-2}{3(x+4)},\quad x\neq -5$$

**21.**
$$\dfrac{\left(\dfrac{6x^2-17x+5}{3x^2+3x}\right)}{\left(\dfrac{3x-1}{3x+1}\right)} = \dfrac{6x^2-17x+5}{3x^2+3x}\cdot\dfrac{3x+1}{3x-1}$$

$$= \dfrac{\cancel{(3x-1)}(2x-5)(3x+1)}{3x(x+1)\cancel{(3x-1)}}$$

$$= \dfrac{(2x-5)(3x+1)}{3x(x+1)},\quad x\neq\dfrac{1}{3},\ x\neq-\dfrac{1}{3}$$

**25.** $\dfrac{\left(1+\dfrac{3}{y}\right)}{y} = \dfrac{\left(1+\dfrac{3}{y}\right)\cdot y}{(y)\cdot y} = \dfrac{1(y)+\dfrac{3}{y}(y)}{y\cdot y} = \dfrac{y+3}{y^2}$

**Alternate method:**

$\dfrac{\left(1+\dfrac{3}{y}\right)}{y} = \dfrac{\left(\dfrac{y}{y}+\dfrac{3}{y}\right)}{y} = \dfrac{\left(\dfrac{y+3}{y}\right)}{\left(\dfrac{y}{1}\right)} = \dfrac{y+3}{y}\div\dfrac{y}{1} = \dfrac{y+3}{y}\cdot\dfrac{1}{y} = \dfrac{(y+3)(1)}{(y)(y)} = \dfrac{y+3}{y^2}$

**27.** $\dfrac{\left(\dfrac{x}{2}\right)}{\left(2+\dfrac{3}{x}\right)} = \dfrac{\left(\dfrac{x}{2}\right)\cdot 2x}{\left(2+\dfrac{3}{x}\right)\cdot 2x} = \dfrac{\left(\dfrac{x}{2}\right)\cdot 2x}{\left(2\cdot 2x+\dfrac{3}{x}\cdot 2x\right)} = \dfrac{x\cdot x}{4x+3(2)} = \dfrac{x^2}{4x+6},\quad x\neq 0$

**Alternate method:**

$\dfrac{\left(\dfrac{x}{2}\right)}{\left(2+\dfrac{3}{x}\right)} = \dfrac{\left(\dfrac{x}{2}\right)}{\left(\dfrac{2x}{x}+\dfrac{3}{x}\right)} = \dfrac{\left(\dfrac{x}{2}\right)}{\left(\dfrac{2x+3}{x}\right)} = \dfrac{x}{2}\div\dfrac{2x+3}{x} = \dfrac{x}{2}\cdot\dfrac{x}{2x+3} = \dfrac{x\cdot x}{2(2x+3)} = \dfrac{x^2}{4x+6},\quad x\neq 0$

**29.** $\dfrac{\left(\dfrac{x}{3}-\dfrac{4}{3}\right)}{\left(5+\dfrac{1}{x}\right)} = \dfrac{\left(\dfrac{x}{3}-\dfrac{4}{3}\right)}{\left(\dfrac{5x}{x}+\dfrac{1}{x}\right)} = \dfrac{\left(\dfrac{x-4}{3}\right)}{\left(\dfrac{5x+1}{x}\right)} = \dfrac{x-4}{3}\cdot\dfrac{x}{5x+1} = \dfrac{x(x-4)}{3(5x+1)},\quad x\neq 0$

**Alternate method:**

$\dfrac{\left(\dfrac{x}{3}-\dfrac{4}{3}\right)}{\left(5+\dfrac{1}{x}\right)} = \dfrac{\left(\dfrac{x}{3}-\dfrac{4}{3}\right)}{\left(5+\dfrac{1}{x}\right)}\cdot\dfrac{3x}{3x} = \dfrac{\dfrac{x}{3}(3x)-\dfrac{4}{3}(3x)}{5(3x)+\dfrac{1}{x}(3x)} = \dfrac{x^2-4x}{15x+3} = \dfrac{x(x-4)}{3(5x+1)},\quad x\neq 0$

**31.** $\dfrac{\left(\dfrac{x}{4}-\dfrac{4}{x}\right)}{(x-4)} = \dfrac{\left(\dfrac{x}{4}-\dfrac{4}{x}\right)\cdot 4x}{(x-4)\cdot 4x} = \dfrac{\left(\dfrac{x}{4}\cdot 4x-\dfrac{4}{x}\cdot 4x\right)}{(x-4)\cdot 4x} = \dfrac{x\cdot x-4\cdot 4}{4x(x-4)} = \dfrac{x^2-16}{4x(x-4)} = \dfrac{(x-4)(x+4)}{4x(x-4)} = \dfrac{x+4}{4x},\quad x\neq 4$

**Alternate method:**

$\dfrac{\left(\dfrac{x}{4}-\dfrac{4}{x}\right)}{(x-4)} = \dfrac{\left(\dfrac{x}{4}-\dfrac{4}{x}\right)}{(x-4)} = \dfrac{\left(\dfrac{x^2}{4x}-\dfrac{16}{4x}\right)}{(x-4)} = \dfrac{x^2-16}{4x}\div(x-4) = \dfrac{x^2-16}{4x}\cdot\dfrac{1}{x-4} = \dfrac{(x+4)(x-4)(1)}{4x(x-4)} = \dfrac{x+4}{4x},\quad x\neq 4$

**33.** $\dfrac{\left(\dfrac{10}{x+1}\right)}{\left(\dfrac{1}{2x+2}+\dfrac{3}{x+1}\right)} = \dfrac{\left(\dfrac{10}{x+1}\right)}{\left(\dfrac{1}{2(x+1)}+\dfrac{2(3)}{2(x+1)}\right)}$

$$= \dfrac{\left(\dfrac{10}{x+1}\right)}{\left(\dfrac{7}{2(x+1)}\right)} = \dfrac{10}{x+1} \div \dfrac{7}{2(x+1)} = \dfrac{10}{x+1} \cdot \dfrac{2(x+1)}{7} = \dfrac{10(2)\cancel{(x+1)}}{\cancel{(x+1)}(7)} = \dfrac{20}{7}, \quad x \neq -1$$

**Alternate method:**

$$\dfrac{\left(\dfrac{10}{x+1}\right)}{\left(\dfrac{1}{2x+2}+\dfrac{3}{x+1}\right)} = \dfrac{\left(\dfrac{10}{x+1}\right)}{\left(\dfrac{1}{2(x+1)}+\dfrac{3}{x+1}\right)}$$

$$= \dfrac{\left(\dfrac{10}{x+1}\right)\cdot 2(x+1)}{\left(\dfrac{1}{2(x+1)}+\dfrac{3}{x+1}\right)\cdot 2(x+1)} = \dfrac{\left(\dfrac{10}{x+1}\right)\cdot 2(x+1)}{\dfrac{1}{2(x+1)}\cdot 2(x+1)+\dfrac{3}{x+1}\cdot 2(x+1)} = \dfrac{10\cdot 2}{1+3\cdot 2} = \dfrac{20}{7}, \quad x \neq -1$$

**35.** $\dfrac{\left(\dfrac{1}{x}-\dfrac{1}{x+1}\right)}{\left(\dfrac{1}{x+1}\right)} = \dfrac{\left(\dfrac{1}{x}-\dfrac{1}{x+1}\right)\cdot(x)(x+1)}{\left(\dfrac{1}{x+1}\right)\cdot(x)(x+1)}$

$$= \dfrac{\left(\dfrac{1}{x}\right)(x)(x+1)-\left(\dfrac{1}{x+1}\right)(x)(x+1)}{\left(\dfrac{1}{x+1}\right)(x)(x+1)} = \dfrac{(x+1)-x}{x} = \dfrac{x+1-x}{x} = \dfrac{1}{x}, \quad x \neq -1$$

**Alternate method:**

$$\dfrac{\left(\dfrac{1}{x}-\dfrac{1}{x+1}\right)}{\left(\dfrac{1}{x+1}\right)} = \dfrac{\left[\dfrac{1(x+1)}{x(x+1)}-\dfrac{1(x)}{(x+1)(x)}\right]}{\left(\dfrac{1}{x+1}\right)}$$

$$= \dfrac{\left[\dfrac{(x+1)-x}{x(x+1)}\right]}{\left(\dfrac{1}{x+1}\right)} = \dfrac{1}{x(x+1)} \div \dfrac{1}{x+1} = \dfrac{1}{x(x+1)} \cdot \dfrac{x+1}{1} = \dfrac{1\cancel{(x+1)}}{x\cancel{(x+1)}} = \dfrac{1}{x}, \quad x \neq -1$$

**37.** $\dfrac{\left(\dfrac{x}{5}+\dfrac{x}{6}\right)}{2} = \dfrac{\left(\dfrac{6x}{30}+\dfrac{5x}{30}\right)}{2} = \dfrac{\left(\dfrac{6x+5x}{30}\right)}{2} = \dfrac{\left(\dfrac{11x}{30}\right)}{\left(\dfrac{2}{1}\right)} = \dfrac{11x}{30} \div \dfrac{2}{1} = \dfrac{11x}{30} \cdot \dfrac{1}{2} = \dfrac{11x}{60}$

**Alternate method:**

$$\dfrac{\left(\dfrac{x}{5}+\dfrac{x}{6}\right)}{2} = \dfrac{\left(\dfrac{x}{5}+\dfrac{x}{6}\right)\cdot 30}{2\cdot 30} = \dfrac{\left(\dfrac{x}{5}\right)(30)+\left(\dfrac{x}{6}\right)(30)}{60} = \dfrac{6x+5x}{60} = \dfrac{11x}{60}$$

**39.** We begin by finding the difference between $x/6$ and $x/9$, and we divide this difference by 4.

$$\frac{\dfrac{x}{6} - \dfrac{x}{9}}{4} = \frac{\dfrac{3x}{6(3)} - \dfrac{2x}{9(2)}}{4} = \frac{\dfrac{3x}{18} - \dfrac{2x}{18}}{4} = \frac{\dfrac{x}{18}}{4} = \frac{x}{18} \div 4 = \frac{x}{18} \cdot \frac{1}{4} = \frac{x}{72}$$

To find $x_1$:  $\dfrac{x}{9} + \dfrac{x}{72} = \dfrac{8x}{9(8)} + \dfrac{x}{72}$

$$= \frac{8x + x}{72}$$

$$= \frac{9x}{72}$$

$$= \frac{\cancel{9}x}{\cancel{9}(8)}$$

$$= \frac{x}{8}$$

To find $x_2$:  $\dfrac{x}{8} + \dfrac{x}{72} = \dfrac{9x}{9(8)} + \dfrac{x}{72}$

$$= \frac{9x + x}{72}$$

$$= \frac{10x}{72}$$

$$= \frac{\cancel{2}(5x)}{\cancel{2}(36)}$$

$$= \frac{5x}{36}$$

To find $x_3$:  $\dfrac{5x}{36} + \dfrac{x}{72} = \dfrac{(5x)(2)}{36(2)} + \dfrac{x}{72}$

$$= \frac{10x + x}{72}$$

$$= \frac{11x}{72}$$

The three numbers are $\dfrac{x}{8}$, $\dfrac{5x}{36}$, and $\dfrac{11x}{72}$.

**41.**

$$\frac{1}{\dfrac{1}{R_1} + \dfrac{1}{R_2}} = \frac{1 \cdot R_1 R_2}{\left(\dfrac{1}{R_1} + \dfrac{1}{R_2}\right) \cdot R_1 R_2} = \frac{R_1 R_2}{\dfrac{1}{R_1} \cdot R_1 R_2 + \dfrac{1}{R_2} \cdot R_1 R_2} = \frac{R_1 R_2}{R_2 + R_1}$$

**Alternate method:**

$$\frac{1}{\dfrac{1}{R_1} + \dfrac{1}{R_2}} = \frac{1}{\dfrac{1 R_2}{R_1 R_2} + \dfrac{1 R_1}{R_2 R_1}} = \frac{1}{\dfrac{R_2 + R_1}{R_1 R_2}} = 1 \div \frac{R_2 + R_1}{R_1 R_2} = 1 \cdot \frac{R_1 R_2}{R_2 + R_1} = \frac{R_1 R_2}{R_2 + R_1}$$

**43.** Method 1: Combine the numerator into a single fraction; combine the denominator into a single fraction; divide by inverting the denominator and multiplying.

Method 2: Multiply the numerator and denominator of the complex fraction by the least common denominator for all fractions in the numerator and denominator of the original complex fraction; then simplify the results.

Answers about preferences will vary.

**45.** $\left(12x^2 + 4x - 8\right) + \left(-x^2 - 6x + 5\right) = 12x^2 + 4x - 8 - x^2 - 6x + 5 = 11x^2 - 2x - 3$

**47.** $-4x\left(-6x^2 + 4x - 2\right) = 24x^3 - 16x^2 + 8x$

**49.** $\dfrac{8x^2 + 4x}{x} = \dfrac{8x^2}{x} + \dfrac{4x}{x} = 8x + 4, \quad x \neq 0$

**51.** $\dfrac{x}{5} + \dfrac{3x}{5} = \dfrac{x + 3x}{5} = \dfrac{4x}{5}$

**53.** $\dfrac{6}{5x} - \dfrac{2}{3x} = \dfrac{6(3)}{5x(3)} - \dfrac{2(5)}{3x(5)}$

$$= \frac{18}{15x} - \frac{10}{15x} = \frac{18 - 10}{15x} = \frac{8}{15x}$$

**55.** $\dfrac{8 - 3x}{x + 2} + \dfrac{4 + 2x}{x + 2} = \dfrac{8 - 3x + 4 + 2x}{x + 2}$

$$= \frac{12 - x}{x + 2} \text{ or } \frac{-x + 12}{x + 2} \text{ or } -\frac{x - 12}{x + 2}$$

## Section 7.5   Rational Equations and Applications

**1.** (a) $\dfrac{0}{5} - \dfrac{3}{0} \overset{?}{=} \dfrac{1}{10}$

$0 - \dfrac{3}{0} \neq \dfrac{1}{10}$

$\dfrac{3}{0}$ is undefined.

0 *is not* a solution.

(b) $\dfrac{-1}{5} - \dfrac{3}{-1} \overset{?}{=} \dfrac{1}{10}$

$-\dfrac{1}{5} + 3 \overset{?}{=} \dfrac{1}{10}$

$-\dfrac{1}{5} + \dfrac{15}{5} \overset{?}{=} \dfrac{1}{10}$

$\dfrac{14}{5} \neq \dfrac{1}{10}$

−1 *is not* a solution.

(c) $\dfrac{(1/6)}{5} - \dfrac{3}{(1/6)} \overset{?}{=} \dfrac{1}{10}$

$\dfrac{1}{6}\left(\dfrac{1}{5}\right) - 3\left(\dfrac{6}{1}\right) \overset{?}{=} \dfrac{1}{10}$

$\dfrac{1}{30} - 18 \neq \dfrac{1}{10}$

$\dfrac{1}{6}$ *is not* a solution.

(d) $\dfrac{6}{5} - \dfrac{3}{6} \overset{?}{=} \dfrac{1}{10}$

$\dfrac{6}{5} - \dfrac{1}{2} \overset{?}{=} \dfrac{1}{10}$

$\dfrac{12}{10} - \dfrac{5}{10} \overset{?}{=} \dfrac{1}{10}$

$\dfrac{7}{10} \neq \dfrac{1}{10}$

6 *is not* a solution.

**3.** (a) $\dfrac{5}{2(-1/2)} - \dfrac{4}{(-1/2)} \overset{?}{=} 3$

$\dfrac{5}{-1} - \dfrac{4}{1} \cdot \dfrac{-2}{1} \overset{?}{=} 3$

$-5 + 8 \overset{?}{=} 3$

$3 = 3$

$-\dfrac{1}{2}$ *is a* solution.

(b) $\dfrac{5}{2(4)} - \dfrac{4}{4} \overset{?}{=} 3$

$\dfrac{5}{8} - 1 \overset{?}{=} 3$

$\dfrac{5}{8} - \dfrac{8}{8} \overset{?}{=} 3$

$-\dfrac{3}{8} \neq 3$

4 *is not* a solution.

(c) $\dfrac{5}{2(0)} - \dfrac{4}{0} \overset{?}{=} 3$

$\dfrac{5}{0} - \dfrac{4}{0} \neq 3$

$\dfrac{5}{0}$ and $\dfrac{4}{0}$ are undefined.

0 *is not* a solution.

(d) $\dfrac{5}{2(1/4)} - \dfrac{4}{(1/4)} \overset{?}{=} 3$

$\dfrac{5}{(1/2)} - \dfrac{4}{(1/4)} \overset{?}{=} 3$

$\dfrac{5}{1}\left(\dfrac{2}{1}\right) - \dfrac{4}{1}\left(\dfrac{4}{1}\right) \overset{?}{=} 3$

$10 - 16 \overset{?}{=} 3$

$-6 \neq 3$

$\dfrac{1}{4}$ *is not* a solution.

**5.** $x = 4$

**7.** All real numbers

**9.** $x = -1$

**11.** $x = 7$

**13.** $\dfrac{z}{3} - \dfrac{2z}{8} = 1$   The least common denominator is 24.

$24\left(\dfrac{z}{3} - \dfrac{2z}{8}\right) = 24(1)$   Multiply both sides by 24.

$8z - 3(2z) = 24$

$8z - 6z = 24$

$2z = 24$

$z = 12$

**15.** $\dfrac{t}{3} = 25 - \dfrac{t}{6}$   The least common denominator is 6.

$6\left(\dfrac{t}{3}\right) = 6\left(25 - \dfrac{t}{6}\right)$   Multiply both sides by 6.

$2t = 6(25) - 6\left(\dfrac{t}{6}\right)$

$2t = 150 - t$

$3t = 150$

$t = 50$

**17.** $\dfrac{5x}{7} - \dfrac{2x}{3} = \dfrac{1}{2}$     The least common denominator is 42.

$42\left(\dfrac{5x}{7} - \dfrac{2x}{3}\right) = 42\left(\dfrac{1}{2}\right)$     Multiply both sides by 42.

$6(5x) - 14(2x) = 21$

$30x - 28x = 21$

$2x = 21$

$x = \dfrac{21}{2}$

**19.** $\dfrac{a+3}{4} - \dfrac{a-1}{6} = \dfrac{4}{3}$     The least common denominator is 12.

$12\left(\dfrac{a+3}{4} - \dfrac{a-1}{6}\right) = 12\left(\dfrac{4}{3}\right)$     Multiply both sides by 12.

$3(a+3) - 2(a-1) = 4(4)$

$3a + 9 - 2a + 2 = 16$

$a + 11 = 16$

$a = 5$

**21.** $\dfrac{x-4}{3} + \dfrac{2x+1}{4} = \dfrac{5}{6}$     The least common denominator is 12.

$12\left(\dfrac{x-4}{3} + \dfrac{2x+1}{4}\right) = 12\left(\dfrac{5}{6}\right)$     Multiply both sides by 12.

$4(x-4) + 3(2x+1) = 10$

$4x - 16 + 6x + 3 = 10$

$10x - 13 = 10$

$10x = 23$

$x = \dfrac{23}{10}$

**23.** $2 - \dfrac{4}{x} = 1$     The least common denominator is $x$.

$x\left(2 - \dfrac{4}{x}\right) = x(1)$     Multiply both sides by $x$.

$2x - 4 = x, \ \ x \neq 0$

$x - 4 = 0$

$x = 4$

**25.** $3 - \dfrac{16}{a} = \dfrac{5}{3}$     The least common denominator is $3a$.

$3a\left(3 - \dfrac{16}{a}\right) = 3a\left(\dfrac{5}{3}\right)$     Multiply both sides by $3a$.

$3a(3) - 3a\left(\dfrac{16}{a}\right) = 3a\left(\dfrac{5}{3}\right)$

$9a - 48 = 5a, \ \ a \neq 0$

$-48 = -4a$

$12 = a$

**27.**  $\dfrac{3}{x} + \dfrac{1}{4} = \dfrac{2}{x}$          The least common denominator is $4x$.

$4x\left(\dfrac{3}{x} + \dfrac{1}{4}\right) = 4x\left(\dfrac{2}{x}\right)$          Multiply both sides by $4x$.

$\quad 4(3) + x(1) = 4(2), \ x \neq 0$

$\qquad\quad 12 + x = 8$

$\qquad\qquad\quad x = -4$

**29.**  $\dfrac{6}{12x} + \dfrac{3}{4} = \dfrac{2}{3x}$          The least common denominator is $12x$.

$12x\left(\dfrac{6}{12x} + \dfrac{3}{4}\right) = 12x\left(\dfrac{2}{3x}\right)$          Multiply both sides by $12x$.

$\qquad 6 + 3x(3) = 4(2), \ x \neq 0$

$\qquad\quad 6 + 9x = 8$

$\qquad\qquad 9x = 2$

$\qquad\qquad\ \ x = \dfrac{2}{9}$

**31.**  $\dfrac{10}{y + 3} + \dfrac{10}{3} = 6$          The least common denominator is $3(y + 3)$.

$3(y + 3)\left(\dfrac{10}{y + 3} + \dfrac{10}{3}\right) = 3(y + 3)(6)$          Multiply both sides by $3(y + 3)$.

$3(y + 3)\left(\dfrac{10}{y + 3}\right) + 3(y + 3)\left(\dfrac{10}{3}\right) = 18(y + 3)$

$\qquad\qquad 30 + 10(y + 3) = 18(y + 3), \quad y \neq -3$

$\qquad\qquad 30 + 10y + 30 = 18y + 54$

$\qquad\qquad\quad 10y + 60 = 18y + 54$

$\qquad\qquad\quad -8y + 60 = 54$

$\qquad\qquad\qquad\quad -8y = -6$

$\qquad\qquad\qquad\qquad y = \dfrac{-6}{-8}$

$\qquad\qquad\qquad\qquad y = \dfrac{3}{4}$

**33.**  $\dfrac{1}{x - 4} + 2 = \dfrac{2x}{x - 4}$          The least common denominator is $x - 4$.

$(x - 4)\left(\dfrac{1}{x - 4} + 2\right) = \left(\dfrac{2x}{x - 4}\right)(x - 4)$          Multiply both sides by $x - 4$.

$\qquad 1 + 2(x - 4) = 2x, \quad x \neq 4$

$\qquad\quad 1 + 2x - 8 = 2x$

$\qquad\qquad 2x - 7 = 2x$

$\qquad 2x - 7 - 2x = 2x - 2x$

$\qquad\qquad\quad -7 \neq 0$

The original equation has no solution.

**35.**          $\dfrac{6x}{x-11} + 1 = \dfrac{3}{x-11}$          The least common denominator is $x - 11$.

$$(x-11)\left(\dfrac{6x}{x-11} + 1\right) = \left(\dfrac{3}{x-11}\right)(x-11)$$   Multiply both sides by $x - 11$.

$$6x + 1(x-11) = 3, \quad x \neq 11$$

$$6x + x - 11 = 3$$

$$7x - 11 = 3$$

$$7x = 14$$

$$x = 2$$

**37.**          $\dfrac{7x}{x+1} = \dfrac{5}{x-3} + 7$          The least common denominator is $(x+1)(x-3)$.

$$(x+1)(x-3)\left(\dfrac{7x}{x+1}\right) = (x+1)(x-3)\left(\dfrac{5}{x-3} + 7\right)$$   Multiply both sides by $(x+1)(x-3)$.

$$(x-3)(7x) = (x+1)(5) + 7(x+1)(x-3), \quad x \neq -1, x \neq 3$$

$$7x^2 - 21x = 5x + 5 + 7\left(x^2 - 2x - 3\right)$$

$$7x^2 - 21x = 5x + 5 + 7x^2 - 14x - 21$$

$$7x^2 - 21x = 7x^2 - 9x - 16$$

$$-21x = -9x - 16$$

$$-12x = -16$$

$$x = \dfrac{-16}{-12}$$

$$x = \dfrac{4}{3}$$

**39.**          $\dfrac{4}{x+2} - \dfrac{1}{x} = \dfrac{1}{x}$          The least common denominator is $x(x+2)$.

$$x(x+2)\left(\dfrac{4}{x+2} - \dfrac{1}{x}\right) = x(x+2)\left(\dfrac{1}{x}\right)$$   Multiply both sides by $x(x+2)$.

$$x(x+2)\left(\dfrac{4}{x+2}\right) - x(x+2)\left(\dfrac{1}{x}\right) = x(x+2)\left(\dfrac{1}{x}\right)$$

$$4x - (x+2) = x + 2, \quad x \neq 0, x \neq -2$$

$$4x - x - 2 = x + 2$$

$$3x - 2 = x + 2$$

$$2x - 2 = 2$$

$$2x = 4$$

$$x = 2$$

**41.**     $10 - \dfrac{13}{x} = 4 + \dfrac{5}{x}$          The least common denominator is $x$.

$$x\left(10 - \dfrac{13}{x}\right) = x\left(4 + \dfrac{5}{x}\right)$$   Multiply both sides by $x$.

$$10x - 13 = 4x + 5, \quad x \neq 0$$

$$6x - 13 = 5$$

$$6x = 18$$

$$x = \dfrac{18}{6}$$

$$x = 3$$

**43.**

$$\frac{2}{x-3} + \frac{1}{x} = \frac{x-1}{x-3}$$ The least common denominator is $x(x-3)$.

$$x(x-3)\left(\frac{2}{x-3} + \frac{1}{x}\right) = \left(\frac{x-1}{x-3}\right)x(x-3)$$ Multiply both sides by $x(x-3)$.

$$2x + (x-3) = (x-1)x, \quad x \neq 0, x \neq 3$$

$$2x + x - 3 = x^2 - x$$

$$3x - 3 = x^2 - x$$

$$0 = x^2 - 4x + 3$$

$$0 = (x-3)(x-1)$$

$$x - 3 = 0 \Rightarrow x = 3 \quad \text{(Extraneous)}$$

$$x - 1 = 0 \Rightarrow x = 1$$

**45.**

$$\frac{3}{x(x-3)} + \frac{4}{x} = \frac{1}{x-3}$$ The least common denominator is $x(x-3)$.

$$x(x-3)\left[\frac{3}{x(x-3)} + \frac{4}{x}\right] = x(x-3)\left(\frac{1}{x-3}\right)$$ Multiply both sides by $x(x-3)$.

$$x(x-3)\left[\frac{3}{x(x-3)}\right] + x(x-3)\left(\frac{4}{x}\right) = x(x-3)\left(\frac{1}{x-3}\right)$$

$$3 + (x-3)4 = x, \quad x \neq 0, x \neq 3$$

$$3 + 4x - 12 = x$$

$$4x - 9 = x$$

$$-9 = -3x$$

$$\frac{-9}{-3} = x$$

$$3 = x \quad \text{(Extraneous)}$$

This solution is extraneous because substituting 3 for $x$ in the original equation results in division by 0.
So, the equation has *no solution*.

**47.**

$$\frac{1}{x+3} + \frac{4}{x+4} = -\frac{x}{(x+3)(x+4)}$$ The least common denominator is $(x+3)(x+4)$.

$$(x+3)(x+4)\left(\frac{1}{x+3} + \frac{4}{x+4}\right) = (x+3)(x+4)\left[-\frac{x}{(x+3)(x+4)}\right]$$ Multiply both sides by $(x+3)(x+4)$.

$$(x+4) + 4(x+3) = -x, \quad x \neq -3, x \neq -4$$

$$x + 4 + 4x + 12 = -x$$

$$5x + 16 = -x$$

$$16 = -6x$$

$$\frac{16}{-6} = x$$

$$-\frac{8}{3} = x$$

**49.**
$$\frac{1}{x-3} + \frac{1}{x+3} = \frac{10}{x^2-9}$$ The common denominator is $(x-3)(x+3)$.

$$(x-3)(x+3)\left(\frac{1}{x-3} + \frac{1}{x+3}\right) = (x-3)(x+3)\left(\frac{10}{x^2-9}\right)$$ Multiply both sides by $(x-3)(x+3)$.

$$(x-3)(x+3)\left(\frac{1}{x-3}\right) + (x-3)(x+3)\left(\frac{1}{x+3}\right) = (x-3)(x+3)\left[\frac{10}{(x-3)(x+3)}\right]$$

$$x+3+x-3 = 10, \quad x \neq 3, x \neq -3$$

$$2x = 10$$

$$x = \frac{10}{2}$$

$$x = 5$$

**51.**
$$\frac{3}{x-3} + \frac{2}{x-5} = \frac{4}{x^2-8x+15}$$ The least common denominator is $(x-3)(x-5)$.

$$(x-3)(x-5)\left(\frac{3}{x-3} + \frac{2}{x-5}\right) = \left[\frac{4}{(x-3)(x-5)}\right](x-3)(x-5)$$ Multiply both sides by $(x-3)(x-5)$.

$$3(x-5) + 2(x-3) = 4, \quad x \neq 3, x \neq 5$$

$$3x - 15 + 2x - 6 = 4$$

$$5x - 21 = 4$$

$$5x = 25$$

$$x = 5 \quad \text{(Extraneous)}$$

The original equation has no solution.

**53.**
$$x + 4 = \frac{-4}{x}$$

$$x(x+4) = x\left(\frac{-4}{x}\right)$$ Multiply both sides by $x$.

$$x^2 + 4x = -4, \quad x \neq 0$$

$$x^2 + 4x + 4 = 0$$

$$(x+2)^2 = 0$$

$$x + 2 = 0 \Rightarrow x = -2$$

**55.**
$$\frac{20-x}{x} = x$$

$$x\left(\frac{20-x}{x}\right) = x(x)$$ Multiply both sides by $x$.

$$20 - x = x^2, \quad x \neq 0$$

$$0 = x^2 + x - 20$$

$$0 = (x+5)(x-4)$$

$$0 = x + 5 \Rightarrow x = -5$$

$$0 = x - 4 \Rightarrow x = 4$$

**57.**
$$2y = \frac{y+6}{y+1}$$

$$(y+1)2y = (y+1)\left(\frac{y+6}{y+1}\right)$$ Multiply both sides by $(y+1)$.

$$2y^2 + 2y = y + 6, \quad y \neq -1$$

$$2y^2 + y - 6 = 0$$

$$(2y-3)(y+2) = 0$$

$$2y - 3 = 0 \Rightarrow 2y = 3 \Rightarrow y = \frac{3}{2}$$

$$y + 2 = 0 \Rightarrow y = -2$$

**59.** $x + \dfrac{1}{x} = \dfrac{5}{2}$      The least common denominator is $2x$.

$$2x\left(x + \dfrac{1}{x}\right) = 2x\left(\dfrac{5}{x}\right) \quad \text{Multiply both sides by } 2x.$$

$$2x(x) + 2x\left(\dfrac{1}{x}\right) = 2x\left(\dfrac{5}{2}\right)$$

$$2x^2 + 2 = 5x, \ \ x \neq 0$$

$$2x^2 - 5x + 2 = 0$$

$$(2x - 1)(x - 2) = 0$$

$$2x - 1 = 0 \Rightarrow 2x = 1 \Rightarrow x = \dfrac{1}{2}$$

$$x - 2 = 0 \Rightarrow x = 2$$

**61.** $\dfrac{x + 3}{x^2 - 9} + \dfrac{4}{x - 3} = -2$      The least common denominator is $(x + 3)(x - 3)$.

$$(x + 3)(x - 3)\left(\dfrac{x + 3}{x^2 - 9} + \dfrac{4}{x - 3}\right) = (x + 3)(x - 3)(-2) \quad \text{Multiply both sides by } (x + 3)(x - 3).$$

$$x + 3 + 4(x + 3) = -2(x^2 - 9), \quad x \neq -3, x \neq 3$$

$$x + 3 + 4x + 12 = -2x^2 + 18$$

$$5x + 15 = -2x^2 + 18$$

$$2x^2 + 5x - 3 = 0$$

$$(2x - 1)(x + 3) = 0$$

$$2x - 1 = 0 \Rightarrow 2x = 1 \Rightarrow x = \dfrac{1}{2}$$

$$x + 3 = 0 \Rightarrow x = -3 \quad \text{(Extraneous)}$$

**63.** $\dfrac{x}{2} = \dfrac{1 + \dfrac{3}{x}}{1 + \dfrac{1}{x}}$

$$x\left(1 + \dfrac{1}{x}\right) = 2\left(1 + \dfrac{3}{x}\right), \quad x \neq -1 \quad \text{Cross-multiply.}$$

$$x + 1 = 2 + \dfrac{6}{x}$$

$$x(x + 1) = x\left(2 + \dfrac{6}{x}\right)$$

$$x^2 + x = 2x + 6, \ \ x \neq 0$$

$$x^2 - x - 6 = 0$$

$$(x - 3)(x + 2) = 0$$

$$x - 3 = 0 \Rightarrow x = 3$$

$$x + 2 = 0 \Rightarrow x = -2$$

**65. (a)** The $x$-intercept appears to be $(4, 0)$.

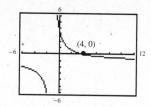

**(b)**
$$0 = \frac{5}{x + 1} - 1$$
$$(x + 1)(0) = (x + 1)\left(\frac{5}{x + 1} - 1\right)$$
$$0 = 5 - 1(x + 1), \quad x \neq -1$$
$$0 = 5 - x - 1$$
$$0 = 4 - x$$
$$x = 4$$

This confirms the $x$-intercept of $(4, 0)$.

**67. (a)** The $x$-intercepts appear to be $(-2, 0)$ and $(2, 0)$.

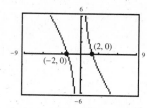

**(b)**
$$0 = \frac{6}{x} - \frac{3x}{2}$$
$$2x(0) = 2x\left(\frac{6}{x} - \frac{3x}{2}\right)$$
$$0 = 2(6) - x(3x), \quad x \neq 0$$
$$0 = 12 - 3x^2$$
$$3x^2 - 12 = 0$$
$$3(x^2 - 4) = 0$$
$$3(x + 2)(x - 2) = 0$$
$$3 \neq 0$$
$$x + 2 = 0 \Rightarrow x = -2$$
$$x - 2 = 0 \Rightarrow x = 2$$

This confirms that the $x$-intercepts are $(-2, 0)$ and $(2, 0)$.

**69. (a)** The $x$-intercept appears to be $(7, 0)$.

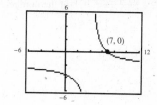

**(b)**
$$0 = 2\left(\frac{3}{x - 4} - 1\right)$$
$$0 = \frac{6}{x - 4} - 2$$
$$(x - 4)(0) = (x - 4)\left(\frac{6}{x - 4} - 2\right)$$
$$0 = 6 - 2(x - 4), \quad x \neq 4$$
$$0 = 6 - 2x + 8$$
$$0 = 14 - 2x$$
$$2x = 14$$
$$x = 7$$

This confirms that the $x$-intercept is $(7, 0)$.

**71. (a)** The $x$-intercepts appear to be $(2, 0)$ and $(-3, 0)$.

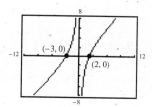

**(b)**
$$0 = x + \frac{x - 6}{x}$$
$$x(0) = x\left(x + \frac{x - 6}{x}\right)$$
$$0 = x^2 + x - 6, \quad x \neq 0$$
$$(x + 3)(x - 2) = 0$$
$$x + 3 = 0 \Rightarrow x = -3$$
$$x - 2 = 0 \Rightarrow x = 2$$

This confirms that the $x$-intercepts are $(2, 0)$ and $(-3, 0)$.

**73.** *Verbal Model:*  $\boxed{\text{A number}}$ + $\boxed{\begin{array}{l}\text{The reciprocal}\\\text{of the number}\end{array}}$ = $\boxed{\dfrac{10}{3}}$

*Labels:*    Number = $x$

   Reciprocal = $\dfrac{1}{x}$

*Equation:*    $x + \dfrac{1}{x} = \dfrac{10}{3}$

$$3x\left(x + \dfrac{1}{x}\right) = 3x\left(\dfrac{10}{x}\right)$$

$$3x(x) + 3x\left(\dfrac{1}{x}\right) = 3x\left(\dfrac{10}{3}\right)$$

$$3x^2 + 3 = 10x, \quad x \neq 0$$

$$3x^2 - 10x + 3 = 0$$

$$(3x - 1)(x - 3) = 0$$

$$3x - 1 = 0 \Rightarrow 3x = 1 \Rightarrow x = \dfrac{1}{3}$$

$$x - 3 = 0 \Rightarrow x = 3$$

The number is $\dfrac{1}{3}$ or 3.

**75.** *Verbal Model:*  $3 \cdot \boxed{\text{A number}}$ + $25 \cdot \boxed{\begin{array}{l}\text{The reciprocal}\\\text{of the number}\end{array}}$ = 20

*Labels:*    The number = $x$

   Reciprocal = $\dfrac{1}{x}$

*Equation:*    $3x + 25\left(\dfrac{1}{x}\right) = 20$

$$3x + \dfrac{25}{x} = 20$$

$$x\left(3x + \dfrac{25}{x}\right) = 20x$$

$$3x^2 + 25 = 20x, \quad x \neq 0$$

$$3x^2 - 20x + 25 = 0$$

$$(3x - 5)(x - 5) = 0$$

$$3x - 5 = 0 \Rightarrow 3x = 5 \Rightarrow x = \dfrac{5}{3}$$

$$x - 5 = 0 \Rightarrow x = 5$$

The number is $\dfrac{5}{3}$ or 5.

**77.** *Verbal Model:*  $\boxed{\begin{array}{c}\text{Time for trip}\\\text{of first car}\end{array}} = \boxed{\begin{array}{c}\text{Time for trip}\\\text{of second car}\end{array}}$

*Labels:*  First car: Distance = 440 miles, Rate = $x$ mph, Time = $\dfrac{440}{x}$ hours

Second car: Distance = 416 miles, Rate = $x - 3$ mph, Time = $\dfrac{416}{x-3}$

*Equation:*
$$\frac{440}{x} = \frac{416}{x-3}$$
$$440(x-3) = 416x, \quad x \neq 0, 3$$
$$440x - 1320 = 416x$$
$$24x = 1320$$
$$x = 55$$
$$x - 3 = 55 - 3 = 52$$

The first car has an average speed of 55 miles per hour and the second car has an average speed of 52 miles per hour.

**79.** *Verbal Model:*  $\boxed{\text{Time for first skater}} = \boxed{\text{Time for second skater}}$

*Labels:*  First skater:

| | |
|---|---|
| Distance = 3192 | (meters) |
| Rate = $x$ | (meters per second) |
| Time = $\dfrac{3192}{x}$ | (seconds) |

Second skater:

| | |
|---|---|
| Distance = 2880 | (meters) |
| Rate = $x - 1.3$ | (meters per second) |
| Time = $\dfrac{2880}{x-1.3}$ | (seconds) |

*Equation:*
$$\frac{3192}{x} = \frac{2880}{x-1.3}$$
$$2880x = 3192(x-1.3), \quad x \neq 0, x \neq 1.3 \quad \text{Cross-multiply.}$$
$$2880x = 3192x - 4149.6$$
$$-312x = -4149.6$$
$$x = \frac{-4149.6}{-312}$$
$$x = 13.3 \text{ and } x - 1.3 = 12$$

The first skater travels at an average speed of 13.3 meters per second, and the second skater travels at an average speed of 12 meters per second.

**81.**

|  | Person #1 | Person #2 | Together |
|---|---|---|---|
| (a) | 4 days | 4 days | 2 days |
| (b) | 4 hours | 6 hours | $2\frac{2}{5}$ hours |
| (c) | 4 hours | $2\frac{1}{2}$ hours | $1\frac{7}{13}$ hours |

*Verbal Model:* $\boxed{\text{Rate for Person \#1}}$ + $\boxed{\text{Rate for Person \#2}}$ = $\boxed{\text{Rate together}}$

(a) *Labels:*  Person #1: Time = 4 days, Rate = $\dfrac{1}{4}$ task per day

Person #2: Time = 4 days, Rate = $\dfrac{1}{4}$ task per day

Together: Time = $x$ days, Rate = $\dfrac{1}{x}$ task per day

*Equation:*
$$\frac{1}{4} + \frac{1}{4} = \frac{1}{x}$$
$$4x\left(\frac{1}{4} + \frac{1}{4}\right) = 4x\left(\frac{1}{x}\right)$$
$$4x\left(\frac{1}{4}\right) + 4x\left(\frac{1}{4}\right) = 4x\left(\frac{1}{x}\right)$$
$$x + x = 4, \quad x \neq 0$$
$$2x = 4$$
$$x = 2$$

So, 2 days are required to complete the task when the two people work together.

(b) *Labels:*  Person #1: Time = 4 hours, Rate = $\dfrac{1}{4}$ task per hour

Person #2: Time = 6 hours, Rate = $\dfrac{1}{6}$ task per hour

Together: Time = $x$ hours, Rate = $\dfrac{1}{x}$ task per hour

*Equation:*
$$\frac{1}{4} + \frac{1}{6} = \frac{1}{x}$$
$$12x\left(\frac{1}{4} + \frac{1}{6}\right) = 12x\left(\frac{1}{x}\right)$$
$$12x\left(\frac{1}{4}\right) + 12x\left(\frac{1}{6}\right) = 12x\left(\frac{1}{x}\right)$$
$$3x + 2x = 12, \quad x \neq 0$$
$$5x = 12$$
$$x = \frac{12}{5}$$

So, $\dfrac{12}{5}$ or $2\dfrac{2}{5}$ hours are required to complete the task when the two people work together.

(c) *Labels:*     Person #1: Time $= 4$ hours, Rate $= \dfrac{1}{4}$ task per hour

Person #2: Time $= 2\dfrac{1}{2}$ hours, Rate $= 1 \Big/ \left(2\dfrac{1}{2}\right)$ task per hour or $\dfrac{2}{5}$ task per hour

Note: $\dfrac{1}{2\frac{1}{2}} = \dfrac{1}{\frac{5}{2}} = 1 \div \dfrac{5}{2} = 1 \cdot \dfrac{2}{5} = \dfrac{2}{5}$

Together: Time $= x$ hours, Rate $= \dfrac{1}{x}$ task per hour

*Equation:*
$$\frac{1}{4} + \frac{2}{5} = \frac{1}{x}$$
$$20x\left(\frac{1}{4} + \frac{2}{5}\right) = 20x\left(\frac{1}{x}\right)$$
$$5x + 8x = 20, \quad x \neq 0$$
$$13x = 20$$
$$x = \frac{20}{13}$$

So, $\dfrac{20}{13}$ or $1\dfrac{7}{13}$ hours are required to complete the task when the two people work together.

**83.** *Verbal Model:*   $\boxed{\text{Rate for first person}} \; + \; \boxed{\text{Rate for second person}} \; = \; \boxed{\text{Rate for first person and second person}}$

*Labels:*     Time for first person $= 4$

Rate for first person $= \dfrac{1}{4}$

Time for second person $= t$

Rate for second person $= \dfrac{1}{t}$

Time for first person and second person $= 1$

Rate for first person and second person $= \dfrac{1}{1} = 1$

*Equation:*
$$\frac{1}{4} + \frac{1}{t} = 1$$
$$4t\left(\frac{1}{4} + \frac{1}{t}\right) = 4t(1)$$
$$t + 4 = 4t, \quad t \neq 0$$
$$4 = 3t$$
$$\frac{4}{3} = t$$

It would take the second person $\dfrac{4}{3}$ hour to accomplish the task working alone.

**85.** *Verbal Model:*   $\boxed{\text{Rate for first person (slower)}} + \boxed{\text{Rate for second person (faster)}} = \boxed{\text{Rate for both persons}}$

*Labels:*   First person:   Time $= 1.5x$   (minutes)

Rate $= \dfrac{1}{1.5x}$   (yard per minute)

Second person:   Time $= x$   (minutes)

Rate $= \dfrac{1}{x}$   (yard per minute)

Both persons:   Time $= 30$   (minutes)

Rate $= \dfrac{1}{30}$   (yard per minute)

*Equation:*

$$\frac{1}{1.5x} + \frac{1}{x} = \frac{1}{30}$$

$$30x\left(\frac{1}{1.5x} + \frac{1}{x}\right) = \left(\frac{1}{30}\right)30x$$

$$20 + 30 = x$$

$$50 = x \text{ and } 1.5x = 75$$

The first person (the slower one) would take 75 minutes to rake the front lawn.

**87.** *Verbal Model:*   $\boxed{\text{Future batting average}} = \boxed{\text{Total hits}} \div \boxed{\text{Total times at bat}}$

*Labels:*   Future batting average $= 0.275$

Number of consecutive hits $= x$

Total hits $= x + 6$

Total times at bat $= x + 35$

*Equation:*

$$0.275 = \frac{x + 6}{x + 35}$$

$$(x + 35)(0.275) = (x + 35)\left(\frac{x + 6}{x + 35}\right)$$

$$0.275x + 9.625 = x + 6, \quad x \neq -35$$

$$-0.725x + 9.625 = 6$$

$$-0.725x = -3.625$$

$$x = 5$$

The softball player must get 5 successive hits to obtain a batting average of 0.275.

**89.** *Verbal Model:*   $\boxed{\text{Serve percentage}} = \boxed{\text{Successful serves}} \div \boxed{\text{Total serves}}$

   *Labels:*         Current serves $= 32$

               Current successful serves $= 26$

               Additional consecutive successful serves $= x$

   *Equation:*                $0.900 = \dfrac{x + 26}{x + 32}$

$$(x + 32)(0.900) = \left(\dfrac{x + 26}{x + 32}\right)(x + 32)$$

$$0.9x + 28.8 = x + 26$$

$$28.8 = 0.1x + 26$$

$$2.8 = 0.1x$$

$$\dfrac{2.8}{0.1} = x$$

$$28 = x$$

To obtain a serve percentage of 0.900, the player must serve the ball successfully for the next 28 consecutive times.

**91.**                $N = \dfrac{250(5 + 3t)}{25 + t}$

$$125 = \dfrac{250(5 + 3t)}{25 + t}$$

$$125(25 + t) = 250(5 + 3t), \quad t \neq -25$$

$$3125 + 125t = 1250 + 750t$$

$$3125 = 1250 + 625t$$

$$1875 = 625t$$

$$3 = t$$

So, it takes 3 years for the herd to increase to 125 deer.

**93.** *Verbal Model:* | Distance traveled by car | = | Distance traveled by truck |

*Labels:*

Speed of car $= x$            (miles per hour)

Speed of truck $= x - 10$       (miles per hour)

Time car travels $= \dfrac{100}{x}$,         (hours)

Time truck travels $= \dfrac{100}{x} + \dfrac{1}{3}$     (hours)    Note: 20 minutes $= \dfrac{1}{3}$ hour

Distance traveled by car $= 100$      (miles)

Distance traveled by truck $=$ (speed of truck)(time truck travels) $= (x - 10)\left(\dfrac{100}{x} + \dfrac{1}{3}\right)$

*Equation:*

$$100 = (x - 10)\left(\frac{100}{x} + \frac{1}{3}\right)$$

$$100 = \frac{100(x - 10)}{x} + \frac{x - 10}{3}$$

$$3x(100) = 3x\left(\frac{100(x - 10)}{x} + \frac{x - 10}{3}\right)$$

$$300x = 300(x - 10) + x(x - 10), \quad x \neq 0$$

$$300x = 300x - 3000 + x^2 - 10x$$

$$x^2 - 10x - 3000 = 0$$

$$(x - 60)(x + 50) = 0$$

$$x - 60 = 0 \Rightarrow x = 60 \text{ and } x - 10 = 50$$

$$x + 50 = 0 \Rightarrow x = -50 \quad \text{(Extraneous)}$$

The speed of the car is 60 miles per hour, and the speed of the truck is 50 miles per hour.

**95.** *Verbal Model:* | Time for trip to meeting | = | Time for first portion of return trip |

*Formula:*      Distance $=$ (Rate)(Time) $\Rightarrow$ Time $= \dfrac{\text{Distance}}{\text{Rate}}$

*Labels:*

Wind speed $= w$                        (miles per hour)

Trip to meeting $=$ Distance $= 1500$ miles

                Rate $= 600 + w$      (miles per hour)

Return trip: Distance $= 1500 - 300 = 1200$ miles

                Rate $= 600 - w$      (miles per hour)

*Equation:*

$$\frac{1500}{600 + w} = \frac{1200}{600 - w}$$

$$1500(600 - w) = 1200(600 + w), \quad w \neq 600, w \neq -600$$

$$900{,}000 - 1500w = 720{,}000 + 1200w$$

$$900{,}000 - 2700w = 720{,}000$$

$$-2700w = -180{,}000$$

$$w = \frac{-180{,}000}{-2700}$$

$$w = \frac{200}{3} \text{ or } w = 66\frac{2}{3}$$

The speed of the wind is $66\dfrac{2}{3}$ mph.

**97.** The error is neglecting to multiply the right-hand side of the equation by the common denominator of $2(x + 5)$.

$$\frac{12}{x + 5} + \frac{1}{2} = 2$$

$$2(x + 5)\left(\frac{12}{x + 5} + \frac{1}{2}\right) = 2(2)(x + 5)$$

$$2(12) + 1(x + 5) = 4(x + 5), \quad x \neq -5$$

$$24 + x + 5 = 4x + 20$$

$$29 + x = 4x + 20$$

$$29 = 3x + 20$$

$$9 = 3x$$

$$3 = x$$

**99.**

$$\frac{6}{x - 3} + \frac{4x}{x + 3} = \frac{8x^2}{x^2 - 9}$$

$$(x + 3)(x - 3)\left(\frac{6}{x - 3} + \frac{4x}{x + 3}\right) = \left[\frac{8x^2}{(x + 3)(x - 3)}\right](x + 3)(x - 3)$$

$$6(x + 3) + 4x(x - 3) = 8x^2, \quad x \neq 3, x \neq -3$$

$$6x + 18 + 4x^2 - 12x = 8x^2$$

$$4x^2 - 6x + 18 = 8x^2$$

$$0 = 4x^2 + 6x - 18$$

$$0 = 2(2x^2 + 3x - 9)$$

$$0 = 2(2x - 3)(x + 3)$$

$$0 \neq 2$$

$$2x - 3 = 0 \Rightarrow x = \frac{3}{2}$$

$$x + 3 = 0 \Rightarrow x = -3 \quad \text{(Extraneous)}$$

True. The equation has "trial solutions" of $\frac{3}{2}$ and $-3$. By checking them, you can conclude that $-3$ is an extraneous solution because it results in division by zero.

**101.**

| | |
|---|---|
| $\dfrac{3 + x}{x} = \dfrac{7 + 2x}{2x}$ | First equation |
| $\dfrac{3}{x} + \dfrac{x}{x} = \dfrac{7}{2x} + \dfrac{2x}{2x}$ | Addition rule: $\dfrac{a + b}{c} = \dfrac{a}{c} + \dfrac{b}{c}$ |
| $\dfrac{3}{x} + 1 = \dfrac{7}{2x} + 1$ | Simplify fractions. |
| $\dfrac{3}{x} + 1 - 1 = \dfrac{7}{2x} + 1 - 1$ | Subtract 1 from each side. |
| $\dfrac{3}{x} = \dfrac{7}{2x}$ | Simplify to obtain the second equation. |

True. The first equation is equivalent to the second.

Note: Neither equation has a solution because the trial solution results in division by zero.

$$\frac{3}{x} = \frac{7}{2x}$$

$$3(2x) = x(7) \quad \text{Cross-multiply.}$$

$$6x = 7x$$

$$6x - 6x = 7x - 6x$$

$$0 = x \quad \text{(Extraneous)}$$

**103.** Because $x - 1$, the expression $x - 1 = 0$. In Step 4, where both sides of the equation are divided by $x - 1$, each side of the equation is divided by 0, but division by zero is undefined.

**105.** $2t^3 + 5t^2 - 6t - 15 = t^2(2t + 5) - 3(2t + 5)$
$$= (2t + 5)(t^2 - 3)$$

**107.** $6y^3 + 3y^2 - 2y - 1 = 3y^2(2y + 1) - 1(2y + 1)$
$$= (2y + 1)(3y^2 - 1)$$

**109.** $4x^2 - x^3 - 8 + 2x = x^2(4 - x) - 2(4 - x)$
$$= (4 - x)(x^2 - 2)$$
$$\text{or} -(x - 4)(x^2 - 2)$$

**111.** $\dfrac{\left(\dfrac{x^5}{12}\right)}{\left(\dfrac{x^2}{54}\right)} = \dfrac{x^5}{12} \cdot \dfrac{54}{x^2}$

$$= \frac{6 \cdot 9 \cdot x^2 \cdot x^3}{2 \cdot 6 \cdot x^2}$$

$$= \frac{(\cancel{6})(9)(\cancel{x^2})(x^3)}{(2)(\cancel{6})(\cancel{x^2})}$$

$$= \frac{9x^3}{2}, \quad x \neq 0$$

**113.** $\dfrac{\left[\dfrac{(2y)^3}{15x}\right]}{\left[\dfrac{22y^2}{(3x)^2}\right]} = \dfrac{(2y)^3}{15x} \cdot \dfrac{(3x)^2}{22y^2}$

$$= \frac{8y^3 \cdot 9x^2}{15x \cdot 22y^2}$$

$$= \frac{2 \cdot 4 \cdot 3 \cdot 3 \cdot y \cdot y^2 \cdot x \cdot x}{3 \cdot 5 \cdot 2 \cdot 11 \cdot x \cdot y^2}$$

$$= \frac{(\cancel{2})(4)(\cancel{3})(3)(y)(\cancel{y^2})(\cancel{x})(x)}{(\cancel{3})(5)(\cancel{2})(11)(\cancel{x})(\cancel{y^2})}$$

$$= \frac{12xy}{55}, \quad x \neq 0, y \neq 0$$

**115.** $\dfrac{\left(\dfrac{x^2 + 3x - 18}{x + 5}\right)}{(2x - 6)} = \dfrac{\left(\dfrac{x^2 + 3x - 18}{x + 5}\right)}{\left(\dfrac{2x - 6}{1}\right)}$

$$= \frac{x^2 + 3x - 18}{x + 5} \cdot \frac{1}{2x - 6}$$

$$= \frac{(x + 6)(x - 3)(1)}{(x + 5)(2)(x - 3)}$$

$$= \frac{(x + 6)(\cancel{x - 3})}{(x + 5)(2)(\cancel{x - 3})}$$

$$= \frac{x + 6}{2(x + 5)}, \quad x \neq 3$$

# Review Exercises for Chapter 7

**1.** The denominator is zero when $x - 5 = 0$ or $x = 5$. So, the domain is all real values of $x$ such that $x \neq 5$.

**3.** The denominator $t^2 - 3t + 2 = (t - 2)(t - 1)$ is zero when $t = 2$ or 1. So, the domain is all real values of $t$ such that $t \neq 2$ and $t \neq 1$.

**5.** $\dfrac{2x}{x + 4}$

(a) $x = 0 \Rightarrow \dfrac{2x}{x + 4} = \dfrac{2(0)}{0 + 4} = \dfrac{0}{4} = 0$

(b) $x = 2 \Rightarrow \dfrac{2x}{x + 4} = \dfrac{2(2)}{2 + 4} = \dfrac{4}{6} = \dfrac{2}{3}$

(c) $x = -3 \Rightarrow \dfrac{2x}{x + 4} = \dfrac{2(-3)}{-3 + 4} = \dfrac{-6}{1} = -6$

(d) $x = -4 \Rightarrow \dfrac{2x}{x + 4} = \dfrac{2(-4)}{-4 + 4} = \dfrac{-8}{0}$, undefined; Division by zero is undefined.

**7.** $\dfrac{x-1}{x^2+4}$

(a) $x = 2 \Rightarrow \dfrac{x-1}{x^2+4} = \dfrac{2-1}{2^2+4} = \dfrac{1}{8}$

(b) $x = 1 \Rightarrow \dfrac{x-1}{x^2+4} = \dfrac{1-1}{1^2+4} = \dfrac{0}{5} = 0$

(c) $x = -2 \Rightarrow \dfrac{x-1}{x^2+4} = \dfrac{-2-1}{(-2)^2+4} = -\dfrac{3}{8}$

(d) $x = -5 \Rightarrow \dfrac{x-1}{x^2+4} = \dfrac{-5-1}{(-5)^2+4} = -\dfrac{6}{29}$

**9.** $\dfrac{6t}{18} = \dfrac{\cancel{6}(t)}{\cancel{6}(3)} = \dfrac{t}{3}$

**11.** $\dfrac{4x^5}{x^2} = \dfrac{\cancel{x^2}(4x^3)}{\cancel{x^2}(1)} = 4x^3, \qquad x \neq 0$

**13.** $\dfrac{7x^2y}{21xy^2} = \dfrac{(\cancel{7})(\cancel{x})(\cancel{x})(x)}{(\cancel{7})(3)(\cancel{x})(\cancel{y})(y)} = \dfrac{x}{3y}, \qquad x \neq 0$

**15.** $\dfrac{3b-6}{4b-8} = \dfrac{3(b-2)}{4(b-2)} = \dfrac{3\cancel{(b-2)}}{4\cancel{(b-2)}} = \dfrac{3}{4}, \quad b \neq 2$

**17.** $\dfrac{4x-4y}{y-x} = \dfrac{4(x-y)}{y-x} = \dfrac{4\cancel{(x-y)}}{-1\cancel{(x-y)}} = -4, \quad x \neq y$

**19.** $\dfrac{x^2-9}{x^2-x-6} = \dfrac{(x+3)\cancel{(x-3)}}{\cancel{(x-3)}(x+2)} = \dfrac{x+3}{x+2}, \, x \neq 3$

**21.** $\dfrac{1-x^3}{x^2-1} = \dfrac{(-1)(x^3-1)}{x^2-1}$

$\qquad = \dfrac{(-1)\cancel{(x-1)}(x^2+x+1)}{(x+1)\cancel{(x-1)}}$

$\qquad = -\dfrac{x^2+x+1}{x+1}, \quad x \neq 1$

**23.** $\dfrac{x^2-3xy-18y^2}{x^2+4xy+3y^2} = \dfrac{(x-6y)\cancel{(x+3y)}}{(x+y)\cancel{(x+3y)}}$

$\qquad = \dfrac{x-6y}{x+y}, \quad x \neq -3y$

**25.** $\dfrac{x(x-4)+7(x-4)}{x^2+7x} = \dfrac{(x-4)\cancel{(x+7)}}{x\cancel{(x+7)}}$

$\qquad = \dfrac{x-4}{x}, \quad x \neq -7$

**27.**

| $x$ | 1 | 1.5 | 2 | 2.5 | 3 |
|---|---|---|---|---|---|
| $\dfrac{x-2}{x^2-4}$ | $\dfrac{1}{3}$ | $\dfrac{2}{7}$ | Undefined | $\dfrac{2}{9}$ | $\dfrac{1}{5}$ |
| $\dfrac{1}{x+2}$ | $\dfrac{1}{3}$ | $\dfrac{2}{7}$ | $\dfrac{1}{4}$ | $\dfrac{2}{9}$ | $\dfrac{1}{5}$ |

$\dfrac{x-2}{x^2-4} = \dfrac{\cancel{(x-2)}}{(x+2)\cancel{(x-2)}} = \dfrac{1}{x+2}, \; x \neq 2$

The expressions are equal for all values of $x$ except 2. When $x = 2$, the first expression is undefined.

**29.** Probability $= \dfrac{\text{Area of shaded region}}{\text{Total area of figure}}$

$\qquad = \dfrac{x(x-3)}{x(2x+1)} = \dfrac{\cancel{x}(x-3)}{\cancel{x}(2x+1)} = \dfrac{x-3}{2x+1}$

**31.** $\dfrac{x^2}{6} \cdot \dfrac{2x}{x^3} = \dfrac{\cancel{(x^2)}(2)\cancel{(x)}}{3(2)\cancel{(x^2)}\cancel{(x)}} = \dfrac{1}{3}, \quad x \neq 0$

**33.** $\dfrac{5x^2y}{4} \cdot \dfrac{6x}{10y^3} = \dfrac{5(6)x^3y}{4(10)y^3} = \dfrac{(\cancel{5})(\cancel{2})(3)(x^3)\cancel{y}}{(4)(\cancel{5})(\cancel{2})(y^2)\cancel{y}} = \dfrac{3x^3}{4y^2}$

**35.** $\dfrac{x-5}{6} \cdot \dfrac{3}{x-5} = \dfrac{3(x-5)}{3(2)(x-5)}$

$\qquad = \dfrac{\cancel{3}\cancel{(x-5)}(1)}{\cancel{3}(2)\cancel{(x-5)}}$

$\qquad = \dfrac{1}{2}, \quad x \neq 5$

**37.** $\dfrac{2-x}{x+3} \cdot \dfrac{4x+12}{x^2-4} = \dfrac{(2-x)(4)(x+3)}{(x+3)(x+2)(x-2)}$

$\qquad = \dfrac{-1\cancel{(x-2)}(4)\cancel{(x+3)}}{\cancel{(x+3)}(x+2)\cancel{(x-2)}}$

$\qquad = -\dfrac{4}{x+2}, \quad x \neq -3, x \neq 2$

**39.** $\dfrac{x+6}{x-8} \cdot (4x) = \dfrac{x+6}{x-8} \cdot \dfrac{4x}{1}$

$\qquad = \dfrac{(x+6)(4x)}{x-8}$

$\qquad = \dfrac{4x(x+6)}{x-8}$

**41.** $\dfrac{x^2-36}{6}\cdot\dfrac{3}{x^2-12x+36}=\dfrac{(x+6)(x-6)}{6}\cdot\dfrac{3}{(x-6)(x-6)}=\dfrac{3(x+6)(x-6)}{3(2)(x-6)(x-6)}=\dfrac{x+6}{2(x-6)}$

**43.** $\dfrac{5}{8}\div\dfrac{u}{v}=\dfrac{5}{8}\cdot\dfrac{v}{u}=\dfrac{5v}{8u},\quad v\neq 0$

**45.** $10y^2\div\dfrac{y}{5}=\dfrac{10y^2}{1}\cdot\dfrac{5}{y}=\dfrac{50y^2}{y}=\dfrac{50y(y)}{y}=50y,\quad y\neq 0$

**47.** $\dfrac{3(x+7)}{21x^2}\div\dfrac{12(x+7)^2}{9x}=\dfrac{3(x+7)}{21x^2}\cdot\dfrac{9x}{12(x+7)^2}$

$=\dfrac{3(3)(3)(x)(x+7)}{3(7)(4)(3)(x)(x)(x+7)(x+7)}$

$=\dfrac{3(3)(3)(x)(x+7)}{3(7)(4)(3)(x)(x)(x+7)(x+7)}$

$=\dfrac{3}{28x(x+7)}$

**49.** $\dfrac{9-x^2}{2x+6}\div\dfrac{2x^2-5x-3}{6x-4}=\dfrac{(-1)(x^2-9)}{2x+6}\cdot\dfrac{6x-4}{2x^2-5x-3}$

$=-\dfrac{(x+3)(x-3)(2)(3x-2)}{2(x+3)(2x+1)(x-3)}$

$=-\dfrac{(x+3)(x-3)(2)(3x-2)}{2(x+3)(2x+1)(x-3)}$

$=-\dfrac{3x-2}{2x+1},\quad x\neq-3,x\neq3,x\neq\dfrac{2}{3}$

**51.** $\dfrac{x^2-16}{x+2}\div\dfrac{x+4}{6x+12}=\dfrac{x^2-16}{x+2}\cdot\dfrac{6x+12}{x+4}$

$=\dfrac{(x+4)(x-4)(6)(x+2)}{(x+2)(x+4)}$

$=\dfrac{(x+4)(x-4)(6)(x+2)}{(x+2)(x+4)}$

$=6(x-4),\quad x\neq-2,x\neq-4$

**53.** $\dfrac{x^2-8x}{x-1}\div\dfrac{x^2-16x+64}{x^2-1}=\dfrac{x^2-8x}{x-1}\cdot\dfrac{x^2-1}{x^2-16x+64}=\dfrac{x(x-8)(x+1)(x-1)}{(x-1)(x-8)(x-8)}=\dfrac{x(x+1)}{x-8},\quad x\neq1,x\neq-1$

**55.** $\left(\dfrac{3x^2}{8}\cdot\dfrac{x-7}{x}\right)\div\dfrac{x}{32}=\dfrac{3x^2(x-7)}{8x}\cdot\dfrac{32}{x}=\dfrac{3(8)(4)(x)(x)(x-7)}{(8)(x)(x)}=\dfrac{3(8)(4)(x)(x)(x-7)}{(8)(x)(x)}=12(x-7),\quad x\neq0$

**57.** $\dfrac{5x}{8}-\dfrac{3x}{8}=\dfrac{2x}{8}=\dfrac{2(x)}{2(4)}=\dfrac{x}{4}$

**59.** $\dfrac{4x-5}{x+2}+\dfrac{2x+1}{x+2}=\dfrac{4x-5+2x+1}{x+2}=\dfrac{6x-4}{x+2}$

**61.** $\dfrac{5t+1}{2t-3}-\dfrac{2t-7}{2t-3}=\dfrac{5t+1-(2t-7)}{2t-3}$

$=\dfrac{5t+1-2t+7}{2t-3}$

$=\dfrac{3t+8}{2t-3}$

**63.** $\dfrac{x-5}{x^2+2x-3}+\dfrac{3x+14}{x^2+2x-3}=\dfrac{x-5+3x+14}{x^2+2x-3}$

$\qquad\qquad\qquad\qquad\quad=\dfrac{4x+9}{x^2+2x-3}$

**65.** $20x^2=(2)(2)(5)(x)(x)$

$\quad\;\; 24=(2)(2)(2)(3)$

$\quad\; 30x^3=(2)(3)(5)(x)(x)(x)$

The different factors are 2, 3, 5, and $x$. Using the highest powers of these factors, we conclude that the least common multiple is $2^3\cdot 3\cdot 5\cdot x^3=120x^3$.

**67.** $2(x-2)=2\cdot(x-2)$

$\quad 12(x-2)=2\cdot 2\cdot 3\cdot(x-2)$

The different factors are 2, 3, and $x-2$. Using the highest powers of the factors, we conclude that the least common multiple is $2^2(3)(x-2)$ or $12(x-2)$.

**69.** $x^2-16=(x+4)(x-4)$

$\quad x(x-4)$

The different factors are $x$, $x+4$, and $x-4$. Using the highest powers of the factors, we conclude that the least common multiple is $x(x+4)(x-4)$ or $x(x^2-16)$.

**71.** $\dfrac{7}{4x}=\dfrac{7(3x^2)}{4x(3x^2)}=\dfrac{7(3x^2)}{12x^3}$

The missing factor is $3x^2$.

**73.** $\dfrac{x-3}{x-1}=\dfrac{(x-3)(x+1)}{(x-1)(x+1)}=\dfrac{(x-3)(x+1)}{x^2-1}$

The missing factor is $x+1$.

**75.** $\dfrac{5t}{16}-\dfrac{5t}{24}=\dfrac{5t(3)}{16(3)}-\dfrac{5t(2)}{24(2)}=\dfrac{15t}{48}-\dfrac{10t}{48}=\dfrac{5t}{48}$

**77.** $\dfrac{5}{2x}+\dfrac{7}{3x}=\dfrac{5(3)}{2x(3)}+\dfrac{7(2)}{3x(2)}=\dfrac{15}{6x}+\dfrac{14}{6x}=\dfrac{29}{6x}$

**79.** $\dfrac{1}{x+2}-\dfrac{1}{x+1}=\dfrac{1(x+1)}{(x+2)(x+1)}-\dfrac{1(x+2)}{(x+1)(x+2)}$

$\qquad\qquad\qquad\quad=\dfrac{1(x+1)-1(x+2)}{(x+2)(x+1)}$

$\qquad\qquad\qquad\quad=\dfrac{x+1-x-2}{(x+2)(x+1)}$

$\qquad\qquad\qquad\quad=\dfrac{-1}{(x+2)(x+1)}$

$\qquad\qquad\quad$ or $-\dfrac{1}{(x+2)(x+1)}$

**81.** $\dfrac{4x}{x-1}-\dfrac{3}{1-x}=\dfrac{4x}{x-1}-\dfrac{(-1)(3)}{(-1)(1-x)}$

$\qquad\qquad\qquad=\dfrac{4x}{x-1}-\dfrac{-3}{x-1}$

$\qquad\qquad\qquad=\dfrac{4x-(-3)}{x-1}$

$\qquad\qquad\qquad=\dfrac{4x+3}{x-1}$

**83.** $\dfrac{2x-1}{5-3x}-\dfrac{3x+2}{3x-5}=\dfrac{2x-1}{5-3x}-\dfrac{(-1)(3x+2)}{(-1)(3x-5)}$

$\qquad\qquad\qquad\qquad=\dfrac{2x-1}{5-3x}-\dfrac{-3x-2}{5-3x}$

$\qquad\qquad\qquad\qquad=\dfrac{2x-1-(-3x-2)}{5-3x}$

$\qquad\qquad\qquad\qquad=\dfrac{2x-1+3x+2}{5-3x}$

$\qquad\qquad\qquad\qquad=\dfrac{5x+1}{5-3x}$ or $-\dfrac{5x+1}{3x-5}$

**85.** $\dfrac{1}{x+4}-\dfrac{x-1}{x^2+4x+4}=\dfrac{1(x^2+4x+4)}{(x+4)(x^2+4x+4)}-\dfrac{(x-1)(x+4)}{(x-1)(x^2+4x+4)}$

$\qquad\qquad\qquad\qquad\qquad=\dfrac{x^2+4x+4}{(x+4)(x+2)^2}-\dfrac{x^2+3x-4}{(x-1)(x+2)^2}$

$\qquad\qquad\qquad\qquad\qquad=\dfrac{x^2+4x+4-x^2-3x+4}{(x+4)(x+2)^2}$

$\qquad\qquad\qquad\qquad\qquad=\dfrac{x+8}{(x+4)(x+2)^2}$

**87.** $\dfrac{x-7}{x^2-16}-\dfrac{3}{4-x}=\dfrac{x-7}{(x+4)(x-4)}-\dfrac{3(-1)}{(4-x)(-1)}$

$=\dfrac{x-7}{(x+4)(x-4)}+\dfrac{3}{x-4}$

$=\dfrac{x-7}{(x+4)(x-4)}+\dfrac{3(x+4)}{(x-4)(x+4)}$

$=\dfrac{x-7+3x+12}{(x+4)(x-4)}$

$=\dfrac{4x+5}{(x+4)(x-4)}$

**89.** $x-1+\dfrac{1}{x+2}+\dfrac{1}{x-1}=\dfrac{x-1}{1}+\dfrac{1}{x+2}+\dfrac{1}{x-1}$

$=\dfrac{(x-1)(x+2)(x-1)}{1(x+2)(x-1)}+\dfrac{1(x-1)}{(x+2)(x-1)}+\dfrac{1(x+2)}{(x-1)(x+2)}$

$=\dfrac{(x-1)^2(x+2)}{(x+2)(x-1)}+\dfrac{(x-1)}{(x+2)(x-1)}+\dfrac{(x+2)}{(x+2)(x-1)}$

$=\dfrac{(x-1)^2(x+2)+(x-1)+(x+2)}{(x+2)(x-1)}$

$=\dfrac{(x^2-2x+1)(x+2)+(x-1)+(x+2)}{(x+2)(x-1)}$

$=\dfrac{x^3+2x^2-2x^2-4x+x+2+x-1+x+2}{(x+2)(x-1)}$

$=\dfrac{x^3-x+3}{(x+2)(x-1)}$

**91.** $2x+\dfrac{3}{2(x-4)}-\dfrac{1}{2(x+2)}=\dfrac{2x(2)(x-4)(x+2)}{2(x-4)(x+2)}+\dfrac{3(x+2)}{2(x-4)(x+2)}-\dfrac{1(x-4)}{2(x-4)(x+2)}$

$=\dfrac{2x(2)(x-4)(x+2)+3(x+2)-(x-4)}{2(x-4)(x+2)}$

$=\dfrac{4x(x^2+2x-4x-8)+3x+6-x+4}{2(x-4)(x+2)}$

$=\dfrac{4x(x^2-2x-8)+2x+10}{2(x-4)(x+2)}$

$=\dfrac{4x^3-8x^2-32x+2x+10}{2(x-4)(x+2)}$

$=\dfrac{4x^3-8x^2-30x+10}{2(x-4)(x+2)}$

$=\dfrac{2(2x^3-4x^2-15x+5)}{2(x-4)(x+2)}$

$=\dfrac{2x^3-4x^2-15x+5}{(x-4)(x+2)}$

**93.** $\dfrac{\left(\dfrac{7}{18}\right)}{\left(\dfrac{3}{28}\right)} = \dfrac{7}{18} \cdot \dfrac{28}{3} = \dfrac{7 \cdot 7 \cdot 2 \cdot 2}{2 \cdot 9 \cdot 3} = \dfrac{7 \cdot 7 \cdot 2 \cdot \cancel{2}}{\cancel{2} \cdot 9 \cdot 3} = \dfrac{98}{27}$

**95.** $\dfrac{5x}{\left(\dfrac{x}{y}\right)} = \dfrac{5x}{1} \div \dfrac{x}{y} = \dfrac{5x}{1} \cdot \dfrac{y}{x} = \dfrac{5xy}{x} = \dfrac{\cancel{x}(5y)}{\cancel{x}} = 5y, \quad x \neq 0, y \neq 0$

**97.** $\dfrac{\left(\dfrac{x^2 + x}{x^2 + x - 12}\right)}{\left(\dfrac{4x + 4}{x^2 - 6x - 9}\right)} = \dfrac{x^2 + x}{x^2 + x - 12} \div \dfrac{4x + 4}{x^2 - 6x + 9}$

$\qquad = \dfrac{x(x + 1)}{(x + 4)(x - 3)} \cdot \dfrac{(x - 3)(x - 3)}{4(x + 1)}$

$\qquad = \dfrac{x\cancel{(x + 1)}\cancel{(x - 3)}(x - 3)}{4(x + 4)\cancel{(x - 3)}\cancel{(x + 1)}}$

$\qquad = \dfrac{x(x - 3)}{4(x + 4)}, \quad x \neq 3, x \neq -1$

**99.** $\dfrac{\left(\dfrac{2}{x} + 2\right)}{\left(1 - \dfrac{1}{x}\right)} = \dfrac{\left(\dfrac{2}{x} + \dfrac{2x}{x}\right)}{\left(\dfrac{x}{x} - \dfrac{1}{x}\right)}$

$\qquad = \dfrac{\left(\dfrac{2x + 2}{x}\right)}{\left(\dfrac{x - 1}{x}\right)}$

$\qquad = \dfrac{2x + 2}{x} \div \dfrac{x - 1}{x}$

$\qquad = \dfrac{2x + 2}{x} \cdot \dfrac{x}{x - 1}$

$\qquad = \dfrac{\cancel{x}(2x + 2)}{\cancel{x}(x - 1)} = \dfrac{2x + 2}{x - 1}, \quad x \neq 0$

**101.** $\dfrac{\left(\dfrac{1}{x + 1} - \dfrac{1}{4}\right)}{x - 3} = \dfrac{\left[\dfrac{4(1)}{4(x + 1)} - \dfrac{1(x + 1)}{4(x + 1)}\right]}{x - 3}$

$\qquad = \dfrac{\left[\dfrac{4 - (x + 1)}{4(x + 1)}\right]}{x - 3} = \dfrac{\left[\dfrac{4 - x - 1}{4(x + 1)}\right]}{x - 3}$

$\qquad = \dfrac{\left[\dfrac{3 - x}{4(x + 1)}\right]}{x - 3}$

$\qquad = \dfrac{3 - x}{4(x + 1)} \div (x - 3)$

$\qquad = \dfrac{-1(x - 3)}{4(x + 1)} \cdot \dfrac{1}{x - 3}$

$\qquad = -\dfrac{\cancel{(x - 3)}(1)}{4(x + 1)\cancel{(x - 3)}}$

$\qquad = -\dfrac{1}{4(x + 1)}, \quad x \neq 3$

**103.** $\dfrac{\left(\dfrac{1}{x + 3}\right)}{\left(\dfrac{4}{6x + 18} + \dfrac{2}{x + 3}\right)} = \dfrac{\left(\dfrac{1}{x + 3}\right)}{\left[\dfrac{4}{6(x + 3)} + \dfrac{2}{x + 3}\right]} \cdot \dfrac{6(x + 3)}{6(x + 3)} = \dfrac{6}{4 + 2(6)} = \dfrac{6}{4 + 12} = \dfrac{6}{16} = \dfrac{3}{8}, \quad x \neq -3$

**105.** $\dfrac{\left(\dfrac{x}{3} + \dfrac{5x}{12}\right)}{2} = \dfrac{\left(\dfrac{4x}{12} + \dfrac{5x}{12}\right)}{2} = \dfrac{9x}{12} \cdot \dfrac{1}{2} = \dfrac{9x}{24} = \dfrac{3x}{8}$

The average of the two real numbers is $\dfrac{3x}{8}$.

**107.**   $\dfrac{x}{6} + \dfrac{5x}{8} = 19$

$24\left(\dfrac{x}{6} + \dfrac{5x}{8}\right) = (19)(24)$

$4x + 15x = 456$

$19x = 456$

$x = \dfrac{456}{19}$

$x = 24$

**109.**   $\dfrac{t+1}{9} = \dfrac{2}{3} - t$

$9\left(\dfrac{t+1}{9}\right) = 9\left(\dfrac{2}{3} - t\right)$

$t + 1 = 3(2) - 9t$

$t + 1 = 6 - 9t$

$10t + 1 = 6$

$10t = 5$

$t = \dfrac{5}{10}$

$t = \dfrac{1}{2}$

**111.**   $\dfrac{2}{x} + \dfrac{3}{8} = \dfrac{5}{x}$

$8x\left(\dfrac{2}{x} + \dfrac{3}{8}\right) = \left(\dfrac{5}{x}\right)(8x)$

$16 + 3x = 40, \quad x \neq 0$

$3x = 24$

$x = 8$

**113.**   $\dfrac{7}{x} - 2 = \dfrac{3}{x} + 6$

$x\left(\dfrac{7}{x} - 2\right) = x\left(\dfrac{3}{x} + 6\right)$

$7 - 2x = 3 + 6x, \quad x \neq 0$

$7 - 8x = 3$

$-8x = -4$

$x = \dfrac{-4}{-8}$

$x = \dfrac{1}{2}$

**115.**   $\dfrac{x}{x+3} - \dfrac{3x}{x^2 - 9} = 1$

$(x-3)(x+3)\left(\dfrac{x}{x+3} - \dfrac{3x}{x^2-9}\right) = (x-3)(x+3)(1)$

$(x-3)(x) - 3x = (x-3)(x+3), \quad x \neq 3, x \neq -3$

$x^2 - 3x - 3x = x^2 - 9$

$x^2 - 6x = x^2 - 9$

$-6x = -9$

$x = \dfrac{-9}{-6}$

$x = \dfrac{3}{2}$

**117.**   $\dfrac{t}{t-4} + \dfrac{3}{t-2} = 0$

$(t-4)(t-2)\left(\dfrac{t}{t-4} + \dfrac{3}{t-2}\right) = (t-4)(t-2)(0)$

$(t-4)(t-2)\left(\dfrac{t}{t-4}\right) + (t-4)(t-2)\left(\dfrac{3}{t-2}\right) = 0$

$t(t-2) + 3(t-4) = 0, \quad t \neq 4, t \neq 2$

$t^2 - 2t + 3t - 12 = 0$

$t^2 + t - 12 = 0$

$(t+4)(t-3) = 0$

$t + 4 = 0 \Rightarrow t = -4$

$t - 3 = 0 \Rightarrow t = 3$

**119.**

$$\frac{2}{x-5} + \frac{x}{x-8} = \frac{-6}{x^2 - 13x + 40}$$

$$(x-5)(x-8)\left(\frac{2}{x-5} + \frac{x}{x-8}\right) = \left(\frac{-6}{x^2 - 13x + 40}\right)(x-5)(x-8)$$

$$2(x-8) + x(x-5) = -6, \quad x \neq 5, x \neq 8$$

$$2x - 16 + x^2 - 5x = -6$$

$$x^2 - 3x - 16 = -6$$

$$x^2 - 3x - 10 = 0$$

$$(x-5)(x+2) = 0$$

$$x - 5 = 0 \Rightarrow x = 5 \quad \text{(Extraneous)}$$

$$x + 2 = 0 \Rightarrow x = -2$$

**121. (a)**

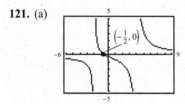

The *x*-intercept appears to be $\left(-\frac{1}{2}, 0\right)$.

**(b)**

$$0 = \frac{1}{x+2} + \frac{3}{x-4}$$

$$(x+2)(x-4)(0) = (x+2)(x-4)\left(\frac{1}{x+2} + \frac{3}{x-4}\right)$$

$$0 = 1(x-4) + 3(x+2), \quad x \neq -2, 4$$

$$0 = x - 4 + 3x + 6$$

$$0 = 4x + 2$$

$$-2 = 4x$$

$$\frac{-2}{4} = x$$

$$-\frac{1}{2} = x$$

This confirms that the *x*-intercept is $\left(-\frac{1}{2}, 0\right)$.

**123.** *Verbal Model:* $\boxed{\text{A number}}$ + $\boxed{\text{The reciprocal of the number}}$ = $\dfrac{41}{20}$

*Labels:*  The number $= x$

The reciprocal of the number $= \dfrac{1}{x}$

*Equation:*
$$x + \frac{1}{x} = \frac{41}{20}$$
$$20x\left(x + \frac{1}{x}\right) = 20x\left(\frac{41}{20}\right)$$
$$20x^2 + 20 = 41x, \quad x \neq 0$$
$$20x^2 - 41x + 20 = 0$$
$$(5x - 4)(4x - 5) = 0$$
$$5x - 4 = 0 \Rightarrow 5x = 4 \Rightarrow x = \frac{4}{5}$$
$$4x - 5 = 0 \Rightarrow 4x = 5 \Rightarrow x = \frac{5}{4}$$

The number is $\dfrac{4}{5}$ or $\dfrac{5}{4}$.

**125.** *Verbal Model:* $\boxed{\text{Time for first car}}$ = $\boxed{\text{Time for second car}}$

*Labels:*  First car:  Distance $= 159$  (miles)

Rate $= x$  (miles per hour)

Time $= \dfrac{159}{x}$  (hours)

Second car:  Distance $= 174$  (miles)

Rate $= x + 5$  (miles per hour)

Time $= \dfrac{174}{x + 5}$  (hours)

*Equation:*
$$\frac{159}{x} = \frac{174}{x + 5}$$
$$174x = 159(x + 5), \quad x \neq 0, x \neq -5 \quad \text{Cross-multiply.}$$
$$174x = 159x + 795$$
$$15x = 795$$
$$x = \frac{795}{15}$$
$$x = 53 \text{ and } x + 5 = 58$$

The first car travels at an average speed of 53 miles per hour, and the second car travels at an average speed of 58 miles per hour.

**127.** *Verbal Model:*  | Rate for first person | + | Rate for second person | = | Rate for both persons |

*Labels:*   First person:   Time $= 3$    (hours)

Rate $= \dfrac{1}{3}$    (hours)

Second person   Time $= x$    (hours)

Rate $= \dfrac{1}{x}$    (job per hour)

Both persons   Time $= 1$    (hours)

Rate $= \dfrac{1}{1}$    (job per hour)

*Equation:*
$$\frac{1}{3} + \frac{1}{x} = 1$$
$$3x\left(\frac{1}{3} + \frac{1}{x}\right) = (1)3x$$
$$x + 3 = 3x$$
$$3 = 2x$$
$$\frac{3}{2} = x$$

The second person would take $1\dfrac{1}{2}$ hours to complete the landscaping job.

**129.** *Verbal Model:*  | Batting average | = | Total hits | ÷ | Total times at bat |

*Labels:*   Current times at bat $= 40$

Current hits $= 40(0.300) = 12$

Additional consecutive hits $= x$

Total hits $= x + 12$

Total times at bat $= x + 40$

*Equation:*
$$0.440 = \frac{x + 12}{x + 40}$$
$$0.440(x + 40) = \left(\frac{x + 12}{x + 40}\right)(x + 40)$$
$$0.44x + 17.6 = x + 12, \quad x \ne -40$$
$$17.6 = 0.56x + 12$$
$$5.6 = 0.56x$$
$$10 = x$$

The player must successfully hit the ball for the next 10 consecutive times at bat.

# Chapter Test for Chapter 7

**1.** The denominator $x^2 - 81 = (x + 9)(x - 9)$ is zero when $x = 9$ or $x = -9$. So, the domain is all real values of $x$ such that $x \ne 9$ and $x \ne -9$.

**2.** $\dfrac{2x^2}{x + 1} = \dfrac{2x^2[(x)(x + 1)]}{(x + 1)[(x)(x + 1)]} = \dfrac{2x^2(x^2 + x)}{x(x + 1)^2}$

The missing factor is $x(x + 1)$ or $x^2 + x$.

**3.** $\dfrac{3y(2y - 1)}{6y^3(2y - 1)^2} = \dfrac{3(y)(2y - 1)}{2(3)(y)(y^2)(2y - 1)(2y - 1)}$

$= \dfrac{\cancel{3}\,(\cancel{y})\,(\cancel{2y - 1})}{2(\cancel{3})\,(\cancel{y})\,(y^2)(\cancel{2y - 1})(2y - 1)}$

$= \dfrac{1}{2y^2(2y - 1)}$

**4.** $\dfrac{x^2 - 64}{x^2 - 3x - 40} = \dfrac{(x+8)(x-8)}{(x-8)(x+5)} = \dfrac{(x+8)\cancel{(x-8)}}{\cancel{(x-8)}(x+5)} = \dfrac{x+8}{x+5}, \quad x \ne 8$

**5.** $\dfrac{18x}{5} \cdot \dfrac{15}{3x^3} = \dfrac{18x \cdot 15}{5 \cdot 3x^3} = \dfrac{18\,\cancel{(x)}\,\cancel{(5)}\,\cancel{(3)}}{\cancel{(5)}\,\cancel{(3)}\,\cancel{(x)}(x^2)} = \dfrac{18}{x^2}$

**6.** $(x+2)^2 \cdot \dfrac{x-2}{x^3 + 2x^2} = \dfrac{(x+2)^2}{1} \cdot \dfrac{x-2}{x^2(x+2)} = \dfrac{(x+2)(x+2)(x-2)}{x^2(x+2)} = \dfrac{(x+2)(x-2)}{x^2}, \quad x \ne -2$

**7.** $\dfrac{3x^2}{4} \div \dfrac{9x^3}{10} = \dfrac{3x^2}{4} \cdot \dfrac{10}{9x^3} = \dfrac{3x^2 \cdot 10}{4 \cdot 9x^3} = \dfrac{\cancel{(3)}\,\cancel{(x^2)}\,(2)(5)}{(2)(2)\cancel{(3)}(3)\cancel{(x^2)}(x)} = \dfrac{5}{6x}$

**8.** $\left[\left(\dfrac{x}{x-3}\right)^2 \cdot \dfrac{x^2}{x^2 - 3x}\right] \div (x-3)^5 = \left[\dfrac{x^2}{(x-3)^2} \cdot \dfrac{x^2}{x(x-3)}\right] \div \dfrac{(x-3)^5}{1}$

$\qquad = \dfrac{x^4}{(x-3)^3(x)} \cdot \dfrac{1}{(x-3)^5}$

$\qquad = \dfrac{x^4}{(x)(x-3)^8}$

$\qquad = \dfrac{\cancel{(x)}(x^3)}{\cancel{(x)}(x-3)^8}$

$\qquad = \dfrac{x^3}{(x-3)^8}, \quad x \ne 0$

**9.** $\dfrac{3}{x+2} + 6 = \dfrac{3}{x+2} + \dfrac{6(x+2)}{1(x+2)} = \dfrac{3 + 6(x+2)}{x+2} = \dfrac{3 + 6x + 12}{x+2} = \dfrac{6x+15}{x+2}$

**10.** $\dfrac{2}{x+1} - \dfrac{2x}{x^2 + 2x + 1} = \dfrac{2}{x+1} - \dfrac{2x}{(x+1)^2} = \dfrac{2(x+1)}{(x+1)(x+1)} - \dfrac{2x}{(x+1)^2} = \dfrac{2(x+1) - 2x}{(x+1)^2} = \dfrac{2x + 2 - 2x}{(x+1)^2} = \dfrac{2}{(x+1)^2}$

**11.** $\dfrac{16}{\left(\dfrac{2}{x} + 8\right)} = \dfrac{16}{\left(\dfrac{2}{x} + 8\right)} \cdot \dfrac{(x)}{(x)}$

$\qquad = \dfrac{16x}{2 + 8x}$

$\qquad = \dfrac{2(8x)}{2(1 + 4x)}$

$\qquad = \dfrac{\cancel{2}(8x)}{\cancel{2}(1 + 4x)}$

$\qquad = \dfrac{8x}{4x + 1}, \quad x \ne 0$

**12.** $\dfrac{\left(\dfrac{3}{x+1} - \dfrac{3}{x-1}\right)}{\left(\dfrac{5}{x^2 - 1}\right)} = \dfrac{\left(\dfrac{3}{x+1} - \dfrac{3}{x-1}\right) \cdot (x+1)(x-1)}{\left[\dfrac{5}{(x+1)(x-1)}\right] \cdot (x+1)(x-1)}$

$\qquad = \dfrac{3(x-1) - 3(x+1)}{5}$

$\qquad = \dfrac{3x - 3 - 3x - 3}{5}$

$\qquad = \dfrac{-6}{5}, \quad x \ne -1, x \ne 1$

$\qquad = -\dfrac{6}{5}, \quad x \ne -1, x \ne 1$

**13.** $\dfrac{\left(\dfrac{t}{t-5}\right)}{\left(\dfrac{t^2}{5-t}\right)} = \dfrac{t}{t-5} \div \dfrac{t^2}{5-t} = \dfrac{t}{t-5} \cdot \dfrac{5-t}{t^2} = \dfrac{t \cdot (5-t)}{(t-5)(t^2)} = \dfrac{\cancel{(t)}(-1)\cancel{(t-5)}}{\cancel{(t-5)}\cancel{(t^2)}(t)} = -\dfrac{1}{t}, \quad t \ne 5$

**14.** $\dfrac{\left(9x - \dfrac{1}{x}\right)}{\left(\dfrac{1}{x} - 3\right)} = \dfrac{\left(9x - \dfrac{1}{x}\right) \cdot x}{\left(\dfrac{1}{x} - 3\right) \cdot x} = \dfrac{9x^2 - 1}{1 - 3x} = \dfrac{(3x + 1)(3x - 1)}{-1(3x - 1)} = \dfrac{3x + 1}{-1} = -3x - 1, \quad x \neq 0, x \neq \dfrac{1}{3}$

**15. (a)** $\dfrac{1}{4} + \dfrac{2}{1} \overset{?}{=} \dfrac{3}{2}$

$\dfrac{1}{4} + \dfrac{2(4)}{1(4)} \overset{?}{=} \dfrac{3(2)}{2(2)}$

$\dfrac{1}{4} + \dfrac{8}{4} \overset{?}{=} \dfrac{6}{4}$

$\dfrac{9}{4} \neq \dfrac{6}{4}$

1 *is not* a solution.

**(b)** $\dfrac{2}{4} + \dfrac{2}{2} \overset{?}{=} \dfrac{3}{2}$

$\dfrac{1}{2} + \dfrac{2}{2} \overset{?}{=} \dfrac{3}{2}$

$\dfrac{3}{2} = \dfrac{3}{2}$

2 *is* a solution.

**(c)** $\dfrac{-\dfrac{1}{2}}{4} + \dfrac{2}{-\dfrac{1}{2}} \overset{?}{=} \dfrac{3}{2}$

$\left(-\dfrac{1}{2} \div 4\right) + \left(2 \div -\dfrac{1}{2}\right) \overset{?}{=} \dfrac{3}{2}$

$\left(-\dfrac{1}{2} \cdot \dfrac{1}{4}\right) + \left(\dfrac{2}{1} \cdot -\dfrac{2}{1}\right) \overset{?}{=} \dfrac{3}{2}$

$-\dfrac{1}{8} + \dfrac{-4}{1} \overset{?}{=} \dfrac{3}{2}$

$-\dfrac{1}{8} + \dfrac{-4(8)}{1(8)} \overset{?}{=} \dfrac{3(4)}{2(4)}$

$-\dfrac{1}{8} - \dfrac{32}{8} \overset{?}{=} \dfrac{12}{8}$

$-\dfrac{33}{8} \neq \dfrac{12}{8}$

$-\dfrac{1}{2}$ *is not* a solution.

**(d)** $\dfrac{4}{4} + \dfrac{2}{4} \overset{?}{=} \dfrac{3}{2}$

$1 + \dfrac{1}{2} \overset{?}{=} \dfrac{3}{2}$

$\dfrac{3}{2} = \dfrac{3}{2}$

4 *is* a solution.

**16.** $5 + \dfrac{t}{3} = t + 2$

$3\left(5 + \dfrac{t}{3}\right) = 3(t + 2)$

$3(5) + 3\left(\dfrac{t}{3}\right) = 3(t + 2)$

$15 + t = 3t + 6$

$15 = 2t + 6$

$9 = 2t$

$\dfrac{9}{2} = t$

**17.** $\dfrac{5}{x + 1} - \dfrac{1}{x} = \dfrac{6}{x}$

$x(x + 1)\left(\dfrac{5}{x + 1} - \dfrac{1}{x}\right) = \left(\dfrac{6}{x}\right)(x)(x + 1)$

$5x - (x + 1) = 6(x + 1), \quad x \neq 0, x \neq -1$

$5x - x - 1 = 6x + 6$

$4x - 1 = 6x + 6$

$-1 = 2x + 6$

$-7 = 2x$

$-\dfrac{7}{2} = x$

**18.**

$$\frac{x-3}{x-2} + \frac{x+1}{x+3} = \frac{2x^2-15}{x^2+x-6}$$

$$(x-2)(x+3)\left(\frac{x-3}{x-2} + \frac{x+1}{x+3}\right) = \left[\frac{2x^2-15}{(x-2)(x+3)}\right](x-2)(x+3)$$

$$(x+3)(x-3) + (x-2)(x+1) = 2x^2-15, \quad x \neq 2, x \neq -3$$

$$x^2 - 9 + x^2 - x - 2 = 2x^2 - 15$$

$$2x^2 - x - 11 = 2x^2 - 15$$

$$-x - 11 = -15$$

$$-x = -4$$

$$x = 4$$

**19.** *Verbal Model:* $\boxed{\text{Time for van}} = \boxed{\text{Time for car}}$

*Labels:*

Van: Distance = 54 (miles)

Rate = $x - 6$ (miles per hour)

Time = $\dfrac{54}{x-6}$ (hours)

Car: Distance = 63 (miles)

Rate = $x$ (miles per hour)

Time = $\dfrac{63}{x}$ (hours)

*Equation:*

$$\frac{54}{x-6} = \frac{63}{x}$$

$$63(x-6) = 54x, \quad x \neq 0, x \neq 6 \quad \text{Cross-multiply.}$$

$$63x - 378 = 54x$$

$$9x - 378 = 0$$

$$9x = 378$$

$$x = \frac{378}{9}$$

$$x = 42 \text{ and } x - 6 = 36$$

The van travels at an average speed of 36 miles per hour, and the car travels at an average speed of 42 miles per hour.

# C H A P T E R  8
# Systems of Linear Equations and Inequalities

# CHAPTER 8
## Systems of Linear Equations and Inequalities

### Section 8.1  Solving Systems of Equations by Graphing

**1.** (a)  $(2, 3)$

$$2 + 3(3) \stackrel{?}{=} 11$$
$$2 + 9 = 11$$

$$-2 + 3(3) \stackrel{?}{=} 7$$
$$-2 + 9 = 7$$

$(2, 3)$ *is* a solution.

(b)  $(5, 4)$

$$5 + 3(4) \stackrel{?}{=} 11$$
$$5 + 12 \neq 11$$

$$-5 + 3(4) \stackrel{?}{=} 7$$
$$-5 + 12 = 7$$

$(5, 4)$ is *not* a solution.

**3.** (a)  $(5, -3)$

$$2(5) - 3(-3) \stackrel{?}{=} -8$$
$$10 + 9 \neq -8$$

$$5 + (-3) \stackrel{?}{=} 1$$
$$5 - 3 \neq 1$$

$(5, -3)$ is *not* a solution.

(b)  $(-1, 2)$

$$2(-1) - 3(2) \stackrel{?}{=} -8$$
$$-2 - 6 = -8$$

$$-1 + 2 \stackrel{?}{=} 1$$
$$-1 + 2 = 1$$

$(-1, 2)$ *is* a solution.

**5.** (a)  $(0, -1)$

$$5(0) - 3(-1) \stackrel{?}{=} 3$$
$$0 + 3 = 3$$

$$-10(0) + 6(-1) \stackrel{?}{=} -6$$
$$0 - 6 = -6$$

$(0, -1)$ *is* a solution.

(b)  $(3, 4)$

$$5(3) - 3(4) \stackrel{?}{=} 3$$
$$15 - 12 = 3$$

$$-10(3) + 6(4) \stackrel{?}{=} -6$$
$$-30 + 24 = -6$$

$(3, 4)$ *is* a solution.

**7.** The graph indicates that the solution is $(2, 0)$.

**9.** The graph indicates that the solution is $(-1, -1)$.

**11.** The graph indicates that the system has infinitely many solutions. The two lines coincide so there are infinitely many points of intersection.

**13.** The graph indicates that the solution is $(-12, -10)$.

**15.** The graph indicates that the system has no solution. The lines are parallel, so there is no point of intersection.

**17.** The solution is $(1, 2)$.

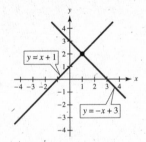

**19.** The solution is $(2, 0)$.

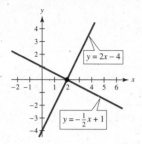

**21.** The solution is $(3, 0)$.

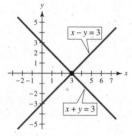

**23.** There is no solution. The system is inconsistent.

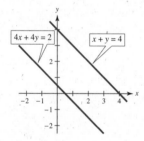

**25.** The solution is $(7, -2)$.

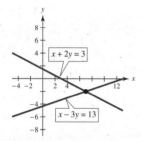

**27.** The solution is $(2, -1)$.

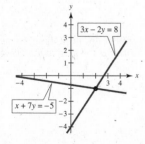

**29.** The solution is $(1, 1)$.

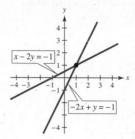

**31.** There are infinitely many solutions. The solution set consists of all ordered pairs $(x, y)$ such that

$$x - 2y = 4.$$

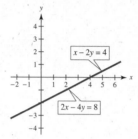

**33.** The solution is $(0, -3)$.

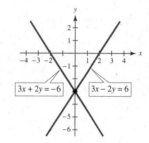

**35.** The solution is $(5, 4)$.

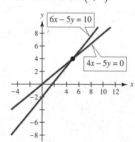

**37.** There are infinitely many solutions. The solution set consists of all ordered pairs $(x, y)$ such that

$$2x + 5y = 5.$$

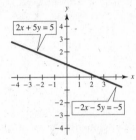

**39.** There are infinitely many solutions. The solution set consists of all ordered pairs $(x, y)$ such that

$$4x + 5y = 20.$$

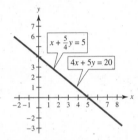

**41.** There is no solution. The system is inconsistent.

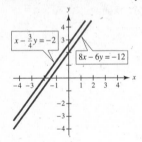

**43.** $y_1 = 2x - 1$

$y_2 = -3x + 9$

The solution is $(2, 3)$.

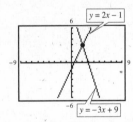

**45.** $y_1 = x - 1$

$y_2 = -2x + 8$

The solution is $(3, 2)$.

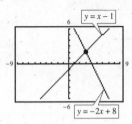

**47.** $2x - 3y = -12$ $\qquad$ $-8x + 12y = -12$

$\qquad -3y = -2x - 12$ $\qquad$ $12y = 8x - 12$

$\qquad y = \frac{2}{3}x + 4$ $\qquad$ $y = \frac{2}{3}x - 1$

$\qquad m = \frac{2}{3}$ $\qquad$ $m = \frac{2}{3}$

These two lines are parallel; they have the *same* slope and *different* y-intercepts. So, the system has *no* solution. The system is inconsistent.

**49.** $-x + 4y = 7$ $\qquad$ $3x - 12y = -21$

$\qquad 4y = x + 7$ $\qquad$ $-12y = -3x - 21$

$\qquad y = \frac{1}{4}x + \frac{7}{4}$ $\qquad$ $y = \frac{1}{4}x + \frac{7}{4}$

$\qquad m = \frac{1}{4}$ $\qquad$ $m = \frac{1}{4}$

These two lines coincide; they have the *same* slope and the *same* y-intercept. So, the system has *infinitely many* solutions.

**51.** $-2x + 3y = 4$ $\qquad$ $2x + 3y = 8$

$\qquad 3y = 2x + 4$ $\qquad$ $3y = -2x + 8$

$\qquad y = \frac{2}{3}x + \frac{4}{3}$ $\qquad$ $y = -\frac{2}{3}x + \frac{8}{3}$

$\qquad m = \frac{2}{3}$ $\qquad$ $m = -\frac{2}{3}$

These two lines do *not* have the same slope, so the two lines intersect. The system has *one* solution.

**53.** $-6x + 8y = 9$ $\qquad$ $3x - 4y = -6$

$\qquad 8y = 6x + 9$ $\qquad$ $-4y = -3x - 6$

$\qquad y = \frac{6}{8}x + \frac{9}{8}$ $\qquad$ $y = \frac{3}{4}x + \frac{6}{4}$

$\qquad y = \frac{3}{4}x + \frac{9}{8}$ $\qquad$ $y = \frac{3}{4}x + \frac{3}{2}$

$\qquad m = \frac{3}{4}$ $\qquad$ $m = \frac{3}{4}$

These two lines are parallel; they have the *same* slope and *different* y-intercepts. So, the system has no solution. The system is inconsistent.

**55.** The solution is $(11, 9)$.

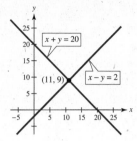

**57.** (a)  *Verbal Model:*

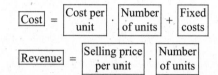

$$\boxed{\text{Cost}} = \boxed{\begin{array}{c}\text{Cost per}\\\text{unit}\end{array}} \cdot \boxed{\begin{array}{c}\text{Number}\\\text{of units}\end{array}} + \boxed{\begin{array}{c}\text{Fixed}\\\text{costs}\end{array}}$$

$$\boxed{\text{Revenue}} = \boxed{\begin{array}{c}\text{Selling price}\\\text{per unit}\end{array}} \cdot \boxed{\begin{array}{c}\text{Number}\\\text{of units}\end{array}}$$

   *Labels:*

Cost $= C$

Cost per unit $=$ \$16.75

Number of units $= x$

Fixed costs $=$ \$400.00

Revenue $= R$

Selling price per units $=$ \$23

   *Equations:*
$$\begin{cases} C = 16.75x + 400 \\ R = 23x \end{cases}$$

(b)

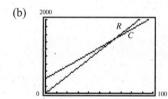

The graphs intersect at $(64, 1472)$. The solution indicates that the break-even point is reached when 64 feeders are produced and sold. Both the cost and the revenue are \$1472 when $x$ is 64, so the sales cover the costs. Sales over 64 feeders will generate profit.

**59.**  $x - 200y = -200$       $x - 199y = 198$

$-200y = -x - 200$       $-199y = -x + 198$

$y = \frac{1}{200}x + 1$       $y = \frac{1}{199}x - \frac{198}{199}$

$m = \frac{1}{200}$       $m = \frac{1}{199}$

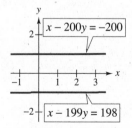

The graphical method is difficult to apply because the slopes of the two lines are nearly equal, and therefore, the lines *appear* to be parallel on the portion of the graph pictured. The point where the lines intersect is far from this portion of the graph. (Using techniques found in the next sections of this chapter, we can determine that the solution of this system is $(79,400, 398)$; these are the coordinates of the point of intersection of the two lines.)

**61.** False. A consistent system has either one solution or infinitely many solutions.

**63.** *Sample answers:*

$$\begin{cases} x + y = 0 \\ x + y = 1 \end{cases} \qquad \begin{cases} 2x - 3y = 5 \\ -4x + 6y = 7 \end{cases}$$

**65.** $\dfrac{2}{3} + \dfrac{1}{3} = \dfrac{2 + 1}{3} = \dfrac{3}{3} = 1$

**67.** $\dfrac{2}{7} - \dfrac{1}{3} = \dfrac{2(3)}{7(3)} - \dfrac{1(7)}{3(7)} = \dfrac{6 - 7}{21} = -\dfrac{1}{21}$

**69.**
$$x - 6 = 5x$$
$$x - x - 6 = 5x - x$$
$$-6 = 4x$$
$$\dfrac{-6}{4} = \dfrac{4x}{4}$$
$$-\dfrac{3}{2} = x$$

**71.**
$$y - 3(4y - 2) = 1$$
$$y - 12y + 6 = 1$$
$$-11y + 6 - 6 = 1 - 6$$
$$-11y = -5$$
$$y = \dfrac{-5}{-11}$$
$$y = \dfrac{5}{11}$$

**73.**
$$\dfrac{x}{5} + \dfrac{2x}{5} = 3$$
$$5\left(\dfrac{x}{5} + \dfrac{2x}{5}\right) = 3 \cdot 5$$
$$x + 2x = 15$$
$$3x = 15$$
$$x = 5$$

**75.**
$$\dfrac{x - 3}{x + 1} = \dfrac{4}{3}$$
$$3(x - 3) = 4(x + 1), \quad x \neq -1 \quad \text{Cross-multiply}$$
$$3x - 9 = 4x + 4$$
$$3x - 4x - 9 = 4x - 4x + 4$$
$$-x - 9 + 9 = 4 + 9$$
$$-x = 13$$
$$x = -13$$

## Section 8.2   Solving Systems of Equations by Substitution

**1.**
$$\begin{cases} x - y = 0 \Rightarrow x = y \\ x + y = 2 \end{cases}$$
$$y + y = 2 \qquad \text{Replace } x \text{ by } y \text{ in second equation.}$$
$$2y = 2$$
$$y = 1 \Rightarrow x = 1 \qquad \text{Replace } y \text{ by 1 in revised first equation.}$$
$$(1, 1)$$

**3.**
$$\begin{cases} -2x + y = 4 \\ -x + y = 3 \Rightarrow y = x + 3 \end{cases}$$
$$-2x + x + 3 = 4 \qquad \text{Replace } y \text{ by } x + 3 \text{ in first equation.}$$
$$-x + 3 = 4$$
$$-x = 1$$
$$x = -1 \Rightarrow y = -1 + 3 \qquad \text{Replace } x \text{ by } -1 \text{ in revised second equation.}$$
$$y = 2$$
$$(-1, 2)$$

**5.** $\begin{cases} -x + y = 1 \Rightarrow y = 1 + x \\ x - y = 1 \end{cases}$

$\qquad x - (1 + x) = 1$  $\qquad\qquad\qquad$ Replace $y$ by $1 + x$ in second equation.

$\qquad x - 1 - x = 1$

$\qquad\qquad -1 = 1$ $\qquad\qquad\qquad$ False

The system has *no* solution. The system is inconsistent.

**7.** $\begin{cases} 2x - y = 2 \Rightarrow -y = 2 - 2x \Rightarrow y = -2 + 2x \\ 4x + 3y = 9 \end{cases}$

$\qquad 4x + 3(-2 + 2x) = 9$  $\qquad\qquad\qquad$ Replace $y$ by $-2 + 2x$ in second equation.

$\qquad 4x - 6 + 6x = 9$

$\qquad\quad 10x - 6 = 9$

$\qquad\qquad 10x = 15$

$\qquad\qquad x = \frac{3}{2} \Rightarrow y = -2 + 2\left(\frac{3}{2}\right)$ $\qquad$ Replace $x$ by $\frac{2}{3}$ in revised first equation.

$\qquad\qquad\qquad\qquad y = -2 + 3$

$\qquad\qquad\qquad\qquad y = 1$

$\left(\frac{3}{2}, 1\right)$

**9.** $\begin{cases} 2x + y = 3 \Rightarrow y = -2x + 3 \\ 4x + 2y = 6 \end{cases}$

$\qquad 4x + 2(-2x + 3) = 6$ $\qquad\qquad$ Replace $y$ with $-2x + 3$.

$\qquad 4x - 4x + 6 = 6$

$\qquad\qquad\qquad 6 = 6$

There are infinitely many solutions. The solution set consists of all points $(x, y)$ such that $2x + y = 3$.

**11.** $\begin{cases} y = 2x - 1 \\ y = -x + 5 \end{cases}$

$\qquad 2x - 1 = -x + 5$ $\qquad\qquad$ Replace $y$ by $2x - 1$ in second equation.

$\qquad 3x - 1 = 5$

$\qquad\quad 3x = 6$

$\qquad\quad x = 2 \Rightarrow y = 2(2) - 1$ $\qquad$ Replace $x$ by $2$ in first equation.

$\qquad\qquad\qquad y = 3$

$(2, 3)$

**13.** $\begin{cases} x = 4y - 5 \\ x = 3y \end{cases}$

$\qquad 4y - 5 = 3y$ $\qquad\qquad\qquad$ Replace $x$ by $4y - 5$ in second equation.

$\qquad y - 5 = 0$

$\qquad\quad y = 5 \Rightarrow x = 3(5)$ $\qquad$ Replace $y$ by $5$ in second equation.

$\qquad\qquad\quad x = 15$

$(15, 5)$

**15.** $\begin{cases} 2x \quad\quad = \quad 8 \Rightarrow x = 4 \\ x - 2y = 12 \end{cases}$

$\qquad 4 - 2y = 12$ $\qquad\qquad\qquad$ Replace $x$ by 4 in second equation.

$\qquad\quad -2y = 8$

$\qquad\qquad y = -4$

$(4, -4)$

**17.** $\begin{cases} x - y = 0 \Rightarrow x = y \\ 2x + y = 0 \end{cases}$

$\qquad 2(y) + y = 0$ $\qquad\qquad\qquad$ Replace $x$ by $y$ in second equation.

$\qquad 2y + y = 0$

$\qquad\quad 3y = 0$

$\qquad\quad y = 0 \Rightarrow x = 0$ $\qquad$ Replace $y$ by 0 in revised first equation.

$(0, 0)$

**19.** $\begin{cases} x - 2y = -10 \Rightarrow x = 2y - 10 \\ 3x - \ y = \ 0 \end{cases}$

$\qquad 3(2y - 10) - y = 0$ $\qquad\qquad\qquad$ Replace $x$ by $2y - 10$ in second equation.

$\qquad\quad 6y - 30 - y = 0$

$\qquad\qquad 5y - 30 = 0$

$\qquad\qquad\quad 5y = 30$

$\qquad\qquad\quad y = 6 \Rightarrow x = 2(6) - 10$ $\qquad$ Replace $y$ by 6 in revised first equation.

$\qquad\qquad\qquad\qquad\quad x = 12 - 10$

$\qquad\qquad\qquad\qquad\quad x = 2$

$(2, 6)$

**21.** $\begin{cases} 2x - y = -2 \\ 4x + y = \ \ 5 \Rightarrow y = -4x + 5 \end{cases}$

$\qquad 2x - (-4x + 5) = -2$ $\qquad\qquad\qquad$ Replace $y$ by $-4x + 5$ in first equation.

$\qquad 2x + 4x - 5 = -2$

$\qquad\qquad 6x - 5 = -2$

$\qquad\qquad\quad 6x = 3$

$\qquad\qquad\quad x = \frac{3}{6}$

$\qquad\qquad\quad x = \frac{1}{2} \Rightarrow y = -4\left(\frac{1}{2}\right) + 5$ $\qquad$ Replace $x$ by $\frac{1}{2}$ in revised second equation.

$\qquad\qquad\qquad\qquad\quad y = -2 + 5$

$\qquad\qquad\qquad\qquad\quad y = 3$

$\left(\frac{1}{2}, 3\right)$

**23.** $\begin{cases} x + 2y = \phantom{-}1 \Rightarrow x = -2y + 1 \\ 5x - 4y = -23 \end{cases}$

$5(-2y + 1) - 4y = -23$           Replace $x$ by $-2y + 1$ in second equation.

$-10y + 5 - 4y = -23$

$-14y + 5 = -23$

$-14y = -28$

$y = 2 \Rightarrow x = -2(2) + 1$      Replace $y$ by 2 in revised first equation.

$x = -4 + 1$

$x = -3$

$(-3, 2)$

**25.** $\begin{cases} \phantom{-}8x + 4y = -2 \Rightarrow 4y = -8x - 2 \Rightarrow y = -2x - \dfrac{1}{2} \\ -12x + 5y = -8 \end{cases}$

$-12x + 5\left(-2x - \dfrac{1}{2}\right) = -8$         Replace $y$ by $-2x - \dfrac{1}{2}$ in second equation.

$-12x - 10x - \dfrac{5}{2} = -8$

$-22x - \dfrac{5}{2} = -8$

$2\left(-22x - \dfrac{5}{2}\right) = (-8)2$

$-44x - 5 = -16$

$-44x = -11$

$x = \dfrac{-11}{-44}$

$x = \dfrac{1}{4} \Rightarrow y = -2\left(\dfrac{1}{4}\right) - \dfrac{1}{2}$    Replace $x$ by $\dfrac{1}{4}$ in revised first equation.

$y = -\dfrac{1}{2} - \dfrac{1}{2}$

$y = -1$

$\left(\dfrac{1}{4}, -1\right)$

**27.** $\begin{cases} 5x + 3y = 11 \\ \phantom{5}x - 5y = \phantom{1}5 \Rightarrow x = 5y + 5 \end{cases}$

$5(5y + 5) + 3y = 11$          Replace $x$ by $5y + 5$ in first equation.

$25y + 25 + 3y = 11$

$28y + 25 = 11$

$28y = -14$

$y = -\dfrac{1}{2} \Rightarrow x = 5\left(-\dfrac{1}{2}\right) + 5$    Replace $y$ by $-\dfrac{1}{2}$ in revised second equation.

$x = -\dfrac{5}{2} + \dfrac{10}{2}$

$x = \dfrac{5}{2}$

$\left(\dfrac{5}{2}, -\dfrac{1}{2}\right)$

**29.** $\begin{cases} 5x + 2y = 0 \\ x - 3y = 0 \Rightarrow x = 3y \end{cases}$

$5(3y) + 2y = 0$       Replace $x$ with $3y$.

$15y + 2y = 0$

$17y = 0$

$y = 0 \Rightarrow x = 3(0)$    Replace $y$ with 0 in revised second equation.

$x = 0$

$(0, 0)$

**31.** $\begin{cases} 2x + 5y = -4 \\ 3x - y = 11 \Rightarrow -y = -3x + 11 \Rightarrow y = 3x - 11 \end{cases}$

$2x + 5(3x - 11) = -4$      Replace $y$ with $3x - 11$ in the first equation.

$2x + 15x - 55 = -4$

$17x - 55 = -4$

$17x = 51$

$x = 3 \Rightarrow y = 3(3) - 11$    Replace $x$ with 3 in revised second equation.

$y = -2$

$(3, -2)$

**33.** $\begin{cases} 4x - y = 2 \Rightarrow -y = -4x + 2 \Rightarrow y = 4x - 2 \\ 2x - \frac{1}{2}y = 1 \end{cases}$

$2x - \frac{1}{2}(4x - 2) = 1$     Replace $y$ by $4x - 2$ in second equation.

$2x - 2x + 1 = 1$

$1 = 1$

The system has *infinitely many* solutions. The solution set consists of all ordered pairs $(x, y)$ such that $4x - y = 2$.

**35.** $\begin{cases} \frac{1}{5}x + \frac{1}{2}y = 8 \\ 2x + y = 20 \Rightarrow y = -2x + 20 \end{cases}$

$\frac{1}{5}x + \frac{1}{2}(-2x + 20) = 8$     Replace $y$ by $-2x + 20$ in first equation.

$\frac{1}{5}x - x + 10 = 8$

$-\frac{4}{5}x + 10 = 8$

$-\frac{4}{5}x = -2$

$-4x = -10$

$x = \frac{-10}{-4}$

$x = \frac{5}{2} \Rightarrow y = -2\left(\frac{5}{2}\right) + 20$    Replace $x$ by $\frac{5}{2}$ in revised second equation.

$y = -5 + 20$

$y = 15$

$\left(\frac{5}{2}, 15\right)$

**37.** $\begin{cases} -5x + 4y = 14 \Rightarrow 4y = 5x + 14 \Rightarrow y = \frac{5}{4}x + \frac{7}{2} \\ 5x - 4y = 4 \end{cases}$

$5x - 4\left(\frac{5}{4}x + \frac{7}{2}\right) = 4$      Replace $y$ by $\frac{5}{4}x + \frac{7}{2}$ in second equation.

$5x - 5x - 14 = 4$

$-14 = 4$      False

The system has *no* solution. The system is inconsistent.

**39.** $\begin{cases} x - 4y = 2 \Rightarrow x = 4y + 2 \\ 5x + 2.5y = 10 \end{cases}$

$5(4y + 2) + 2.5y = 10$      Replace $x$ by $4y + 2$ in second equation.

$20y + 10 + 2.5y = 10$

$22.5y + 10 = 10$

$22.5y = 0$

$y = 0 \Rightarrow x = 4(0) + 2$      Replace $y$ by 0 in revised first equation.

$x = 0 + 2$

$x = 2$

$(2, 0)$

**41.** $\begin{cases} -6x + 1.5y = 6 \\ 8x - 2y = -8 \Rightarrow -2y = -8x - 8 \Rightarrow y = 4x + 4 \end{cases}$

$-6x + 1.5(4x + 4) = 6$      Replace $y$ by $4x + 4$ in first equation.

$-6x + 6x + 6 = 6$

$6 = 6$

The system has *infinitely many* solutions. The solution set consists of all ordered pairs $(x, y)$ such that $y = 4x + 4$.

**43.** $\begin{cases} \dfrac{x}{3} - \dfrac{y}{4} = 2 \Rightarrow 4x - 3y = 24 \\ \dfrac{x}{2} + \dfrac{y}{6} = 3 \Rightarrow 3x + y = 18 \Rightarrow y = -3x + 18 \end{cases}$

$4x - 3(-3x + 18) = 24$      Replace $y$ by $-3x + 18$ in revised first equation.

$4x + 9x - 54 = 24$

$13x - 54 = 24$

$13x = 78$

$x = 6 \Rightarrow y = -3(6) + 18$      Replace $x$ by 6 in revised second equation.

$y = -18 + 18$

$y = 0$

$(6, 0)$

**45.** $\begin{cases} \dfrac{x}{4} + \dfrac{y}{2} = 1 \Rightarrow x + 2y = 4 \Rightarrow x = -2y + 4 \\ \dfrac{x}{2} - \dfrac{y}{3} = 1 \Rightarrow 3x - 2y = 6 \end{cases}$

$3(-2y + 4) - 2y = 6$         Replace $x$ by $-2y + 4$ in revised second equation.

$-6y + 12 - 2y = 6$

$-8y + 12 = 6$

$-8y = -6$

$y = \dfrac{-6}{-8}$

$y = \dfrac{3}{4} \Rightarrow x = -2\left(\dfrac{3}{4}\right) + 4$     Replace $y$ by $\dfrac{3}{4}$ in revised first equation.

$x = -\dfrac{3}{2} + \dfrac{8}{2}$

$x = \dfrac{5}{2}$

$\left(\dfrac{5}{2}, \dfrac{3}{4}\right)$

**47.** $\begin{cases} 2(x - 5) = y + 2 \Rightarrow 2x - 10 = y + 2 \Rightarrow 2x - 12 = y \\ 3x = 4(y + 2) \Rightarrow 3x = 4y + 8 \end{cases}$

$3x = 4(2x - 12) + 8$         Replace $y$ by $2x - 12$ in revised second equation.

$3x = 8x - 48 + 8$

$3x = 8x - 40$

$-5x = -40$

$x = 8 \Rightarrow y = 2(8) - 12$     Replace $x$ by $8$ in revised first equation.

$y = 16 - 12$

$y = 4$

$(8, 4)$

**49.** $\begin{cases} y = \frac{1}{4}x + \frac{19}{4} \\ y = \frac{8}{5}x - 2 \end{cases}$

$\frac{1}{4}x + \frac{19}{4} = \frac{8}{5}x - 2$   Replace $y$ by $\frac{1}{4}x + \frac{19}{4}$ in second equation.

$20\left(\frac{1}{4}x + \frac{19}{4}\right) = 20\left(\frac{8}{5}x - 2\right)$

$5x + 95 = 32x - 40$

$95 = 27x - 40$

$135 = 27x$

$\frac{135}{27} = x$

$5 = x \Rightarrow y = \frac{1}{4} \cdot 5 + \frac{19}{4}$   Replace $x$ by 5 in first equation.

$\qquad\qquad\quad y = \frac{5}{4} + \frac{19}{4}$

$\qquad\qquad\quad y = \frac{24}{4}$

$\qquad\qquad\quad y = 6$

$(5, 6)$

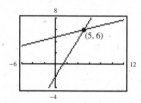

**51.** $\begin{cases} 3x + 2y = 12 \\ x - y = 3 \Rightarrow x = y + 3 \end{cases}$

$3(y + 3) + 2y = 12$   Replace $x$ by $y + 3$ in first equation.

$3y + 9 + 2y = 12$

$5y + 9 = 12$

$5y = 3$

$y = \frac{3}{5} \Rightarrow x = \frac{3}{5} + 3$   Replace $y$ by $\frac{3}{5}$ in revised second equation.

$\qquad\qquad\quad x = \frac{3}{5} + \frac{15}{5}$

$\qquad\qquad\quad x = \frac{18}{5}$

$\left(\frac{18}{5}, \frac{3}{5}\right)$

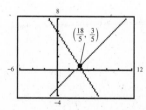

**53.** $\begin{cases} 5x + 3y = 15 \\ 2x - 3y = 6 \end{cases} \Rightarrow -3y = -2x + 6$

$$y = \frac{-2x}{-3} + \frac{6}{-3}$$

$$y = \frac{2}{3}x - 2$$

$5x + 3\left(\dfrac{2}{3}x - 2\right) = 15$       Replace $y$ by $\dfrac{2}{3}x - 2$ in first equation.

$5x + 2x - 6 = 15$

$7x - 6 = 15$

$7x = 21$

$x = 3 \Rightarrow y = \dfrac{2}{3}(3) - 2$       Replace $x$ by 3 in revised second equation.

$$y = 2 - 2$$

$$y = 0$$

$(3, 0)$

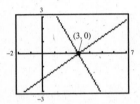

**55.** $(2, 1)$

*Sample answers:*

$\begin{cases} x - 2y = 0 \\ x + y = 3 \end{cases}$       $\begin{cases} 3x + 2y = 8 \\ 5x - 4y = 6 \end{cases}$

**57.** $\left(\frac{3}{5}, -3\right)$

*Sample answers:*

$\begin{cases} -5x - y = 0 \\ 5x + 2y = -3 \end{cases}$       $\begin{cases} 10x + y = 3 \\ 5x + 4y = -9 \end{cases}$

**59.** $\begin{cases} x + y = 40 \Rightarrow x = -y + 40 \\ x - y = 10 \end{cases}$

$(-y + 40) - y = 10$       Replace $x$ by $-y + 40$ in second equation.

$-y + 40 - y = 10$

$-2y + 40 = 10$

$-2y = -30$

$y = 15 \Rightarrow x = -15 + 40$       Replace $y$ by 15 in revised first equation.

$x = 25$

$(25, 15)$

The two numbers are 25 and 15.

**61.** $\begin{cases} x + y = 15{,}000 \Rightarrow x = -y + 15{,}000 \\ 0.05x + 0.08y = 900 \end{cases}$

$0.05(-y + 15{,}000) + 0.08y = 900$         Replace $x$ by $-y + 15{,}000$ in second equation.

$-0.05y + 750 + 0.08y = 900$

$0.03y + 750 = 900$

$0.03y = 150$

$y = \dfrac{150}{0.03}$

$y = 5000 \Rightarrow x = -5000 + 15{,}000$         Replace $y$ by 5000 in revised first equation.

$x = 10{,}000$

$(10{,}000,\ 5000)$

So, \$10,000 is invested at 5%, and \$5000 is invested at 8%.

**63.** $\begin{cases} x + y = 6 \Rightarrow x = 6 - y \\ 16.95x + 7.50y = 63.90 \end{cases}$

$16.95(6 - y) + 7.50y = 63.90$

$101.70 - 16.95y + 7.50y = 63.90$

$101.70 - 9.45y = 63.90$

$-9.45y = -37.80$

$y = \dfrac{-37.80}{-9.45}$

$y = 4 \Rightarrow x = 6 - 4$

$x = 2$

So, 2 adults attended the dinner.

**65.** $\begin{cases} y = 18{,}000 + 0.26x \\ y = 22{,}000 + 0.22x \end{cases}$

$18{,}000 + 0.26x = 22{,}000 + 0.22x$         Replace $y$ by $18{,}000 + 0.26x$ in second equation.

$18{,}000 + 0.04x = 22{,}000$

$0.04x = 4000$

$x = \dfrac{4000}{0.04}$

$x = 100{,}000$

The costs of the two models will be the same after they are driven 100,000 miles.

**67.** First, find the point of intersection of the two lines:

$$\begin{cases} x - 2y = 3 \\ 3x + y = 16 \Rightarrow y = -3x + 16 \end{cases}$$

$$x - 2(-3x + 16) = 3 \qquad\qquad \text{Replace } y \text{ by } -3x + 16 \text{ in first equation.}$$

$$x + 6x - 32 = 3$$

$$7x - 32 = 3$$

$$7x = 35$$

$$x = 5 \Rightarrow y = -3(5) + 16 \qquad \text{Replace } x \text{ by } 5 \text{ in revised second equation.}$$

$$y = -15 + 16$$

$$y = 1$$

$$(5, 1)$$

Then, find the equation of the line through $(5, 1)$ with slope 2:

$$y - y_1 = m(x - x_1)$$

$$y - 1 = 2(x - 5)$$

$$y - 1 = 2x - 10$$

$$y = 2x - 9$$

The equation of the line in slope-intercept form is $y = 2x - 9$. The equation in general form is $2x - y - 9 = 0$.

**69.** The substitution method yields exact solutions.

**71.** $\begin{cases} x + by = 1 \Rightarrow x = -by + 1 \\ x + 2y = 2 \end{cases}$

$$-by + 1 + 2y = 2 \qquad\qquad \text{Replace } x \text{ by } -by + 1 \text{ in second equation.}$$

$$-by + 2y + 1 = 2$$

$$y(-b + 2) + 1 = 2$$

The variable $y$ will "drop out" of this equation when $-b + 2 = 0$.

$$-b + 2 = 0 \Rightarrow -b = -2 \Rightarrow b = 2$$

So, when $b = 2$, the equation becomes $1 = 2$, a *false* statement. The system is inconsistent when $b = 2$.

**73.** $\begin{cases} -6x + y = 4 \\ 2x + by = 3 \Rightarrow 2x = -by + 3 \Rightarrow x = -\dfrac{b}{2}y + \dfrac{3}{2} \end{cases}$

$$-6\left(-\frac{b}{2}y + \frac{3}{2}\right) + y = 4 \qquad \text{Replace } x \text{ by } -\frac{b}{2}y + \frac{3}{2} \text{ in first equation.}$$

$$3by - 9 + y = 4$$

$$3by + y - 9 = 4$$

$$y(3b + 1) - 9 = 4$$

The variable will "drop out" of this equation when $3b + 1 = 0$.

$$3b + 1 = 0 \Rightarrow 3b = -1 \Rightarrow b = -\frac{1}{3}$$

So, when $b = -\frac{1}{3}$, the equation becomes $-9 = 4$, which is a *false* statement. The system is inconsistent when $b = -\frac{1}{3}$.

**75.** $x(3 - x) - 2(3 - x) = (3 - x)(x - 2)$

**77.** $4y^2 - 20y + 25 = (2y)^2 - 2(2y)(5) + 5^2$
$$= (2y - 5)^2$$

**79.** $14 - 2x = x + 2$
$$14 - 3x = 2$$
$$-3x = -12$$
$$x = 4$$

**81.** $z^2 - 4z - 12 = 0$
$$(z - 6)(z + 2) = 0$$
$$z - 6 = 0 \Rightarrow z = 6$$
$$z + 2 = 0 \Rightarrow z = -2$$

**83.**   $3x + 4y = 10$         $3x + 4y = -1$
    $4y = -3x + 10$      $4y = -3x - 1$
    $y = -\frac{3}{4}x + \frac{10}{4}$       $y = -\frac{3}{4}x - \frac{1}{4}$
    $y = -\frac{3}{4}x + \frac{5}{2}$

The lines are parallel. The system has no solution; it is inconsistent.

**85.**   $5x + y = -3$         $x + 2y = -6$
    $y = -5x - 3$         $2y = -x - 6$
                $y = -\frac{1}{2}x - 3$

The solution is $(0, -3)$.

# Section 8.3   Solving Systems of Equations by Elimination

**1.**   $\begin{cases} 2x + y = 4 \\ x - y = 2 \end{cases}$
    $3x\ \ \ \ \ = 6$
    $x\ \ \ \ \ = 2$

$2 - y = 2$     Replace $x$ by 2 in second equation.
    $-y = 0$
     $y = 0$

$(2, 0)$

**3.** $\begin{cases} x - y = 0 \\ 3x - 2y = -1 \end{cases} \Rightarrow \begin{array}{l} -2x + 2y = 0 \\ \underline{3x - 2y = -1} \\ x \qquad = -1 \end{array}$   Multiply equation by $-2$.

$-1 - y = 0$   Replace $x$ by $-1$ in first equation.

$-y = 1$

$y = -1$

$(-1, -1)$

**5.** $\begin{cases} x - y = 4 \\ x + y = 12 \end{cases}$

$2x \qquad = 16$

$x \qquad = 8$

$8 + y = 12$   Replace $x$ by 8 in second equation.

$y = 4$

$(8, 4)$

**9.** $\begin{cases} 3x - 5y = 1 \\ 2x + 5y = 9 \end{cases}$

$5x \qquad = 10$

$x \qquad = 2$

$3(2) - 5y = 1$   Replace $x$ by 2 in first equation.

$6 - 5y = 1$

$-5y = -5$

$y = 1$

$(2, 1)$

**7.** $\begin{cases} -x + 2y = 12 \\ x + 6y = 20 \end{cases}$

$8y = 32$

$y = 4$

$-x + 2(4) = 12$   Replace $y$ by 4 in first equation.

$-x + 8 = 12$

$-x = 4$

$x = -4$

$(-4, 4)$

**11.** $\begin{cases} 2a + 5b = 3 \\ 2a + b = 9 \end{cases} \Rightarrow \begin{array}{l} 2a + 5b = 3 \\ \underline{-2a - b = -9} \\ 4b = -6 \end{array}$   Multiply equation by $-1$.

$b = \dfrac{-6}{4}$

$b = -\dfrac{3}{2}$

$2a - \dfrac{3}{2} = 9$   Replace $b$ by $-\dfrac{3}{2}$ in second equation.

$2\left(2a - \dfrac{3}{2}\right) = 9 \cdot 2$

$4a - 3 = 18$

$4a = 21$

$a = \dfrac{21}{4}$

$\left(\dfrac{21}{4}, -\dfrac{3}{2}\right)$

**13.** $\begin{cases} -x + 2y = 6 \Rightarrow -2x + 4y = 12 \\ 2x + 5y = 6 \Rightarrow \underline{\phantom{-}2x + 5y = \phantom{0}6} \end{cases}$   Multiply equation by 2.

$$9y = 18$$
$$y = \phantom{0}2$$

$-x + 2(2) = 6$   Replace $y$ by 2 in first equation.

$$-x + 4 = 6$$
$$-x = 2$$
$$x = -2$$

$(-2, 2)$

**15.** $\begin{cases} 3x - 4y = 11 \Rightarrow 9x - 12y = \phantom{-}33 \\ 2x + 3y = -4 \Rightarrow \underline{8x + 12y = -16} \end{cases}$   Multiply equation by 3.
   Multiply equation by 4.

$$17x \phantom{0000} = \phantom{0}17$$
$$x \phantom{00000} = \phantom{00}1$$

$2(1) + 3y = -4$   Replace $x$ with 1 in second equation.

$$2 + 3y = -4$$
$$3y = -6$$
$$y = -2$$

$(1, -2)$

**17.** $\begin{cases} \phantom{-}3x + 2y = -1 \Rightarrow \phantom{-}6x + \phantom{0}4y = -2 \\ -2x + 7y = \phantom{-}9 \Rightarrow \underline{-6x + 21y = 27} \end{cases}$   Multiply equation by 2.
   Multiply equation by 3.

$$25y = 25$$
$$y = \phantom{0}1$$

$3x + 2(1) = -1$   Replace $y$ with 1 in first equation.

$$3x + 2 = -1$$
$$3x = -3$$
$$x = -1$$

$(-1, 1)$

**19.** $\begin{cases} 3x - 4y = 1 \Rightarrow \phantom{0}9x - 12y = 3 \\ 4x + 3y = 1 \Rightarrow \underline{16x + 12y = 4} \end{cases}$   Multiply equation by 3.
   Multiply equation by 4.

$$25x \phantom{0000} = 7$$
$$x \phantom{00000} = \tfrac{7}{25}$$

$4\left(\tfrac{7}{25}\right) + 3y = 1$   Replace $x$ with $\tfrac{7}{25}$ in second equation.

$$\tfrac{28}{25} + 3y = 1$$
$$25\left(\tfrac{28}{25} + 3y\right) = 25(1)$$
$$28 + 75y = 25$$
$$75y = -3$$
$$y = -\tfrac{3}{75}$$
$$y = -\tfrac{1}{25}$$

$\left(\tfrac{7}{25}, -\tfrac{1}{25}\right)$

**21.** $\begin{cases} 3x + 2y = 10 \\ 2x + 5y = 3 \end{cases} \Rightarrow \begin{array}{r} 6x + 4y = 20 \\ -6x - 15y = -9 \end{array}$  Multiply equation by 2.
Multiply equation by −3.

$$\begin{array}{r} -11y = 11 \\ y = -1 \end{array}$$

$3x + 2(-1) = 10$  Replace $y$ with −1 in first equation.

$$3x - 2 = 10$$
$$3x = 12$$
$$x = 4$$

$(4, -1)$

**23.** $\begin{cases} 5u + 6v = 14 \\ 3u + 5v = 7 \end{cases} \Rightarrow \begin{array}{r} 15u + 18v = 42 \\ -15u - 25v = -35 \end{array}$  Multiply equation by 3.
Multiply equation by −5.

$$\begin{array}{r} -7v = 7 \\ v = -1 \end{array}$$

$5u + 6(-1) = 14$  Replace $v$ by −1 in first equation.

$$5u - 6 = 14$$
$$5u = 20$$
$$u = 4$$

$(4, -1)$

**25.** $\begin{cases} 2x - 4y = 1 \\ 5x - 10y = -3 \end{cases} \Rightarrow \begin{array}{r} 10x - 20y = 5 \\ -10x + 20y = 6 \end{array}$  Multiply equation by 5.
Multiply equation by −2.

$$0 = 11 \qquad \text{False}$$

The system has *no* solution; it is inconsistent.

**27.** $\begin{cases} -3x - 12y = 3 \\ 5x + 20y = -5 \end{cases} \Rightarrow \begin{array}{r} -x - 4y = 1 \\ x + 4y = -1 \end{array}$  Multiply equation by $\frac{1}{3}$.
Multiply equation by $\frac{1}{5}$.

$$0 = 0$$

The system has *infinitely many* solutions. The solution set consists of all ordered pairs $(x, y)$ such that $x + 4y = -1$.

**29.** $\begin{cases} 6r + 5s = 3 \\ \frac{3}{2}r - \frac{5}{4}s = \frac{3}{4} \end{cases} \Rightarrow \begin{array}{r} 6r + 5s = 3 \\ -6r + 5s = -3 \end{array}$  
Multiply equation by −4.

$$\begin{array}{r} 10s = 0 \\ s = 0 \end{array}$$

$6r + 5(0) = 3$  Replace $s$ by 0 in first equation.

$$6r = 3$$
$$r = \tfrac{1}{2}$$

$\left(\tfrac{1}{2}, 0\right)$

**31.** $\begin{cases} \frac{1}{2}s - t = \frac{3}{2} \Rightarrow s - 2t = 3 \\ 4s + 2t = 27 \Rightarrow 4s + 2t = 27 \end{cases}$   Multiply equation by 2.

$$\underline{\phantom{4s + 2t = 27}}$$
$$5s \quad\quad = 30$$
$$s \quad\quad = 6$$

$4(6) + 2t = 27$   Replace $s$ by 6 in second equation.

$24 + 2t = 27$

$2t = 3$

$t = \frac{3}{2}$

$\left(6, \frac{3}{2}\right)$

**33.** $\begin{cases} 0.4a + 0.7b = 3 \Rightarrow -0.8a - 1.4b = -6 \\ 0.8a + 1.4b = 6 \Rightarrow 0.8a + 1.4b = 6 \end{cases}$   Multiply equation by $-2$.

$$\underline{\phantom{0.8a + 1.4b = 6}}$$
$$0 = 0$$

The system has *infinitely many* solutions. The solution set consists of all ordered pairs $(a, b)$ such that $4a + 7b = 30$.

**35.** $\begin{cases} 0.02x - 0.05y = -0.19 \\ 0.03x + 0.04y = 0.52 \end{cases}$

Multiply both equations by 100.

$2x - 5y = -19 \Rightarrow 8x - 20y = -76$   Multiply equation by 4.

$3x + 4y = 52 \Rightarrow 15x + 20y = 260$   Multiply equation by 5.

$$\underline{\phantom{15x + 20y = 260}}$$
$$23x \quad\quad = 184$$
$$x \quad\quad = 8$$

$2(8) - 5y = -19$   Replace $x$ by 8 in first revised equation.

$16 - 5y = -19$

$-5y = -35$

$y = 7$

$(8, 7)$

**37.** $\begin{cases} 3x + 2y = 10 \\ x - 2y = 14 \end{cases}$

$4x \quad\quad = 24$

$x \quad\quad = 6$

$6 - 2y = 14$   Replace $x$ by 6 in second equation.

$-2y = 8$

$y = -4$

$(6, -4)$

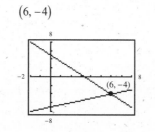

**39.** $\begin{cases} 7x + 8y = 6 \\ 3x - 4y = 10 \end{cases} \Rightarrow \begin{aligned} 7x + 8y &= 6 \\ 6x - 8y &= 20 \end{aligned}$   Multiply equation by 2.

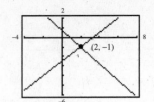

$$\overline{\hspace{2cm}}$$
$$13x \quad\;\; = 26$$
$$x \quad\;\;\; = 2$$

$7(2) + 8y = 6$   Replace $x$ by 2 in first equation.

$14 + 8y = 6$

$8y = -8$

$y = -1$

The solution is $(2, -1)$.

**41.** $\begin{cases} 5x + 2y = 7 \\ 3x - 6y = -3 \end{cases} \Rightarrow \begin{aligned} 15x + 6y &= 21 \\ 3x - 6y &= -3 \end{aligned}$   Multiply equation by 3.

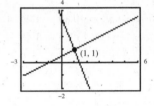

$$\overline{\hspace{2cm}}$$
$$18x \quad\;\;\; = 18$$
$$x \quad\;\;\;\; = 1$$

$5(1) + 2y = 7$   Replace $x$ by 1 in first equation.

$5 + 2y = 7$

$2y = 2$

$y = 1$

The solution is $(1, 1)$.

**43.** $\begin{cases} x - y = 2 \\ \phantom{x -} y = 3 \end{cases}$

$x - 3 = 2$   Replace $y$ by 3 in first equation.

$x = 5$

$(5, 3)$

**45.** $\begin{cases} 6x + 21y = 132 \\ 6x - 4y = 32 \end{cases} \Rightarrow \begin{aligned} 6x + 21y &= 132 \\ -6x + 4y &= -32 \end{aligned}$   Multiply equation by $-1$.

$$\overline{\hspace{2cm}}$$
$$25y = 100$$
$$y = 4$$

$6x + 21(4) = 132$   Replace $y$ by 4 in first equation.

$6x + 84 = 132$

$6x = 48$

$x = 8$

$(8, 4)$

**47.** $\begin{cases} -4x + 3y = 11 \Rightarrow -12x + 9y = 33 \\ \phantom{-}3x - 10y = 15 \Rightarrow \phantom{-}12x - 40y = 60 \end{cases}$   Multiply equation by 3.

Multiply equation by 4.

$$-31y = 93$$
$$y = -3$$

$-4x + 3(-3) = 11$   Replace $y$ by $-3$ in first equation.

$$-4x - 9 = 11$$
$$-4x = 20$$
$$x = -5$$

$(-5, -3)$

**49.** $\begin{cases} 0.1x - 0.1y = 0 \Rightarrow 3x - 3y = 0 \\ 0.8x + 0.3y = 1.5 \Rightarrow 8x + 3y = 15 \end{cases}$   Multiply equation by 30.

Multiply equation by 10.

$$11x \phantom{+ 3y} = 15$$
$$x \phantom{+ 3y} = \tfrac{15}{11}$$

$3\left(\tfrac{15}{11}\right) - 3y = 0$   Replace $x$ by $\tfrac{15}{11}$ in revised first equation.

$$\tfrac{45}{11} - 3y = 0$$
$$11\left(\tfrac{45}{11} - 3y\right) = 0 \cdot 11$$
$$45 - 33y = 0$$
$$-33y = -45$$
$$y = \tfrac{45}{33}$$
$$y = \tfrac{15}{11}$$

$\left(\tfrac{15}{11}, \tfrac{15}{11}\right)$

**51.** $\begin{cases} -\dfrac{x}{4} + y = 1 \Rightarrow -x + 4y = 4 \\ \dfrac{x}{4} + \dfrac{y}{2} = 1 \Rightarrow \phantom{-}x + 2y = 4 \end{cases}$   Multiply equation by 4.

Multiply equation by 4.

$$6y = 8$$
$$y = \frac{4}{3}$$

$x + 2\left(\dfrac{4}{3}\right) = 4$   Replace $y$ by $\dfrac{4}{3}$ in revised second equation.

$$x + \frac{8}{3} = 4$$
$$3x + 8 = 12$$
$$3x = 4$$
$$x = \frac{4}{3}$$

$\left(\dfrac{4}{3}, \dfrac{4}{3}\right)$

**53.** $\begin{cases} 3(x+5) - 7 = 2(3-2y) \Rightarrow 3x + 15 - 7 = 6 - 4y \Rightarrow 3x + 4y = -2 \\ \quad\quad 2x + 1 = 4(y+2) \Rightarrow \quad\quad 2x + 1 = 4y + 8 \Rightarrow 2x - 4y = \phantom{-}7 \end{cases}$

$$\phantom{xxxxxxxxxxxxxxxxxxxxxxxxxxxxxxxxxxxxxxxxxxxxx} 5x \phantom{xx} = 5$$
$$\phantom{xxxxxxxxxxxxxxxxxxxxxxxxxxxxxxxxxxxxxxxxxxxxx} x \phantom{xxx} = 1$$

$3(1) + 4y = -2$    Replace $x$ by 1 in revised first equation.

$\phantom{x}3 + 4y = -2$

$\phantom{xxx} 4y = -5$

$\phantom{xxxx} y = -\frac{5}{4}$

$\left(1, -\frac{5}{4}\right)$

**55.** $\begin{cases} 213x + 632y = 3799 \\ 275x + 816y = 4905 \end{cases}$

$\phantom{xx} 58{,}575x + 173{,}800y = \phantom{-}1{,}044{,}725$    Multiply equation by 275.

$\phantom{xx} {-}58{,}575x - 173{,}808y = -1{,}044{,}765$    Multiply equation by $-213$.

$\phantom{xxxxxxxxxxxxxxxx} -8y = \phantom{xxxx} -40$

$\phantom{xxxxxxxxxxxxxxxxxx} y = \phantom{xxxxx} 5$

$213x + 632(5) = 3799$    Replace $y$ by 5 in first equation.

$\phantom{x}213x + 3160 = 3799$

$\phantom{xxxxx}213x = 639$

$\phantom{xxxxxxx} x = 3$

So, student tickets cost \$3 and general admission tickets cost \$5.

**57.** $\begin{cases} x + \phantom{xx} y = 10{,}000 \\ 0.075x + 0.10y = \phantom{xxx} 850 \end{cases}$

$\phantom{xx} -75x - \phantom{x} 75y = -750{,}000$    Multiply equation by $-75$.

$\phantom{xxx} 75x + 100y = \phantom{-} 850{,}000$    Multiply equation by 1000.

$\phantom{xxxxxxx} 25y = \phantom{-} 100{,}000$

$\phantom{xxxxxxxxx} y = \dfrac{100{,}000}{25}$

$\phantom{xxxxxxxxx} y = \phantom{xx} 4000$

The smallest amount that you can invest at 10% is \$4000.

**59.** $\begin{cases} x + y = 82 \\ x - y = 14 \end{cases}$

$2x \phantom{xxx} = 96$

$\phantom{x}x \phantom{xxxx} = 48$

$48 + y = 82$    Replace $x$ by 48 in first equation.

$\phantom{xxxx} y = 34$

The two numbers are 48 and 34.

**61.** $\begin{cases} x + \phantom{x}y = \phantom{x}18.52 \Rightarrow -9x - 9y = -166.68 \\ 19.3x + 9y = 277.92 \Rightarrow 19.3x + 9y = \phantom{-}277.92 \end{cases}$

$$\phantom{xxxxxxxxxxxxxxxxxxxxxxxxx} 10.3x \phantom{xxxx} = 111.24$$
$$\phantom{xxxxxxxxxxxxxxxxxxxxxxxxxxxxx} x \phantom{xxxx} = \frac{111.24}{10.3}$$
$$\phantom{xxxxxxxxxxxxxxxxxxxxxxxxxxxxx} x \phantom{xxxx} = 10.8$$

$10.8 + y = 18.52$

$\phantom{xxxx} y = 7.72$

The gold in the bracelet has a volume of 10.8 cubic centimeters and a weight of $19.3(10.8) = 208.44$ grams.

The copper in the bracelet has a volume of 7.72 cubic centimeters and a weight of $9(7.72) = 69.48$ grams.

$$\frac{208.44}{208.44 + 69.48} = \frac{208.44}{277.92} = 0.75 = 75\%$$

Yes, the bracelet is 18k gold.

**63.** Find the intersection of the two lines:

$$\begin{cases} 3x + 4y = 7 \\ 5x - 4y = 1 \end{cases}$$

$$8x \quad\quad = 8$$

$$x \quad\quad = 1$$

$3(1) + 4y = 7$      Replace $x$ by 1 in first equation.

$$3 + 4y = 7$$

$$4y = 4$$

$$y = 1$$

$$(1, 1)$$

Then find the equation of the line:

$$y - y_1 = m(x - x_1)$$

$$y - 1 = \tfrac{1}{3}(x - 1)$$

$$y - 1 = \tfrac{1}{3}x - \tfrac{1}{3}$$

$$y = \tfrac{1}{3}x - \tfrac{1}{3} + \tfrac{3}{3}$$

$$y = \tfrac{1}{3}x + \tfrac{2}{3}$$

**65.** $\begin{cases} 0.3x - 0.2y = 0.9 \\ 0.7x + 0.2y = 1.1 \end{cases}$

Multiply both sides of each equation by 10 to clear the system of decimals.

$$3x - 2y = 9$$

$$7x + 2y = 11$$

**67.** $\begin{cases} y = 3x + 4 \\ y = -x + 8 \end{cases}$

**69.**

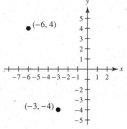

$$m = \frac{-4 - 4}{-3 - (-6)} = \frac{-8}{3} = -\frac{8}{3}$$

**71.**

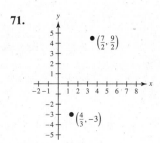

$$m = \frac{-3 - \dfrac{9}{2}}{\dfrac{4}{3} - \dfrac{7}{2}}$$

$$= \frac{\dfrac{6}{2} - \dfrac{9}{2}}{\dfrac{8}{6} - \dfrac{21}{6}}$$

$$= \frac{-\dfrac{15}{2}}{-\dfrac{13}{6}}$$

$$= -\frac{15}{2} \cdot -\frac{6}{13}$$

$$= \frac{15(3)\cancel{(2)}}{\cancel{(2)}(13)}$$

$$= \frac{45}{13}$$

**73.**

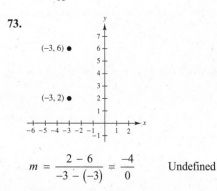

$$m = \frac{2 - 6}{-3 - (-3)} = \frac{-4}{0} \quad\quad \text{Undefined}$$

The slope is undefined.

**75.** $x \le 3$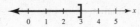

**77.** $x + 5 < 6$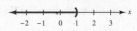
$x < 1$

**79.** $\begin{cases} y = x \\ x + 3y = 20 \end{cases}$

$y + 3y = 20$      Replace $x$ by $y$ in second equation.

$\quad 4y = 20$

$\quad\quad y = 5 \Rightarrow 5 = x$    Replace $y$ by 5 in first equation.

$(5, 5)$

**81.** $\begin{cases} 2x + y = 5 \Rightarrow y = -2x + 5 \\ 5x + 3y = 12 \end{cases}$

$5x + 3(-2x + 5) = 12$      Replace $y$ by $-2x + 5$ in second equation.

$5x - 6x + 15 = 12$

$-x + 15 = 12$

$-x = -3$

$x = 3 \Rightarrow y = -2(3) + 5$    Replace $x$ by 3 in revised first equation.

$y = -6 + 5$

$y = -1$

$(3, -1)$

# Mid-Chapter Quiz for Chapter 8

**1.** $(4, 2)$

$3(4) + 4(2) \overset{?}{=} 4$        $5(4) - 3(2) \overset{?}{=} 14$

$12 + 8 \overset{?}{=} 4$           $20 - 6 \overset{?}{=} 14$

$20 \ne 4$              $14 = 14$

The ordered pair $(4, 2)$ is *not* a solution because
substituting $x = 4$ and $y = 2$ into the equation
$3x + 4y = 4$ yields $20 = 4$ which is a contradiction.

**2.**    $2x - 3y = 7$         $3x + 5y = 1$

$2(2) - 3(-1) \overset{?}{=} 7$     $3(2) + 5(-1) \overset{?}{=} 1$

$4 + 3 = 7$          $6 - 5 = 1$

Yes, $(2, -1)$ is a solution.

**3.** The graph indicates that the solution of the system is
$(3, 2)$.

**4.** The graph indicates that the solution of the system is
$(4, 1)$.

**5.** The graph indicates that the solution of the system is
$(4, -1)$.

**6.** The solution is $(-3, 11)$.

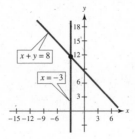

**7.** The solution is $(2, 2)$.

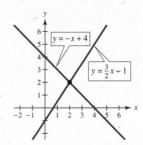

**8.** The solution is $(-1, 4)$.

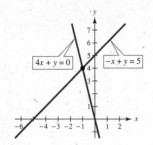

**9.** $\begin{cases} x - y = 4 \\ \phantom{x -} y = 2 \end{cases}$

$\phantom{xxx}x - 2 = 4$ \qquad Replace $y$ by 2 in first equation.

$\phantom{xxxxxx}x = 6$

The solution is $(6, 2)$.

**10.** $\begin{cases} y = -\frac{2}{3}x + 5 \\ y = \phantom{-}2x - 3 \end{cases}$

$\phantom{xxx}2x - 3 = -\frac{2}{3}x + 5$ \qquad Replace $y$ by $2x - 3$ in first equation.

$\phantom{x}3(2x - 3) = 3\left(-\frac{2}{3}x + 5\right)$

$\phantom{xxx}6x - 9 = -2x + 15$

$\phantom{xxx}8x - 9 = 15$

$\phantom{xxxxx}8x = 24$

$\phantom{xxxxxxx}x = 3 \Rightarrow y = 2(3) - 3$ \qquad Replace $x$ by 3 in second equation.

$\phantom{xxxxxxxxxxxx}y = 3$

$(3, 3)$

The solution is $(3, 3)$.

**11.** $\begin{cases} 2x - \phantom{x}y = -7 \Rightarrow -y = -2x - 7 \Rightarrow y = 2x + 7 \\ 4x + 3y = 16 \end{cases}$

$\phantom{x}4x + 3(2x + 7) = 16$ \qquad Replace $y$ by $2x + 7$ in second equation.

$\phantom{x}4x + 6x + 21 = 16$

$\phantom{xxx}10x + 21 = 16$

$\phantom{xxxxx}10x = -5$

$\phantom{xxxxxxx}x = -\frac{1}{2} \Rightarrow y = 2\left(-\frac{1}{2}\right) + 7$ \qquad Replace $x$ by $-\frac{1}{2}$ in revised first equation.

$\phantom{xxxxxxxxxxxx}y = -1 + 7$

$\phantom{xxxxxxxxxxxx}y = 6$

$\left(-\frac{1}{2}, 6\right)$

The solution is $\left(-\frac{1}{2}, 6\right)$.

**12.** $\begin{cases} x + y = -2 \\ x - y = \phantom{-}4 \end{cases}$

$\overline{\phantom{xxxxxxxxxxx}}$

$2x \phantom{xxxx} = \phantom{-}2$

$\phantom{x}x \phantom{xxxxx} = \phantom{-}1 \Rightarrow 1 + y = -2$

$\phantom{xxxxxxxxxxxxx}y = -3$

$(1, -3)$

**13.** $\begin{cases} 2x + \phantom{x}y = \phantom{x}1 \Rightarrow -10x - 5y = -5 \\ 6x + 5y = 13 \Rightarrow \phantom{-}6x + 5y = 13 \end{cases}$

$\phantom{xxxxxxxxxxxxxxxx}\overline{\phantom{xxxxxxxxxxxx}}$

$\phantom{xxxxxxxxxxxxxxx}-4x \phantom{xxxx} = \phantom{-}8$

$\phantom{xxxxxxxxxxxxxxxxx}x \phantom{xxxx} = -2$

$\phantom{xxxxxx}2(-2) + y = 1$

$\phantom{xxxxxxxx}-4 + y = 1$

$\phantom{xxxxxxxxxxxx}y = 5$

$(-2, 5)$

**14.** $\begin{cases} -x + 3y = 10 \Rightarrow -9x + 27y = 90 \\ 9x - 4y = 5 \Rightarrow \underline{\phantom{-}9x - 4y = 5} \end{cases}$

$$23y = 95$$
$$y = \tfrac{95}{23}$$

$$-x + 3\left(\tfrac{95}{23}\right) = 10$$

$$-x + \tfrac{285}{23} = 10$$

$$23\left(-x + \tfrac{285}{23}\right) = 23(10)$$

$$-23x + 285 = 230$$

$$-23x = -55$$

$$x = \tfrac{55}{23}$$

$\left(\tfrac{55}{23}, \tfrac{95}{23}\right)$

**15.** $(-1, 9)$

*Sample answers:*

$\begin{cases} 2x + y = 7 \\ -5x - 2y = -13 \end{cases}$  $\begin{cases} 4x + y = 5 \\ 7x + 2y = 11 \end{cases}$

**16.** $(3, 0)$

*Sample answers:*

$\begin{cases} -x + y = -3 \\ 4x - 3y = 12 \end{cases}$  $\begin{cases} 2x + 3y = 6 \\ -x + 7y = -3 \end{cases}$

**17.** $\begin{cases} 5x + ky = 3 \Rightarrow 5x = -ky + 3 \Rightarrow x = \dfrac{-ky}{5} + \dfrac{3}{5} \\ 10x - 4y = 1 \end{cases}$

$$10\left(\dfrac{-ky}{5} + \dfrac{3}{5}\right) - 4y = 1 \qquad \text{Replace } x \text{ by } \dfrac{-ky}{5} + \dfrac{3}{5} \text{ in second equation.}$$

$$-2ky + 6 - 4y = 1$$

$$-(2k + 4)y + 6 = 1$$

The variable $y$ will "drop out" of this equation when $2k + 4 = 0$.

$$2k + 4 = 0 \Rightarrow 2k = -4 \Rightarrow k = -2$$

So, when $k = -2$, the equation becomes $6 = 1$, which is a *false* statement. The system is inconsistent when $k = -2$.

**18.** $\begin{cases} 8x - 5y = 16 \\ kx - 0.5y = 3 \Rightarrow -0.5y = -kx + 3 \Rightarrow y = 2kx - 6 \end{cases}$

$$8x - 5(2kx - 6) = 16 \qquad \text{Replace } y \text{ by } 2kx - 6 \text{ in first equation.}$$

$$8x - 10kx + 30 = 16$$

$$(8 - 10k)x + 30 = 16$$

The variable $x$ will "drop out" of this equation when $8 - 10k = 0$.

$$8 - 10k = 0 \Rightarrow 8 = 10k \Rightarrow 0.8 = k$$

So, when $k = 0.8$, the equation becomes $30 = 16$, which is a *false* statement. The system is inconsistent when $k = 0.8$.

**19.** $\begin{cases} x + y = 50 \Rightarrow y = -x + 50 \\ x - y = 22 \end{cases}$

$$x - (-x + 50) = 22 \qquad \text{Replace } y \text{ by } -x + 50 \text{ in the second equation.}$$

$$x + x - 50 = 22$$

$$2x - 50 = 22$$

$$2x = 72$$

$$x = 36 \Rightarrow y = -36 + 50 \qquad \text{Replace } x \text{ by } 36 \text{ in revised first equation.}$$

$$y = 14$$

$(36, 14)$

The two numbers are 36 and 14.

**20.** $\begin{cases} x + y = 32 \\ \quad x = 4y + 2 \end{cases}$

$\quad\quad (4y + 2) + y = 32$  ———  Replace $x$ by $4y + 2$ in first equation.

$\quad\quad\quad\quad 5y + 2 = 32$

$\quad\quad\quad\quad\quad 5y = 30$

$\quad\quad\quad\quad\quad y = 6 \Rightarrow x = 4(6) + 2$  ———  Replace $y$ by 6 in second equation.

$\quad\quad\quad\quad\quad\quad\quad\quad x = 26$

$(26, 6)$

The price of the book was $26, and the price of the calendar was $6.

## Section 8.4   Applications of Systems of Linear Equations

**1. (a)**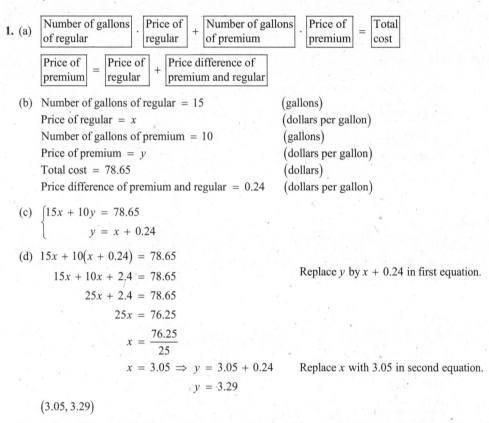

**(b)** Number of gallons of regular $= 15$ \quad (gallons)

Price of regular $= x$ \quad (dollars per gallon)

Number of gallons of premium $= 10$ \quad (gallons)

Price of premium $= y$ \quad (dollars per gallon)

Total cost $= 78.65$ \quad (dollars)

Price difference of premium and regular $= 0.24$ \quad (dollars per gallon)

**(c)** $\begin{cases} 15x + 10y = 78.65 \\ \quad\quad y = x + 0.24 \end{cases}$

**(d)** $15x + 10(x + 0.24) = 78.65$

$\quad\quad 15x + 10x + 2.4 = 78.65$  ———  Replace $y$ by $x + 0.24$ in first equation.

$\quad\quad\quad\quad 25x + 2.4 = 78.65$

$\quad\quad\quad\quad\quad\quad 25x = 76.25$

$\quad\quad\quad\quad\quad\quad x = \dfrac{76.25}{25}$

$\quad\quad\quad\quad\quad x = 3.05 \Rightarrow y = 3.05 + 0.24$  ———  Replace $x$ with 3.05 in second equation.

$\quad\quad\quad\quad\quad\quad\quad\quad y = 3.29$

$(3.05, 3.29)$

Regular gasoline costs $3.05 per gallon and premium gasoline costs $3.29 per gallon.

**3.** *Verbal Model:*   $\boxed{\text{Larger number}} + \boxed{\text{Smaller number}} = 67$

$\boxed{\text{Larger number}} - \boxed{\text{Smaller number}} = 17$

*Labels:*   Larger number $= x$

Smaller number $= y$

*System of Equations:*   $\begin{cases} x + y = 67 \\ x - y = 17 \end{cases}$

*Solving by Elimination:*   $\begin{cases} x + y = 67 \\ \underline{x - y = 17} \end{cases}$

$\qquad\qquad 2x \quad\;\; = 84$

$\qquad\qquad x \quad\;\;\; = 42$

$42 + y = 67$   Replace $x$ by 42 in first equation.

$\qquad\quad y = 25$

The two numbers are 42 and 25.

**5.** *Verbal Model:*   $\boxed{\text{Larger number}} + \boxed{\text{Smaller number}} = 132$

$\boxed{\text{Larger number}} = 2 \cdot \boxed{\text{Smaller number}} + 6$

*Labels:*   Larger number $= x$

Smaller number $= y$

*System of Equations:*   $\begin{cases} x + y = 132 \\ x = 2y + 6 \end{cases}$

*Solving by Substitution:*   $\begin{cases} x + y = 132 \\ x = 2y + 6 \end{cases}$

$(2y + 6) + y = 132$   Replace $x$ by $2y + 6$ in first equation.

$\qquad 3y + 6 = 132$

$\qquad\quad 3y = 126$

$\qquad\quad\; y = 42$

$\qquad\quad\; x = 2(42) + 6$   Replace $y$ by 42 in second equation.

$\qquad\quad\; x = 90$

The two numbers are 90 and 42.

**7.** *Verbal Model:*   $\boxed{\text{Larger number}} + 2 \cdot \boxed{\text{Smaller number}} = 100$

$\boxed{\text{Larger number}} - \boxed{\text{Smaller number}} = 10$

*Labels:*   Larger number $= x$

Smaller number $= y$

*System of Equations:*   $\begin{cases} x + 2y = 100 \\ x - y = 10 \end{cases}$

*Solving by Elimination:*   $\begin{cases} x + 2y = 100 \Rightarrow \quad x + 2y = 100 \\ x - y = 10 \Rightarrow \underline{-x + y = -10} \end{cases}$   Multiply by $-1$.

$\qquad\qquad\qquad\qquad\qquad 3y = 90$

$\qquad\qquad\qquad\qquad\qquad\; y = 30$

$\qquad\qquad\qquad\quad x - 30 = 10$   Replace $y$ by 30 in first equation.

$\qquad\qquad\qquad\quad x \qquad\; = 40$

The two numbers are 40 and 30.

**9.** *Verbal Model:*   Number of dimes + Number of quarters = Number of coins

Value of dimes + Value of quarters = Value of coins

*Labels:*   Number of dimes $= x$

Number of quarters $= y$

Number of coins $= 21$

Value of dimes $= 0.10x$     (dollars)

Value of quarters $= 0.25y$     (dollars)

Value of coins $= 4.05$     (dollars)

*System of Equations:*   $\begin{cases} x + y = 21 \\ 0.10x + 0.25y = 4.05 \end{cases}$

*Solving by Elimination:*   $\begin{cases} x + y = 21 \Rightarrow -10x - 10y = -210 \\ 0.10x + 0.25y = 4.05 \Rightarrow \underline{\quad 10x + 25y = 405} \end{cases}$     Multiply equation by $-10$.

Multiply equation by 100.

$$15y = 195$$
$$y = 13$$

$$x + 13 = 21$$
$$x = 8$$

There are 8 dimes and 13 quarters.

**11.** *Verbal Model:*   Number of nickels + Number of quarters = Number of coins

Value of nickels + Value of quarters = Value of coins

*Labels:*   Number of nickels $= x$

Number of quarters $= y$

Number of coins $= 35$

Value of nickels $= 0.05x$     (dollars)

Value of quarters $= 0.25y$     (dollars)

Total value of coins $= 5.75$     (dollars)

*System of Equations:*   $\begin{cases} x + y = 35 \\ 0.05x + 0.25y = 5.75 \end{cases}$

*Solving by Elimination:*   $\begin{cases} x + y = 35 \Rightarrow -5x - 5y = -175 \\ 0.05x + 0.25y = 5.75 \Rightarrow \underline{\quad 5x + 25y = 575} \end{cases}$     Multiply equation by $-5$.

Multiply equation by 100.

$$20y = 400$$
$$y = 20$$

$$x + 20 = 35$$
$$x = 15$$

$$(15, 20)$$

There are 15 nickels and 20 quarters.

**13.** *Verbal Model:* $\boxed{\text{Number of nickels}} + \boxed{\text{Number of dimes}} = \boxed{\text{Number of coins}}$

$\boxed{\text{Value of nickels}} + \boxed{\text{Value of dimes}} = \boxed{\text{Value of coins}}$

*Labels:*

Number of nickels $= x$

Number of dimes $= y$

Number of coins $= 44$

Value of nickels $= 0.05x$    (dollars)

Value of dimes $= 0.10y$    (dollars)

Total value of coins $= 3.00$    (dollars)

*System of Equations:*
$$\begin{cases} x + y = 44 \\ 0.05x + 0.10y = 3.00 \end{cases}$$

*Solving by Elimination:*
$$\begin{cases} x + y = 44 & \Rightarrow -5x - 5y = -220 \\ 0.05x + 0.10y = 3.00 & \Rightarrow \underline{\ \ 5x + 10y = 300\ \ } \end{cases}$$

Multiply equation by $-5$.

Multiply equation by 100.

$$5y = 80$$
$$y = 16 \Rightarrow x + 16 = 44$$
$$x = 28$$

$(28, 16)$

There are 28 nickels and 16 dimes in the cash register.

**15.** *Verbal Model:* $2 \cdot \boxed{\begin{array}{c}\text{Length of}\\\text{rectangle}\end{array}} + 2 \cdot \boxed{\begin{array}{c}\text{Width of}\\\text{rectangle}\end{array}} = 40$

$\boxed{\begin{array}{c}\text{Length of}\\\text{rectangle}\end{array}} = \boxed{\begin{array}{c}\text{Width of}\\\text{rectangle}\end{array}} + 4$

*Labels:*

Length of rectangle $= l$    (feet)

Width of rectangle $= w$    (feet)

*System of Equations:*
$$\begin{cases} 2l + 2w = 40 \\ l = w + 4 \end{cases}$$

*Solving by Substitution:*
$$\begin{cases} 2l + 2w = 40 \\ l = w + 4 \end{cases}$$

$2(w + 4) + 2w = 40$                Replace $l$ by $w + 4$ in first equation.

$2w + 8 + 2w = 40$

$4w + 8 = 40$

$4w = 32$

$w = 8 \Rightarrow l = 8 + 4$          Replace $w$ by 8 in second equation.

$l = 12$

The length of the rectangle is 12 feet and the width of the rectangle is 8 feet.

**17.** *Verbal Model:*

$$2 \cdot \boxed{\begin{array}{c}\text{Length of}\\\text{rectangle}\end{array}} + 2 \cdot \boxed{\begin{array}{c}\text{Width of}\\\text{rectangle}\end{array}} = 16$$

$$\boxed{\begin{array}{c}\text{Width of}\\\text{rectangle}\end{array}} = \tfrac{1}{3} \cdot \boxed{\begin{array}{c}\text{Length of}\\\text{rectangle}\end{array}} \cdot$$

*Labels:*

Length of rectangle $= l$   (yards)

Width of rectangle $= w$   (yards)

*System of Equations:*

$$\begin{cases} 2l + 2w = 16 \\ \quad\ w = \tfrac{1}{3}l \end{cases}$$

*Solving by Substitution:*

$$\begin{cases} 2l + 2w = 16 \\ \quad\ w = \tfrac{1}{3}l \end{cases}$$

$$2l + 2\left(\tfrac{1}{3}l\right) = 16 \qquad \text{Replace } w \text{ by } \tfrac{1}{3}l \text{ in first equation.}$$

$$2l + \tfrac{2}{3}l = 16$$

$$6l + 2l = 48$$

$$8l = 48$$

$$l = 6 \Rightarrow w = \tfrac{1}{3}(6)$$

$$w = 2$$

The length of the rectangle is 6 yards and the width is 2 yards.

**19.** *Verbal Model:*

$$2 \cdot \boxed{\begin{array}{c}\text{Length of}\\\text{rectangle}\end{array}} + 2 \cdot \boxed{\begin{array}{c}\text{Width of}\\\text{rectangle}\end{array}} = 35.2$$

$$\boxed{\begin{array}{c}\text{Length of}\\\text{rectangle}\end{array}} = 120\% \cdot \boxed{\begin{array}{c}\text{Width of}\\\text{rectangle}\end{array}}$$

*Labels:*

Length of rectangle $= l$   (meters)

Width of rectangle $= w$   (meters)

*System of Equations:*

$$\begin{cases} 2l + 2w = 35.2 \\ \quad\ l = 1.20w \end{cases}$$

*Solving by Substitution:*

$$\begin{cases} 2l + 2w = 35.2 \\ \quad\ l = 1.20w \end{cases}$$

$$2(1.20w) + 2w = 35.2 \qquad \text{Replace } l \text{ by } 1.20\,w \text{ in first equation.}$$

$$2.40w + 2w = 35.2$$

$$4.40w = 35.2$$

$$w = 8 \text{ and } l = 1.20(8)$$

$$l = 9.6$$

The length of the rectangle is 9.6 meters and the width is 8 meters.

**21.** *Verbal Model:*   Wholesale cost + Markup = Selling price

Markup = 0.45 · Wholesale cost

*Labels:*   Wholesale cost = $C$   (dollars)

Markup = $M$   (dollars)

Selling price = 108.75   (dollars)

*System of Equations:*   $\begin{cases} C + M = 108.75 \\ \quad\; M = 0.45C \end{cases}$

*Solving by Substitution:*   $\begin{cases} C + M = 108.75 \\ \quad\; M = 0.45C \end{cases}$

$C + 0.45C = 108.75$   Replace $M$ by $0.45C$ in first equation.

$1.45C = 108.75$

$C = 75$

The wholesale cost of the watch is $75.00.

**23.** *Verbal Model:*   Wholesale cost + Markup = Selling price

Markup = 0.3 · Wholesale cost

*Labels:*   Wholesale cost = $C$   (dollars)

Markup = $M$   (dollars)

Selling price = 359   (dollars)

*System of Equations:*   $\begin{cases} C + M = 359 \\ \quad\; M = 0.3C \end{cases}$

*Solving by Substitution:*   $\begin{cases} C + M = 359 \\ \quad\; M = 0.3C \end{cases}$

$C + 0.3C = 359$   Replace $M$ by $0.3C$ in first equation.

$1.3C = 359$

$C \approx 276.15$

The wholesale cost of the air conditioner is $276.15.

**25.** *Verbal Model:*   List price − Discount = Sale price

Discount = Discount rate · List price

*Labels:*   List price = $L$   (dollars)

Discount = $D$   (dollars)

Sale price = 280   (dollars)

Discount rate = 0.30   (percent in decimal form)

*System of Equations:*   $\begin{cases} L - D = 280 \\ \quad\; D = 0.30L \end{cases}$

*Solving by Substitution:*   $\begin{cases} L - D = 280 \\ \quad\; D = 0.30L \end{cases}$

$L - 0.30L = 280$   Replace $D$ by $0.30L$ in first equation.

$0.70L = 280$

$L = \dfrac{280}{0.70}$

$L = 400$

The list price of the treadmill is $400.

**27.** *Verbal Model:*

$$\boxed{\text{Distance one person drives}} + \boxed{\text{Distance other person drives}} = \boxed{\text{Total distance}}$$

$$\boxed{\text{Distance one person drives}} = \boxed{2} \cdot \boxed{\text{Distance other person drives}}$$

*Labels:*

Distance one person drives $= x$    (miles)

Distance other person drives $= y$    (miles)

Total distance $= 450$    (miles)

*System of Equations:*
$$\begin{cases} x + y = 450 \\ x = 2y \end{cases}$$

*Solving by Substitution:*
$$\begin{cases} x + y = 450 \\ x = 2y \end{cases}$$

$2y + y = 450$     Replace $x$ by $2y$ in first equation.

$3y = 450$

$y = \frac{450}{3}$

$y = 150 \Rightarrow x = 2(150)$    Replace $y$ by 150 in second equation.

$x = 300$

$(300, 150)$

One driver drove 300 miles and the other driver drove 150 miles.

**29.** *Verbal Model:*

$$\boxed{\text{Total annual pay}} = \boxed{36,000} + \boxed{1.5\%} \cdot \boxed{\text{Total sales}}$$

$$\boxed{\text{Total annual pay}} = \boxed{36,000} + \boxed{2\%} \cdot \boxed{\text{Total sales}}$$

*Labels:*

Total annual pay $= x$    (dollars)

Total sales $= y$    (dollars)

*System of Equations:*
$$\begin{cases} x = 36,000 + 0.015y \\ x = 32,000 + 0.02y \end{cases}$$

*Solving by Substitution:*
$$\begin{cases} x = 36,000 + 0.015y \\ x = 32,000 + 0.02y \end{cases}$$

$32,000 + 0.02y = 36,000 + 0.015y$    Replace $x$ by $32,000 + 0.02y$ in first equation.

$32,000 + 0.005y = 36,000$

$0.005y = 4000$

$y = \dfrac{4000}{0.005}$

$y = 800,000$

Your total sales would need to be $800,000 to earn the same amount at each job.

**31.** *Verbal Model:*

$$\boxed{\text{Total monthly cost}} = \boxed{80} + \boxed{0.80} \cdot \boxed{\text{Number of minutes over 1500}}$$

$$\boxed{\text{Total monthly cost}} = \boxed{105} + \boxed{0.30} \cdot \boxed{\text{Number of minutes over 1500}}$$

*Labels:*

Total monthly cost $= x$ (dollars)

Number of minutes over 1500 $= y$ (minutes)

*System of Equations:*

$$\begin{cases} x = 80 + 0.80y \\ x = 105 + 0.30y \end{cases}$$

*Solving by Substitution:*

$$\begin{cases} x = 80 + 0.80y \\ x = 105 + 0.30y \end{cases}$$

$80 + 0.80y = 105 + 0.30y$    Replace $x$ by $80 + 0.80y$ in second equation.

$80 + 0.50y = 105$

$0.50y = 25$

$y = \dfrac{25}{0.50}$

$y = 50$

If you used 50 minutes over the free 1500 minutes, the two plans would cost the same.

**33.** *Verbal Model:*

$$\boxed{\begin{array}{c}\text{Number of}\\ \text{adult tickets}\end{array}} + \boxed{\begin{array}{c}\text{Number of}\\ \text{children's tickets}\end{array}} = \boxed{\begin{array}{c}\text{Total number}\\ \text{of tickets}\end{array}}$$

$$\boxed{\begin{array}{c}\text{Receipts from}\\ \text{adult tickets}\end{array}} + \boxed{\begin{array}{c}\text{Receipts from}\\ \text{children's tickets}\end{array}} = \boxed{\text{Total receipts}}$$

*Labels:*

Number of adult tickets $= x$ (tickets)

Number of children's tickets $= y$ (tickets)

Total number of tickets $= 500$ (tickets)

Receipts from adult tickets $= 18x$ (dollars)

Receipts from children's tickets $= 12.50y$ (dollars)

Total receipts $= 8312.50$ (dollars)

*System of Equations:*

$$\begin{cases} x + y = 500 \\ 18x + 12.50y = 8312.50 \end{cases}$$

*Solving by Substitution:*

$$\begin{cases} x + y = 500 \quad \Rightarrow x = -y + 500 \\ 18x + 12.50y = 8312.50 \end{cases}$$

$18(-y + 500) + 12.50y = 8312.50$    Replace $x$ by $-y + 500$ in second equation.

$-18y + 9000 + 12.50y = 8312.50$

$-5.50y + 9000 = 8312.50$

$-5.50y = -687.50$

$y = \dfrac{-787.50}{-5.50}$

$y = 125 \Rightarrow x = -125 + 500$    Replace $y$ by 125 in revised first equation.

$x = 375$

$(375, 125)$

So, 375 adult tickets and 125 children's tickets were sold for the dinner.

**35.** *Verbal Model:*     $\boxed{\text{Cost of large truck}} + 4 \cdot \boxed{\text{Cost of small truck}} = \boxed{118,000}$

   $2 \cdot \boxed{\text{Cost of large truck}} + 2 \cdot \boxed{\text{Cost of small truck}} = \boxed{107,000}$

*Labels:*     Cost of large truck $= x$     (dollars)

   Cost of small truck $= y$     (dollars)

*System of Equations:*     $\begin{cases} x + 4y = 118,000 \\ 2x + 2y = 107,000 \end{cases}$

*Solving by Elimination:*

$$\begin{cases} x + 4y = 118,000 \Rightarrow & x + 4y = \phantom{-}118,000 \\ 2x + 2y = 107,000 \Rightarrow & -4x - 4y = -214,000 \end{cases} \quad \text{Multiply equation by } -2.$$

$$\begin{aligned} -3x \phantom{+4y} &= -96,000 \\ x &= \phantom{-}32,000 \end{aligned}$$

$32,000 + 4y = 118,000$     Replace $x$ by 32,000 in first equation.

$4y = 86,000$

$y = 21,500$

The cost of the large truck is \$32,000 and the cost of the small truck is \$21,500.

**37.** *Verbal Model:*     $\boxed{\begin{array}{l}\text{Number of gallons}\\\text{of regular}\end{array}} \cdot \boxed{\begin{array}{l}\text{Price of}\\\text{regular}\end{array}} + \boxed{\begin{array}{l}\text{Number of gallons}\\\text{of premium}\end{array}} \cdot \boxed{\begin{array}{l}\text{Price of}\\\text{premium}\end{array}} = \boxed{\begin{array}{l}\text{Total}\\\text{cost}\end{array}}$

   $\boxed{\begin{array}{l}\text{Price of}\\\text{premium}\end{array}} = \boxed{\begin{array}{l}\text{Price of}\\\text{regular}\end{array}} + \boxed{\begin{array}{l}\text{Price difference of}\\\text{premium and regular}\end{array}}$

*Labels:*     Number of gallons of regular $= 8$     (gallons)

   Price of regular $= x$     (dollars per gallon)

   Number of gallons of premium $= 12$     (gallons)

   Price of premium $= y$     (dollars per gallon)

   Total cost $= 65.40$     (dollars)

   Price difference of premium and regular $= 0.15$     (dollars per gallon)

*System of Equations:*     $\begin{cases} 8x + 12y = 65.40 \\ \phantom{8x + 12}y = x + 0.15 \end{cases}$

*Solving by Substitution:*     $\begin{cases} 8x + 12y = 65.40 \\ \phantom{8x + 12}y = x + 0.15 \end{cases}$

$8x + 12(x + 0.15) = 65.40$     Replace $y$ by $x + 0.15$ in first equation.

$8x + 12x + 1.80 = 65.40$

$20x + 1.80 = 65.40$

$20x = 63.60$

$x = \dfrac{63.60}{20}$

$x = 3.18 \Rightarrow y = 3.18 + 0.15$     Replace $x$ with 3.18 in second equation.

$y = 3.33$

$(3.18, 3.33)$

Regular gasoline costs \$3.18 per gallon and premium gasoline costs \$3.33 per gallon.

**39.** *Verbal Model:*

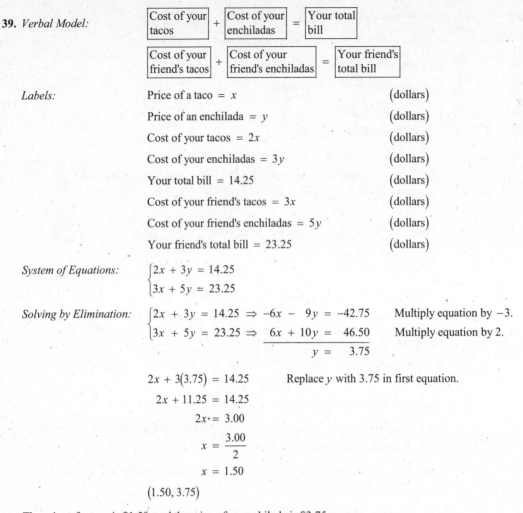

$$\boxed{\text{Cost of your tacos}} + \boxed{\text{Cost of your enchiladas}} = \boxed{\text{Your total bill}}$$

$$\boxed{\text{Cost of your friend's tacos}} + \boxed{\text{Cost of your friend's enchiladas}} = \boxed{\text{Your friend's total bill}}$$

*Labels:*

| | |
|---|---|
| Price of a taco $= x$ | (dollars) |
| Price of an enchilada $= y$ | (dollars) |
| Cost of your tacos $= 2x$ | (dollars) |
| Cost of your enchiladas $= 3y$ | (dollars) |
| Your total bill $= 14.25$ | (dollars) |
| Cost of your friend's tacos $= 3x$ | (dollars) |
| Cost of your friend's enchiladas $= 5y$ | (dollars) |
| Your friend's total bill $= 23.25$ | (dollars) |

*System of Equations:*
$$\begin{cases} 2x + 3y = 14.25 \\ 3x + 5y = 23.25 \end{cases}$$

*Solving by Elimination:*

$$\begin{aligned} 2x + 3y &= 14.25 \Rightarrow -6x - 9y = -42.75 \quad &\text{Multiply equation by } -3. \\ 3x + 5y &= 23.25 \Rightarrow \underline{\phantom{-}6x + 10y = \phantom{-}46.50} \quad &\text{Multiply equation by } 2. \\ & \phantom{= 23.25 \Rightarrow} \quad y = \phantom{4}3.75 \end{aligned}$$

$$2x + 3(3.75) = 14.25 \qquad \text{Replace } y \text{ with 3.75 in first equation.}$$
$$2x + 11.25 = 14.25$$
$$2x = 3.00$$
$$x = \frac{3.00}{2}$$
$$x = 1.50$$

$$(1.50, 3.75)$$

The price of a taco is \$1.50, and the price of an enchilada is \$3.75.

**41.** *Verbal Model:*

$$\boxed{\text{Boat speed in still water}} - \boxed{\text{Current speed}} = \boxed{\text{Upstream speed}}$$

$$\boxed{\text{Boat speed in still water}} + \boxed{\text{Current speed}} = \boxed{\text{Downstream speed}}$$

*Labels:*

Boat speed in still water $= x$ (miles per hour)

Current speed $= y$ (miles per hour)

Upstream speed $= \dfrac{10 \text{ miles}}{\frac{1}{2} \text{ hours}} = \dfrac{10}{1} \cdot \dfrac{2}{1} = 20$ (miles per hour)

Downstream speed $= \dfrac{10 \text{ miles}}{\frac{1}{3} \text{ hours}} = \dfrac{10}{1} \cdot \dfrac{3}{1} = 30$ (miles per hour)

*System of Equations:*
$$\begin{cases} x - y = 20 \\ x + y = 30 \end{cases}$$

*Solving by Elimination:*
$$\begin{cases} x - y = 20 \\ x + y = 30 \end{cases}$$
$$2x \quad = 50$$
$$x \quad = 25$$

$$25 + y = 30 \qquad \text{Replace } x \text{ by 25 in second equation.}$$
$$y = 5$$

The speed of the current is 5 miles per hour.

**43.** *Verbal Model:*

$$\boxed{\text{Plane speed in still air}} - \boxed{\text{Wind speed}} = \boxed{\text{Speed with headwind}}$$

$$\boxed{\text{Plane speed in still air}} + \boxed{\text{Wind speed}} = \boxed{\text{Speed with tailwind}}$$

*Labels:*

Plane speed in still air $= x$ (miles per hour)

Wind speed $= y$ (miles per hour)

Speed with headwind $= \dfrac{2100 \text{ miles}}{3.5 \text{ hour}} = 600$ (miles per hour)

Speed with tailwind $= \dfrac{2100 \text{ miles}}{3 \text{ hours}} = 700$ (miles per hour)

*System of Equations:*
$$\begin{cases} x - y = 600 \\ x + y = 700 \end{cases}$$

*Solving by Elimination:*
$$\begin{cases} x - y = 600 \\ x + y = 700 \end{cases}$$
$$2x \quad = 1300$$
$$x \quad = 650$$

$$650 + y = 700 \quad \text{Replace } x \text{ by 650 in second equation.}$$
$$y = 50$$

The speed of the plane in still air is 650 miles per hour, and the speed of the wind is 50 miles per hour.

**45.** *Verbal Model:*

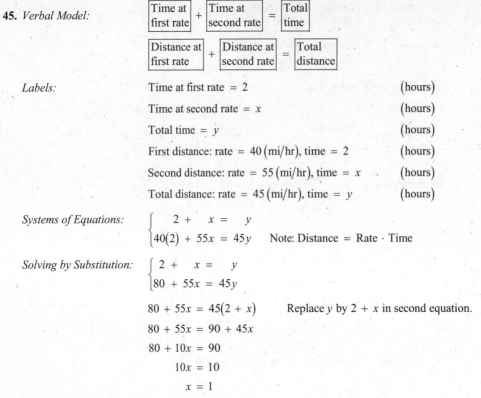

$$\boxed{\begin{array}{c}\text{Time at}\\\text{first rate}\end{array}} + \boxed{\begin{array}{c}\text{Time at}\\\text{second rate}\end{array}} = \boxed{\begin{array}{c}\text{Total}\\\text{time}\end{array}}$$

$$\boxed{\begin{array}{c}\text{Distance at}\\\text{first rate}\end{array}} + \boxed{\begin{array}{c}\text{Distance at}\\\text{second rate}\end{array}} = \boxed{\begin{array}{c}\text{Total}\\\text{distance}\end{array}}$$

*Labels:*

| | |
|---|---|
| Time at first rate $= 2$ | (hours) |
| Time at second rate $= x$ | (hours) |
| Total time $= y$ | (hours) |
| First distance: rate $= 40\,(\text{mi/hr})$, time $= 2$ | (hours) |
| Second distance: rate $= 55\,(\text{mi/hr})$, time $= x$ | (hours) |
| Total distance: rate $= 45\,(\text{mi/hr})$, time $= y$ | (hours) |

*Systems of Equations:*

$$\begin{cases} 2 + x = y \\ 40(2) + 55x = 45y \end{cases} \quad \text{Note: Distance} = \text{Rate} \cdot \text{Time}$$

*Solving by Substitution:*

$$\begin{cases} 2 + x = y \\ 80 + 55x = 45y \end{cases}$$

$80 + 55x = 45(2 + x)$      Replace $y$ by $2 + x$ in second equation.

$80 + 55x = 90 + 45x$

$80 + 10x = 90$

$10x = 10$

$x = 1$

The van must travel for 1 hour at 55 miles per hour.

**47.** *Verbal Model:*

$$\boxed{\text{Time of first runner}} + \boxed{\text{Time of second runner}} = \boxed{\text{Total time}}$$

$$\boxed{\text{Distance of first runner}} + \boxed{\text{Distance of second runner}} = \boxed{\text{Total distance}}$$

*Labels:*

Time of first runner $= x$ (minutes)

Time of second runner $= y$ (minutes)

Total time $= 41$ (minutes)

Distance of first runner $= 110x$ (meters)

Distance of second runner $= 160y$ (meters)

Total distance $= 5660$ (meters)

*System of Equations:*

$$\begin{cases} x + y = 41 \\ 110x + 160y = 5660 \end{cases}$$

*Solving by Substitution:*

$$\begin{cases} x + y = 41 \Rightarrow x = -y + 41 \\ 110x + 160y = 5660 \end{cases}$$

$110(-y + 41) + 160y = 5660$   Replace $x$ by $-y + 41$ in second equation.

$-110y + 4510 + 160y = 5660$

$50y + 4510 = 5660$

$50y = 1150$

$y = \frac{1150}{50}$

$y = 23 \Rightarrow x = -23 + 41$   Replace $y$ by 23 in revised first equation.

$x = 18$

$(18, 23)$

The first runner ran 18 minutes and the second runner ran 23 minutes.

**49.** *Verbal Model:*

$$\boxed{\text{Liters of } 35\% \text{ solution}} + \boxed{\text{Liters of } 60\% \text{ solution}} = \boxed{\text{Liters of } 50\% \text{ solution}}$$

$$\boxed{\text{Alcohol in } 35\% \text{ solution}} + \boxed{\text{Alcohol in } 60\% \text{ solution}} = \boxed{\text{Alcohol in } 50\% \text{ solution}}$$

*Labels:*

Number of liters of 35% solution $= x$

Number of liters of 60% solution $= y$

Number of liters of 50% solution $= 10$

Alcohol in 35% solution $= 0.35x$ (liters)

Alcohol in 60% solution $= 0.60y$ (liters)

Alcohol in 50% solution $= 0.50(10)$ (liters)

*System of Equations:*

$$\begin{cases} x + y = 10 \\ 0.35x + 0.60y = 0.50(10) \end{cases}$$

*Solving System by Elimination:*

$$\begin{cases} x + y = 10 \\ 0.35x + 0.60y = 5 \end{cases}$$

$$\begin{array}{rl} -35x - 35y = -350 & \text{Multiply first equation by } -35. \\ 35x + 60y = 500 & \text{Multiply second equation by } 100. \\ \hline 25y = 150 & \\ y = 6 & \end{array}$$

$$x + 6 = 10 \qquad \text{Replace } y \text{ by 6 in first equation.}$$
$$x = 4$$

So, 4 liters of the 35% solution and 6 liters of the 60% solution must be used.

**51.** *Verbal Model:*

$$\boxed{\text{Pounds of } \$4.25/\text{lb nuts}} + \boxed{\text{Pounds of } \$6.55/\text{lb nuts}} = \boxed{\text{Pounds of mixture}}$$

$$\boxed{\text{Value of } \$4.25/\text{lb nuts}} = \boxed{\text{Value of } \$6.55/\text{lb nuts}} = \boxed{\text{Value of mixture}}$$

*Labels:*

Pounds of \$4.25/lb nuts $= x$

Pounds of \$6.55/lb nuts $= y$

Pounds of mixture $= 10$

Value of \$4.25/lb nuts $= 4.25x$ (dollars)

Value of \$6.55/lb nuts $= 6.55y$ (dollars)

Value of mixture $= 5.86(10) = 58.60$ (dollars)

*System of Equations:*

$$\begin{cases} x + y = 10 \\ 4.25x + 6.55y = 58.60 \end{cases}$$

*Solving by Elimination:*

$$\begin{array}{ll} \begin{cases} x + y = 10 & \Rightarrow -425x - 425y = -4250 \\ 4.25x + 6.55y = 58.60 & \Rightarrow \underline{425x + 655y = 5860} \end{cases} & \begin{array}{l}\text{Multiply by } -425. \\ \text{Multiply by } 100.\end{array} \\ \qquad\qquad\qquad\qquad\qquad 230y = 1610 & \\ \qquad\qquad\qquad\qquad\qquad\quad y = 7 & \end{array}$$

$$x + 7 = 10 \qquad \text{Replace } y \text{ by 7 in first equation.}$$
$$x = 3$$

So, 3 pounds of the \$4.25/lb walnuts and 7 pounds of the \$6.55/lb cashews were used in the mixture.

**53.** *Verbal Model:*

$$\boxed{\text{Tons of }\$110/\text{ton hay}} + \boxed{\text{Tons of }\$60/\text{ton hay}} = \boxed{\text{Total tons of hay}}$$

$$\boxed{\text{Value of }\$110/\text{ton hay}} + \boxed{\text{Value of }\$60/\text{ton hay}} = \boxed{\text{Total value of hay}}$$

*Labels:*

Tons of $110/ton hay $= x$

Tons of $60/ton hay $= y$

Total tons of hay $= 100$

Value of $110/ton hay $= 110x$       (dollars)

Value of $60/ton hay $= 60y$       (dollars)

Total value of hay $= 100(75) = 7500$       (dollars)

*System of Equations:*

$$\begin{cases} x + y = 100 \\ 110x + 60y = 7500 \end{cases}$$

*Solving by Elimination:*

$$\begin{cases} x + y = 100 \Rightarrow -60x - 60y = -6000 \\ 110x + 60y = 7500 \Rightarrow 110x + 60y = \phantom{-}7500 \end{cases}$$     Multiply by $-60$.

$$\begin{aligned} 50x &= 1500 \\ x &= 30 \end{aligned}$$

$30 + y = 100$    Replace $x$ by 30 in first equation.

$$y = 70$$

So, 30 tons of $110/ton hay and 70 tons of $60/ton hay are required.

**55.** *Verbal Model:*

$$\boxed{\text{Amount invested at 6\%}} + \boxed{\text{Amount invested at 9.5\%}} = \boxed{\text{Total investment}}$$

$$\boxed{\text{Interest at 6\%}} + \boxed{\text{Interest at 9.5\%}} = \boxed{\text{Total interest}}$$

*Labels:*

Amount invested at 8% $= x$       (dollars)

Amount invested at 9.5% $= y$       (dollars)

Total investment $= 24,000$       (dollars)

Interest at 6% $= 0.06x$       (dollars)

Interest at 9.5% $= 0.095y$       (dollars)

Total interest $= 2000$       (dollars)

*System of Equations:*

$$\begin{cases} x + y = 24{,}000 \\ 0.06x + 0.095y = 2000 \end{cases}$$

*Solving by Elimination:*

$$\begin{cases} x + y = 24{,}000 \\ 0.06x + 0.095y = 2000 \end{cases}$$

$$\begin{aligned} -60x - 60y &= -1{,}440{,}000 \quad &&\text{Multiply first equation by } -60. \\ 60x + 95y &= \phantom{-}2{,}000{,}000 \quad &&\text{Multiply second equation by 1000.} \\ \hline 35y &= \phantom{-}560{,}000 \\ y &= \phantom{-}16{,}000 \end{aligned}$$

So, at least $16,000 should be invested at 9.5% to meet your goal.

**57.** *Verbal Model:*

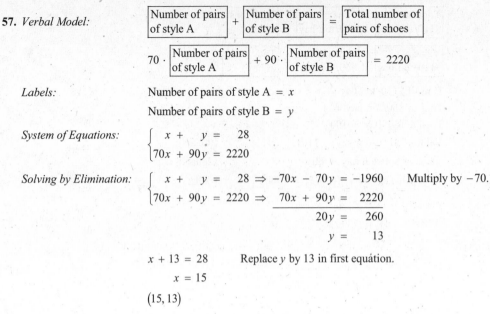

$$70 \cdot \boxed{\begin{array}{l}\text{Number of pairs}\\\text{of style A}\end{array}} + 90 \cdot \boxed{\begin{array}{l}\text{Number of pairs}\\\text{of style B}\end{array}} = 2220$$

*Labels:*

Number of pairs of style A $= x$

Number of pairs of style B $= y$

*System of Equations:*

$$\begin{cases} x + y = 28 \\ 70x + 90y = 2220 \end{cases}$$

*Solving by Elimination:*

$$\begin{cases} x + y = 28 \Rightarrow -70x - 70y = -1960 \qquad \text{Multiply by } -70. \\ 70x + 90y = 2220 \Rightarrow \underline{70x + 90y = 2220} \\ \phantom{70x + 90y = 2220 \Rightarrow} 20y = 260 \\ \phantom{70x + 90y = 2220 \Rightarrow} y = 13 \end{cases}$$

$$x + 13 = 28 \qquad \text{Replace } y \text{ by 13 in first equation.}$$
$$x = 15$$

$$(15, 13)$$

There are 15 pairs of Style A shoes and 13 pairs of Style B shoes.

**59.** *Verbal Model:*

$$\boxed{\begin{array}{l}\text{Minutes of}\\\text{walking}\end{array}} + \boxed{\begin{array}{l}\text{Minutes of}\\\text{jogging}\end{array}} = 30$$

$$0.05 \cdot \boxed{\begin{array}{l}\text{Minutes of}\\\text{walking}\end{array}} + 0.1 \cdot \boxed{\begin{array}{l}\text{Minutes of}\\\text{jogging}\end{array}} = 2.5$$

*Labels:*

Minutes of walking $= x$

Minutes of jogging $= y$

*System of Equations:*

$$\begin{cases} x + y = 30 \\ 0.05x + 0.1y = 2.5 \end{cases}$$

*Solving by Elimination:*

$$\begin{cases} x + y = 30 \Rightarrow -0.05x - 0.05y = -1.5 \qquad \text{Multiply by } -0.05. \\ 0.05x + 0.1y = 2.5 \Rightarrow \underline{0.05x + 0.1y = 2.5} \\ \phantom{0.05x + 0.1y = 2.5 \Rightarrow} 0.05y = 1 \\ \phantom{0.05x + 0.1y = 2.5 \Rightarrow} y = 20 \end{cases}$$

$$x + 20 = 30$$
$$x = 10$$

$$(10, 20)$$

The number of minutes spent walking was 10 minutes.

**61.**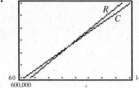

$x \approx 96$

$R = 8950(96) = \$859,200$

Algebraic check:   $R = C$

$$8950x = 7650x + 125,000$$
$$1300x = 125,000$$
$$x \approx 96$$

**63.**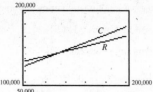

$x \approx 133,333.33$   or   $x \approx 133,333$

$R = 0.85(133,333) = \$113,333.05$

Algebraic check:   $R = C$

$$0.85x = 0.55x + 40,000$$
$$0.30x = 40,000$$
$$x \approx 133,333.33 \text{ or } x \approx 133,333$$

**65.** $y = mx + b$

$(2, -1)$: $-1 = m(2) + b \Rightarrow 2m + b = -1 \Rightarrow -2m - b = 1$
$(6, 1)$:   $1 = m(6) + b \Rightarrow 6m + b = \phantom{-}1 \Rightarrow \dfrac{6m + b = 1}{}$

$$4m \quad\;\; = 2$$
$$m \quad\;\; = \tfrac{1}{2}$$

$$2\left(\tfrac{1}{2}\right) + b = -1$$
$$1 + b = -1$$
$$b = -2$$

$\left(\tfrac{1}{2}, -2\right)$

The equation of the line is $y = \tfrac{1}{2}x - 2$.

**67.** $y = mx + b$

$(-3, 6)$: $6 = m(-3) + b \Rightarrow -3m + b = 6 \Rightarrow 3m - b = -6$
$(5, 2)$:   $2 = m(5)\phantom{-} + b \Rightarrow \phantom{-}5m + b = 2 \Rightarrow \dfrac{5m + b = \phantom{-}2}{}$

$$8m \quad\;\; = -4$$
$$m \quad\;\; = -\tfrac{1}{2}$$

$$-3\left(-\tfrac{1}{2}\right) + b = 6$$
$$\tfrac{3}{2} + b = 6$$
$$b = \tfrac{9}{2}$$

$\left(-\tfrac{1}{2}, \tfrac{9}{2}\right)$

The equation of the line is $y = -\tfrac{1}{2}x + \tfrac{9}{2}$.

**69.** *Verbal Model:*    | Pounds of cashews | + | Pounds of peanuts | = | Pounds of mixed nuts |

   | Value of cashews | + | Value of peanuts | = | Value of mixed nuts |

*Labels:*        Pounds of cashews $= x$        (pounds)

            Pounds of mixed nuts $= y$        (pounds)

*System of Equations:*   $\begin{cases} x + 15 \;=\; y \\ 7x + 37.5 = 4y \end{cases}$

**71.** The interest rates were not written correctly in decimal form; $5\% = 0.05$ and $8\% = 0.08$.

    *Labels:*         Amount invested at $5\% = x$    (dollars)

                       Amount invested at $8\% = y$    (dollars)

    *System:*
$$\begin{cases} x + y = 9000 \\ 0.05x + 0.08y = 645 \end{cases}$$

**73.** $m_1 = \dfrac{4-7}{4-8}$        $m_2 = \dfrac{5-1}{-1-2}$

     $= \dfrac{-3}{-4}$           $= \dfrac{4}{-3}$

     $= \dfrac{3}{4}$           $= -\dfrac{4}{3}$

    The lines are perpendicular because $m_1 = -\dfrac{1}{m_2}$.

**75.** $m_1 = \dfrac{-2-0}{7-12}$        $m_2 = \dfrac{9-7}{-5-0}$

     $= \dfrac{-2}{-5}$           $= \dfrac{2}{-5}$

     $= \dfrac{2}{5}$           $= -\dfrac{2}{5}$

    The lines are neither parallel nor perpendicular because $m_1 \neq m_2$ and $m_1 \neq -\dfrac{1}{m_2}$.

**77.**
$$\begin{cases} 3x + 3y = 7 \Rightarrow \phantom{-}3x + \phantom{-}3y = \phantom{-}7 \\ 3x + 5y = 3 \Rightarrow -3x - \phantom{-}5y = -3 \end{cases}$$
                                       Multiply equation by $-1$.
$$-2y = 4$$
$$y = -2$$

    $3x + 3(-2) = 7$         Replace $y$ by $-2$ in first equation.

       $3x - 6 = 7$

           $3x = 13$

            $x = \dfrac{13}{3}$

    $\left(\dfrac{13}{3}, -2\right)$

**79.**
$$\begin{cases} 6x + 5y = 19 \Rightarrow \phantom{-}6x + \phantom{-}5y = \phantom{-}19 \\ 2x + 3y = 5 \Rightarrow -6x - \phantom{-}9y = -15 \end{cases}$$
                                         Multiply equation by $-3$.
$$-4y = 4$$
$$y = -1$$

    $2x + 3(-1) = 5$         Replace $y$ by $-1$ in second equation.

       $2x - 3 = 5$

           $2x = 8$

            $x = 4$

    $(4, -1)$

## Section 8.5  Systems of Linear Inequalities

    **1.** Graph (c)

    **2.** Graph (b)

    **3.** Graph (f)

    **4.** Graph (e)

    **5.** Graph (a)

    **6.** Graph (d)

**7.**

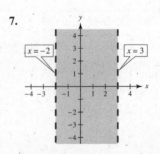

**9.**

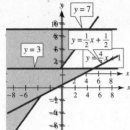

**11.**

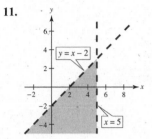

**13.**

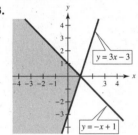

**15.**

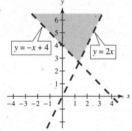

**17.**

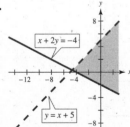

**19.**

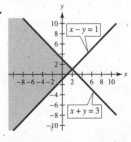

**21.**

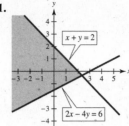

**23.**

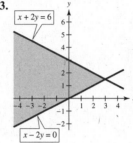

**25.**

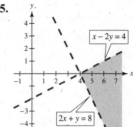

**27.**

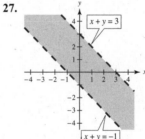

**29.**

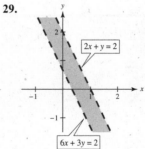

**31.**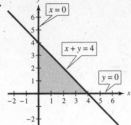

**33.**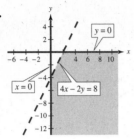

**35.**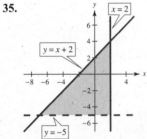

**37.**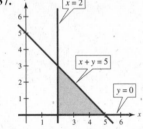

**39.**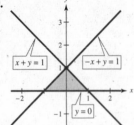

**41.**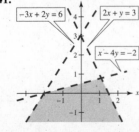

**43.**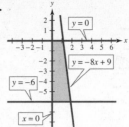

**45.**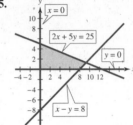

**47.**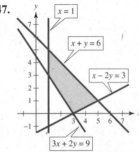

**49.**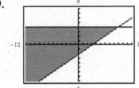

**51.**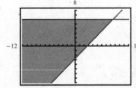

**53.**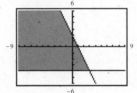

**55.** Vertical line through $(1, 3)$ and $(1, -5)$: $x = 1$

Vertical line through $(8, 3)$ and $(8, -5)$: $x = 8$

Horizontal line through $(1, -5)$ and $(8, -5)$: $y = -5$

Horizontal line through $(1, 3)$ and $(8, 3)$: $y = 3$

System of inequalities:

Region on and to the right of $x = 1$ $\qquad \begin{cases} x \geq 1 \\ x \leq 8 \\ y \geq -5 \\ y \leq 3 \end{cases}$

Region on and to the left of $x = 8$

Region on and above $y = -5$

Region on and below $y = 3$

**57.** Line through $(-6, 3)$ and $(4, 12)$: $m = \dfrac{12 - 3}{4 - (-6)} = \dfrac{9}{10}$

$$y - 3 = \frac{9}{10}(x + 6) \Rightarrow y = \frac{9}{10}x + \frac{42}{5}$$

Line through $(-6, 3)$ and $(3, 9)$: $m = \dfrac{9 - 3}{3 - (-6)} = \dfrac{6}{9} = \dfrac{2}{3}$

$$y - 3 = \frac{2}{3}(x + 6) \Rightarrow y = \frac{2}{3}x + 7$$

Line through $(3, 9)$ and $(4, 12)$: $m = \dfrac{12 - 9}{4 - 3} = \dfrac{3}{1} = 3$

$$y - 9 = 3(x - 3) \Rightarrow y = 3x$$

Systems of inequalities:

Region on and below the line $y = \dfrac{9}{10}x + \dfrac{42}{5}$ $\qquad \begin{cases} y \leq \dfrac{9}{10}x + \dfrac{42}{5} \\ y \geq 3x \\ y \geq \dfrac{2}{3}x + 7 \end{cases}$

Region on and above the line $y = 3x$

Region on and above the line $y = \dfrac{2}{3}x + 7$

**59.** *Verbal Model:*   $\boxed{\begin{array}{l}\text{Amount invested} \\ \text{in Account X}\end{array}} + \boxed{\begin{array}{l}\text{Amount invested} \\ \text{in Account Y}\end{array}} \leq 20{,}000$

$\boxed{\begin{array}{l}\text{Amount invested} \\ \text{in Account X}\end{array}} \geq 5000$

$\boxed{\begin{array}{l}\text{Amount invested} \\ \text{in Account Y}\end{array}} \geq 2\boxed{\begin{array}{l}\text{Amount invested} \\ \text{in Account X}\end{array}}$

*Labels:*   Amount invested in Account X $= x$   (dollars)

Amount invested in Account Y $= y$   (dollars)

*System of Inequalities:* $\begin{cases} x + y \leq 20{,}000 \\ x \qquad \geq 5000 \\ \qquad y \geq 2x \end{cases}$

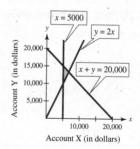

**61.** *Verbal Model:*

$$\boxed{\text{Number of } \$15 \text{ tickets}} + \boxed{\text{Number of } \$25 \text{ tickets}} \geq 15{,}000$$

$$15 \cdot \boxed{\text{Number of } \$15 \text{ tickets}} + 25 \cdot \boxed{\text{Number of } \$25 \text{ tickets}} \geq 275{,}000$$

$$\boxed{\text{Number of } \$15 \text{ tickets}} \geq 8000$$

$$\boxed{\text{Number of } \$25 \text{ tickets}} \geq 4000$$

*Labels:*    Number of $15 tickets $= x$

Number of $25 tickets $= y$

*System of Inequalities:*
$$\begin{cases} x + y \geq 15{,}000 \\ 15x + 25y \geq 275{,}000 \\ x \geq 8000 \\ y \geq 4000 \end{cases}$$

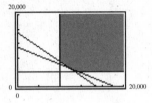

**63.** Horizontal line through $(0, 0)$ and $(90, 0)$: $y = 0$

Horizontal line through $(70, -10)$ and $(90, -10)$: $y = -10$

Vertical line through $(90, -10)$ and $(90, 0)$: $x = 90$

Line through $(0, 0)$ and $(70, -10)$: $m = \dfrac{-10 - 0}{70 - 0} = \dfrac{-10}{70} = -\dfrac{1}{7}$ and $y$-intercept: $(0, 0) \Rightarrow y = -\dfrac{1}{7}x$

System of inequalities:

Region on and below $y = 0$:

Region on and above $y = -10$:

Region on and to the left of $x = 90$:

Region on and above $y = -\dfrac{1}{7}x$:

$$\begin{cases} y \leq 0 \\ y \geq -10 \\ x \leq 90 \\ y \geq -\dfrac{1}{7}x \end{cases}$$

**65.** The graph of a linear equation splits the *xy*-plane into two parts, each of which is a half-plane.
Here are some examples of half-planes:

$y < 5$

$x > -4$

$y < 3x + 2$

**67.** Yes. The solution of a system of linear inequalities is a single point if the system consists of two pairs of inequalities graphed as solid lines, each having different inequalities but the same corresponding equations.

Here are examples:

$$\begin{cases} y \geq x \\ y \leq x \\ y \leq 3 \\ y \geq 3 \end{cases} \qquad \begin{cases} 2x + y \leq 4 \\ 2x + y \geq 4 \\ x + y \geq 0 \\ x + y \leq 0 \end{cases} \qquad \begin{cases} y \leq 0 \\ y \geq 0 \\ x \geq 0 \\ x \leq 0 \end{cases}$$

The solution is
the point $(3, 3)$.

The solution is
the point $(4, -4)$.

The solution is
the point $(0, 0)$.

**69.** $(3) \cdot (3) \cdot (3) \cdot (3) = (3)^4$

**73.** $(-4)^4 = (-4) \cdot (-4) \cdot (-4) \cdot (-4)$

**71.** $\left(\frac{1}{2}\right) \cdot \left(\frac{1}{2}\right) \cdot \left(\frac{1}{2}\right) \cdot \left(\frac{1}{2}\right) \cdot \left(\frac{1}{2}\right) \cdot \left(\frac{1}{2}\right) = \left(\frac{1}{2}\right)^6$

**75.** $\left(-\frac{3}{4}\right)^2 = \left(-\frac{3}{4}\right) \cdot \left(-\frac{3}{4}\right)$

**77.** $2x + 4y = 8$

$$4y = -2x + 8$$

$$y = \frac{-2}{4}x + \frac{8}{4}$$

$$y = -\frac{1}{2}x + 2$$

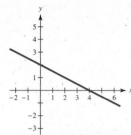

**79.** $0.3x - 0.2y = 0.8$

$$10(0.3x - 0.2y) = (0.8)10$$

$$3x - 2y = 8$$

$$-2y = -3x + 8$$

$$y = \frac{-3}{-2}x + \frac{8}{-2}$$

$$y = \frac{3}{2}x - 4$$

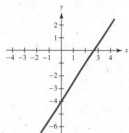

# Review Exercises for Chapter 8

**1.** (a) $(2, -1)$

$$3(2) - 5(-1) \stackrel{?}{=} 11$$

$$6 + 5 = 11$$

$$-2 + 2(-1) = -4$$

$$-2 - 2 = -4$$

$(2, -1)$ *is* a solution.

(b) $(3, -2)$

$$3(3) - 5(-2) \stackrel{?}{=} 11$$

$$9 + 10 \neq 11$$

$(3, -2)$ *is not* a solution.

**3.** (a) $(0.5, -0.7)$

$$0.2(0.5) + 0.4(-0.7) \stackrel{?}{=} 5$$

$$0.01 - 0.28 \neq 5$$

$(0.5, -0.7)$ *is not* a solution.

(b) $(15, 5)$

$$0.2(15) - 0.4(5) \stackrel{?}{=} 5$$

$$3 + 2 = 5$$

$$15 + 3(5) \stackrel{?}{=} 30$$

$$15 + 15 \stackrel{?}{=} 30$$

$(15, 5)$ *is* a solution.

**5.** The solution is $(1, 2)$.

$$\begin{cases} 2(1) + 2 = 4 \\ 2(1) - 2 = 0 \end{cases}$$

**7.** The solution is $(0, -3)$.

$$\begin{cases} 3(0) - 2(-3) = 6 \\ 2(0) - (-3) = 3 \end{cases}$$

**9.**

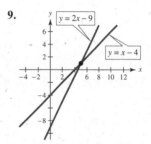

The solution is $(5, 1)$.

**11.**

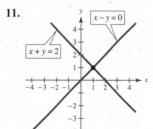

The solution is $(1, 1)$.

**13.**

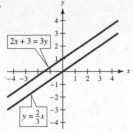

There is *no* solution. The system is inconsistent.

**15.**

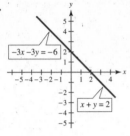

The system has *infinitely many* solutions. The solution set consists of all ordered pairs $(x, y)$ such that

$$x + y = 2.$$

**17.**

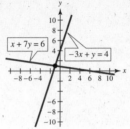

The solution is $(-1, 1)$.

**19.**

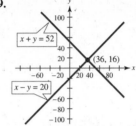

The solution is $(36, 16)$.

**21.**
$$\begin{cases} y = 2x \\ y = x + 4 \end{cases}$$

$2x = x + 4$     Replace $y$ by $2x$ in second equation.

$x = 4 \Rightarrow y = 2(4)$     Replace $x$ by 4 in first equation.

$\qquad\qquad y = 8$

$(4, 8)$

**23.**
$$\begin{cases} x = 3y - 2 \\ x = 6 - y \end{cases}$$

$3y - 2 = 6 - y$     Replace $x$ by $3y - 2$ in second equation.

$4y - 2 = 6$

$\qquad 4y = 8$     Replace $x$ by $3y - 2$ in second equation.

$\qquad\quad y = 2$

$x = 3(2) - 2$     Replace $y$ by 2 in first equation.

$x = 6 - 2$

$x = 4$

$(4, 2)$

**25.** $\begin{cases} x - 2y = 6 \Rightarrow x = 2y + 6 \\ 3x + 2y = 10 \end{cases}$

$\quad 3(2y + 6) + 2y = 10 \quad$ Replace $x$ by $2y + 6$ in second equation.

$\quad\quad 6y + 18 + 2y = 10$

$\quad\quad\quad 8y + 18 = 10$

$\quad\quad\quad\quad 8y = -8$

$\quad\quad\quad\quad\quad y = -1$

$\quad x = 2(-1) + 6 \quad\quad\quad$ Replace $y$ by $-1$ in revised first equation.

$\quad x = -2 + 6$

$\quad x = 4$

$\quad (4, -1)$

**27.** $\begin{cases} 2x - y = 2 \Rightarrow -y = -2x + 2 \Rightarrow y = 2x - 2 \\ 6x + 8y = 39 \end{cases}$

$\quad 6x + 8(2x - 2) = 39 \quad\quad\quad\quad$ Replace $y$ by $2x - 2$ in second equation.

$\quad 6x + 16x - 16 = 39$

$\quad\quad 22x - 16 = 39$

$\quad\quad\quad 22x = 55$

$\quad\quad\quad\quad x = \frac{55}{22}$

$\quad\quad\quad\quad x = \frac{5}{2} \Rightarrow y = 2\left(\frac{5}{2}\right) - 2 \quad$ Replace $x$ by $\frac{5}{2}$ in revised first equation.

$\quad\quad\quad\quad\quad\quad\quad\quad y = 5 - 2$

$\quad\quad\quad\quad\quad\quad\quad\quad y = 3$

$\quad \left(\frac{5}{2}, 3\right)$

**29.** $\begin{cases} \frac{3}{5}x - y = 8 \Rightarrow -y = -\frac{3}{5}x + 8 \Rightarrow y = \frac{3}{5}x - 8 \\ 2x - 3y = 25 \end{cases}$

$\quad 2x - 3\left(\frac{3}{5}x - 8\right) = 25 \quad\quad\quad$ Replace $y$ by $\frac{3}{5}x - 8$ in second equation.

$\quad 2x - \frac{9}{5}x + 24 = 25$

$\quad 5\left(2x - \frac{9}{5}x + 24\right) = 25 \cdot 5$

$\quad 10x - 9x + 120 = 125$

$\quad\quad x + 120 = 125$

$\quad\quad\quad x = 5 \Rightarrow y = \frac{3}{5}(5) - 8 \quad$ Replace $x$ by $5$ in revised first equation.

$\quad\quad\quad\quad\quad\quad y = 3 - 8$

$\quad\quad\quad\quad\quad\quad y = -5$

$\quad (5, -5)$

**31.** $\begin{cases} x = y + 3 \\ x = y + 1 \end{cases}$

$\quad y + 3 = y + 1 \quad$ Replace $x$ by $y + 3$ in second equation.

$\quad\quad 3 = 1 \quad\quad\quad$ False

So, there is *no* solution. The system is inconsistent.

**33.** $\begin{cases} -6x + y = -3 \Rightarrow y = 6x - 3 \\ 12x - 2y = 6 \end{cases}$

$12x - 2(6x - 3) = 6$       Replace $y$ by $6x - 3$ in second equation.

$12x - 12x + 6 = 6$

$6 = 6$       True

This system has infinitely many solutions. The solution set consists of all points $(x, y)$ such that $y = 6x - 3$.

**35.** $\begin{cases} x + y = 12{,}000 \Rightarrow y = 12{,}000 - x \\ 0.05x + 0.10y = 800 \end{cases}$

$0.05x + 0.10(12{,}000 - x) = 800$       Replace $y$ by $12{,}000 - x$ in second equation.

$0.05x + 1200 - 0.10x = 800$

$1200 - 0.05x = 800$

$-0.05x = -400$

$x = \dfrac{-400}{-0.05}$

$x = 8000$    and    $y = 12{,}000 - 8000$

$y = 4000$

$(8000, 4000)$

So, $8000 is invested at 5% and $4000 is invested at 10%.

**37.** $\begin{cases} x + y = 1510 \Rightarrow y = 1510 - x \\ 5x + 3y = 6138 \end{cases}$

$5x + 3(1510 - x) = 6138$       Replace $y$ by $1510 - x$ in second equation.

$5x + 4530 - 3x = 6138$

$2x + 4530 = 6138$

$2x = 1608$

$x = 804$    and    $y = 1510 - 804$

$y = 706$

$(804, 706)$

So, 804 adult tickets and 706 children's tickets were sold.

**39.** $\begin{cases} 3x - y = 5 \\ 2x + y = 5 \end{cases}$

$5x = 10$

$x = 2$

$2(2) + y = 5$       Replace $x$ by 2 in second equation.

$4 + y = 5$

$y = 1$

$(2, 1)$

**41.** $\begin{cases} 5x + 4y = 2 \\ -x + y = -22 \end{cases} \Rightarrow \begin{array}{l} 5x + 4y = 2 \\ -5x + 5y = -110 \end{array}$     Multiply equation by 5.

$$\overline{\phantom{-5x +} 9y = -108}$$

$$y = -12$$

$-x + (-12) = -22$     Replace $y$ by $-12$ in second equation.

$$-x = -10$$

$$x = 10$$

$(10, -12)$

**43.** $\begin{cases} 8x - 6y = 4 \\ -4x + 3y = -2 \end{cases} \Rightarrow \begin{array}{l} 8x - 6y = 4 \\ -8x + 6y = -4 \end{array}$     Multiply equation by 2.

$$\overline{\phantom{-8x + 6y} 0 = 0}$$

The system has *infinitely many* solutions. The solution set consists of all ordered pairs $(x, y)$ such that $4x - 3y = 2$.

**45.** $\begin{cases} \dfrac{2}{3}x + \dfrac{1}{12}y = \dfrac{3}{4} \\ 3x - 4y = 2 \end{cases} \Rightarrow \begin{array}{l} 32x + 4y = 36 \\ 3x - 4y = 2 \end{array}$     Multiply equation by 48.

$$\overline{\phantom{32x +} 35x = 38}$$

$$x = \frac{38}{35}$$

$3\left(\dfrac{38}{35}\right) - 4y = 2$     Replace $x$ by $\dfrac{38}{35}$ in second equation.

$$35 \cdot 3\left(\frac{38}{35}\right) - 35 \cdot 4y = 2 \cdot 35$$

$$114 - 140y = 70$$

$$-140y = -44$$

$$y = \frac{-44}{-140}$$

$$y = \frac{11}{35}$$

$\left(\dfrac{38}{35}, \dfrac{11}{35}\right)$

**47.** $\begin{cases} 0.2x + 0.1y = 0.03 \\ 0.3x - 0.1y = -0.13 \end{cases} \Rightarrow \begin{array}{l} 20x + 10y = 3 \\ 30x - 10y = -13 \end{array}$    Multiply by 100. <br> Multiply by 100.

$$\overline{\phantom{30x -} 50x = -10}$$

$$x = \frac{-10}{50}$$

$$x = -0.2$$

$20(-0.2) + 10y = 3$     Replace $x$ by $-0.2$ in revised first equation.

$$-4 + 10y = 3$$

$$10y = 7$$

$$y = 0.7$$

$(-0.2, 0.7)$ or $\left(-\dfrac{1}{5}, \dfrac{7}{10}\right)$

**49.** $\begin{cases} 6x - 5y = 0 \\ \quad\quad y = 6 \end{cases}$

$6x - 5(6) = 0$     Replace $y$ by 6 in first equation.

$6x - 30 = 0$

$\quad\quad 6x = 30$

$\quad\quad\; x = 5$

$(5, 6)$

**51.** $\begin{cases} -x + 4y = 4 \\ \;\; x + \; y = 6 \end{cases}$

$\quad\quad\quad 5y = 10$

$\quad\quad\quad\; y = 2$

$x + 2 = 6$     Replace $y$ by 2 in second equation.

$\quad\; x = 4$

$(4, 2)$

**53.** $\begin{cases} x - \; y = 0 \Rightarrow x = y \\ x - 6y = 5 \end{cases}$

$y - 6y = 5$     Replace $x$ by $y$ in second equation.

$\quad -5y = 5$

$\quad\quad y = -1$

$x = -1$     Replace $y$ by $-1$ in revised first equation.

$(-1, -1)$

**55.** $\begin{cases} 5x + 8y = 8 \\ \; x - 8y = 16 \end{cases}$

$6x \quad\quad\; = 24$

$\; x \quad\quad\; = 4$

$5(4) + 8y = 8$     Replace $x$ by 4 in first equation.

$20 + 8y = 8$

$\quad\quad 8y = -12$

$\quad\quad\; y = -\frac{3}{2}$

$\left(4, -\frac{3}{2}\right)$

**57.** $\begin{cases} 6x - 3y = 27 \Rightarrow \quad 6x - 3y = \quad 27 \\ -2x + \; y = -9 \Rightarrow -6x + 3y = -27 \end{cases}$     Multiply equation by 3.

$\quad\quad\quad\quad\quad\quad\quad\quad 0 = \quad 0$

The system has infinitely many solutions. The solution set consists of all ordered pairs $(x, y)$ such that $-2x + y = -9$.

**59.** $\begin{cases} 2x + 6y = 16 \Rightarrow \quad 2x + 6y = 16 \\ 2x + 3y = \; 7 \Rightarrow -2x - 3y = -7 \end{cases}$     Multiply equation by $-1$.

$\quad\quad\quad\quad\quad\quad\quad\quad 3y = \quad 9$

$\quad\quad\quad\quad\quad\quad\quad\quad\; y = \quad 3$

$2x + 6(3) = 16$     Replace $y$ by 3 in first equation.

$2x + 18 = 16$

$\quad\quad 2x = -2$

$\quad\quad\; x = -1$

$(-1, 3)$

**61.** $\begin{cases} \frac{1}{5}x + \frac{3}{2}y = 2 \\ 2x + 13y = 20 \end{cases}$ $\Rightarrow$ $\begin{array}{r} -2x - 15y = -20 \\ 2x + 13y = 20 \\ \hline -2y = 0 \\ y = 0 \end{array}$ Multiply equation by $-10$.

$2x + 13(0) = 20$  Replace $y$ by 0 in second equation.

$2x = 20$

$x = 10$

$(10, 0)$

**63.** $\begin{cases} \frac{1}{3}x + \frac{4}{7}y = 3 \\ 2x + 3y = 15 \end{cases}$ $\Rightarrow$ $\begin{array}{r} 7x + 12y = 63 \\ -8x - 12y = -60 \\ \hline -x = 3 \\ x = -3 \end{array}$ Multiply equation by 21.

Multiply equation by $-4$.

$2(-3) + 3y = 15$  Replace $x$ by $-3$ in second equation.

$-6 + 3y = 15$

$3y = 21$

$y = 7$

$(-3, 7)$

**65.** $\begin{cases} 1.2s + 4.2t = -1.7 \\ 3.0s - 1.8t = 1.9 \end{cases}$

Multiply both equations by 10.

$12s + 42t = -17 \Rightarrow \begin{array}{r} -60s - 210t = 85 \\ 60s - 36t = 38 \\ \hline -246t = 123 \\ t = -\frac{123}{246} \\ t = -\frac{1}{2} \end{array}$ Multiply equation by $-5$.

$30s - 18t = 19 \Rightarrow$  Multiply equation by 2.

$30s - 18\left(-\frac{1}{2}\right) = 19$  Replace $t$ by $-\frac{1}{2}$ in second equation.

$30s + 9 = 19$

$30s = 10$

$s = \frac{1}{3}$

$\left(\frac{1}{3}, -\frac{1}{2}\right)$

**67.** *Verbal Model:* $\boxed{\text{Number of dimes}} + \boxed{\text{Number of quarters}} = \boxed{\text{Number of coins}}$

$0.10 \cdot \boxed{\text{Number of dimes}} + 0.25 \cdot \boxed{\text{Number of quarters}} = \boxed{\text{Total value}}$

*Labels:* Number of dimes $= x$

Number of quarters $= y$

Number of coins $= 15$

Total value $= \$2.85$

*System of Equations:*
$$\begin{cases} x + y = 15 \\ 0.10x + 0.25y = 2.85 \end{cases}$$

*Solving by Elimination:*
$$\begin{cases} x + y = 15 \Rightarrow -0.10x - 0.10y = -1.5 \qquad \text{Multiply by } -0.10. \\ 0.10x + 0.25y = 2.85 \Rightarrow \underline{\phantom{-}0.10x + 0.25y = \phantom{-}2.85} \\ \phantom{0.10x + 0.25y = 2.85 \Rightarrow} 0.15y = 1.35 \\ \phantom{0.10x + 0.25y = 2.85 \Rightarrow} y = 9 \end{cases}$$

$x + 9 = 15$    Replace $y$ by 9 in first equation.

$\phantom{x + 9} x = 6$

$(6, 9)$

There are 6 dimes and 9 quarters.

**69.** *Verbal Model:* $2 \cdot \boxed{\text{Width of rectangle}} + 2 \cdot \boxed{\text{Height of rectangle}} = \boxed{\text{Perimeter of rectangle}}$

$\boxed{\text{Height of rectangle}} = \frac{2}{3} \cdot \boxed{\text{Width of rectangle}}$

*Labels:* Height of rectangle $= h$        (inches)

Width of rectangle $= w$        (inches)

Perimeter of rectangle $= 120$    (inches)

*System of Equations:*
$$\begin{cases} 2w + 2h = 120 \\ h = \frac{2}{3}w \end{cases}$$

*Solving by Substitution:*
$$\begin{cases} 2w + 2h = 120 \\ h = \frac{2}{3}w \end{cases}$$

$2w + 2\left(\frac{2}{3}w\right) = 120$        Replace $h$ by $\frac{2}{3}$ in first equation.

$2w + \frac{4}{3}w = 120$

$3\left(2w + \frac{4}{3}w\right) = 3(120)$

$6w + 4w = 360$

$10w = 360$

$w = 36 \qquad h = \frac{2}{3}(36)$    Replace $w$ by 36 in second equation.

$\phantom{w = 36 \qquad} h = 24$

$(36, 24)$

The width of the rectangle is 36 inches, and the height of the rectangle is 24 inches.

**71.** *Verbal Model:*   $\boxed{\text{Wholesale cost}} + \boxed{\text{Markup}} = \boxed{\text{Selling price}}$

$\boxed{\text{Markup}} = \boxed{\text{Markup rate}} \cdot \boxed{\text{Wholesale cost}}$

*Labels:*

| | |
|---|---|
| Wholesale cost $= C$ | (dollars) |
| Markup $= M$ | (dollars) |
| Selling price $= 109$ | (dollars) |
| Markup rate $= 0.40$ | (percent in decimal form) |

*System of Equations:*   $\begin{cases} C + M = 109 \\ \quad M = 0.40C \end{cases}$

*Solving by Substitution:*   $\begin{cases} C + M = 109 \\ \quad M = 0.40C \end{cases}$

$C + 0.40C = 109$     Replace $M$ by $0.40C$ in first equation.

$1.40C = 109$

$C = \dfrac{109}{1.40}$

$C \approx 77.857$

The wholesale cost of the DVD player is approximately $77.86.

**73.** *Verbal Model:*   $\boxed{\begin{array}{l}\text{Number of gallons}\\\text{of gasoline}\end{array}} \cdot \boxed{\begin{array}{l}\text{Price of}\\\text{gasoline}\end{array}} + \boxed{\begin{array}{l}\text{Number of gallons}\\\text{of diesel}\end{array}} \cdot \boxed{\begin{array}{l}\text{Price of}\\\text{diesel}\end{array}} = \boxed{\begin{array}{l}\text{Total}\\\text{cost}\end{array}}$

$\boxed{\begin{array}{l}\text{Price of}\\\text{diesel}\end{array}} = \boxed{\begin{array}{l}\text{Price of}\\\text{gasoline}\end{array}} + \boxed{\begin{array}{l}\text{Price difference of}\\\text{diesel and gasoline}\end{array}}$

*Labels:*

| | |
|---|---|
| Number of gallons of gasoline $= 2$ | (gallons) |
| Price of gasoline $= x$ | (dollars per gallon) |
| Number of gallons of diesel $= 5$ | (gallons) |
| Price of diesel $= y$ | (dollars per gallon) |
| Total cost $= 22.03$ | (dollars) |
| Price difference of diesel and gasoline $= 0.08$ | (dollars per gallon) |

*System of Equations:*   $\begin{cases} 2x + 5y = 22.03 \\ \quad\quad y = x + 0.08 \end{cases}$

*Solving by Substitution:*   $\begin{cases} 2x + 5y = 22.03 \\ \quad\quad y = x + 0.08 \end{cases}$

$2x + 5(x + 0.08) = 22.03$     Replace $y$ by $x + 0.08$ in first equation.

$2x + 5x + 0.40 = 22.03$

$7x + 0.40 = 22.03$

$7x = 21.63$

$x = \dfrac{21.63}{7}$

$x = 3.09 \Rightarrow y = 3.09 + 0.08$     Replace $x$ with 3.09 in second equation.

$y = 3.17$

$(3.09, 3.17)$

Gasoline costs $3.09 per gallon and diesel fuel costs $3.17 per gallon.

**75.** *Verbal Model:*

$$\boxed{\begin{array}{c}\text{Time at}\\\text{first rate}\end{array}} + \boxed{\begin{array}{c}\text{Time at}\\\text{second rate}\end{array}} = \boxed{\begin{array}{c}\text{Total}\\\text{time}\end{array}}$$

$$\boxed{\begin{array}{c}\text{Distance at}\\\text{first rate}\end{array}} + \boxed{\begin{array}{c}\text{Distance at}\\\text{second rate}\end{array}} = \boxed{\begin{array}{c}\text{Total}\\\text{distance}\end{array}}$$

*Labels:*

Time at first rate $= 3$     (hours)

Time at second rate $= x$     (hours)

Total time $= y$     (hours)

First distance: rate $= 40$     (miles per hour);     time $= 3$     (hours)

Second distance: rate $= 55$     (miles per hour);     time $= x$     (hours)

Total distance: rate $= 49$     (miles per hour);     time $= y$     (hours)

*System of Equations:*
$$\begin{cases} 3 + x = y \\ 40(3) + 55x = 49y \end{cases}$$

*Solving by Substitution:*
$$\begin{cases} 3 + x = y \\ 40(3) + 55x = 49y \end{cases}$$

$40(3) + 55x = 49(3 + x)$     Replace $y$ by $3 + x$ in second equation.

$120 + 55x = 147 + 49x$

$120 + 6x = 147$

$6x = 27$

$x = \dfrac{27}{6}$

$x = \dfrac{9}{2}$ or $4.5$

The car must travel for 4.5 hours at 55 miles per hour.

**77.**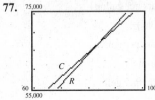

$x = 83$

$R = 800(83) = \$66,400$

Algebraic check:     $R = C$

$800x = 650x + 12,500$

$150x = 12,500$

$x = \dfrac{12,500}{150}$

$x \approx 83$

**79.**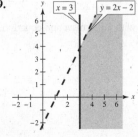

**81.**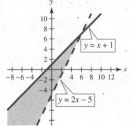

**83.**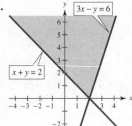

**85.**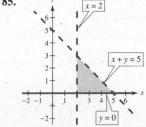

**87.**

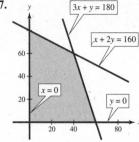

**89.** Line through $(1, 5)$ and $(6, 10)$;  $m = \dfrac{10 - 5}{6 - 1} = \dfrac{5}{5} = 1$;  $y - 5 = 1(x - 1) \Rightarrow y = x + 4$

Line through $(6, 10)$ and $(8, 6)$;  $m = \dfrac{6 - 10}{8 - 6} = \dfrac{-4}{2} = -2$;  $y - 10 = -2(x - 6) \Rightarrow y = -2x + 22$

Line through $(8, 6)$ and $(3, 1)$;  $m = \dfrac{1 - 6}{3 - 8} = \dfrac{-5}{-5} = 1$;  $y - 1 = 1(x - 3) \Rightarrow y = x - 2$

Line through $(3, 1)$ and $(1, 5)$;  $m = \dfrac{5 - 1}{1 - 3} = \dfrac{4}{-2} = -2$;  $y - 5 = -2(x - 1) \Rightarrow y = -2x + 7$

Region on or below $y = x + 4$:  $y \le x + 4$

Region on or below $y = -2x + 22$:  $y \le -2x + 22$

Region on or above $y = x - 2$:  $y \ge x - 2$

Region on or above $y = -2x + 7$  $y \ge -2x + 7$

System of inequalities: $\begin{cases} y \le x + 4 \\ y \le -2x + 22 \\ y \ge x - 2 \\ y \ge -2x + 7 \end{cases}$

**91.** *Verbal Model:*

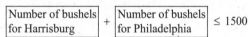

$\boxed{\begin{array}{c}\text{Number of bushels} \\ \text{for Harrisburg}\end{array}} + \boxed{\begin{array}{c}\text{Number of bushels} \\ \text{for Philadelphia}\end{array}} \le 1500$

$\boxed{\begin{array}{c}\text{Number of bushels} \\ \text{for Harrisburg}\end{array}} \ge 400$

$\boxed{\begin{array}{c}\text{Number of bushels} \\ \text{for Philadelphia}\end{array}} \ge 600$

*Labels:*  Number of bushels for Harrisburg $= x$

Number of bushels for Philadelphia $= y$

*System of Equations:* $\begin{cases} x + y \le 1500 \\ x \ge 400 \\ y \ge 600 \end{cases}$

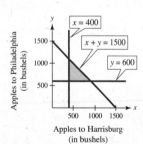

**93.** *Verbal Model:*

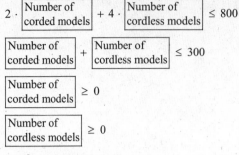

$$2 \cdot \boxed{\begin{array}{c}\text{Number of}\\\text{corded models}\end{array}} + 4 \cdot \boxed{\begin{array}{c}\text{Number of}\\\text{cordless models}\end{array}} \le 800$$

$$\boxed{\begin{array}{c}\text{Number of}\\\text{corded models}\end{array}} + \boxed{\begin{array}{c}\text{Number of}\\\text{cordless models}\end{array}} \le 300$$

$$\boxed{\begin{array}{c}\text{Number of}\\\text{corded models}\end{array}} \ge 0$$

$$\boxed{\begin{array}{c}\text{Number of}\\\text{cordless models}\end{array}} \ge 0$$

*Labels:*    Number of corded models $= x$

Number of cordless models $= y$

*System of Equations:*
$$\begin{cases} 2x + 4y \le 800 \\ x + y \le 300 \\ x \ge 0 \\ y \ge 0 \end{cases}$$

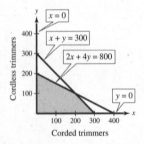

# Chapter Test for Chapter 8

**1.**

| $x - 6y = -19$ | $4x - 5y = 0$ |
|---|---|
| $3 - 6(-2) \overset{?}{=} -19$ | $4(3) - 5(-2) \overset{?}{=} 0$ |
| $3 + 12 \overset{?}{=} -19$ | $12 + 10 \overset{?}{=} 0$ |
| $15 \ne -19$ | $22 \ne 0$ |

$(3, -2)$ *is not* a solution because it does not satisfy the two equations.

| $x - 6y = -19$ | $4x - 5y = 0$ |
|---|---|
| $5 - 6(4) \overset{?}{=} -19$ | $4(5) - 5(4) \overset{?}{=} 0$ |
| $5 - 24 \overset{?}{=} -19$ | $20 - 20 \overset{?}{=} 0$ |
| $-19 = -19$ | $0 = 0$ |

$(5, 4)$ *is* a solution of the system because it satisfies both equations.

**2.**

| $3x + 4y = 16$ | $3x - 4y = 8$ |
|---|---|
| $4y = -3x + 16$ | $-4y = -3x + 8$ |
| $y = -\frac{3}{4}x + 4$ | $y = \frac{3}{4}x - 2$ |
| $m = -\frac{3}{4}$ | $m = \frac{3}{4}$ |

These two lines do *not* have the same slope, so the two lines intersect. The system has *one* solution.

**3.**

| $x - 2y = -4$ | $x - 2y = 2$ |
|---|---|
| $-2y = -x - 4$ | $-2y = -x + 2$ |
| $y = \frac{1}{2}x + 2$ | $y = \frac{1}{2}x - 1$ |
| $m = \frac{1}{2}$ | $m = \frac{1}{2}$ |

These two lines are parallel; they have the *same* slope and *different* $y$-intercepts. So, the system has *no* solution. The system is inconsistent.

**4.**

| $2x - y = 5$ | $-4x + 2y = -10$ |
|---|---|
| $-y = -2x + 5$ | $2y = 4x - 10$ |
| $y = 2x - 5$ | $y = 2x - 5$ |
| $m = 2$ | $m = 2$ |

These two lines coincide; they have the *same* slope and the *same* $y$-intercept. So, the system has *infinitely many* solutions.

**5.**

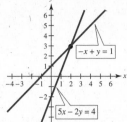

The solution is $(2, 3)$.

**7.**

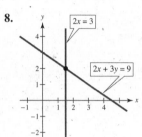

The solution is $(3, -5)$.

**6.**

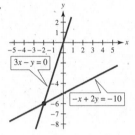

The solution is $(-2, -6)$.

**8.**

The solution is $\left(\frac{3}{2}, 2\right)$.

**9.** $\begin{cases} x + 5y = 10 \Rightarrow x = -5y + 10 \\ -2x - 10y = 1 \end{cases}$

$-2(-5y + 10) - 10y = 1$   Replace $x$ by $-5y + 10$ in second equation.

$10y - 20 - 10y = 1$

$-20 \neq 1$

This system has *no* solution. It is an inconsistent system.

**10.** $\begin{cases} x + 3y = 15 \Rightarrow x = -3y + 15 \\ -2x + 5y = 14 \end{cases}$

$-2(-3y + 15) + 5y = 14$   Replace $x$ by $-3y + 15$ in second equation.

$6y - 30 + 5y = 14$

$11y = 44$

$y = 4 \Rightarrow x = -3(4) + 15$   Replace $y$ by 4 in revised first equation.

$x = -12 + 15$

$x = 3$

$(3, 4)$

**11.** $\begin{cases} y = 14 - 5x \\ x = y - 2 \end{cases}$

$x = (14 - 5x) - 2$   Replace $y$ by $14 - 5x$ in second equation.

$x = 12 - 5x$

$6x = 12$

$x = 2 \Rightarrow y = 14 - 5(2)$   Replace $x$ by 2 in first equation.

$y = 14 - 10$

$y = 4$

$(2, 4)$

**12.** $\begin{cases} x + y = 8 \\ 2x - y = -2 \end{cases}$

$3x \quad\;\; = 6$

$x \quad\;\; = 2$

$2 + y = 8$      Replace $x$ by 2 in first equation.

$y = 6$

$(2, 6)$

**13.** $\begin{cases} 7x + 6y = 36 \Rightarrow 14x + 12y = 72 \\ 5x - 4y = 5 \Rightarrow 15x - 12y = 15 \end{cases}$      Multiply equation by 2.

Multiply equation by 3.

$29x \quad\quad = 87$

$x \quad\quad = 3$

$7(3) + 6y = 36$      Replace $x$ by 3 in first equation.

$21 + 6y = 36$

$6y = 15$

$y = \frac{15}{6}$

$y = \frac{5}{2}$

$\left(3, \frac{5}{2}\right)$

**14.** $\begin{cases} \frac{1}{2}x - \frac{1}{4}y = 1 \Rightarrow -4x + 2y = -8 \\ 4x + 5y = 22 \Rightarrow 4x + 5y = 22 \end{cases}$      Multiply equation by $-8$.

$7y = 14$

$y = 2$ and $4x + 5(2) = 22$      Replace $y$ by 2 in second equation.

$4x + 10 = 22$

$4x = 12$

$x = 3$

$(3, 2)$

**15.**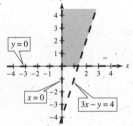

**16.**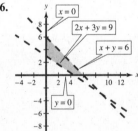

**17.** *Verbal Model:*   | Liters of 30% solution | + | Liters of 5% solution | = | Liters of 20% solution |

| Acid in 30% solution | + | Acid in 5% solution | = | Acid in 20% solution |

*Labels:*

Liters of 30% solution $= x$

Liters of 5% solution $= y$

Liters of 20% solution $= 20$

Acid in 30% solution $= 0.30x$          (liters)

Acid in 5% solution $= 0.05y$          (liters)

Acid in 20% solution $= 0.20(20) = 4$          (liters)

*System of Equations:*   $\begin{cases} x + y = 20 \\ 0.30x + 0.05y = 4 \end{cases}$

*Solving by Elimination:*   $\begin{cases} x + y = 20 \Rightarrow \phantom{-}5x + 5y = \phantom{-}100 \quad \text{Multiply by 5.} \\ 0.30x + 0.05y = 4 \Rightarrow \underline{-30x - 5y = -400} \quad \text{Multiply by } -100. \end{cases}$

$$-25x \phantom{- 5y} = -300$$
$$x \phantom{- 5y} = 12$$

$x + y = 20$          Replace $x$ by 12 in first equation.

$12 + y = 20$

$y = 8$

$(12, 8)$

The mixture contains 12 liters of the 30% solution and 8 liters of the 5% solution.

**18.** *Verbal Model:*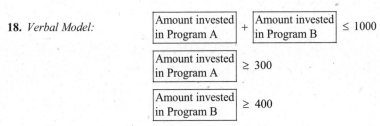

*Labels:*

Amount invested in Program A $= x$          (dollars)

Amount invested in Program B $= y$          (dollars)

*System of Inequalities:*   $\begin{cases} x + y \leq 1000 \\ x \geq 300 \\ y \geq 400 \end{cases}$

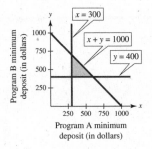

# CHAPTER 9
## Radical Expressions and Equations

# CHAPTER 9
## Radical Expressions and Equations

### Section 9.1   Roots and Radicals

1. Number: 16

   Positive square root: 4

   Negative square root: −4

3. Number: 36

   Positive square root: 6

   Negative square root: −6

5. Number: $\frac{9}{49}$

   Positive square root: $\frac{3}{7}$

   Negative square root: $-\frac{3}{7}$

7. Number: $\frac{81}{16}$

   Positive square root: $\frac{9}{4}$

   Negative square root: $-\frac{9}{4}$

9. A negative number has no square root in the real numbers.

11. Number: 0.01

    Positive square root: 0.1

    Negative square root: −0.1

13. A negative number has no square root in the real numbers.

15. The positive number 27 has one cube root, 3, because $3^3 = 27$.

17. The negative number −8 has one cube root, −2, because $(-2)^3 = -8$.

19. The positive number 1 has two real fourth roots, 1 and −1, because $1^4 = 1$ and $(-1)^4 = 1$.

21. The negative number −81 has no fourth root because no real number can be multiplied by itself four times to obtain −81.

23. The positive number $\frac{1}{8}$ has one cube root, $\frac{1}{2}$, because $\left(\frac{1}{2}\right)^3 = \frac{1}{8}$.

25. $\sqrt{100} = 10$

27. $-\sqrt{100} = -10$

29. A negative number has no square root in the real numbers.

31. $\sqrt{9} = 3$

33. $-\sqrt{169} = -13$

35. $-\sqrt{\frac{1}{9}} = -\frac{1}{3}$

37. $\sqrt{\frac{49}{64}} = \frac{7}{8}$

39. $-\sqrt{\frac{81}{121}} = -\frac{9}{11}$

41. $\sqrt{0.16} = 0.4$

43. $\sqrt{0.04} = 0.2$

45. $\sqrt[3]{8} = 2$

47. $\sqrt[3]{-125} = -5$

49. $\sqrt[3]{216} = 6$

51. $\sqrt[4]{10,000} = 10$

53. $\sqrt[3]{\frac{8}{27}} = \frac{2}{3}$

55. $\sqrt[5]{-1} = -1$

57. $-\sqrt[6]{64} = -2$

59. Irrational

    (15 *is not* a perfect square.)

61. Rational

    (49 *is* a perfect square; $-\sqrt{49} = -7$.)

63. Irrational

    (24 *is not* a perfect square.)

65. Rational

    (400 *is* a perfect square; $\sqrt{400} = 20$.)

**67.** Rational

$\left(\frac{36}{25} \text{ is a perfect square; } -\sqrt{\frac{36}{25}} = -\frac{6}{5}.\right)$

**69.** Irrational

$\left(0.18 \text{ is not a perfect square.}\right)$

**71.** Rational

$\left(1.21 \text{ is a perfect square; } \sqrt{1.21} = 1.1.\right)$

**73.** $\sqrt{43} \approx 6.557$

**75.** The negative number $-12$ has no square root in the real numbers.

**77.** $-\sqrt{137} \approx -11.705$

**79.** The negative number $-632$ has no square root in the real numbers.

**81.** $\sqrt{2580} \approx 50.794$

**83.** $-\sqrt{517.8} \approx -22.755$

**85.** $-\sqrt{\frac{15}{24}} \approx -0.791$

**87.** $\sqrt{\frac{95}{6}} \approx 3.979$

**89.** $16 - \sqrt{92.6} \approx 6.377$

**91.** $2 + 4\sqrt{7} \approx 12.583$

**93.** $\dfrac{-5 + \sqrt{49}}{4} = 0.5$

**95.** $\dfrac{-4 - 3\sqrt{2}}{12} \approx -0.687$

**97.** The number 55 lies between two integers that are perfect squares, $49 = 7^2$ and $64 = 8^2$. Therefore, $\sqrt{55}$ is between 7 and 8. Since 55 is closer to 49 than to 64, we could estimate $\sqrt{55}$ to be 7.4. Using a calculator, we find $\sqrt{55} \approx 7.416$.

**99.** The number 70 lies between two integers that are perfect squares, $64 = 8^2$ and $81 = 9^2$. Since 70 is closer to 64 than to 81, we could estimate $\sqrt{70}$ to be 8.4. Using a calculator, we find $\sqrt{70} \approx 8.367$.

**101.** The number 130 lies between two integers that are perfect squares; $121 = 11^2$ and $144 = 12^2$. Since 130 is closer to 121 than to 144, we could estimate $\sqrt{130}$ to be 11.4. Using a calculator, we find $\sqrt{130} \approx 11.402$.

**103.** The number 300 lies between two integers that are perfect squares, $289 = 17^2$ and $324 = 18^2$. Since 300 is closer to 289 than to 324, we could estimate $\sqrt{300}$ to be 17.3. Using a calculator, we find $\sqrt{300} \approx 17.321$.

**105.** (a) $\sqrt{5(3) - (-1)} = \sqrt{15 + 1}$

$= \sqrt{16}$

$= 4$

(b) $\sqrt{5(-1) - (4)} = \sqrt{-5 - 4}$

$= \sqrt{-9}$

A negative number has no square root in the real numbers.

**107.** (a) $\sqrt{1^2 - 4(3)} = \sqrt{1 - 12}$

$= \sqrt{-11}$

A negative number has no square root in the real numbers.

(b) $\sqrt{(4)^2 - 4(-2)} = \sqrt{16 + 8}$

$= \sqrt{24}$

$\approx 4.90$

**109.** (a) $\sqrt{(5)^2 - 4(4)(1)} = \sqrt{25 - 16} = \sqrt{9} = 3$

(b) $\sqrt{7^2 - 4(-2)(3)} = \sqrt{49 + 24} = \sqrt{73} \approx 8.54$

**111.** When $v = \frac{3}{4}$, $0.03\sqrt{v} = 0.03\sqrt{\frac{3}{4}} \approx 0.026$.

The particle size is about 0.026 inch.

**113.** $\sqrt{l^2 + w^2 + h^2} = \sqrt{12^2 + 9^2 + 3^2}$

$= \sqrt{144 + 81 + 9}$

$= \sqrt{234}$

$\approx 15.30$

**115.** $\sqrt{571{,}536} = 756$

The dimensions of the square base of the pyramid are 756 feet by 756 feet.

**117. (a)**

| $x$ | 0 | 1 | 2 | 4 | 6 | 8 |
|---|---|---|---|---|---|---|
| $\sqrt{x}$ | 0 | 1 | 1.41 | 2 | 2.45 | 2.83 |

| $x$ | 10 | 12 | 14 | 16 | 18 | 20 |
|---|---|---|---|---|---|---|
| $\sqrt{x}$ | 3.16 | 3.46 | 3.74 | 4 | 4.24 | 4.47 |

**(b)**

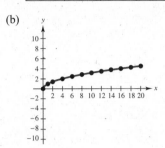

**119. (a)** $\left(\sqrt{8.2}\right)^2 = 8.2$

**(b)** $\left(\sqrt{142}\right)^2 = 142$

**(c)** $\left(\sqrt{22}\right)^2 = 22$

**(d)** $\left(\sqrt{850}\right)^2 = 850$

**121.** $x < 0$

$\sqrt{x^2} \neq x$ for negative values of $x$. For example,

$\sqrt{(-4)^2} = \sqrt{16} = 4$.

**123. (a)**

$0^2 = 0 \qquad 4^2 = 16 \qquad 8^2 = 64$

$1^2 = 1 \qquad 5^2 = 25 \qquad 9^2 = 81$

$2^2 = 4 \qquad 6^2 = 36$

$3^2 = 9 \qquad 7^2 = 49$

Writing the squares of the integers from 0 to 9, we see that all the last digits of these squares are from this list: 0, 1, 4, 5, 6, 9.

Since all integers end in one of the 10 integers from 0 to 9, *every* perfect square integer must have a last digit from the list 0, 1, 4, 5, 6, 9.

**(b)** No, 5,788,942,862 could not be a perfect square because its last digit, 2, *is not* on the list of possible last digits. (The list is 0, 1, 4, 5, 6, 9.)

**125.**

$$\frac{4}{x-3} + \frac{6}{x-3} = -1$$

$$(x-3)\left(\frac{4}{x-3} + \frac{6}{x-3}\right) = -1(x-3)$$

$$4 + 6 = -x + 3$$

$$10 = -x + 3$$

$$7 = -x$$

$$-7 = x$$

**127.**

$$\frac{x-1}{8} + \frac{x+5}{8} = \frac{x-6}{8}$$

$$8\left(\frac{x-1}{8} + \frac{x+5}{8}\right) = \left(\frac{x-6}{8}\right) \cdot 8$$

$$x - 1 + x + 5 = x - 6$$

$$2x + 4 = x - 6$$

$$x + 4 = -6$$

$$x = -10$$

**129.**

$$\frac{x}{2} = 10 - \frac{x}{3}$$

$$6\left(\frac{x}{2}\right) = \left(10 - \frac{x}{3}\right) \cdot 6$$

$$3x = 60 - 2x$$

$$5x = 60$$

$$x = \frac{60}{5}$$

$$x = 12$$

**131.**

$$\frac{2}{3x} - \frac{1}{5} = \frac{4}{15x}$$

$$15x\left(\frac{2}{3x} - \frac{1}{5}\right) = \left(\frac{4}{15x}\right) \cdot 15x$$

$$10 - 3x = 4$$

$$-3x = -6$$

$$x = \frac{-6}{-3}$$

$$x = 2$$

**133.**

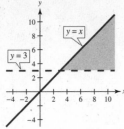

**135.**

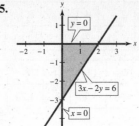

# Section 9.2   Simplifying Radicals

**1.** $\sqrt{3} \cdot \sqrt{5} = \sqrt{15}$

**3.** $\sqrt{10} \cdot \sqrt{7} = \sqrt{70}$

**5.** $\sqrt{5} \cdot \sqrt{6} = \sqrt{30}$

**7.** $\sqrt{2} \cdot \sqrt{x} = \sqrt{2x}$

**9.** $\sqrt{2x} \cdot \sqrt{3y} = \sqrt{6xy}$

**11.** $\sqrt{4 \cdot 15} = \sqrt{4} \cdot \sqrt{15} = 2\sqrt{15}$

**13.** $\sqrt{64 \cdot 11} = \sqrt{64} \cdot \sqrt{11} = 8\sqrt{11}$

**15.** $\sqrt{8} = \sqrt{4 \cdot 2} = \sqrt{4} \cdot \sqrt{2} = 2\sqrt{2}$

**17.** $\sqrt{45} = \sqrt{9 \cdot 5} = \sqrt{9} \cdot \sqrt{5} = 3\sqrt{5}$

**19.** $\sqrt{128} = \sqrt{64 \cdot 2} = \sqrt{64} \cdot \sqrt{2} = 8\sqrt{2}$

**21.** $\sqrt[3]{-24} = \sqrt[3]{-8 \cdot 3} = \sqrt[3]{-8} \cdot \sqrt[3]{3} = -2\sqrt[3]{3}$

**23.** $\sqrt[3]{375} = \sqrt[3]{125 \cdot 3} = \sqrt[3]{125} \cdot \sqrt[3]{3} = 5\sqrt[3]{3}$

**25.** $\sqrt[4]{48} = \sqrt[4]{16 \cdot 3} = \sqrt[4]{16} \cdot \sqrt[4]{3} = 2\sqrt[4]{3}$

**27.** $\sqrt[4]{162} = \sqrt[4]{81 \cdot 2} = \sqrt[4]{81} \cdot \sqrt[4]{2} = 3\sqrt[4]{2}$

**29.** $\sqrt{4x^2} = \sqrt{4 \cdot x^2} = \sqrt{4} \cdot \sqrt{x^2} = 2|x|$

**31.** $\sqrt{64x^3} = \sqrt{64 \cdot x^2 \cdot x}$
$= \sqrt{64} \cdot \sqrt{x^2} \cdot \sqrt{x}$
$= 8x\sqrt{x}$

Note: If $x$ were negative, $x^3$ would be negative and the original radical would be undefined in the real numbers. So, we can assume that the variable is nonnegative and absolute value signs are not necessary.

**33.** $\sqrt{84x^2} = \sqrt{4 \cdot 21 \cdot x^2}$
$= \sqrt{4} \cdot \sqrt{21} \cdot \sqrt{x^2}$
$= 2 \cdot \sqrt{21} \cdot |x|$
$= 2|x|\sqrt{21}$

**35.** $\sqrt{x^6} = \sqrt{\left(x^3\right)^2} = |x^3|$

**37.** $\sqrt{20x^6} = \sqrt{4 \cdot 5 \cdot x^6}$
$= \sqrt{4} \cdot \sqrt{5} \cdot \sqrt{\left(x^3\right)^2}$
$= 2 \cdot \sqrt{5} \cdot |x^3|$
$= 2|x^3|\sqrt{5}$

**39.** $\sqrt{x^2y^3} = \sqrt{x^2 \cdot y^2 \cdot y}$
$= \sqrt{x^2} \cdot \sqrt{y^2} \cdot \sqrt{y}$
$= |x|y\sqrt{y}$

**41.** $\sqrt{180x^5y^8} = \sqrt{36 \cdot 5 \cdot x^4 \cdot x \cdot y^8}$
$= \sqrt{36} \cdot \sqrt{5} \cdot \sqrt{x^4} \cdot \sqrt{x} \cdot \sqrt{y^8}$
$= 6 \cdot \sqrt{5} \cdot x^2 \cdot \sqrt{x} \cdot y^4$
$= 6x^2y^4\sqrt{5x}$

**43.** $\sqrt[3]{27a^4} = \sqrt[3]{27 \cdot a^3 \cdot a}$
$= \sqrt[3]{27} \cdot \sqrt[3]{a^3} \cdot \sqrt[3]{a}$
$= 3a\sqrt[3]{a}$

**45.** $\sqrt[4]{t^7} = \sqrt[4]{t^4 \cdot t^3} = \sqrt[4]{t^4} \cdot \sqrt[4]{t^3} = t\sqrt[4]{t^3}$

**47.** $\sqrt[4]{16y^5} = \sqrt[4]{16 \cdot y^4 \cdot y}$
$= \sqrt[4]{16} \cdot \sqrt[4]{y^4} \cdot \sqrt[4]{y}$
$= 2y\sqrt[4]{y}$

**49.** $\dfrac{\sqrt{28}}{\sqrt{7}} = \sqrt{\dfrac{28}{7}} = \sqrt{4} = 2$

**51.** $\dfrac{\sqrt{72}}{\sqrt{9}} = \dfrac{\sqrt{36}\cdot\sqrt{2}}{3}$ or $\dfrac{\sqrt{72}}{\sqrt{9}} = \sqrt{8}$

$\qquad = \dfrac{6\sqrt{2}}{3} \qquad\qquad = \sqrt{4\cdot 2}$

$\qquad\qquad\qquad\qquad\quad = \sqrt{4}\cdot\sqrt{2}$

$\qquad = 2\sqrt{2} \qquad\qquad = 2\sqrt{2}$

**53.** $\dfrac{\sqrt{35}}{\sqrt{16}} = \dfrac{\sqrt{35}}{4}$

**55.** $\dfrac{\sqrt{48}}{\sqrt{64}} = \dfrac{\sqrt{16}\cdot\sqrt{3}}{8} = \dfrac{4\sqrt{3}}{8} = \dfrac{\sqrt{3}}{2}$

**57.** $\sqrt{\dfrac{13}{49}} = \dfrac{\sqrt{13}}{\sqrt{49}} = \dfrac{\sqrt{13}}{7}$

**59.** $\sqrt{\dfrac{66}{88}} = \sqrt{\dfrac{22\cdot 3}{22\cdot 4}} = \sqrt{\dfrac{3}{4}} = \dfrac{\sqrt{3}}{\sqrt{4}} = \dfrac{\sqrt{3}}{2}$

**61.** $\sqrt{\dfrac{3x^2}{27}} = \sqrt{\dfrac{3x^2}{3(9)}} = \sqrt{\dfrac{x^2}{9}} = \dfrac{\sqrt{x^2}}{\sqrt{9}} = \dfrac{|x|}{3}$

**63.** $\sqrt{\dfrac{12x^2}{25}} = \dfrac{\sqrt{12x^2}}{\sqrt{25}}$

$\qquad = \dfrac{\sqrt{4\cdot 3\cdot x^2}}{5}$

$\qquad = \dfrac{\sqrt{4}\cdot\sqrt{3}\cdot\sqrt{x^2}}{5}$

$\qquad = \dfrac{2|x|\sqrt{3}}{5}$

**65.** $\sqrt{\dfrac{x^6}{16y^2}} = \dfrac{\sqrt{x^6}}{\sqrt{16y^2}} = \dfrac{\sqrt{(x^3)^2}}{\sqrt{16}\cdot\sqrt{y^2}} = \dfrac{|x^3|}{4|y|}$

**67.** $\sqrt{\dfrac{11x^2}{44y^6}} = \sqrt{\dfrac{11\cdot x^2}{11\cdot 4\cdot y^6}} = \dfrac{\sqrt{x^2}}{\sqrt{4}\cdot\sqrt{y^6}} = \dfrac{|x|}{2|y^3|}$

**69.** $\sqrt{\dfrac{9u^5}{48u^7}} = \sqrt{\dfrac{3(3)(u^5)}{3(16)(u^5)(u^2)}}$

$\qquad = \sqrt{\dfrac{3}{16\cdot u^2}}$

$\qquad = \dfrac{\sqrt{3}}{\sqrt{16\cdot u^2}}$

$\qquad = \dfrac{\sqrt{3}}{4|u|}$

**71.** $\sqrt{\dfrac{1}{3}} = \dfrac{\sqrt{1}}{\sqrt{3}} = \dfrac{1}{\sqrt{3}} = \dfrac{1}{\sqrt{3}}\cdot\dfrac{\sqrt{3}}{\sqrt{3}} = \dfrac{\sqrt{3}}{\sqrt{9}} = \dfrac{\sqrt{3}}{3}$

**73.** $\dfrac{1}{\sqrt{7}} = \dfrac{1}{\sqrt{7}}\cdot\dfrac{\sqrt{7}}{\sqrt{7}} = \dfrac{\sqrt{7}}{\sqrt{49}} = \dfrac{\sqrt{7}}{7}$

**75.** $\dfrac{5}{\sqrt{10}} = \dfrac{5}{\sqrt{10}}\cdot\dfrac{\sqrt{10}}{\sqrt{10}} = \dfrac{5\sqrt{10}}{\sqrt{100}} = \dfrac{5\sqrt{10}}{10} = \dfrac{\sqrt{10}}{2}$

**77.** $\dfrac{\sqrt{2}}{\sqrt{3}} = \dfrac{\sqrt{2}}{\sqrt{3}}\cdot\dfrac{\sqrt{3}}{\sqrt{3}} = \dfrac{\sqrt{6}}{\sqrt{9}} = \dfrac{\sqrt{6}}{3}$

**79.** $\sqrt{\dfrac{2}{45}} = \dfrac{\sqrt{2}}{\sqrt{9\cdot 5}}\cdot\dfrac{\sqrt{5}}{\sqrt{5}} = \dfrac{\sqrt{10}}{\sqrt{3^2\cdot 5^2}} = \dfrac{\sqrt{10}}{3\cdot 5} = \dfrac{\sqrt{10}}{15}$

**81.** $\sqrt{\dfrac{11}{8}} = \dfrac{\sqrt{11}}{\sqrt{8}}\cdot\dfrac{\sqrt{2}}{\sqrt{2}} = \dfrac{\sqrt{22}}{\sqrt{16}} = \dfrac{\sqrt{22}}{4}$

**83.** $\sqrt{\dfrac{12}{54}} = \sqrt{\dfrac{6\cdot 2}{6\cdot 9}} = \dfrac{\sqrt{2}}{\sqrt{9}} = \dfrac{\sqrt{2}}{3}$

**85.** $\dfrac{1}{\sqrt{y}} = \dfrac{1}{\sqrt{y}}\cdot\dfrac{\sqrt{y}}{\sqrt{y}} = \dfrac{\sqrt{y}}{\sqrt{y^2}} = \dfrac{\sqrt{y}}{y}$

**87.** $\sqrt{\dfrac{5}{x}} = \dfrac{\sqrt{5}}{\sqrt{x}} = \dfrac{\sqrt{5}}{\sqrt{x}}\cdot\dfrac{\sqrt{x}}{\sqrt{x}} = \dfrac{\sqrt{5x}}{\sqrt{x^2}} = \dfrac{\sqrt{5x}}{x}$

**89.** $\sqrt{\dfrac{3}{16x^5}} = \dfrac{\sqrt{3}}{\sqrt{16x^5}}\cdot\dfrac{\sqrt{x}}{\sqrt{x}}$

$\qquad = \dfrac{\sqrt{3x}}{\sqrt{16x^6}}$

$\qquad = \dfrac{\sqrt{3x}}{\sqrt{16}\cdot\sqrt{(x^3)^2}}$

$\qquad = \dfrac{\sqrt{3x}}{4x^3}$

**91.** $\dfrac{\sqrt{2t}}{\sqrt{8r}} = \sqrt{\dfrac{2(t)}{2(4)(r)}}$

$\qquad = \sqrt{\dfrac{t}{4r}}$

$\qquad = \dfrac{\sqrt{t}}{\sqrt{4r}}\cdot\dfrac{\sqrt{r}}{\sqrt{r}}$

$\qquad = \dfrac{\sqrt{rt}}{\sqrt{4r^2}}$

$\qquad = \dfrac{\sqrt{rt}}{2r}$

**93.** $\dfrac{\sqrt{12x^3}}{\sqrt{3y}} = \sqrt{\dfrac{\cancel{3}(4)x^3}{\cancel{3}(y)}}$

$= \sqrt{\dfrac{4x^3}{y}}$

$= \dfrac{\sqrt{4x^3}}{\sqrt{y}} \cdot \dfrac{\sqrt{y}}{\sqrt{y}}$

$= \dfrac{\sqrt{4 \cdot x^2 \cdot x \cdot y}}{y}$

$= \dfrac{2x\sqrt{xy}}{y}$

**95.** $\dfrac{4}{\sqrt[3]{9}} = \dfrac{4}{\sqrt[3]{9}} \cdot \dfrac{\sqrt[3]{3}}{\sqrt[3]{3}} = \dfrac{4\sqrt[3]{3}}{\sqrt[3]{27}} = \dfrac{4\sqrt[3]{3}}{3}$

**97.** $\dfrac{7}{\sqrt[3]{3}} = \dfrac{7}{\sqrt[3]{3}} \cdot \dfrac{\sqrt[3]{9}}{\sqrt[3]{9}} = \dfrac{7\sqrt[3]{9}}{\sqrt[3]{27}} = \dfrac{7\sqrt[3]{9}}{3}$

**99.** $\dfrac{18}{\sqrt[3]{16}} = \dfrac{18}{\sqrt[3]{4^2}}$

$= \dfrac{18}{\sqrt[3]{4^2}} \cdot \dfrac{\sqrt[3]{4}}{\sqrt[3]{4}}$

$= \dfrac{18\sqrt[3]{4}}{\sqrt[3]{4^3}}$

$= \dfrac{18\sqrt[3]{4}}{4}$

$= \dfrac{\cancel{2} \cdot 9\sqrt[3]{4}}{\cancel{2} \cdot 2}$

$= \dfrac{9\sqrt[3]{4}}{2}$

**101.** $\sqrt[3]{\dfrac{1}{x^2}} = \dfrac{\sqrt[3]{1}}{\sqrt[3]{x^2}} \cdot \dfrac{\sqrt[3]{x}}{\sqrt[3]{x}} = \dfrac{\sqrt[3]{x}}{\sqrt[3]{x^3}} = \dfrac{\sqrt[3]{x}}{x}$

**103.** $\sqrt[3]{\dfrac{1}{8y^2}} = \dfrac{\sqrt[3]{1}}{\sqrt[3]{8y^2}} \cdot \dfrac{\sqrt[3]{y}}{\sqrt[3]{y}} = \dfrac{\sqrt[3]{y}}{\sqrt[3]{8y^3}} = \dfrac{\sqrt[3]{y}}{2y}$

**105.**

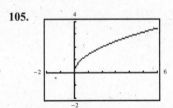

The two graphs are the same.

$\sqrt{2}\sqrt{x} = \sqrt{2x}$

The two expressions are equivalent.

**107.**

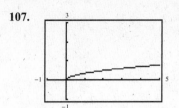

The two graphs are the same.

$\dfrac{\sqrt{x}}{\sqrt{8}} = \dfrac{\sqrt{x}}{\sqrt{8}} \cdot \dfrac{\sqrt{2}}{\sqrt{2}} = \dfrac{\sqrt{2x}}{\sqrt{16}} = \dfrac{\sqrt{2x}}{4} = \dfrac{1}{4}\sqrt{2x}$

The two expressions are equivalent.

**109.** $>$

$\sqrt{160} > 12$

**111.** $>$

$4\sqrt{2} > 5$

**113.** $<$

$5\sqrt{6} < 6\sqrt{5}$

**115.** Area $= (\text{Length})(\text{Width})$

$= \left(\sqrt{20}\right)\left(\sqrt{10}\right)$

$= \sqrt{200}$

$= \sqrt{100} \cdot \sqrt{2}$

$= 10\sqrt{2}$

$\approx 14.14$ square units

**117.** Area $= \frac{1}{2}(\text{Base})(\text{Height})$

$= \frac{1}{2}\left(6\sqrt{3}\right)\left(4\sqrt{2}\right)$

$= \frac{1}{2}\left(24\sqrt{6}\right)$

$= 12\sqrt{6}$

$\approx 29.39$ square units

**119.** $S = \pi r\sqrt{r^2 + h^2}$

$= \pi(4)\sqrt{4^2 + 8^2}$

$= 4\pi\sqrt{80}$

$= 4\pi\sqrt{16 \cdot 5}$

$= 4\pi \cdot 4\sqrt{5}$

$= 16\pi\sqrt{5}$

$\approx 112.4$

The lateral surface area is about 112.4 square feet.

**121.** $S = \pi r\sqrt{r^2 + h^2}$

$= \pi(2)\sqrt{2^2 + 5^2}$

$= 2\pi\sqrt{29}$

$\approx 33.8$

The lateral surface area is about 33.8 square meters.

**123.** (a) $P = \sqrt{d^3} = \sqrt{d^2 \cdot d} = d\sqrt{d}$

The simplified form of the formula is $P = d\sqrt{d}$.

(b) For $d = 9.54$, $P = 9.54\sqrt{9.54}$

$\approx 9.54(3.089)$

$\approx 29.47.$

Saturn's orbital period is approximately 29.5 years.

(c) For $d = 0.72$, $P = 0.72\sqrt{0.72}$

$\approx 0.72(0.85)$

$\approx 0.61.$

Venus's orbital period is approximately 0.6 year.

**125.** (a) $\sqrt{x^2} = x$, False

For $x = -2$, $\sqrt{(-2)^2} = \sqrt{4} \neq -2.$

(b) $\sqrt{x^2} = |x|$, True

(c) $\sqrt[3]{x^3} = |x|$, False

For $x = -3$, $\sqrt[3]{(-3)^3} = \sqrt[3]{-27} \neq |-3|.$

(d) $\sqrt[3]{x^3} = x$, True

# Mid-Chapter Quiz for Chapter 9

**1.** $\sqrt{121} = 11$

**2.** $-\sqrt{0.25} = -0.5$

**3.** $\sqrt[3]{-8} = -2$

**4.** $\sqrt[4]{-16}$

This is not a real number. There is no real number that can be multiplied by itself four times to obtain $-16$.

**5.** $\sqrt{-\frac{1}{49}}$

This is not a real number. There is no real number that can be multiplied by itself to obtain $-\frac{1}{49}$.

**6.** $-\sqrt{1.44} = -1.2$

**7.** Irrational

5 is not a perfect square.

**127.** False, because $\dfrac{\sqrt{50}}{\sqrt{2}} = \sqrt{\dfrac{50}{2}} = \sqrt{25} = 5.$

**129.** False, because $\sqrt[3]{72x^4} = \sqrt[3]{8 \cdot 9 \cdot x^3 \cdot x} = 2x\sqrt[3]{9x}.$

**131.** $\dfrac{\left(\dfrac{x}{6}\right)}{\left(2 - \dfrac{5}{x}\right)} = \dfrac{\left(\dfrac{x}{6}\right)}{\left(\dfrac{2x}{x} - \dfrac{5}{x}\right)}$

$= \dfrac{\left(\dfrac{x}{6}\right)}{\left(\dfrac{2x - 5}{x}\right)}$

$= \dfrac{x}{6} \cdot \dfrac{x}{2x - 5}$

$= \dfrac{x^2}{6(2x - 5)}, \; x \neq 0$

**133.** $\sqrt{81} = 9$

**135.** $\sqrt{-49}$

A negative number has no square root in the real numbers.

**137.** $\sqrt[3]{-64} = -4$

**139.** $-\sqrt[4]{16} = -2$

**8.** $\sqrt{\dfrac{3}{4}} = \dfrac{\sqrt{3}}{\sqrt{4}} = \dfrac{\sqrt{3}}{2}$

The number is irrational because 3 is not a perfect square.

**9.** $\sqrt{900} = 30$

The number is rational because 900 is a perfect square.

**10.** $\sqrt{61} \approx 7.810$

**11.** $13 + \sqrt{27.4} \approx 13 + 5.235 \approx 18.235$

**12.** $\dfrac{7 - \sqrt{16}}{5} = \dfrac{7 - 4}{5} = \dfrac{3}{5} = 0.6$

**13.** $\sqrt{5} \cdot \sqrt{19} = \sqrt{5 \cdot 19} = \sqrt{95}$

**14.** $\sqrt{5x} \cdot \sqrt{6y} = \sqrt{30xy}$

**15.** $\dfrac{\sqrt{42}}{\sqrt{6}} = \sqrt{\dfrac{42}{6}} = \sqrt{7}$

**16.** $\sqrt{50} = \sqrt{25 \cdot 2} = \sqrt{25} \cdot \sqrt{2} = 5\sqrt{2}$

**17.** $\sqrt{72x^2} = \sqrt{36x^2}\sqrt{2} = 6|x|\sqrt{2}$

**18.** $\sqrt{45b^8} = \sqrt{9 \cdot 5 \cdot b^8} = \sqrt{9} \cdot \sqrt{5} \cdot \sqrt{\left(b^4\right)^2} = 3b^4\sqrt{5}$

**19.** $\sqrt[3]{64x^3} = \sqrt[3]{64} \cdot \sqrt[3]{x^3} = 4x$

**20.** $\sqrt[4]{x^9} = \sqrt[4]{x^8 \cdot x} = \sqrt[4]{\left(x^2\right)^4} \cdot \sqrt[4]{x} = x^2\sqrt[4]{x}$

**21.** $\sqrt{18u^5v^2} = \sqrt{9 \cdot 2 \cdot u^4 \cdot u \cdot v^2}$

$\qquad = \sqrt{9} \cdot \sqrt{2} \cdot \sqrt{\left(u^2\right)^2} \cdot \sqrt{u} \cdot \sqrt{v^2}$

$\qquad = 3u^2|v|\sqrt{2u}$

**22.** $\dfrac{\sqrt{576}}{\sqrt{18}} = \sqrt{\dfrac{576}{18}} = \sqrt{32} = \sqrt{16} \cdot \sqrt{2} = 4\sqrt{2}$

**23.** $\dfrac{\sqrt{63x^2}}{\sqrt{64}} = \dfrac{\sqrt{9 \cdot 7 \cdot x^2}}{\sqrt{64}} = \dfrac{3|x|\sqrt{7}}{8}$

**24.** $\sqrt{\dfrac{90b^4}{2b^2}} = \sqrt{45b^2} = \sqrt{9b^2}\sqrt{5} = 3|b|\sqrt{5}, \quad b \neq 0$

**25.** $\sqrt{\dfrac{3}{2}} = \dfrac{\sqrt{3}}{\sqrt{2}} \cdot \dfrac{\sqrt{2}}{\sqrt{2}} = \dfrac{\sqrt{6}}{\sqrt{4}} = \dfrac{\sqrt{6}}{2}$

**26.** $\dfrac{2}{\sqrt{12}} = \dfrac{2}{\sqrt{4}\sqrt{3}} = \dfrac{2}{2\sqrt{3}} = \dfrac{1}{\sqrt{3}} \cdot \dfrac{\sqrt{3}}{\sqrt{3}} = \dfrac{\sqrt{3}}{\sqrt{9}} = \dfrac{\sqrt{3}}{3}$

**27.** $\sqrt[3]{\dfrac{1}{9}} = \dfrac{\sqrt[3]{1}}{\sqrt[3]{9}} \cdot \dfrac{\sqrt[3]{3}}{\sqrt[3]{3}} = \dfrac{\sqrt[3]{3}}{\sqrt[3]{27}} = \dfrac{\sqrt[3]{3}}{3}$

**28.** $\dfrac{4a}{\sqrt{2a}} = \dfrac{4a}{\sqrt{2a}} \cdot \dfrac{\sqrt{2a}}{\sqrt{2a}}$

$\qquad = \dfrac{4a\sqrt{2a}}{\sqrt{4a^2}} = \dfrac{4a\sqrt{2a}}{2a} = 2\sqrt{2a}, \quad a \neq 0$

**29.** $\sqrt{\dfrac{5a^3}{4a}} = \sqrt{\dfrac{5a^2}{4}} = \dfrac{\sqrt{a^2}\sqrt{5}}{\sqrt{4}} = \dfrac{a\sqrt{5}}{2}, \quad a \neq 0$

**30.** $\dfrac{\sqrt[3]{x}}{\sqrt[3]{27x^4}} = \sqrt[3]{\dfrac{\cancel{x}}{27x^3(\cancel{x})}}$

$\qquad = \sqrt[3]{\dfrac{1}{27x^3}}$

$\qquad = \dfrac{\sqrt[3]{1}}{\sqrt[3]{27x^3}}$

$\qquad = \dfrac{1}{\sqrt[3]{27}\sqrt[3]{x^3}}$

$\qquad = \dfrac{1}{3x}$

**31.** $\sqrt{l^2 + w^2 + h^2} = \sqrt{15^2 + 8^2 + 4^2}$

$\qquad = \sqrt{225 + 64 + 16}$

$\qquad = \sqrt{305}$

$\qquad \approx 17.46$

**32.** $\sqrt{361} = 19$

The dimensions of the entire room are 19 feet by 19 feet. The area rug is a square with each side of length $19 - 4$ feet. The area of the rug is $15^2$ or 225 square feet.

# Section 9.3 Operations with Radical Expressions

**1.** $3\sqrt{5} - \sqrt{5} = (3 - 1)\sqrt{5} = 2\sqrt{5}$

**3.** $10\sqrt{11} + 8\sqrt{11} = (10 + 8)\sqrt{11} = 18\sqrt{11}$

**5.** $\frac{2}{5}\sqrt{3} - \frac{6}{5}\sqrt{3} = \left(\frac{2}{5} - \frac{6}{5}\right)\sqrt{3} = -\frac{4}{5}\sqrt{3}$

**9.** $9\sqrt{17} + 7\sqrt{2} - 11\sqrt{17} + \sqrt{2} = (9 - 11)\sqrt{17} + (7 + 1)\sqrt{2} = -2\sqrt{17} + 8\sqrt{2}$

**11.** $4\sqrt[3]{5} + 2\sqrt[3]{5} = (4 + 2)\sqrt[3]{5} = 6\sqrt[3]{5}$

**13.** $4\sqrt[4]{8} - 9\sqrt[4]{8} = (4 - 9)\sqrt[4]{8} = -5\sqrt[4]{8}$

**15.** $9\sqrt[3]{7} + 3\sqrt[3]{7} - 4\sqrt[3]{7} = (9 + 3 - 4)\sqrt[3]{7} = 8\sqrt[3]{7}$

**7.** $\sqrt{3} - 5\sqrt{7} - 12\sqrt{3} = \sqrt{3} - 12\sqrt{3} - 5\sqrt{7}$

$\qquad = (1 - 12)\sqrt{3} - 5\sqrt{7}$

$\qquad = -11\sqrt{3} - 5\sqrt{7}$

**17.** $8\sqrt[4]{6} - 3\sqrt[4]{6} + 5\sqrt[4]{6} = (8 - 3 + 5)\sqrt[4]{6}$

$\qquad = 10\sqrt[4]{6}$

**19.** $5\sqrt{x} - 3\sqrt{x} = (5 - 3)\sqrt{x} = 2\sqrt{x}$

**21.** $4\sqrt{u} - 3 + \sqrt{u} + 8 = (4 + 1)\sqrt{u} + (-3 + 8)$

$\qquad = 5\sqrt{u} + 5$

**23.** $3 + \sqrt[3]{x} + 5 + 4\sqrt[3]{x} = (3 + 5) + (1 + 4)\sqrt[3]{x}$
$$= 8 + 5\sqrt[3]{x}$$

**25.** $3 + 3\sqrt[3]{x^2} + 2 - 8\sqrt[3]{x^2} = 3 + 2 + (3 - 8)\sqrt[3]{x^2}$
$$= 5 - 5\sqrt[3]{x^2}$$

**27.** $8\sqrt[5]{a^3} - 3\sqrt[5]{a^3} + 5 + 6\sqrt[5]{a^3} = (8 - 3 + 6)\sqrt[5]{a^3} + 5$
$$= 11\sqrt[5]{a^3} + 5$$

**29.** $12\sqrt{8} - 3\sqrt{8} = (12 - 3)\sqrt{8}$
$$= 9\sqrt{8}$$
$$= 9\sqrt{4 \cdot 2}$$
$$= 9 \cdot 2\sqrt{2}$$
$$= 18\sqrt{2}$$

**31.** $2\sqrt{45} + 12\sqrt{80} = 2\sqrt{9 \cdot 5} + 12\sqrt{16 \cdot 5}$
$$= 2 \cdot 3\sqrt{5} + 12 \cdot 4\sqrt{5}$$
$$= 6\sqrt{5} + 48\sqrt{5}$$
$$= (6 + 48)\sqrt{5}$$
$$= 54\sqrt{5}$$

**33.** $8\sqrt{54} + \sqrt{50} + \sqrt{2} = 8\sqrt{9 \cdot 6} + \sqrt{25 \cdot 2} + \sqrt{2}$
$$= 8 \cdot 3\sqrt{6} + 5\sqrt{2} + \sqrt{2}$$
$$= 24\sqrt{6} + (5 + 1)\sqrt{2}$$
$$= 24\sqrt{6} + 6\sqrt{2}$$

**35.** $\sqrt{9x} + \sqrt{36x} = \sqrt{9 \cdot x} + \sqrt{36 \cdot x}$
$$= 3\sqrt{x} + 6\sqrt{x}$$
$$= (3 + 6)\sqrt{x}$$
$$= 9\sqrt{x}$$

**37.** $\sqrt{81b} + \sqrt{b} = \sqrt{81 \cdot b} + \sqrt{b}$
$$= 9\sqrt{b} + \sqrt{b}$$
$$= (9 + 1)\sqrt{b}$$
$$= 10\sqrt{b}$$

**39.** $\sqrt{45z} - \sqrt{125z} = \sqrt{9 \cdot 5z} - \sqrt{25 \cdot 5z}$
$$= 3\sqrt{5z} - 5\sqrt{5z}$$
$$= (3 - 5)\sqrt{5z}$$
$$= -2\sqrt{5z}$$

**41.** $3\sqrt{3y} - \sqrt{27y} + \sqrt{y} = 3\sqrt{3y} - \sqrt{9 \cdot 3y} + \sqrt{y}$
$$= 3\sqrt{3y} - 3\sqrt{3y} + \sqrt{y}$$
$$= (3 - 3)\sqrt{3y} + \sqrt{y}$$
$$= \sqrt{y}$$

**43.** $\sqrt{32x^3} - 2\sqrt{8x^3} = \sqrt{16 \cdot 2 \cdot x^2 \cdot x} - 2\sqrt{4 \cdot 2 \cdot x^2 \cdot x} = 4x\sqrt{2x} - 2 \cdot 2x\sqrt{2x} = (4x - 4x)\sqrt{2x} = 0\sqrt{2x} = 0$

**45.** $\sqrt{x^3y} + 4\sqrt{xy} = \sqrt{x^2 \cdot xy} + 4\sqrt{xy} = |x|\sqrt{xy} + 4\sqrt{xy} = (|x| + 4)\sqrt{xy}$

**47.** $\sqrt{\dfrac{a}{4}} - \sqrt{\dfrac{a}{9}} = \dfrac{\sqrt{a}}{\sqrt{4}} - \dfrac{\sqrt{a}}{\sqrt{9}}$

$$= \dfrac{\sqrt{a}}{2} - \dfrac{\sqrt{a}}{3}$$

$$= \left(\dfrac{1}{2} - \dfrac{1}{3}\right)\sqrt{a} \qquad \text{Note: } \dfrac{\sqrt{a}}{2} = \dfrac{1}{2}\sqrt{a}; \dfrac{\sqrt{a}}{3} = \dfrac{1}{3}\sqrt{a}$$

$$= \left(\dfrac{3}{6} - \dfrac{2}{6}\right)\sqrt{a}$$

$$= \dfrac{1}{6}\sqrt{a} \text{ or } \dfrac{\sqrt{a}}{6}$$

**49.** $\sqrt{2} \cdot \sqrt{8} = \sqrt{16} = 4$

**51.** $\sqrt{3} \cdot \sqrt{27} = \sqrt{81} = 9$

**53.** $\sqrt{10} \cdot \sqrt{6} = \sqrt{60} = \sqrt{4 \cdot 15} = 2\sqrt{15}$

**55.** $\sqrt[3]{4} \cdot \sqrt[3]{2} = \sqrt[3]{8} = 2$

**57.** $\sqrt[4]{2} \cdot \sqrt[4]{8} = \sqrt[4]{16} = 2$

**59.** $\sqrt{7}\left(1 - \sqrt{2}\right) = \sqrt{7} \cdot 1 - \sqrt{7} \cdot \sqrt{2} = \sqrt{7} - \sqrt{14}$

338 Chapter 9 Radical Expressions and Equations

**61.** $\sqrt{6}\left(\sqrt{12} + 8\right) = \sqrt{6} \cdot \sqrt{12} + \sqrt{6} \cdot 8$

$\qquad = \sqrt{72} + 8\sqrt{6}$

$\qquad = \sqrt{36 \cdot 2} + 8\sqrt{6}$

$\qquad = 6\sqrt{2} + 8\sqrt{6}$

**63.** $\sqrt[3]{2}\left(\sqrt[3]{4} + 5\right) = \sqrt[3]{2} \cdot \sqrt[3]{4} + \sqrt[3]{2} \cdot 5$

$\qquad = \sqrt[3]{8} + 5\sqrt[3]{2}$

$\qquad = 2 + 5\sqrt[3]{2}$

**69.** $\left(\sqrt{5} + \sqrt{2}\right)\left(\sqrt{5} - \sqrt{3}\right) = \sqrt{5} \cdot \sqrt{5} - \sqrt{5} \cdot \sqrt{3} + \sqrt{2} \cdot \sqrt{5} - \sqrt{2} \cdot \sqrt{3} = 5 - \sqrt{15} + \sqrt{10} - \sqrt{6}$

**71.** $\left(1 + \sqrt{11}\right)\left(1 - \sqrt{11}\right) = 1^2 - \left(\sqrt{11}\right)^2 = 1 - 11 = -10$

**73.** $\left(\sqrt{10} + \sqrt{5}\right)\left(\sqrt{10} - \sqrt{5}\right) = \left(\sqrt{10}\right)^2 - \left(\sqrt{5}\right)^2$

$\qquad = 10 - 5 = 5$

**75.**

$\qquad\qquad\qquad$ F $\qquad$ O $\qquad$ I $\qquad$ L

$\left(\sqrt[3]{4} + 5\right)\left(\sqrt[3]{3} + 2\right) = \sqrt[3]{4 \cdot 3} + 2\sqrt[3]{4} + 5\sqrt[3]{3} + 10$

$\qquad\qquad = \sqrt[3]{12} + 2\sqrt[3]{4} + 5\sqrt[3]{3} + 10$

**77.** $\left(\sqrt{13} + 2\right)^2 = \left(\sqrt{13} + 2\right)\left(\sqrt{13} + 2\right)$

$\qquad = \sqrt{13} \cdot \sqrt{13} + 2\sqrt{13} + 2\sqrt{13} + 4$

$\qquad = 13 + 4\sqrt{13} + 4$

$\qquad = 17 + 4\sqrt{13}$

We could use the pattern for the square of a binomial.

Pattern: $(a + b)^2 = a^2 + 2ab + b^2$

$\left(\sqrt{13} + 2\right)^2 = \left(\sqrt{13}\right)^2 + 2\left(\sqrt{13}\right)(2) + 2^2$

$\qquad = 13 + 4\sqrt{13} + 4$

$\qquad = 17 + 4\sqrt{13}$

**79.** $\left(\sqrt[3]{2} - 1\right)^2 = \left(\sqrt[3]{2} - 1\right)\left(\sqrt[3]{2} - 1\right)$

$\qquad = \sqrt[3]{2 \cdot 2} - \sqrt[3]{2} - \sqrt[3]{2} + 1$

$\qquad = \sqrt[3]{4} + (-1 - 1)\sqrt[3]{2} + 1$

$\qquad = \sqrt[3]{4} - 2\sqrt[3]{2} + 1$

**65.** $\sqrt[4]{2}\left(6 + \sqrt[4]{8}\right) = \sqrt[4]{2} \cdot 6 + \sqrt[4]{2} \cdot \sqrt[4]{8}$

$\qquad = 6\sqrt[4]{2} + \sqrt[4]{16}$

$\qquad = 6\sqrt[4]{2} + 2$

**67.**

$\qquad\qquad\qquad$ F $\qquad$ O $\qquad$ I $\qquad$ L

$\left(\sqrt{2} - 1\right)\left(\sqrt{2} + 3\right) = \sqrt{2 \cdot 2} + 3\sqrt{2} - 1\sqrt{2} - 3$

$\qquad = 2 + (3 - 1)\sqrt{2} - 3$

$\qquad = -1 + 2\sqrt{2}$

**81.** $\left(3 - \sqrt{8}\right)^2 = \left(3 - \sqrt{8}\right)\left(3 - \sqrt{8}\right)$

$\qquad = 9 - 3\sqrt{8} - 3\sqrt{8} + \sqrt{8 \cdot 8}$

$\qquad = 9 + (-3 - 3)\sqrt{8} + 8$

$\qquad = 17 - 6\sqrt{8}$

$\qquad = 17 - 6\sqrt{4 \cdot 2}$

$\qquad = 17 - 6 \cdot 2\sqrt{2}$

$\qquad = 17 - 12\sqrt{2}$

**83.** $\sqrt{x}\left(\sqrt{x} + 5\right) = \sqrt{x} \cdot \sqrt{x} + \sqrt{x} \cdot 5 = x + 5\sqrt{x}$

**85.**

$\qquad\qquad\qquad$ F $\qquad$ O $\qquad$ I $\qquad$ L

$\left(\sqrt{x} + 7\right)\left(\sqrt{x} - 2\right) = \sqrt{x \cdot x} - 2\sqrt{x} + 7\sqrt{x} - 14$

$\qquad\qquad = \sqrt{x^2} + (-2 + 7)\sqrt{x} - 14$

$\qquad\qquad = x + 5\sqrt{x} - 14$

**87.** $\left(3 + \sqrt{x}\right)^2 = \left(3 + \sqrt{x}\right)\left(3 + \sqrt{x}\right)$

$\qquad = 3 \cdot 3 + 3\sqrt{x} + 3\sqrt{x} + \sqrt{x} \cdot \sqrt{x}$

$\qquad = 9 + 6\sqrt{x} + x$

We could use the pattern for the square of a binomial.

Pattern: $(a + b)^2 = a^2 + 2ab + b^2$

$\left(3 + \sqrt{x}\right)^2 = 3^2 + 2(3)\left(\sqrt{x}\right) + \left(\sqrt{x}\right)^2$

$\qquad = 9 + 6\sqrt{x} + x$

**89.** $\left(2\sqrt{x} - 3\right)\left(2\sqrt{x} + 3\right) = 2^2\left(\sqrt{x}\right)^2 - 3^2 = 4x - 9, x \geq 0$

**91.** *Expression* $\qquad$ *Conjugate* $\qquad$ *Product*

$\quad 4 + \sqrt{5} \qquad\qquad 4 - \sqrt{5} \qquad (4)^2 - \left(\sqrt{5}\right)^2 = 16 - 5 = 11$

© 2010 Cengage Learning. All Rights Reserved. May not be scanned, copied or duplicated, or posted to a publicly accessible website, in whole or in part.

**93.** 

| Expression | Conjugate | Product |
|---|---|---|
| $\sqrt{t} - 5$ | $\sqrt{t} + 5$ | $\left(\sqrt{t}\right)^2 - (5)^2 = t - 25, t \geq 0$ |

**95.** 

| Expression | Conjugate | Product |
|---|---|---|
| $\sqrt{15} - \sqrt{7}$ | $\sqrt{15} + \sqrt{7}$ | $\left(\sqrt{15}\right)^2 - \left(\sqrt{7}\right)^2 = 15 - 7 = 8$ |

**97.** 

| Expression | Conjugate | Product |
|---|---|---|
| $\sqrt{u} - \sqrt{2}$ | $\sqrt{u} + \sqrt{2}$ | $\left(\sqrt{u}\right)^2 - \left(\sqrt{2}\right)^2 = u - 2, u \geq 0$ |

**99.**
$$\frac{5}{\sqrt{14} - 2} = \frac{5}{\sqrt{14} - 2} \cdot \frac{\sqrt{14} + 2}{\sqrt{14} + 2}$$
$$= \frac{5\left(\sqrt{14} + 2\right)}{\left(\sqrt{14}\right)^2 - (2)^2}$$
$$= \frac{5\left(\sqrt{14} + 2\right)}{14 - 4}$$
$$= \frac{5\left(\sqrt{14} + 2\right)}{10}$$
$$= \frac{\sqrt{14} + 2}{2}$$

**101.**
$$\frac{16}{\sqrt{11} + 3} = \frac{16}{\sqrt{11} + 3} \cdot \frac{\sqrt{11} - 3}{\sqrt{11} - 3}$$
$$= \frac{16\left(\sqrt{11} - 3\right)}{\left(\sqrt{11}\right)^2 - 3^2}$$
$$= \frac{16\left(\sqrt{11} - 3\right)}{11 - 9}$$
$$= \frac{16\left(\sqrt{11} - 3\right)}{2}$$
$$= 8\left(\sqrt{11} - 3\right)$$
$$= 8\sqrt{11} - 24$$

**103.**
$$\frac{4}{\sqrt{7} - \sqrt{3}} = \frac{4}{\sqrt{7} - \sqrt{3}} \cdot \frac{\sqrt{7} + \sqrt{3}}{\sqrt{7} + \sqrt{3}}$$
$$= \frac{4\left(\sqrt{7} + \sqrt{3}\right)}{\left(\sqrt{7}\right)^2 - \left(\sqrt{3}\right)^2}$$
$$= \frac{4\left(\sqrt{7} + \sqrt{3}\right)}{7 - 3}$$
$$= \frac{4\left(\sqrt{7} + \sqrt{3}\right)}{4}$$
$$= \sqrt{7} + \sqrt{3}$$

**105.**
$$\frac{8\sqrt{6}}{\sqrt{6} + \sqrt{2}} = \frac{8\sqrt{6}}{\sqrt{6} + \sqrt{2}} \cdot \frac{\sqrt{6} - \sqrt{2}}{\sqrt{6} - \sqrt{2}}$$
$$= \frac{8\sqrt{6}\left(\sqrt{6} - \sqrt{2}\right)}{\left(\sqrt{6}\right)^2 - \left(\sqrt{2}\right)^2}$$
$$= \frac{8\sqrt{36} - 8\sqrt{12}}{6 - 2}$$
$$= \frac{8 \cdot 6 - 8\sqrt{4 \cdot 3}}{4}$$
$$= \frac{48 - 8 \cdot 2\sqrt{3}}{4}$$
$$= \frac{48 - 16\sqrt{3}}{4}$$
$$= 12 - 4\sqrt{3}$$

**107.**
$$\frac{\sqrt{5} + 1}{\sqrt{13} + 7} = \frac{\sqrt{5} + 1}{\sqrt{13} + 7} \cdot \frac{\sqrt{13} - 7}{\sqrt{13} - 7}$$
$$= \frac{\sqrt{65} - 7\sqrt{5} + \sqrt{13} - 7}{\left(\sqrt{13}\right)^2 - (7)^2}$$
$$= \frac{\sqrt{65} - 7\sqrt{5} + \sqrt{13} - 7}{13 - 49}$$
$$= \frac{\sqrt{65} - 7\sqrt{5} + \sqrt{13} - 7}{-36}$$
$$= -\frac{\sqrt{65} - 7\sqrt{5} + \sqrt{13} - 7}{36}$$

**109.**
$$\frac{2}{5 - \sqrt{y}} = \frac{2}{5 - \sqrt{y}} \cdot \frac{5 + \sqrt{y}}{5 + \sqrt{y}}$$
$$= \frac{2\left(5 + \sqrt{y}\right)}{(5)^2 - \left(\sqrt{y}\right)^2}$$
$$= \frac{10 + 2\sqrt{y}}{25 - y}$$

**111.** $\dfrac{9}{\sqrt{x}+2} = \dfrac{9}{\sqrt{x}+2} \cdot \dfrac{\sqrt{x}-2}{\sqrt{x}-2}$

$\qquad = \dfrac{9(\sqrt{x}-2)}{(\sqrt{x})^2 - 2^2}$

$\qquad = \dfrac{9\sqrt{x}-18}{x-4}$

**113.** $\dfrac{\sqrt{y}}{7-\sqrt{y}} = \dfrac{\sqrt{y}}{7-\sqrt{y}} \cdot \dfrac{7+\sqrt{y}}{7+\sqrt{y}}$

$\qquad = \dfrac{\sqrt{y}(7+\sqrt{y})}{7^2 - (\sqrt{y})^2}$

$\qquad = \dfrac{7\sqrt{y} + (\sqrt{y})^2}{49-y}$

$\qquad = \dfrac{7\sqrt{y} + y}{49-y}$

**115.** $\dfrac{-2\sqrt{x}}{3\sqrt{x}-\sqrt{2}} = \dfrac{-2\sqrt{x}}{3\sqrt{x}-\sqrt{2}} \cdot \dfrac{3\sqrt{x}+\sqrt{2}}{3\sqrt{x}+\sqrt{2}}$

$\qquad = \dfrac{-2\sqrt{x}(3\sqrt{x}+\sqrt{2})}{(3\sqrt{x})^2 - (\sqrt{2})^2}$

$\qquad = \dfrac{-6\sqrt{x^2} - 2\sqrt{2x}}{9x-2}$

$\qquad = \dfrac{-6x - 2\sqrt{2x}}{9x-2}$

**117.** $\dfrac{\sqrt{x}-5}{\sqrt{x}-1} = \dfrac{\sqrt{x}-5}{\sqrt{x}-1} \cdot \dfrac{\sqrt{x}+1}{\sqrt{x}+1}$

$\qquad = \dfrac{(\sqrt{x}-5)(\sqrt{x}+1)}{(\sqrt{x})^2 - (1)^2}$

$\qquad = \dfrac{\sqrt{x^2} + \sqrt{x} - 5\sqrt{x} - 5}{x-1}$

$\qquad = \dfrac{x - 4\sqrt{x} - 5}{x-1}$

**119.** $\dfrac{x}{\sqrt{x}+\sqrt{y}} = \dfrac{x}{\sqrt{x}+\sqrt{y}} \cdot \dfrac{\sqrt{x}-\sqrt{y}}{\sqrt{x}-\sqrt{y}}$

$\qquad = \dfrac{x(\sqrt{x}-\sqrt{y})}{(\sqrt{x})^2 - (\sqrt{y})^2}$

$\qquad = \dfrac{x(\sqrt{x}-\sqrt{y})}{x-y}$

$\qquad = \dfrac{x\sqrt{x} - x\sqrt{y}}{x-y}$

**121.** $3 - \dfrac{1}{\sqrt{3}} = 3 - \dfrac{1}{\sqrt{3}} \cdot \dfrac{\sqrt{3}}{\sqrt{3}}$

$\qquad = \dfrac{3}{1} - \dfrac{\sqrt{3}}{3} = \dfrac{9}{3} - \dfrac{\sqrt{3}}{3} = \dfrac{9-\sqrt{3}}{3}$

Note: Here is another way to do this exercise.

$3 - \dfrac{1}{\sqrt{3}} = \dfrac{3}{1} - \dfrac{1}{\sqrt{3}} = \dfrac{3\sqrt{3}}{\sqrt{3}} - \dfrac{1}{\sqrt{3}}$

$\qquad = \dfrac{3\sqrt{3}-1}{\sqrt{3}} = \dfrac{3\sqrt{3}-1}{\sqrt{3}} \cdot \dfrac{\sqrt{3}}{\sqrt{3}}$

$\qquad = \dfrac{(3\sqrt{3}-1)\sqrt{3}}{3} = \dfrac{3\sqrt{9}-\sqrt{3}}{3}$

$\qquad = \dfrac{3 \cdot 3 - \sqrt{3}}{3} = \dfrac{9-\sqrt{3}}{3}$

**123.** $\sqrt{50} - \dfrac{6}{\sqrt{2}} = \sqrt{25 \cdot 2} - \dfrac{6}{\sqrt{2}} \cdot \dfrac{\sqrt{2}}{\sqrt{2}}$

$\qquad = 5\sqrt{2} - \dfrac{6\sqrt{2}}{2}$

$\qquad = 5\sqrt{2} - 3\sqrt{2}$

$\qquad = 2\sqrt{2}$

**125.** $\dfrac{4}{\sqrt{3}} + 2 = \dfrac{4}{\sqrt{3}} \cdot \dfrac{\sqrt{3}}{\sqrt{3}} + \dfrac{2}{1}$

$\qquad = \dfrac{4\sqrt{3}}{(\sqrt{3})^2} + \dfrac{2}{1} \cdot \dfrac{3}{3}$

$\qquad = \dfrac{4\sqrt{3}}{3} + \dfrac{6}{3}$

$\qquad = \dfrac{4\sqrt{3}+6}{3}$

**127.** $\sqrt{5} + \sqrt{3} > \sqrt{5+3}$

$\sqrt{5} + \sqrt{3} \approx 3.97$

$\sqrt{5+3} = \sqrt{8} \approx 2.83$

**129.** $5 > \sqrt{3^2 + 2^2}$

$\sqrt{3^2 + 2^2} = \sqrt{13} \approx 3.61$

**131.** Perimeter $= 2(\text{Length}) + 2(\text{Width})$

$$= 2\sqrt{121x} + 2(5\sqrt{x})$$

$$= 2\sqrt{121}\sqrt{x} + 10\sqrt{x}$$

$$= 2 \cdot 11\sqrt{x} + 10\sqrt{x}$$

$$= 22\sqrt{x} + 10\sqrt{x}$$

$$= 32\sqrt{x}$$

Area $= (\text{Length})(\text{Width})$

$$= \left(\sqrt{121x}\right)(5\sqrt{x})$$

$$= 5\sqrt{121}\sqrt{x^2}$$

$$= 5 \cdot 11x$$

$$= 55x$$

**133.** Perimeter $= 2(\text{Length}) + 2(\text{Width})$

$$= 2(2\sqrt{63}) + 2(\sqrt{28} + 7)$$

$$= 4\sqrt{63} + 2\sqrt{28} + 14$$

$$= 4\sqrt{9 \cdot 7} + 2\sqrt{4 \cdot 7} + 14$$

$$= 4 \cdot 3\sqrt{7} + 2 \cdot 2\sqrt{7} + 14$$

$$= 12\sqrt{7} + 4\sqrt{7} + 14$$

$$= 16\sqrt{7} + 14$$

Area $= (\text{Length})(\text{Width})$

$$= (2\sqrt{63})(\sqrt{28} + 7)$$

$$= 2\sqrt{63 \cdot 28} + 14\sqrt{63}$$

$$= 2\sqrt{1764} + 14\sqrt{63}$$

$$= 2(42) + 14\sqrt{9 \cdot 7}$$

$$= 84 + 14 \cdot 3\sqrt{7}$$

$$= 84 + 42\sqrt{7}$$

**135.** $\dfrac{2}{\sqrt{5}-1} = \dfrac{2}{\sqrt{5}-1} \cdot \dfrac{\sqrt{5}+1}{\sqrt{5}+1}$

$$= \dfrac{2(\sqrt{5}+1)}{(\sqrt{5})^2 - (1)^2}$$

$$= \dfrac{2(\sqrt{5}+1)}{5-1}$$

$$= \dfrac{2(\sqrt{5}+1)}{4}$$

$$= \dfrac{\sqrt{5}+1}{2}$$

$$\approx 1.62$$

The Golden Section is about 1.62.

**137.** No, $\sqrt[3]{5}$ and $\sqrt[4]{5}$ are not like radicals because they do not have the same index; the first is a cube root and the second is a fourth root.

**139.** No.

Multiply the numerator and denominator by the conjugate of the denominator.

$$\dfrac{3}{1+\sqrt{5}} = \dfrac{3}{1+\sqrt{5}} \cdot \dfrac{1-\sqrt{5}}{1-\sqrt{5}}$$

$$= \dfrac{3(1-\sqrt{5})}{1^2 - (\sqrt{5})^2}$$

$$= \dfrac{3 - 3\sqrt{5}}{1-5}$$

$$= \dfrac{3 - 3\sqrt{5}}{-4}$$

$$= \dfrac{-3 + 3\sqrt{5}}{4}$$

**141.** The display approaches 1.

These two examples illustrate this, beginning with the numbers 145 and 0.145.

| Example: | 145 | or | 0.145 |
|---|---|---|---|
| | 12.042 | | 0.381 |
| | 3.470 | | 0.617 |
| | 1.863 | | 0.786 |
| | 1.365 | | 0.886 |
| | 1.168 | | 0.941 |
| | 1.081 | | 0.970 |
| | 1.040 | | 0.985 |
| | 1.020 | | 0.992 |
| | 1.010 | | 0.996 |
| | 1.005 | | 0.998 |
| | 1.002 | | 0.999 |
| | etc. | | etc. |

**143.** $x^3 - 27 = x^3 - 3^3$

$$= (x - 3)(x^2 + x \cdot 3 + 3^2)$$

$$= (x - 3)(x^2 + 3x + 9)$$

**145.** $(x + 5)^2 - 144 = (x + 5)^2 - (12)^2$

$$= [(x + 5) + 12][(x + 5) - 12]$$

$$= (x + 5 + 12)(x + 5 - 12)$$

$$= (x + 17)(x - 7)$$

**147.** $\begin{cases} 5x + \phantom{1}y = 20 \Rightarrow y = -5x + 20 \\ 2x - 10y = \phantom{2}1 \end{cases}$

$\quad 2x - 10(-5x + 20) = 1 \quad$ Replace $y$ by $-5x + 20$ in second equation.

$\qquad 2x + 50x - 200 = 1$

$\qquad\quad 52x - 200 = 1$

$\qquad\qquad\quad 52x = 201$

$\qquad\qquad\qquad x = \dfrac{201}{52} \Rightarrow y = -5\left(\dfrac{201}{52}\right) + 20 \quad$ Replace $x$ by $\dfrac{201}{52}$ in revised first equation.

$\qquad\qquad\qquad\qquad\qquad y = \dfrac{-1005}{52} + \dfrac{20(52)}{1(52)}$

$\qquad\qquad\qquad\qquad\qquad y = \dfrac{-1005 + 1040}{52}$

$\qquad\qquad\qquad\qquad\qquad y = \dfrac{35}{52}$

$\left(\dfrac{201}{52}, \dfrac{35}{52}\right)$

**149.** $\begin{cases} 9x - 7y = 39 \Rightarrow 27x - 21y = 117 \quad \text{Multiply equation by 3.} \\ 4x + 3y = -1 \Rightarrow 28x + 21y = \phantom{1}-7 \quad \text{Multiply equation by 7.} \end{cases}$

$\qquad\qquad\qquad\qquad \overline{\phantom{xx}55x \phantom{xxxxxx} = 110}$

$\qquad\qquad\qquad\qquad\qquad x \phantom{xxx} = \frac{110}{55}$

$\qquad\qquad\qquad\qquad\qquad x \phantom{xxx} = 2$

$\quad 4(2) + 3y = -1 \quad$ Replace $x$ by 2 in second equation.

$\qquad 8 + 3y = -1$

$\qquad\quad 3y = -9$

$\qquad\quad\phantom{3}y = -3$

$(2, -3)$

**151.** $\sqrt{\dfrac{2}{3}} = \dfrac{\sqrt{2}}{\sqrt{3}} \cdot \dfrac{\sqrt{3}}{\sqrt{3}} = \dfrac{\sqrt{6}}{\sqrt{3^2}} = \dfrac{\sqrt{6}}{3}$

**153.** $\sqrt{\dfrac{5}{4x^3}} = \dfrac{\sqrt{5}}{\sqrt{4x^3}} \cdot \dfrac{\sqrt{x}}{\sqrt{x}} = \dfrac{\sqrt{5x}}{\sqrt{4x^4}} = \dfrac{\sqrt{5x}}{\sqrt{2^2\left(x^2\right)^2}} = \dfrac{\sqrt{5x}}{2x^2}$

# Section 9.4   Radical Equations and Applications

**1.** (a) $x = -1$

$$\sqrt{-1} - 6 \overset{?}{=} 0$$

$$\sqrt{-1} - 6 \neq 0$$

$\sqrt{-1}$ is not a real number.

$-1$ *is not* a solution.

(b) $x = -36$

$$\sqrt{-36} - 6 \overset{?}{=} 0$$

$$\sqrt{-36} - 6 \neq 0$$

$\sqrt{-36}$ is not a real number.

$-36$ *is not* a solution.

(c) $x = 36$

$$\sqrt{36} - 6 \overset{?}{=} 0$$

$$6 - 6 = 0$$

$36$ *is* a solution.

(d) $x = 6$

$$\sqrt{6} - 6 \overset{?}{=} 0$$

$$\sqrt{6} - 6 \neq 0$$

$6$ *is not* a solution.

**3.** (a) $x = -1$

$$-1 \overset{?}{=} \sqrt{2(-1) + 3}$$

$$-1 \neq \sqrt{1}$$

$-1$ *is not* a solution.

(b) $x = 2$

$$2 \overset{?}{=} \sqrt{2(2) + 3}$$

$$2 \neq \sqrt{7}$$

$2$ *is not* a solution.

(c) $x = 8$

$$8 \overset{?}{=} \sqrt{2(8) + 3}$$

$$8 \neq \sqrt{19}$$

$8$ *is not* a solution.

(d) $x = 3$

$$3 \overset{?}{=} \sqrt{2(3) + 3}$$

$$3 = \sqrt{9}$$

$3$ *is* a solution.

**5.** $\sqrt{x} = 4$

$$\left(\sqrt{x}\right)^2 = 4^2$$

$$x = 16$$

**7.** $\sqrt{x} = 10$

$$\left(\sqrt{x}\right)^2 = (10)^2$$

$$x = 100$$

**9.** $\sqrt{4x} = -6$

$$(4x)^2 = (-6)^2$$

$$4x = 36$$

$$x = 9 \quad \text{Extraneous}$$

The equation has no solution.

**11.** $\sqrt{u} + 3 = 0$

$$\sqrt{u} = -3$$

$$\left(\sqrt{u}\right)^2 = (-3)^2$$

$$u = 9 \quad \text{Extraneous}$$

The equation has no solution.

**13.** $8 - \sqrt{t} = 3$

$$-\sqrt{t} = -5$$

$$\left(-\sqrt{t}\right)^2 = (-5)^2$$

$$t = 25$$

**15.** $\sqrt{x + 4} = 3$

$$\left(\sqrt{x + 4}\right)^2 = 3^2$$

$$x + 4 = 9$$

$$x = 5$$

**17.** $\sqrt{10x} = 100$

$$\left(\sqrt{10x}\right)^2 = (100)^2$$

$$10x = 10,000$$

$$x = 1000$$

**19.** $\sqrt{3x} - 4 = -7$

$$\sqrt{3x} = -3$$

$$\left(\sqrt{3x}\right)^2 = (-3)^2$$

$$3x = 9$$

$$x = 3 \quad \text{Extraneous}$$

The equation has no solution.

**21.** $\sqrt{3x - 2} = 4$

$\left(\sqrt{3x - 2}\right)^2 = 4^2$

$3x - 2 = 16$

$3x = 18$

$x = 6$

**23.** $\sqrt{7x + 5} = 2$

$\left(\sqrt{7x + 5}\right)^2 = (2)^2$

$7x + 5 = 4$

$7x = -1$

$x = -\frac{1}{7}$

**25.** $\sqrt{x - 2} + 1 = 7$

$\sqrt{x - 2} = 6$

$\left(\sqrt{x - 2}\right)^2 = 6^2$

$x - 2 = 36$

$x = 38$

**27.** $\sqrt{4x + 3} - 6 = -5$

$\sqrt{4x + 3} = 1$

$\left(\sqrt{4x + 3}\right)^2 = 1^2$

$4x + 3 = 1$

$4x = -2$

$x = \frac{-2}{4}$

$x = -\frac{1}{2}$

**29.** $\sqrt{1 - 4x} - 3 = 2$

$\sqrt{1 - 4x} = 5$

$\left(\sqrt{1 - 4x}\right)^2 = 5^2$

$1 - 4x = 25$

$-4x = 24$

$x = \frac{24}{-4}$

$x = -6$

**31.** $5\sqrt{x + 1} = 6$

$\left(5\sqrt{x + 1}\right)^2 = 6^2$

$25(x + 1) = 36$

$25x + 25 = 36$

$25x = 11$

$x = \frac{11}{25}$

**33.** $\sqrt{x + 3} = \sqrt{6x - 7}$

$\left(\sqrt{x + 3}\right)^2 = \left(\sqrt{6x - 7}\right)^2$

$x + 3 = 6x - 7$

$-5x + 3 = -7$

$-5x = -10$

$x = \frac{-10}{-5}$

$x = 2$

**35.** $\sqrt{x + 3} = \sqrt{4x - 3}$

$\left(\sqrt{x + 3}\right)^2 = \left(\sqrt{4x - 3}\right)^2$

$x + 3 = 4x - 3$

$-3x + 3 = -3$

$-3x = -6$

$x = 2$

**37.** $\sqrt{5x + 1} = 3\sqrt{x}$

$\left(\sqrt{5x + 1}\right)^2 = \left(3\sqrt{x}\right)^2$

$5x + 1 = 9x$

$1 = 4x$

$\frac{1}{4} = x$

**39.** $\sqrt{3x + 4} = -3\sqrt{x}$

$\left(\sqrt{3x + 4}\right)^2 = \left(-3\sqrt{x}\right)^2$

$3x + 4 = 9x$

$4 = 6x$

$\frac{4}{6} = x$

$\frac{2}{3} = x$    Extraneous

The equation has no solution.

**41.** $\sqrt{3t + 11} - 5\sqrt{t} = 0$

$\sqrt{3t + 11} = 5\sqrt{t}$

$\left(\sqrt{3t + 11}\right)^2 = \left(5\sqrt{t}\right)^2$

$3t + 11 = 25t$

$11 = 22t$

$\frac{11}{22} = t$

$\frac{1}{2} = t$

**43.** $2\sqrt{y + 1} - \sqrt{3y + 6} = 0$

$2\sqrt{y + 1} = \sqrt{3y + 6}$

$\left(2\sqrt{y + 1}\right)^2 = \left(\sqrt{3y + 6}\right)^2$

$4(y + 1) = 3y + 6$

$4y + 4 = 3y + 6$

$y + 4 = 6$

$y = 2$

**45.** $\sqrt{x} = 2 - x$

$\left(\sqrt{x}\right)^2 = (2 - x)^2$

$x = 4 - 4x + x^2$

$0 = 4 - 5x + x^2$

$x^2 - 5x + 4 = 0$

$(x - 4)(x - 1) = 0$

$x - 4 = 0 \Rightarrow x = 4$   Extraneous

$x - 1 = 0 \Rightarrow x = 1$

**47.** $2x + 1 = \sqrt{9x}$

$(2x + 1)^2 = \left(\sqrt{9x}\right)^2$

$(2x)^2 + 2(2x)(1) + 1^2 = 9x$

$4x^2 + 4x + 1 = 9x$

$4x^2 - 5x + 1 = 0$

$(4x - 1)(x - 1) = 0$

$4x - 1 = 0 \Rightarrow 4x = 1 \Rightarrow x = \frac{1}{4}$

$x - 1 = 0 \Rightarrow x = 1$

**49.** $x = \sqrt{20 - x}$

$x^2 = \left(\sqrt{20 - x}\right)^2$

$x^2 = 20 - x$

$x^2 + x - 20 = 0$

$(x + 5)(x - 4) = 0$

$x + 5 = 0 \Rightarrow x = -5$   Extraneous

$x - 4 = 0 \Rightarrow x = 4$

**51.** $x = \sqrt{18 - 3x}$

$x^2 = \left(\sqrt{18 - 3x}\right)^2$

$x^2 = 18 - 3x$

$x^2 + 3x - 18 = 0$

$(x + 6)(x - 3) = 0$

$x + 6 = 0 \Rightarrow x = -6$   Extraneous

$x - 3 = 0 \Rightarrow x = 3$

**53.** $\sqrt{x^2 + 5} = x + 1$

$\left(\sqrt{x^2 + 5}\right)^2 = (x + 1)^2$

$x^2 + 5 = x^2 + 2x + 1$

$5 = 2x + 1$

$4 = 2x$

$2 = x$

**55.** $\sqrt{6x + 7} = x + 2$

$\left(\sqrt{6x + 7}\right)^2 = (x + 2)^2$

$6x + 7 = x^2 + 4x + 4$

$0 = x^2 - 2x - 3$

$0 = (x - 3)(x + 1)$

$x - 3 = 0 \Rightarrow x = 3$

$x + 1 = 0 \Rightarrow x = -1$

**57.**

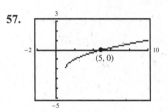

The *x*-intercept appears to be $(5, 0)$.

$y = \sqrt{x - 1} - 2$

$0 = \sqrt{x - 1} - 2$

$2 = \sqrt{x - 1}$

$2^2 = \left(\sqrt{x - 1}\right)^2$

$4 = x - 1$

$5 = x$

This verifies that the *x*-intercept is $(5, 0)$.

**59.**

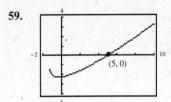

The $x$-intercept appears to be $(5, 0)$.

$$y = x - \sqrt{4x + 5}$$
$$0 = x - \sqrt{4x + 5}$$
$$-x = -\sqrt{4x + 5}$$
$$(-x)^2 = \left(-\sqrt{4x + 5}\right)^2$$
$$x^2 = 4x + 5$$
$$x^2 - 4x - 5 = 0$$
$$(x - 5)(x + 1) = 0$$
$$x - 5 = 0 \Rightarrow x = 5$$
$$x + 1 = 0 \Rightarrow x = -1 \quad \text{Extraneous}$$

This verifies that the $x$-intercept is $(5, 0)$.

**61.** $c^2 = a^2 + b^2$

$$x^2 = 12^2 + 5^2$$
$$x = \sqrt{12^2 + 5^2}$$
$$= \sqrt{144 + 25}$$
$$= \sqrt{169}$$
$$= 13$$

**63.** $c^2 = a^2 + b^2$

$$2^2 = \left(\sqrt{3}\right)^2 + x^2$$
$$4 = 3 + x^2$$
$$1 = x^2$$
$$1 = x$$

**65.** $c^2 = a^2 + b^2$

$$x^2 = 10^2 + (x - 2)^2$$
$$x^2 = 100 + x^2 - 4x + 4$$
$$x^2 = x^2 - 4x + 104$$
$$0 = -4x + 104$$
$$4x = 104$$
$$x = 26$$

**67.** $d = \sqrt{(x_2 - x_1)^2 + (y_2 - y_1)^2}$

$$d = \sqrt{(5 - 1)^2 + (5 - 2)^2}$$
$$= \sqrt{4^2 + 3^2}$$
$$= \sqrt{16 + 9}$$
$$= \sqrt{25}$$
$$= 5$$

**69.** $d = \sqrt{(x_2 - x_1)^2 + (y_2 - y_1)^2}$

$$d = \sqrt{[3 - (-5)]^2 + (-2 - 4)^2}$$
$$= \sqrt{8^2 + (-6)^2}$$
$$= \sqrt{64 + 36}$$
$$= \sqrt{100}$$
$$= 10$$

**71.** $d = \sqrt{(x_2 - x_1)^2 + (y_2 - y_1)^2}$

$$d = \sqrt{(4 - 3)^2 + [6 - (-2)]^2}$$
$$= \sqrt{1^2 + 8^2}$$
$$= \sqrt{1 + 64}$$
$$= \sqrt{65}$$
$$\approx 8.06$$

**73.** $d = \sqrt{(x_2 - x_1)^2 + (y_2 - y_1)^2}$

$$d = \sqrt{[-2 - (-3)]^2 + (6 - 2)^2}$$
$$= \sqrt{1^2 + 4^2}$$
$$= \sqrt{1 + 16}$$
$$= \sqrt{17} \text{ or } d \approx 4.12$$

**75.**

$$t = \sqrt{\frac{d}{16}}$$
$$3 = \sqrt{\frac{d}{16}}$$
$$3^2 = \left(\sqrt{\frac{d}{16}}\right)^2$$
$$9 = \frac{d}{16}$$
$$16 \cdot 9 = d$$
$$144 = d$$

The worker is about 144 feet high.

**77.**
$$v = \sqrt{2gh}$$
$$45 = \sqrt{2 \cdot 32 \cdot h}$$
$$45 = \sqrt{64h}$$
$$(45)^2 = \left(\sqrt{64h}\right)^2$$
$$2025 = 64h$$
$$\frac{2025}{64} = h$$
$$31.64 \approx h$$

The height was approximately 31.64 feet.

**79.**
$$t = 2\pi\sqrt{\frac{L}{32}}$$
$$2 = 2\pi\sqrt{\frac{L}{32}}$$
$$2^2 = \left(2\pi\sqrt{\frac{L}{32}}\right)^2$$
$$4 = 4\pi^2 \cdot \frac{L}{32}$$
$$4 = \frac{\pi^2 L}{8}$$
$$8 \cdot 4 = \pi^3 L$$
$$32 = \pi^3 L$$
$$\frac{32}{\pi^2} = L$$
$$3.24 \approx L$$

The pendulum is about 3.24 feet long.

**81.** $a^2 + b^2 = c^2$ Pythagorean Theorem
$$15^2 + b^2 = 39^2$$
$$225 + b^2 = 1521$$
$$b^2 = 1296$$
$$b = \sqrt{1296}$$
$$b = 36$$

The mast is 36 feet tall.

**83.** $c^2 = a^2 + b^2$
$$c^2 = 30^2 + 60^2 = 900 + 3600 = 4500$$
$$c = \sqrt{4500} = \sqrt{900}\sqrt{5} = 30\sqrt{5} \approx 67.08$$

The length of the diagonal is $30\sqrt{5}$ feet, or about 67.08 feet.

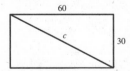

**85.** $c^2 = a^2 + b^2$ Pythagorean Theorem
$$c^2 = 410^2 + 317^2$$
$$c^2 = 268,589$$
$$c = \sqrt{268,589}$$
$$c \approx 518.26$$

The cities are about 518.26 miles apart.

**87.**
$$p = 40 - \sqrt{x - 1}$$
$$34.70 = 40 - \sqrt{x - 1}$$
$$-5.30 = -\sqrt{x - 1}$$
$$5.30 = \sqrt{x - 1}$$
$$(5.30)^2 = \left(\sqrt{x - 1}\right)^2$$
$$28.09 = x - 1$$
$$29.09 = x$$

When the price is $34.70, the demand is about 29 units per day.

**89.** (a)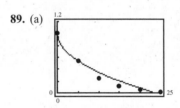

(b)
$$p = 1 - 0.206\sqrt{h}$$
$$0.4 = 1 - 0.206\sqrt{h}$$
$$-0.6 = -0.206\sqrt{h}$$
$$\frac{-0.6}{-0.206} = \sqrt{h}$$
$$\frac{0.6}{0.206} = \sqrt{h}$$
$$\left(\frac{0.6}{0.206}\right)^2 = \left(\sqrt{h}\right)^2$$
$$8.5 \approx h$$

The altitude is about 8.5 kilometers.

**91.** No. The principal square root of a number is positive.

**93.** Yes. The two legs of the right triangle can be of the same length $l$ and the hypotenuse of length $\sqrt{2}l$.

**95.** $y = -5x - 8 \Rightarrow m_1 = -5$
$$y = \frac{1}{5}x \quad \Rightarrow m_2 = \frac{1}{5}$$

The lines are perpendicular because $m_1 = -\dfrac{1}{m_2}$.

**97.** $y = \frac{2}{3}x - \frac{3}{5} \Rightarrow m_1 = \frac{2}{3}$

$y = \frac{2}{3}x + \frac{3}{5} \Rightarrow m_2 = \frac{2}{3}$

The lines are parallel because $m_1 = m_2$.

**99.** $\dfrac{r}{r - 1} \div \dfrac{r^2}{r^2 - 1} = \dfrac{r}{r - 1} \cdot \dfrac{r^2 - 1}{r^2}$

$= \dfrac{r}{r - 1} \cdot \dfrac{(r + 1)(r - 1)}{r^2}$

$= \dfrac{\cancel{r}(r + 1)\cancel{(r - 1)}}{r \cdot \cancel{r}\cancel{(r - 1)}}$

$= \dfrac{r + 1}{r}, \quad r \neq 1$

**101.** $\dfrac{\left(\dfrac{x}{2} - 1\right)}{(x - 2)} = \dfrac{\left(\dfrac{x}{2} - \dfrac{2}{2}\right)}{\left(\dfrac{x - 2}{1}\right)}$

$= \dfrac{x - 2}{2} \cdot \dfrac{1}{x - 2}$

$= \dfrac{1\cancel{(x - 2)}}{2\cancel{(x - 2)}}$

$= \dfrac{1}{2}, \quad x \neq 2$

**103.** $\dfrac{2}{\sqrt{6} - 1} = \dfrac{2}{\sqrt{6} - 1} \cdot \dfrac{\sqrt{6} + 1}{\sqrt{6} + 1}$

$= \dfrac{2\left(\sqrt{6} + 1\right)}{\left(\sqrt{6}\right)^2 - 1^2}$

$= \dfrac{2\left(\sqrt{6} + 1\right)}{6 - 1}$

$= \dfrac{2\sqrt{6} + 2}{5}$

**105.** $\dfrac{\sqrt{5} + 4}{\sqrt{7} - \sqrt{2}} = \dfrac{\sqrt{5} + 4}{\sqrt{7} - \sqrt{2}} \cdot \dfrac{\sqrt{7} + \sqrt{2}}{\sqrt{7} + \sqrt{2}}$

$= \dfrac{\left(\sqrt{5} + 4\right)\left(\sqrt{7} + \sqrt{2}\right)}{\left(\sqrt{7}\right)^2 - \left(\sqrt{2}\right)^2}$

$= \dfrac{\sqrt{35} + \sqrt{10} + 4\sqrt{7} + 4\sqrt{2}}{7 - 2}$

$= \dfrac{\sqrt{35} + \sqrt{10} + 4\sqrt{7} + 4\sqrt{2}}{5}$

# Review Exercises for Chapter 9

**1.** Number: 49

Positive square root: 7

Negative square root: $-7$

**3.** Number: $-4$

A negative number has no square root in the real numbers.

**5.** The negative number $-125$ has one cube root, $-5$, because $(-5)^3 = -125$.

**7.** The positive number 16 has two real fourth roots, 2 and $-2$, because $2^4 = 16$ and $(-2)^4 = 16$.

**9.** $\sqrt{121} = 11$

**11.** $\sqrt{1.44} = 1.2$

**13.** A negative number has no square root in the real numbers.

**15.** $\sqrt[3]{-27} = -3$

**17.** $\sqrt[3]{\dfrac{8}{125}} = \dfrac{\sqrt[3]{8}}{\sqrt[3]{125}} = \dfrac{2}{5}$

**19.** $\sqrt{53} \approx 7.280$

**21.** $\sqrt{\dfrac{3}{20}} \approx 0.387$

**23.** $3 + 2\sqrt{6} \approx 7.899$

**25.** $\dfrac{5 - 3\sqrt{3}}{2} \approx -0.098$

**27.** When $x = -2$ and $y = 3$,

$\sqrt{x^2 y} = \sqrt{(-2)^2 (3)} = \sqrt{12} \approx 3.46$.

**29.** $\sqrt{48} = \sqrt{16 \cdot 3} = 4\sqrt{3}$

**31.** $\sqrt{160} = \sqrt{16 \cdot 10} = 4\sqrt{10}$

**33.** $\sqrt{\dfrac{23}{9}} = \dfrac{\sqrt{23}}{\sqrt{9}} = \dfrac{\sqrt{23}}{3}$

**35.** $\sqrt{\dfrac{20}{9}} = \dfrac{\sqrt{20}}{\sqrt{9}} = \dfrac{\sqrt{4 \cdot 5}}{3} = \dfrac{2\sqrt{5}}{3}$

**37.** $\sqrt[3]{32} = \sqrt[3]{8 \cdot 4} = 2\sqrt[3]{4}$

**39.** $\sqrt[4]{96} = \sqrt[4]{16 \cdot 6} = 2\sqrt[4]{6}$

**41.** $\sqrt{36x^4} = \sqrt{36 \cdot x^4} = 6x^2$

**43.** $\sqrt{4y^3} = \sqrt{4 \cdot y^2 \cdot y} = 2y\sqrt{y}$

**45.** $\sqrt{32a^3b} = \sqrt{16 \cdot 2 \cdot a^2 \cdot ab} = 4a\sqrt{2ab}$

**47.** $\sqrt{x^3y} = \sqrt{x^2 \cdot xy} = |x|\sqrt{xy}$

**49.** $\sqrt[3]{8x^6} = \sqrt[3]{8\left(x^2\right)^3} = 2x^2$

**51.** $\sqrt[4]{81y^5} = \sqrt[4]{3^4 \cdot y^4 \cdot y} = 3y\sqrt[4]{y}$

**53.** $\sqrt{\dfrac{3}{7}} = \dfrac{\sqrt{3}}{\sqrt{7}} \cdot \dfrac{\sqrt{7}}{\sqrt{7}} = \dfrac{\sqrt{21}}{\sqrt{7^2}} = \dfrac{\sqrt{21}}{7}$

**55.** $\dfrac{6}{\sqrt{3}} = \dfrac{6}{\sqrt{3}} \cdot \dfrac{\sqrt{3}}{\sqrt{3}} = \dfrac{6\sqrt{3}}{3} = 2\sqrt{3}$

**57.** $\sqrt{\dfrac{5}{12}} = \dfrac{\sqrt{5}}{\sqrt{12}} \cdot \dfrac{\sqrt{3}}{\sqrt{3}} = \dfrac{\sqrt{15}}{\sqrt{36}} = \dfrac{\sqrt{15}}{6}$

**59.** $\dfrac{3}{\sqrt[3]{2}} = \dfrac{3}{\sqrt[3]{2}} \cdot \dfrac{\sqrt[3]{4}}{\sqrt[3]{4}} = \dfrac{3\sqrt[3]{4}}{\sqrt[3]{8}} = \dfrac{3\sqrt[3]{4}}{2}$

**61.** $\dfrac{4}{\sqrt{x}} = \dfrac{4}{\sqrt{x}} \cdot \dfrac{\sqrt{x}}{\sqrt{x}} = \dfrac{4\sqrt{x}}{\sqrt{x^2}} = \dfrac{4\sqrt{x}}{x}$

**63.** $\sqrt{\dfrac{11a}{b}} = \dfrac{\sqrt{11a}}{\sqrt{b}} \cdot \dfrac{\sqrt{b}}{\sqrt{b}} = \dfrac{\sqrt{11ab}}{b}$

**65.** $\dfrac{\sqrt{6x^2}}{\sqrt{27y^3}} = \dfrac{\sqrt{6x^2}}{\sqrt{27y^3}} \cdot \dfrac{\sqrt{3y}}{\sqrt{3y}}$   or   $\dfrac{\sqrt{6x^2}}{\sqrt{27y^3}} = \sqrt{\dfrac{6x^2}{27y^3}}$

$\qquad = \dfrac{\sqrt{18x^2y}}{\sqrt{81y^4}}$   $\qquad = \sqrt{\dfrac{2x^2}{9y^3}}$

$\qquad = \dfrac{\sqrt{9 \cdot 2 \cdot x^2 \cdot y}}{\sqrt{81\left(y^2\right)^2}}$   $\qquad = \dfrac{\sqrt{2x^2}}{\sqrt{9y^3}} \cdot \dfrac{\sqrt{y}}{\sqrt{y}}$

$\qquad = \dfrac{3x\sqrt{2y}}{9y^2}$   $\qquad = \dfrac{\sqrt{2x^2y}}{\sqrt{9y^4}}$

$\qquad = \dfrac{x\sqrt{2y}}{3y^2}$   $\qquad = \dfrac{\sqrt{2 \cdot x^2 \cdot y}}{\sqrt{9\left(y^2\right)^2}}$

$\qquad\qquad\qquad\qquad = \dfrac{x\sqrt{2y}}{3y^2}$

**67.** $\sqrt[3]{\dfrac{4}{x^2}} = \dfrac{\sqrt[3]{4}}{\sqrt[3]{x^2}} \cdot \dfrac{\sqrt[3]{x}}{\sqrt[3]{x}} = \dfrac{\sqrt[3]{4x}}{\sqrt[3]{x^3}} = \dfrac{\sqrt[3]{4x}}{x}$

**69.** $7\sqrt{2} + 5\sqrt{2} = (7 + 5)\sqrt{2} = 12\sqrt{2}$

**71.** $3\sqrt{5} - 7\sqrt{3} + 11\sqrt{3} = 3\sqrt{5} + (-7 + 11)\sqrt{3}$

$\qquad\qquad\qquad\qquad\quad = 3\sqrt{5} + 4\sqrt{3}$

**73.** $3\sqrt{20} - 10\sqrt{20} = (3 - 10)\sqrt{20}$

$\qquad\qquad\qquad\quad = -7\sqrt{4 \cdot 5}$

$\qquad\qquad\qquad\quad = -7 \cdot 2\sqrt{5}$

$\qquad\qquad\qquad\quad = -14\sqrt{5}$

**75.** $4\sqrt{48} + 2\sqrt{3} - 5\sqrt{12} = 4\sqrt{16 \cdot 3} + 2\sqrt{3} - 5\sqrt{4 \cdot 3}$

$\qquad\qquad\qquad\qquad\quad = 4 \cdot 4\sqrt{3} + 2\sqrt{3} - 5 \cdot 2\sqrt{3}$

$\qquad\qquad\qquad\qquad\quad = 16\sqrt{3} + 2\sqrt{3} - 10\sqrt{3}$

$\qquad\qquad\qquad\qquad\quad = (16 + 2 - 10)\sqrt{3}$

$\qquad\qquad\qquad\qquad\quad = 8\sqrt{3}$

**77.** $\sqrt[4]{4} + 5\sqrt[4]{4} = (1 + 5)\sqrt[4]{4} = 6\sqrt[4]{4}$

**79.** $\sqrt[5]{x} - 8\sqrt[5]{x} = (1 - 8)\sqrt[5]{x} = -7\sqrt[5]{x}$

**81.** $\sqrt{36y} - \sqrt{16y} = 6\sqrt{y} - 4\sqrt{y}$

$$= (6 - 4)\sqrt{y}$$

$$= 2\sqrt{y}$$

**83.** $\sqrt{18x^3} - 3x\sqrt{2x} = \sqrt{9 \cdot 2 \cdot x^2 \cdot x} - 3x\sqrt{2x}$

$$= 3x\sqrt{2x} - 3x\sqrt{2x}$$

$$= 0$$

**85.** $\sqrt{3}\left(\sqrt{6} + 1\right) = \sqrt{18} + 1\sqrt{3} = \sqrt{9 \cdot 2} + \sqrt{3} = 3\sqrt{2} + \sqrt{3}$

**87.** $\sqrt[4]{6}\left(\sqrt[4]{2} - 1\right) = \sqrt[4]{6 \cdot 2} - 1\sqrt[4]{6} = \sqrt[4]{12} - \sqrt[4]{6}$

**89.** $\left(\sqrt{8} + 2\right)\left(3\sqrt{2} - 1\right) = \sqrt{8} \cdot 3\sqrt{2} - \sqrt{8} + 2 \cdot 3\sqrt{2} - 2$

$$= 3\sqrt{16} - \sqrt{4 \cdot 2} + 6\sqrt{2} - 2$$

$$= 3 \cdot 4 - 2\sqrt{2} + 6\sqrt{2} - 2$$

$$= 12 + 4\sqrt{2} - 2$$

$$= 10 + 4\sqrt{2}$$

**91.** $\left(\sqrt{3} - \sqrt{5}\right)\left(\sqrt{3} + \sqrt{5}\right) = \left(\sqrt{3}\right)^2 - \left(\sqrt{5}\right)^2 = 3 - 5 = -2$

**93.** $\left(\sqrt{5} - 2\right)^2 = \left(\sqrt{5} - 2\right)\left(\sqrt{5} - 2\right) = \sqrt{5} \cdot \sqrt{5} - 2\sqrt{5} - 2\sqrt{5} + 4 = 5 - 4\sqrt{5} + 4 = 9 - 4\sqrt{5}$

We could use the pattern for the square of a binomial.

Pattern: $(a - b)^2 = a^2 - 2ab + b^2$

$$\left(\sqrt{5} - 2\right)^2 = \left(\sqrt{5}\right)^2 - 2\left(\sqrt{5}\right)(2) + 2^2 = 5 - 4\sqrt{5} + 4 = 9 - 4\sqrt{5}$$

**95.** $\left(\sqrt[5]{2} + 3\right)^2 = \left(\sqrt[5]{2}\right)^2 + 2\left(\sqrt[5]{2}\right)(3) + 3^2 = \sqrt[5]{4} + 6\sqrt[5]{2} + 9$

**97.** Expression: $\sqrt{x} + 9$

Conjugate: $\sqrt{x} - 9$

Product: $\left(\sqrt{x} + 9\right)\left(\sqrt{x} - 9\right) = \left(\sqrt{x}\right)^2 - 9^2 = x - 81, x \geq 0$

**99.** Expression: $12 - \sqrt{t}$

Conjugate: $12 + \sqrt{t}$

Product: $\left(12 - \sqrt{t}\right)\left(12 + \sqrt{t}\right) = 12^2 - \left(\sqrt{t}\right)^2 = 144 - t, t \geq 0$

**101.** $\dfrac{3}{\sqrt{12}-3} = \dfrac{3}{\sqrt{12}-3} \cdot \dfrac{\sqrt{12}+3}{\sqrt{12}+3}$

$\qquad = \dfrac{3\left(\sqrt{12}+3\right)}{\left(\sqrt{12}\right)^2 - 3^2}$

$\qquad = \dfrac{3\left(\sqrt{12}+3\right)}{12-9}$

$\qquad = \dfrac{3\left(\sqrt{12}+3\right)}{3}$

$\qquad = \sqrt{12}+3$

$\qquad = \sqrt{4\cdot 3}+3$

$\qquad = 2\sqrt{3}+3$

**103.** $\dfrac{\sqrt{x}-3}{\sqrt{x}+3} = \dfrac{\sqrt{x}-3}{\sqrt{x}+3} \cdot \dfrac{\sqrt{x}-3}{\sqrt{x}-3}$

$\qquad = \dfrac{\left(\sqrt{x}-3\right)\left(\sqrt{x}-3\right)}{\left(\sqrt{x}\right)^2 - 3^2}$

$\qquad = \dfrac{\sqrt{x}\cdot\sqrt{x} - 3\sqrt{x} - 3\sqrt{x} + 9}{x-9}$

$\qquad = \dfrac{x - 6\sqrt{x} + 9}{x-9}$

**105.** (a) $x = 4$

$\qquad \sqrt{4} - 4 \overset{?}{=} 0$

$\qquad 2 - 4 \neq 0$

$\qquad$ 4 *is not* a solution.

(b) $x = 8$

$\qquad \sqrt{8} - 4 \overset{?}{=} 0$

$\qquad \sqrt{4\cdot 2} - 4 \overset{?}{=} 0$

$\qquad 2\sqrt{2} - 4 \neq 0$

$\qquad$ 8 *is not* a solution.

(c) $x = 16$

$\qquad \sqrt{16} - 4 \overset{?}{=} 0$

$\qquad 4 - 4 \overset{?}{=} 0$

$\qquad 0 = 0$

$\qquad$ 16 *is* a solution.

(d) $x = -2$

$\qquad \sqrt{-2} - 4 \overset{?}{=} 0$

$\qquad \sqrt{-2} \neq 4$

$\qquad \sqrt{-2}$ is not a real number.

$\qquad -2$ *is not* a solution.

**107.** $\sqrt{y} = 13$

$\qquad \left(\sqrt{y}\right)^2 = 13^2$

$\qquad y = 169$

**109.** $\sqrt{x} + 2 = 0$

$\qquad \sqrt{x} = -2$

$\qquad \left(\sqrt{x}\right)^2 = (-2)^2$

$\qquad x = 4 \quad$ Extraneous

This answer does not check.

$\qquad \sqrt{4} + 2 \overset{?}{=} 0$

$\qquad 2 + 2 \neq 0$

So, the equation has no solution.

**111.** $\sqrt{2t+8} = 5$

$\qquad \left(\sqrt{2t+8}\right)^2 = (5)^2$

$\qquad 2t + 8 = 25$

$\qquad 2t = 17$

$\qquad t = \frac{17}{2}$

**113.** $\sqrt{4x-3} = \sqrt{x+6}$

$\qquad \left(\sqrt{4x-3}\right)^2 = \left(\sqrt{x+6}\right)^2$

$\qquad 4x - 3 = x + 6$

$\qquad 3x - 3 = 6$

$\qquad 3x = 9$

$\qquad x = 3$

**115.** $\sqrt{x-4} = x - 6$

$\qquad \left(\sqrt{x-4}\right)^2 = (x-6)^2$

$\qquad x - 4 = x^2 - 12x + 36$

$\qquad 0 = x^2 - 13x + 40$

$\qquad (x-8)(x-5) = 0$

$\qquad x - 8 = 0 \Rightarrow x = 8$

$\qquad x - 5 = 0 \Rightarrow x = 5 \quad$ Extraneous

**117.** $c^2 = a^2 + b^2$

$\qquad 8^2 = x^2 + 4^2$

$\qquad 64 = x^2 + 16$

$\qquad 48 = x^2$

$\qquad \sqrt{48} = x$

$\qquad \sqrt{16\cdot 3} = x$

$\qquad 4\sqrt{3} = x$ or $x \approx 6.93$

**119.** $d = \sqrt{(-2-1)^2 + (0-4)^2}$

$\phantom{d} = \sqrt{(-3)^2 + (-4)^2}$

$\phantom{d} = \sqrt{9 + 16}$

$\phantom{d} = \sqrt{25}$

$\phantom{d} = 5$

**121.** $d = \sqrt{[-1-(-5)]^2 + [9-(-2)]^2}$

$\phantom{d} = \sqrt{(4)^2 + (11)^2}$

$\phantom{d} = \sqrt{16 + 121}$

$\phantom{d} = \sqrt{137}$

$\phantom{d} = 11.70$

**123.** $d = \sqrt{[2-(-4)]^2 + [-4-(-3)]^2}$

$\phantom{d} = \sqrt{(6)^2 + (-1)^2}$

$\phantom{d} = \sqrt{36 + 1}$

$\phantom{d} = \sqrt{37}$

$\phantom{d} = 6.08$

**125.** $d = \sqrt{(2-7)^2 + (-8-3)^2}$

$\phantom{d} = \sqrt{(-5)^2 + (-11)^2}$

$\phantom{d} = \sqrt{25 + 121}$

$\phantom{d} = \sqrt{146}$

$\phantom{d} \approx 12.08$

**127.**   $v = \sqrt{2gh}$

$80 = \sqrt{2(32)(h)}$

$80 = \sqrt{64h}$

$80 = 8\sqrt{h}$

$10 = \sqrt{h}$

$10^2 = \left(\sqrt{h}\right)^2$

$100 = h$

The object was dropped from a height of 100 feet.

**129.**   $t = 2\pi\sqrt{\dfrac{L}{32}}$

$1.75 = 2\pi\sqrt{\dfrac{L}{32}}$

$(1.75)^2 = \left(2\pi\sqrt{\dfrac{L}{32}}\right)^2$

$(1.75)^2 = 4\pi^2 \cdot \dfrac{L}{32}$

$(1.75)^2 = \dfrac{\pi^2 L}{8}$

$8(1.75)^2 = \pi^2 L$

$\dfrac{8(1.75)^2}{\pi^2} = L$

$2.48 \approx L$

The pendulum is about 2.48 feet long.

**131.** $c^2 = a^2 + b^2$   Pythagorean Theorem

$x^2 = (60)^2 + (100)^2 = 3600 + 10{,}000 = 13{,}600$

$x = \sqrt{13{,}600} = \sqrt{400 \cdot 34} = 20\sqrt{34} \approx 116.6$

The wire is about 116.6 feet long.

**133.** $c^2 = a^2 + b^2$

$c^2 = 20^2 + 42^2$

$c^2 = 400 + 1764$

$c^2 = 2164$

$c = \sqrt{2164}$

$c = \sqrt{4 \cdot 541}$

$c = 2\sqrt{541}$ or $c \approx 46.52$

The length of the diagonal of the pool is approximately 46.52 feet.

## Chapter Test for Chapter 9

**1.** (a) $\sqrt{121} = 11$

  (b) $\sqrt{-36}$

  A negative number has no square root in the real numbers.

**2.** (a) $\sqrt[4]{81} = 3$

  (b) $\sqrt[3]{-64} = -4$

**3.** $\sqrt{28} = \sqrt{4 \cdot 7} = 2\sqrt{7}$

**4.** $\sqrt[3]{54} = \sqrt[3]{27 \cdot 2} = \sqrt[3]{27} \cdot \sqrt[3]{2} = 3\sqrt[3]{2}$

**5.** $\sqrt{32x^2 y^3} = \sqrt{16 \cdot 2 \cdot x^2 \cdot y^2 \cdot y} = 4|x|y\sqrt{2y}$

**6.** $\sqrt{\dfrac{3x^3}{y^4}} = \dfrac{\sqrt{3 \cdot x^2 \cdot x}}{\sqrt{y^4}} = \dfrac{x\sqrt{3x}}{y^2}$

**7.** $\dfrac{5}{\sqrt{15}} = \dfrac{5}{\sqrt{15}} \cdot \dfrac{\sqrt{15}}{\sqrt{15}} = \dfrac{5\sqrt{15}}{15} = \dfrac{\sqrt{15}}{3}$

**8.** $\dfrac{2}{\sqrt[3]{4}} = \dfrac{2}{\sqrt[3]{4}} \cdot \dfrac{\sqrt[3]{2}}{\sqrt[3]{2}} = \dfrac{2\sqrt[3]{2}}{\sqrt[3]{8}} = \dfrac{2\sqrt[3]{2}}{2} = \sqrt[3]{2}$

**9.** $10\sqrt{2} - 7\sqrt{2} = (10 - 7)\sqrt{2} = 3\sqrt{2}$

**10.** $\begin{aligned} 5\sqrt{3x} + 3\sqrt{75x} &= 5\sqrt{3x} + 3\sqrt{25 \cdot 3x} \\ &= 5\sqrt{3x} + 3 \cdot 5\sqrt{3x} \\ &= 5\sqrt{3x} + 15\sqrt{3x} \\ &= 20\sqrt{3x} \end{aligned}$

**11.** $\begin{aligned} 7\sqrt[3]{5} - 6\sqrt[3]{4} + \sqrt[3]{5} &= (7 + 1)\sqrt[3]{5} - 6\sqrt[3]{4} \\ &= 8\sqrt[3]{5} - 6\sqrt[3]{4} \end{aligned}$

**12.** $\begin{aligned} 4\sqrt{2x} - 6\sqrt{32x} + \sqrt{2x^2} &= 4\sqrt{2x} - 6\sqrt{16 \cdot 2x} + \sqrt{2x^2} \\ &= 4\sqrt{2x} - 6 \cdot 4\sqrt{2x} + |x|\sqrt{2} \\ &= 4\sqrt{2x} - 24\sqrt{2x} + |x|\sqrt{2} \\ &= (4 - 24)\sqrt{2x} + |x|\sqrt{2} \\ &= -20\sqrt{2x} + |x|\sqrt{2} \end{aligned}$

**13.** $\begin{aligned} \sqrt{3}\left(2 - \sqrt{12}\right) &= 2\sqrt{3} - \sqrt{3} \cdot \sqrt{12} \\ &= 2\sqrt{3} - \sqrt{36} \\ &= 2\sqrt{3} - 6 \end{aligned}$

**14.** $\sqrt[3]{5}\left(\sqrt[3]{2} + 3\right) = \sqrt[3]{5 \cdot 2} + 3\sqrt[3]{5} = \sqrt[3]{10} + 3\sqrt[3]{5}$

**15.** $\begin{aligned} \left(\sqrt{6} - 3\right)\left(\sqrt{6} + 5\right) &= \sqrt{6} \cdot \sqrt{6} + 5\sqrt{6} - 3\sqrt{6} - 15 \\ &= 6 + 2\sqrt{6} - 15 \\ &= -9 + 2\sqrt{6} \end{aligned}$

**16.** $\begin{aligned} \left(1 - 2\sqrt{x}\right)^2 &= (1)^2 - 2(1)\left(2\sqrt{x}\right) + \left(2\sqrt{x}\right)^2 \\ &= 1 - 4\sqrt{x} + 4x \end{aligned}$

**17.** Number: $\sqrt{3} - 5$

Conjugate: $\sqrt{3} + 5$

Product: $\left(\sqrt{3}\right)^2 - 5^2 = 3 - 25 = -22$

**18.** $\begin{aligned} \dfrac{10}{\sqrt{6} + 1} &= \dfrac{10}{\sqrt{6} + 1} \cdot \dfrac{\sqrt{6} - 1}{\sqrt{6} - 1} \\ &= \dfrac{10\left(\sqrt{6} - 1\right)}{\left(\sqrt{6}\right)^2 - 1^2} \\ &= \dfrac{10\left(\sqrt{6} - 1\right)}{6 - 1} \\ &= \dfrac{10\left(\sqrt{6} - 1\right)}{5} = 2\left(\sqrt{6} - 1\right) \text{ or } 2\sqrt{6} - 2 \end{aligned}$

**19.** $\begin{aligned} \sqrt{y} &= 5 \\ \left(\sqrt{y}\right)^2 &= (5)^2 \\ y &= 25 \end{aligned}$

**20.** $\begin{aligned} 2\sqrt{x + 3} &= 5 \\ \left(2\sqrt{x + 3}\right)^2 &= 5^2 \\ 4(x + 3) &= 25 \\ 4x + 12 &= 25 \\ 4x &= 13 \\ x &= \tfrac{13}{4} \end{aligned}$

**21.** $\begin{aligned} \sqrt{5x - 4} &= \sqrt{3x + 6} \\ \left(\sqrt{5x - 4}\right)^2 &= \left(\sqrt{3x + 6}\right)^2 \\ 5x - 4 &= 3x + 6 \\ 2x - 4 &= 6 \\ 2x &= 10 \\ x &= 5 \end{aligned}$

**22.** $\begin{aligned} 2\sqrt{6y} &= y + 6 \\ \left(2\sqrt{6y}\right)^2 &= (y + 6)^2 \\ 4(6y) &= y^2 + 2 \cdot y \cdot 6 + 6^2 \\ 24y &= y^2 + 12y + 36 \\ 0 &= y^2 - 12y + 36 \\ 0 &= y^2 - 2 \cdot 6 \cdot y + 6^2 \\ 0 &= (y - 6)^2 \\ y - 6 &= 0 \Rightarrow y = 6 \end{aligned}$

**23.** $c^2 = a^2 + b^2$

$c^2 = 6^2 + 4^2$

$c^2 = 36 + 16$

$c^2 = 52$

$c = \sqrt{52}$

$c = \sqrt{4 \cdot 13}$

$c = 2\sqrt{13}$

$c \approx 7.21$

Side $c$ is about 7.21 units long.

**24.** $d = \sqrt{[5 - (-3)]^2 + (2 - 8)^2}$

$= \sqrt{8^2 + (-6)^2}$

$= \sqrt{64 + 36}$

$= \sqrt{100}$

$= 10$

**25.**    $p = 100 - \sqrt{x - 25}$

$90 = 100 - \sqrt{x - 25}$

$-10 = -\sqrt{x - 25}$

$10 = \sqrt{x - 25}$

$10^2 = \left(\sqrt{x - 25}\right)^2$

$100 = x - 25$

$125 = x$

When the price is \$90, the demand is 125 units per day.

# Cumulative Test for Chapters 7–9

**1.** The denominator $x^2 - 4 = (x - 2)(x + 2)$ is zero when $x = 2$ or $x = -2$, so the domain is all real numbers $x$ such that $x \neq 2$ and $x \neq -2$.

**2.** $\dfrac{7}{3x} = \dfrac{7(6x^3)}{3x(6x^3)} = \dfrac{7(6x^3)}{18x^4}$

The missing factor is $(6x^3)$.

**3.** $\dfrac{8 - 2x}{x^2 - 16} = \dfrac{2(4 - x)}{(x - 4)(x + 4)}$

$= \dfrac{2(-1)(x - 4)}{(x - 4)(x + 4)}$

$= -\dfrac{2}{x + 4}, \quad x \neq 4$

**4.** $\dfrac{x^2 - 3x - 10}{x^2 - 4} = \dfrac{(x - 5)(x + 2)}{(x + 2)(x - 2)}$

$= \dfrac{x - 5}{x - 2}, \quad x \neq -2$

**5.** $\dfrac{c}{c - 1} \cdot \dfrac{c^2 + 9c - 10}{c^3} = \dfrac{c(c + 10)(c - 1)}{(c - 1)(c^3)}$

$= \dfrac{c(c + 10)(c - 1)}{(c - 1)(c)(c^2)}$

$= \dfrac{c + 10}{c^2}, \quad c \neq 1$

**6.** $\dfrac{6}{(c - 1)^2} \div \dfrac{8}{c^3 - c^2} = \dfrac{6}{(c - 1)^2} \cdot \dfrac{c^3 - c^2}{8}$

$= \dfrac{6(c^2)(c - 1)}{(c - 1)^2(8)}$

$= \dfrac{2(3)(c^2)(c - 1)}{(c - 1)(c - 1)(2)(4)}$

$= \dfrac{3c^2}{4(c - 1)}, \quad c \neq 0$

**7.** $\dfrac{3}{x - 2} + \dfrac{x}{4 - x^2} = \dfrac{3}{x - 2} + \dfrac{x(-1)}{(4 - x^2)(-1)}$

$= \dfrac{3}{x - 2} + \dfrac{-x}{x^2 - 4}$

$= \dfrac{3(x + 2)}{(x - 2)(x + 2)} + \dfrac{-x}{(x + 2)(x - 2)}$

$= \dfrac{3x + 6 - x}{(x + 2)(x - 2)}$

$= \dfrac{2x + 6}{(x + 2)(x - 2)}$ or $\dfrac{2(x + 3)}{(x + 2)(x - 2)}$

**8.** $\dfrac{5}{x - 2} - \dfrac{2}{x^2} = \dfrac{5 \cdot x^2}{(x - 2) \cdot x^2} - \dfrac{2(x - 2)}{x^2(x - 2)}$

$= \dfrac{5x^2 - 2(x - 2)}{x^2(x - 2)}$

$= \dfrac{5x^2 - 2x + 4}{x^2(x - 2)}$

**9.** $\dfrac{\left(\dfrac{9}{x}\right)}{\left(\dfrac{6}{x}+2\right)} = \dfrac{\left(\dfrac{9}{x}\right)}{\left(\dfrac{6}{x}+\dfrac{2x}{x}\right)}$

$= \dfrac{\left(\dfrac{9}{x}\right)}{\left(\dfrac{6+2x}{x}\right)}$

$= \dfrac{9}{x} \div \dfrac{6+2x}{x}$

$= \dfrac{9}{x} \cdot \dfrac{x}{6+2x}$

$= \dfrac{9(\cancel{x})}{\cancel{x}(6+2x)}$

$= \dfrac{9}{6+2x}, \quad x \neq 0$

**10.** $\dfrac{\left(a-\dfrac{1}{a}\right)}{\left(\dfrac{1}{2}+\dfrac{1}{a}\right)} = \dfrac{\left(a-\dfrac{1}{a}\right)\cdot 2a}{\left(\dfrac{1}{2}+\dfrac{1}{a}\right)\cdot 2a}$

$= \dfrac{(a)(2a)-\left(\dfrac{1}{a}\right)(2a)}{\left(\dfrac{1}{2}\right)(2a)+\left(\dfrac{1}{a}\right)(2a)}$

$= \dfrac{2a^2-2}{a+2}, \quad a \neq 0$

This answer could also be written as $\dfrac{2\left(a^2-1\right)}{a+2}$ or

$\dfrac{2(a+1)(a-1)}{a+2}$.

Alternate Method:

$\dfrac{\left(a-\dfrac{1}{a}\right)}{\left(\dfrac{1}{2}+\dfrac{1}{a}\right)} = \dfrac{\dfrac{a\cdot a}{1\cdot a}-\dfrac{1}{a}}{\dfrac{1\cdot a}{2\cdot a}+\dfrac{1\cdot 2}{a\cdot 2}} = \dfrac{\dfrac{a^2-1}{a}}{\dfrac{a+2}{2a}}$

$= \dfrac{a^2-1}{a} \div \dfrac{a+2}{2a}$

$= \dfrac{(a+1)(a-1)}{a} \cdot \dfrac{2a}{a+2}$

$= \dfrac{(a+1)(a-1)(2)(\cancel{a})}{\cancel{a}(a+2)}$

$= \dfrac{2(a+1)(a-1)}{a+2}, \quad a \neq 0$

**11.** $\dfrac{5}{x}+\dfrac{3}{x} = 24$

$x\left(\dfrac{5}{x}+\dfrac{3}{x}\right) = 24(x)$

$5+3 = 24x$

$8 = 24x$

$\dfrac{1}{3} = x$

**12.** $\dfrac{x}{5}-\dfrac{x}{2} = 3$

$10\left(\dfrac{x}{5}-\dfrac{x}{2}\right) = 10(3)$

$2x-5x = 30$

$-3x = 30$

$x = \dfrac{30}{-3}$

$x = -10$

**13.** $\dfrac{1}{x}-\dfrac{2}{x-9} = 0$

$x(x-9)\left(\dfrac{1}{x}-\dfrac{2}{x-9}\right) = 0 \cdot x(x-9)$

$x-9-2x = 0$

$-x-9 = 0$

$-x = 9$

$x = -9$

**14.** $\dfrac{2x-1}{2x+1} = \dfrac{4}{5}$

$5(2x-1) = 4(2x+1)$

$10x-5 = 8x+4$

$2x-5 = 4$

$2x = 9$

$x = \dfrac{9}{2}$

**15.**

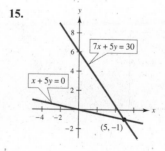

The solution is $(5, -1)$.

**16.**

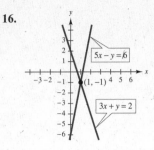

The solution is $(1, -1)$.

**17.**
$$\begin{cases} x - y = 0 \Rightarrow x = y \\ 5x - 3y = 10 \end{cases}$$

$$5(y) - 3y = 10$$
$$5y - 3y = 10$$
$$2y = 10$$
$$y = 5 \text{ and } x = 5$$

$(5, 5)$

**20.**
$$\begin{cases} 2x + y = 4 \Rightarrow 6x + 3y = 12 \quad \text{Multiply equation by 3.} \\ 4x - 3y = 3 \Rightarrow 4x - 3y = 3 \end{cases}$$
$$\overline{\phantom{aaaa}10x \phantom{aaaaa} = 15}$$
$$x \phantom{aaaa} = \frac{15}{10}$$
$$x \phantom{aaaa} = \frac{3}{2}$$

$$2\left(\tfrac{3}{2}\right) + y = 4 \quad \text{Replace } x \text{ by } \tfrac{3}{2} \text{ in first equation.}$$
$$3 + y = 4$$
$$y = 1$$

The solution is $\left(\tfrac{3}{2}, 1\right)$.

**21.**

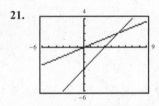

The solution is $(5, 2)$.

**18.**
$$\begin{cases} x + 8y = 6 \Rightarrow x = -8y + 6 \\ 2x + 4y = -3 \end{cases}$$

$$2(-8y + 6) + 4y = -3$$
$$-16y + 12 + 4y = -3$$
$$-12y + 12 = -3$$
$$-12y = -15$$

$$y = \frac{-15}{-12} \qquad x = -8\left(\frac{5}{4}\right) + 6$$
$$y = \frac{5}{4} \qquad x = -10 + 6$$
$$\qquad\qquad x = -4$$

$$\left(-4, \frac{5}{4}\right)$$

**19.**
$$\begin{cases} 4x + 3y = 15 \Rightarrow 4x + 3y = 15 \\ 2x - 5y = 1 \Rightarrow -4x + 10y = -2 \end{cases}$$
$$\overline{\phantom{aaaaaaaaaaaaa}13y = 13}$$
$$\phantom{aaaaaaaaaaaaa}y = 1$$

$$4x + 3(1) = 15$$
$$4x + 3 = 15$$
$$4x = 12$$
$$x = 3$$

$(3, 1)$

**22.** Here are some examples.

(a) $$\begin{cases} x + y = 10 \\ x + y = -4 \end{cases}$$

(b) $$\begin{cases} x - y = 3 \\ -2x + 2y = -6 \end{cases}$$

**23.** $\sqrt{-\tfrac{4}{9}}$

A negative number has no square root in the real numbers.

**24.** $-\sqrt{\dfrac{4}{9}} = -\dfrac{\sqrt{4}}{\sqrt{9}} = -\dfrac{2}{3}$

**25.** $\sqrt{144} = \sqrt{(12)^2} = 12$

**26.** $\sqrt[3]{-125} = -5$

**27.** $-\sqrt{54} = -\sqrt{9 \cdot 6} = -3\sqrt{6}$

**28.** $\sqrt{50x^3} = \sqrt{25 \cdot 2 \cdot x^2 \cdot x} = 5x\sqrt{2x}$

**29.** $\sqrt[3]{32u^4v^6} = \sqrt[3]{8 \cdot 4 \cdot u^3 \cdot u \cdot (v^2)^3} = 2uv^2\sqrt[3]{4u}$

**30.** $\sqrt{\dfrac{32y}{9y^3}} = \sqrt{\dfrac{32}{9y^2}} = \dfrac{\sqrt{16 \cdot 2}}{\sqrt{9y^2}} = \dfrac{4\sqrt{2}}{3|y|}$

**31.** $5\sqrt{x} - 3\sqrt{x} = (5-3)\sqrt{x} = 2\sqrt{x}$

**32.** $\sqrt{7}\left(\sqrt{7} + 2\right) = 7 + 2\sqrt{7}$

**33.** $\left(4 - \sqrt{8}\right)^2 = \left(4 - \sqrt{8}\right)\left(4 - \sqrt{8}\right)$
$= 16 - 8\sqrt{8} + 8$
$= 24 - 8\sqrt{4 \cdot 2}$
$= 24 - 8 \cdot 2\sqrt{2}$
$= 24 - 16\sqrt{2}$

**34.** $\dfrac{8y}{\sqrt{5}-1} = \dfrac{8y}{\sqrt{5}-1} \cdot \dfrac{\sqrt{5}+1}{\sqrt{5}+1}$

$= \dfrac{8y\left(\sqrt{5}+1\right)}{\left(\sqrt{5}\right)^2 - 1^2}$

$= \dfrac{8y\left(\sqrt{5}+1\right)}{5-1}$

$= \dfrac{8y\left(\sqrt{5}+1\right)}{4}$

$= 2y\left(\sqrt{5}+1\right)$ or $2\sqrt{5}\,y + 2y$

**35.** $2\sqrt{y} = 14$
$\sqrt{y} = 7$
$\left(\sqrt{y}\right)^2 = (7)^2$
$y = 49$

**36.** $\sqrt{a-4} = 5$
$\left(\sqrt{a-4}\right)^2 = 5^2$
$a - 4 = 25$
$a = 29$

**37.** $\sqrt{2x+7} = 3\sqrt{x}$
$\left(\sqrt{2x+7}\right)^2 = \left(3\sqrt{x}\right)^2$
$2x + 7 = 9x$
$7 = 7x$
$1 = x$

**38.** $x\left(\sqrt{x} - 2\right) = 0$
$x\sqrt{x} - 2x = 0$
$x\sqrt{x} = 2x$
$\left(x\sqrt{x}\right)^2 = (2x)^2$
$x^2 \cdot x = 4x^2$
$x^3 = 4x^2$
$x^3 - 4x^2 = 0$
$x^2(x-4) = 0$
$x^2 = 0 \Rightarrow x = 0$
$x - 4 = 0 \Rightarrow x = 4$

**39.** $d = \sqrt{(x_2 - x_1)^2 + (y_2 - y_1)^2}$
$d = \sqrt{[1-(-2)]^2 + [3-(-1)]^2}$
$= \sqrt{3^2 + 4^2}$
$= \sqrt{9+16}$
$= \sqrt{25}$
$= 5$

**40.** $d = \sqrt{(x_2 - x_1)^2 + (y_2 - y_1)^2}$
$d = \sqrt{(-3-7)^2 + (0-4)^2}$
$= \sqrt{(-10)^2 + (-4)^2}$
$= \sqrt{100+16}$
$= \sqrt{116}$
$= \sqrt{4 \cdot 29}$
$= 2\sqrt{29}$ or $d \approx 10.77$

**41.** *Verbal Model:*    $\boxed{\text{Time for first half of trip}} + \boxed{\text{Time for second half of trip}} = \dfrac{9}{2}$

*Labels:*    Distance for first half of trip $= 100$     (miles)

Distance for second half of trip $= 100$     (miles)

Speed for first half of trip $= x$     (miles per hour)

Speed for second half of trip $= x + 10$     (miles per hour)

Time for first half of trip $= \dfrac{100}{x}$     (hours)

Time for second half of trip $= \dfrac{100}{x + 10}$     (hours)

*Equation:*

$$\dfrac{100}{x} + \dfrac{100}{x + 10} = \dfrac{9}{2}$$

$$2x(x + 10)\left(\dfrac{100}{x} + \dfrac{100}{x + 10}\right) = 2x(x + 10)\left(\dfrac{9}{2}\right)$$

$$2(x + 10)(100) + 2x(100) = x(x + 10)(9), \quad x \neq 0, x \neq -10$$

$$200(x + 10) + 200x = 9x(x + 10)$$

$$200x + 2000 + 200x = 9x^2 + 90x$$

$$400x + 2000 = 9x^2 + 90x$$

$$0 = 9x^2 - 310x - 2000$$

$$(9x + 50)(x - 40) = 0$$

$$9x + 50 = 0 \Rightarrow 9x = -50 \Rightarrow x = -\dfrac{50}{9} \, (\text{Discard.})$$

$$x - 40 = 0 \Rightarrow x = 40 \text{ and } x + 10 = 50$$

The average speed on the second half of the trip was 50 miles per hour.

**42.** *Verbal model:*    $\boxed{\text{Rate of new employee}} + \boxed{\text{Rate of experienced employee}} = \boxed{\text{Rate when working together}}$

*Labels:*    Experienced employee: time $= x$ (hours), rate $= \dfrac{1}{x}$     (tasks per hour)

New employee: time $= 2x$ (hours), rate $= \dfrac{1}{3}$     (tasks per hour)

Together: time $= 3$ (hours), rate $= \dfrac{1}{3}$     (tasks per hour)

*Equation:*

$$\dfrac{1}{2x} + \dfrac{1}{x} = \dfrac{1}{3}$$

$$6x\left(\dfrac{1}{2x} + \dfrac{1}{x}\right) = 6x\left(\dfrac{1}{3}\right)$$

$$6x\left(\dfrac{1}{2x}\right) + 6x\left(\dfrac{1}{x}\right) = 6x\left(\dfrac{1}{3}\right)$$

$$3 + 6 = 2x, \quad x \neq 0$$

$$9 = 2x$$

$$\dfrac{9}{2} = x$$

The experienced employee can complete the task in $\dfrac{9}{2}$ or $4\dfrac{1}{2}$ hours.

The new employee would take 9 hours to complete the task.

**43.** *Verbal Model:*

| Number of gallons of regular | · | Price of regular | + | Number of gallons of premium | · | Price of premium | = | Total cost |

| Price of premium | = | Price of regular | + | Price difference of premium and regular |

*Labels:*

Number of gallons of regular $= 10$ (gallons)

Price of regular $= x$ (dollars per gallon)

Number of gallons of premium $= 12$ (gallons)

Price of premium $= y$ (dollars per gallon)

Total cost $= 69.94$ (dollars)

Price difference of premium and regular $= 0.20$ (dollars per gallon)

*System:*
$$\begin{cases} 10x + 12y = 69.94 \\ y = x + 0.20 \end{cases}$$

Solving by substitution:
$$\begin{cases} 10x + 12y = 69.94 \\ y = x + 0.20 \end{cases}$$

$10x + 12(x + 0.20) = 69.94$    Replace $y$ by $x + 0.20$ in first equation.

$10x + 12x + 2.40 = 69.94$

$22x + 2.40 = 69.94$

$22x = 67.54$

$x = \dfrac{67.54}{22}$

$x = 3.07 \Rightarrow y = 3.07 + 0.20$    Replace $x$ with 3.07 in second equation.

$y = 3.27$

$(3.07, 3.27)$

Regular gasoline costs $3.07 per gallon and premium gasoline costs $3.27 per gallon.

**44.**

$a^2 + b^2 = c^2$    Pythagorean Theorem

$x^2 + \left(4\sqrt{5}\right)^2 = 12^2$

$x^2 + 16(5) = 144$

$x^2 + 80 = 144$

$x^2 = 64$

$x = 8$

# CHAPTER 10
## Quadratic Equations and Functions

# C H A P T E R  1 0
# Quadratic Equations and Functions

## Section 10.1   Solution by the Square Root Property

**1.** $y^2 - 3y = 0$

$y(y - 3) = 0$

$y = 0$

$y - 3 = 0 \Rightarrow y = 3$

The solutions are 0 and 3.

**3.** $4x^2 + 8x = 0$

$4x(x + 2) = 0$

$4x = 0 \Rightarrow x = 0$

$x + 2 = 0 \Rightarrow x = -2$

The solutions are 0 and $-2$.

**5.** $a^2 - 25 = 0$

$(a + 5)(a - 5) = 0$

$a + 5 = 0 \Rightarrow a = -5$

$a - 5 = 0 \Rightarrow a = 5$

The solutions are $-5$ and 5.

**7.** $9m^2 = 64$

$9m^2 - 64 = 0$

$(3m + 8)(3m - 8) = 0$

$3m + 8 = 0 \Rightarrow 3m = -8 \Rightarrow m = -\frac{8}{3}$

$3m - 8 = 0 \Rightarrow 3m = 8 \Rightarrow m = \frac{8}{3}$

The solutions are $\frac{8}{3}$ and $-\frac{8}{3}$.

**9.** $u(u - 10) + 6(u - 10) = 0$

$(u - 10)(u + 6) = 0$

$u - 10 = 0 \Rightarrow u = 10$

$u + 6 = 0 \Rightarrow u = -6$

The solutions are 10 and $-6$.

**11.** $3z(z + 20) + 12(z + 20) = 0$

$(z + 20)(3z + 12) = 0$

$3(z + 20)(z + 4) = 0$

$3 \neq 0$

$z + 20 = 0 \Rightarrow z = -20$

$z + 4 = 0 \Rightarrow z = -4$

The solutions are $-20$ and $-4$.

**13.** $x^2 - 5x + 6 = 0$

$(x - 2)(x - 3) = 0$

$x - 2 = 0 \Rightarrow x = 2$

$x - 3 = 0 \Rightarrow x = 3$

The solutions are 2 and 3.

**15.** $x^2 + 4x + 4 = 0$

$(x + 2)(x + 2) = 0$

$x + 2 = 0 \Rightarrow x = -2$

The repeated solution is $-2$.

**17.** $16x^2 - 40x + 25 = 0$

$(4x - 5)(4x - 5) = 0$

$(4x - 5)^2 = 0$

$4x - 5 = 0 \Rightarrow 4x = 5 \Rightarrow x = \frac{5}{4}$

The repeated solution is $\frac{5}{4}$.

**19.** $5x^2 - 16x - 16 = 0$

$(5x + 4)(x - 4) = 0$

$5x + 4 = 0 \Rightarrow 5x = -4 \Rightarrow x = -\frac{4}{5}$

$x - 4 = 0 \Rightarrow x = 4$

The solutions are $-\frac{4}{5}$ and 4.

**21.** $6x^2 = -13x + 28$

$6x^2 + 13x - 28 = 0$

$(2x + 7)(3x - 4) = 0$

$2x + 7 = 0 \Rightarrow 2x = -7 \Rightarrow x = -\frac{7}{2}$

$3x - 4 = 0 \Rightarrow 3x = 4 \Rightarrow x = \frac{4}{3}$

The solutions are $-\frac{7}{2}$ and $\frac{4}{3}$.

**23.** $(x - 3)(x + 1) = 5$

$x^2 - 2x - 3 = 5$

$x^2 - 2x - 8 = 0$

$(x - 4)(x + 2) = 0$

$x - 4 = 0 \Rightarrow x = 4$

$x + 2 = 0 \Rightarrow x = -2$

The solutions are 4 and $-2$.

**25.** $x^2 = 64$

$\quad x = \pm\sqrt{64}$

$\quad x = \pm 8$

The solutions are 8 and $-8$.

**27.** $h^2 = 169$

$\quad h = \pm\sqrt{169}$

$\quad h = \pm 13$

The solutions are 13 and $-13$.

**29.** $x^2 = -9$

$\quad x = \pm\sqrt{-9}$

The equation has no real solution.

**31.** $6x^2 = 30$

$\quad x^2 = 5$

$\quad x = \pm\sqrt{5}$

The solutions are $\sqrt{5}$ and $-\sqrt{5}$.

**33.** $7x^2 = 42$

$\quad x^2 = 6$

$\quad x = \pm\sqrt{6}$

The solutions are $\sqrt{6}$ and $-\sqrt{6}$.

**35.** $9x^2 = 49$

$\quad x^2 = \frac{49}{9}$

$\quad x = \pm\sqrt{\frac{49}{9}}$

$\quad x = \pm\frac{7}{3}$

The solutions are $\frac{7}{3}$ and $-\frac{7}{3}$.

**37.** $16y^2 = 25$

$\quad y^2 = \frac{25}{16}$

$\quad y = \pm\sqrt{\frac{25}{16}}$

$\quad y = \pm\frac{5}{4}$

The solutions are $\frac{5}{4}$ and $-\frac{5}{4}$.

**39.** $u^2 - 100 = 0$

$\quad u^2 = 100$

$\quad u = \pm\sqrt{100}$

$\quad u = \pm 10$

The solutions are 10 and $-10$.

**41.** $x^2 + 1 = 0$

$\quad x^2 = -1$

$\quad x = \pm\sqrt{-1}$

The equation has no real solution.

**43.** $2s^2 - 5 = 27$

$\quad 2s^2 = 32$

$\quad s^2 = 16$

$\quad s = \pm\sqrt{16}$

$\quad s = \pm 4$

The solutions are 4 and $-4$.

**45.** $9u^2 - 196 = 0$

$\quad 9u^2 = 196$

$\quad u^2 = \frac{196}{9}$

$\quad u = \pm\sqrt{\frac{196}{9}}$

$\quad u = \pm\frac{14}{3}$

The solutions are $\frac{14}{3}$ and $-\frac{14}{3}$.

**47.** $\quad \frac{1}{2}x^2 - 1 = 3$

$\quad 2\left(\frac{1}{2}x^2 - 1\right) = 2(3)$

$\quad x^2 - 2 = 6$

$\quad x^2 = 8$

$\quad x = \pm\sqrt{8}$

$\quad x = \pm 2\sqrt{2}$

The solutions are $2\sqrt{2}$ and $-2\sqrt{2}$.

**49.** $\quad \frac{1}{3}t^2 - 14 = 2$

$\quad 3\left(\frac{1}{3}t^2 - 14\right) = 3(2)$

$\quad t^2 - 42 = 6$

$\quad t^2 = 48$

$\quad t = \pm\sqrt{48}$

$\quad t = \pm 4\sqrt{3}$

The solutions are $4\sqrt{3}$ and $-4\sqrt{3}$.

**51.** $\quad \frac{1}{4}x^2 + 6 = 2$

$\quad 4\left(\frac{1}{4}x^2 + 6\right) = 4(2)$

$\quad x^2 + 24 = 8$

$\quad x^2 = -16$

$\quad x = \pm\sqrt{-16}$

The equation has no real solution.

**53.** $(x - 3)^2 = 16$

$x - 3 = \pm\sqrt{16}$

$x - 3 = \pm 4$

$x = 3 \pm 4$

$x = 3 + 4 = 7$

$x = 3 - 4 = -1$

The solutions are 7 and −1.

**55.** $(z + 4)^2 = 225$

$z + 4 = \pm\sqrt{225}$

$z + 4 = \pm 15$

$z = -4 \pm 15$

$z = -4 + 15 = 11$

$z = -4 - 15 = -19$

The solutions are 11 and −19.

**57.** $(y - 7)^2 = 6$

$y - 7 = \pm\sqrt{6}$

$y = 7 \pm \sqrt{6}$

The solutions are $7 + \sqrt{6}$ and $7 - \sqrt{6}$.

**59.** $(x + 6)^2 = 3$

$x + 6 = \pm\sqrt{3}$

$x = -6 \pm \sqrt{3}$

The solutions are $-6 + \sqrt{3}$ and $-6 - \sqrt{3}$.

**61.** $(y + 2)^2 = 12$

$y + 2 = \pm\sqrt{12}$

$y + 2 = \pm 2\sqrt{3}$

$y = -2 \pm 2\sqrt{3}$

The solutions are $-2 + 2\sqrt{3}$ and $-2 - 2\sqrt{3}$.

**63.** $(3x + 2)^2 = 9$

$3x + 2 = \pm\sqrt{9}$

$3x + 2 = \pm 3$

$3x = -2 \pm 3$

$x = \dfrac{-2 \pm 3}{3}$

$x = \dfrac{-2 + 3}{3} = \dfrac{1}{3}$

$x = \dfrac{-2 - 3}{3} = \dfrac{-5}{3}$

The solutions are $\dfrac{1}{3}$ and $-\dfrac{5}{3}$.

**65.** $(5x + 2)^2 = 5$

$5x + 2 = \pm\sqrt{5}$

$5x = -2 \pm\sqrt{5}$

$x = \dfrac{-2 \pm\sqrt{5}}{5}$

The solutions are $\dfrac{-2 + \sqrt{5}}{5}$ and $\dfrac{-2 - \sqrt{5}}{5}$.

**67.** $(3x - 4)^2 = 27$

$3x - 4 = \pm\sqrt{27}$

$3x - 4 = \pm 3\sqrt{3}$

$3x = 4 \pm 3\sqrt{3}$

$x = \dfrac{4 \pm 3\sqrt{3}}{3}$

The solutions are $\dfrac{4 + 3\sqrt{3}}{3}$ and $\dfrac{4 - 3\sqrt{3}}{3}$.

**69.** $(2x + 1)^2 = -4$

$2x + 1 = \pm\sqrt{-4}$

The equation has no real solution.

**71.** $4(x + 3)^2 = 25$

$(x + 3)^2 = \dfrac{25}{4}$

$x + 3 = \pm\sqrt{\dfrac{25}{4}}$

$x + 3 = \pm\dfrac{5}{2}$

$x = -3 \pm \dfrac{5}{2}$

$x = -3 + \dfrac{5}{2} = -\dfrac{1}{2}$

$x = -3 - \dfrac{5}{2} = -\dfrac{11}{2}$

The solutions are $-\dfrac{1}{2}$ and $-\dfrac{11}{2}$.

**73.** $8(4x + 3)^2 = 14$

$$(4x + 3)^2 = \frac{14}{8}$$

$$(4x + 3)^2 = \frac{7}{4}$$

$$4x + 3 = \pm\sqrt{\frac{7}{4}}$$

$$4x + 3 = \pm\frac{\sqrt{7}}{2}$$

$$8x + 6 = \pm\sqrt{7}$$

$$8x = -6 \pm \sqrt{7}$$

$$x = \frac{-6 \pm \sqrt{7}}{8}$$

The solutions are $\dfrac{-6 + \sqrt{7}}{8}$ and $\dfrac{-6 - \sqrt{7}}{8}$.

**75.** $3(2x + 3)^2 - 12 = 0$

$$3(2x + 3)^2 = 12$$

$$(2x + 3)^2 = 4$$

$$2x + 3 = \pm\sqrt{4}$$

$$2x + 3 = \pm 2$$

$$2x = -3 \pm 2$$

$$x = \frac{-3 \pm 2}{2}$$

$$x = \frac{-3 + 2}{2} = -\frac{1}{2}$$

$$x = \frac{-3 - 2}{2} = -\frac{5}{2}$$

The solutions are $-\dfrac{1}{2}$ and $-\dfrac{5}{2}$.

**77.** $9(7x + 4)^2 + 1 = 0$

$$9(7x + 4)^2 = -1$$

$$(7x + 4)^2 = -\tfrac{1}{9}$$

$$7x + 4 = \pm\sqrt{-\tfrac{1}{9}}$$

The equation has no real solution.

**79.** $x^2 - 12x + 36 = 49$

$$(x - 6)^2 = 49$$

$$x - 6 = \pm\sqrt{49}$$

$$x - 6 = \pm 7$$

$$x = 6 \pm 7$$

$$x = 6 + 7 = 13$$

$$x = 6 - 7 = -1$$

The solutions are 13 and $-1$.

**81.** $9y^2 + 30y + 25 = 25$

$$(3y + 5)^2 = 25$$

$$3y + 5 = \pm\sqrt{25}$$

$$3y + 5 = \pm 5$$

$$3y = -5 \pm 5$$

$$y = \frac{-5 \pm 5}{3}$$

$$y = \frac{-5 + 5}{3} = 0$$

$$y = \frac{-5 - 5}{3} = -\frac{10}{3}$$

The solutions are 0 and $-\dfrac{10}{3}$.

**83.** $(x + 6)^2 = 16$

$$x + 6 = \pm\sqrt{16}$$

$$x + 6 = \pm 4$$

$$x = -6 \pm 4$$

$$x = -6 + 4 = -2$$

$$x = -6 - 4 = -10 \quad \text{(Extraneous)}$$

If $x = -10$, $x + 6 = -4$, but the length of a side of a square cannot be negative. The solution is $x = -2$.

**85.** $\frac{1}{2}(\text{Base})(\text{Height}) = \text{Area}$

$$\tfrac{1}{2}(8)(x^2 - 6) = 12$$

$$4(x^2 - 6) = 12$$

$$4x^2 - 24 = 12$$

$$4x^2 = 36$$

$$x^2 = 9$$

$$x = \pm\sqrt{9}$$

$$x = \pm 3$$

The solutions are 3 and $-3$. (Note: With either solution, the height of the triangle, $x^2 - 6$, is 3.)

**87.**
$$A = \pi r^2$$

$$10 = \pi r^2$$

$$\frac{10}{\pi} = r^2$$

$$\sqrt{\frac{10}{\pi}} = r$$

$$\sqrt{\frac{10}{3.14}} \approx r$$

$$2\sqrt{\frac{10}{3.14}} \approx d \quad \text{Note: The diameter is twice the radius.}$$

$$3.57 \approx d$$

The diameter of the oil spill is approximately 3.57 miles.

**89.**   $1166.40 = 1000(1 + r)^2$

$1.16640 = (1 + r)^2$

$\sqrt{1.16640} = 1 + r$

$1.08 = 1 + r$

$0.08 = r$

The interest rate is 8%.

**91.**   $R = x\left(5 - \frac{1}{10}x\right), \quad 0 < x < 25$

$60 = x\left(5 - \frac{1}{10}x\right)$

$60 = 5x - \frac{1}{10}x^2$

$10(60) = 10\left(5x - \frac{1}{10}x^2\right)$

$600 = 50x - x^2$

$x^2 - 50x + 600 = 0$

$(x - 30)(x - 20) = 0$

$x - 30 = 0 \Rightarrow x = 30$

$x - 20 = 0 \Rightarrow x = 20$

The domain is $0 < x < 25$. So, the *only* solution is 20.
To produce a revenue of $60, 20 units must be sold.

**93.**   $h = 64 - 16t^2$

$0 = 64 - 16t^2$   The height is 0 when the object

$16t^2 = 64$   reaches the ground.

$t^2 = \frac{64}{16} = 4$

$t = \pm\sqrt{4} = \pm 2$

The answer $t = -2$ is extraneous in this application. So,
it takes 2 seconds for the object to reach the ground.

Note: The equation $0 = 64 - 16t^2$ could also be solved
by factoring.

**95.**   $h = -2.7t^2 + s$

$0 = -2.7t^2 + 5$

$2.7t^2 = 5$

$t^2 = \frac{5}{2.7}$

$t = \pm\sqrt{\frac{5}{2.7}}$   The negative answer is extraneous.

$t \approx 1.36$

It would take approximately 1.36 seconds for each object
to hit the ground.

**97.** False. The equation has two solutions, 6 and −6.

$x^2 = 36$

$x = \pm\sqrt{36}$

$x = \pm 6$

**Check:** $(6)^2 \overset{?}{=} 36 \qquad (-6)^2 \overset{?}{=} 36$

$36 = 36 \qquad\qquad 36 = 36$

**99.** Equation (c), $(x - 2)^2 + 36 = 0$, has no real solutions

because $(x - 2)^2 + 36 \geq 36$ for all real numbers $x$.

**101.** The point $(-4, 10)$ is located 4 units to the left of the
vertical axis and 10 units above the horizontal axis; it is
in Quadrant II.

**103.** The point $(2, -18)$ is located 2 units to the right of the
vertical axis and 18 units below the horizontal axis; it is
in Quadrant IV.

**105.** $\sqrt{51} \approx 7.141$

**107.** $\sqrt{123} \approx 11.091$

**109.** $d = \sqrt{(x_2 - x_1)^2 + (y_2 - y_1)^2}$

$= \sqrt{(5 - 3)^2 + (8 - 2)^2}$

$= \sqrt{2^2 + 6^2}$

$= \sqrt{4 + 36}$

$= \sqrt{40} = 2\sqrt{10} \approx 6.32$

**111.** $d = \sqrt{(x_2 - x_1)^2 + (y_2 - y_1)^2}$

$= \sqrt{[3 - (-8)]^2 + (10 - 4)^2}$

$= \sqrt{11^2 + 6^2}$

$= \sqrt{121 + 36} = \sqrt{157} \approx 12.53$

# Section 10.2   Solution by Completing the Square

**1.** $x^2 + 8x + \boxed{16}$

Half of 8 is 4, and $4^2 = 16$.

$x^2 + 8x + 16 = (x + 4)^2$

**3.** $y^2 - 24y + \boxed{144}$

Half of $-24$ is $-12$, and $(-12)^2 = 144$.

$y^2 - 24y + 144 = (y - 12)^2$

**5.** $t^2 + 3t + \boxed{\frac{9}{4}}$

Half of 3 is $\frac{3}{2}$, and $\left(\frac{3}{2}\right)^2 = \frac{9}{4}$.

$t^2 + 3t + \frac{9}{4} = \left(t + \frac{3}{2}\right)^2$

**7.** $x^2 - x + \boxed{\frac{1}{4}}$

Half of $-1$ is $-\frac{1}{2}$, and $\left(-\frac{1}{2}\right)^2 = \frac{1}{4}$.

$x^2 - x + \frac{1}{4} = \left(x - \frac{1}{2}\right)^2$

**9.** $t^2 - \frac{3}{4}t + \boxed{\frac{9}{64}}$

Half of $-\frac{3}{4}$ is $-\frac{3}{8}$, and $\left(-\frac{3}{8}\right)^2 = \frac{9}{64}$.

$t^2 - \frac{3}{4}t + \frac{9}{64} = \left(t - \frac{3}{8}\right)^2$

**13.**
$$y^2 + 20y = 0$$
$$y^2 + 20y + 10^2 = 100 \quad \text{Half of 20 is 10, and } 10^2 = 100.$$
$$(y + 10)^2 = 100$$
$$y + 10 = \pm\sqrt{100}$$
$$y + 10 = \pm 10$$
$$y = -10 \pm 10$$
$$y = -10 + 10 \Rightarrow y = 0$$
$$y = -10 - 10 \Rightarrow y = -20$$

The solutions are 0 and $-20$.

**15.**
$$x^2 - 18x + 70 = 0$$
$$x^2 - 18x = -70$$
$$x^2 - 18x + (-9)^2 = -70 + 81 \quad \text{Half of } -18 \text{ is } -9, \text{ and } (-9)^2 = 81.$$
$$(x - 9)^2 = 11$$
$$x - 9 = \pm\sqrt{11}$$
$$x = 9 \pm \sqrt{11}$$

The solutions are $9 + \sqrt{11}$ and $9 - \sqrt{11}$.

**17.**
$$y^2 + 4y - 1 = 0$$
$$y^2 + 4y = 1$$
$$y^2 + 4y + (2)^2 = 1 + 4 \quad \text{Half of 4 is 2, and } 2^2 = 4.$$
$$(y + 2)^2 = 5$$
$$y + 2 = \pm\sqrt{5}$$
$$y = -2 \pm \sqrt{5}$$

The solutions are $-2 + \sqrt{5}$ and $-2 - \sqrt{5}$.

**11.**
$$x^2 - 8x = 0$$
$$x^2 - 8x + 4^2 = 16 \quad \text{Half of } -8 \text{ is } -4, \text{ and } (-4)^2 = 16.$$
$$(x - 4)^2 = 16$$
$$x - 4 = \pm\sqrt{16}$$
$$x - 4 = \pm 4$$
$$x = 4 \pm 4$$
$$x = 4 + 4 \Rightarrow x = 8$$
$$x = 4 - 4 \Rightarrow x = 0$$

The solutions are 8 and 0.

**19.** $x^2 - 2x + 3 = 0$

$\quad\quad x^2 - 2x = -3$

$\quad x^2 - 2x + 1 = -3 + 1$   Half of $-2$ is $-1$, and $(-1)^2 = 1$.

$\quad\quad (x - 1)^2 = -2$

$\quad\quad\quad x - 1 = \pm\sqrt{-2}$

The equation has no real solution.

**21.** $x^2 - 8x - 2 = 0$

$\quad\quad x^2 - 8x = 2$

$\quad x^2 - 8x + 16 = 2 + 16$   Half of $-8$ is $-4$, and $(-4)^2 = 16$.

$\quad\quad (x - 4)^2 = 18$

$\quad\quad\quad x - 4 = \pm\sqrt{18}$

$\quad\quad\quad x - 4 = \pm 3\sqrt{2}$

$\quad\quad\quad\quad x = 4 \pm 3\sqrt{2}$

The solutions are $4 + 3\sqrt{2}$ and $4 - 3\sqrt{2}$.

**23.** $y^2 + 14y + 17 = 0$

$\quad\quad y^2 + 14y = -17$

$\quad y^2 + 14y + 49 = -17 + 49$   Half of $14$ is $7$, and $7^2 = 49$.

$\quad\quad (y + 7)^2 = 32$

$\quad\quad\quad y + 7 = \pm\sqrt{32} = \pm 4\sqrt{2}$

$\quad\quad\quad\quad y = -7 \pm 4\sqrt{2}$

The solutions are $-7 + 4\sqrt{2}$ and $-7 - 4\sqrt{2}$.

**25.** $x^2 + 2x - 35 = 0$

$\quad\quad x^2 + 2x = 35$

$\quad x^2 + 2x + 1^2 = 35 + 1$   Half of $2$ is $1$, and $1^2 = 1$.

$\quad\quad (x + 1)^2 = 36$

$\quad\quad\quad x + 1 = \pm\sqrt{36}$

$\quad\quad\quad x + 1 = \pm 6$

$\quad\quad\quad\quad x = -1 \pm 6$

$\quad\quad\quad\quad x = -1 + 6 \Rightarrow x = 5$

$\quad\quad\quad\quad x = -1 - 6 \Rightarrow x = -7$

The solutions are $5$ and $-7$.

**27.** $x^2 - x - 3 = 0$

$$x^2 - x = 3$$

$$x^2 - x + \frac{1}{4} = 3 + \frac{1}{4} \quad \text{Half of } -1 \text{ is } -\frac{1}{2}, \text{ and } \left(-\frac{1}{2}\right)^2 = \frac{1}{4}.$$

$$\left(x - \frac{1}{2}\right)^2 = \frac{12}{4} + \frac{1}{4} = \frac{13}{4}$$

$$x - \frac{1}{2} = \pm\sqrt{\frac{13}{4}} = \pm\frac{\sqrt{13}}{2}$$

$$x = \frac{1}{2} \pm \frac{\sqrt{13}}{2} = \frac{1 \pm \sqrt{13}}{2}$$

The solutions are $\dfrac{1 + \sqrt{13}}{2}$ and $\dfrac{1 - \sqrt{13}}{2}$.

**29.** $t^2 - 13t + 35 = 0$

$$t^2 - 13t = -35$$

$$t^2 - 13t + \left(-\frac{13}{2}\right)^2 = -\frac{140}{4} + \frac{169}{4} \quad \text{Half of } -13 \text{ is } -\frac{13}{2}, \text{ and } \left(-\frac{13}{2}\right)^2 = \frac{169}{4}.$$

$$\left(t - \frac{13}{2}\right)^2 = \frac{29}{4}$$

$$t - \frac{13}{2} = \pm\sqrt{\frac{29}{4}}$$

$$t - \frac{13}{2} = \pm\frac{\sqrt{29}}{2}$$

$$t = \frac{13}{2} \pm \frac{\sqrt{29}}{2}$$

$$t = \frac{13 \pm \sqrt{29}}{2}$$

The solutions are $\dfrac{13 + \sqrt{29}}{2}$ and $\dfrac{13 - \sqrt{29}}{2}$.

**31.** $x^2 + 3x - 4 = 0$

$$x^2 + 3x = 4$$

$$x^2 + 3x + \frac{9}{4} = 4 + \frac{9}{4} \quad \text{Half of } 3 \text{ is } \frac{3}{2}, \text{ and } \left(\frac{3}{2}\right)^2 = \frac{9}{4}.$$

$$\left(x + \frac{3}{2}\right)^2 = \frac{16}{4} + \frac{9}{4} = \frac{25}{4}$$

$$x + \frac{3}{2} = \pm\sqrt{\frac{25}{4}}$$

$$x + \frac{3}{2} = \pm\frac{5}{2}$$

$$x = -\frac{3}{2} \pm \frac{5}{2}$$

$$x = -\frac{3}{2} + \frac{5}{2} = \frac{2}{2} = 1$$

$$x = -\frac{3}{2} - \frac{5}{2} = -\frac{8}{2} = -4$$

The solutions are 1 and $-4$.

**33.** $y^2 - 8y + 48 = 0$

$$y^2 - 8y = -48$$

$$y^2 - 8y + (-4)^2 = -48 + 16 \quad \text{Half of } -8 \text{ is } -4, \text{ and } (-4)^2 = 16.$$

$$(y - 4)^2 = -32$$

$$y - 4 = \pm\sqrt{-32}$$

$$y = 4 \pm \sqrt{-32}$$

The equation has no real solution.

**35.** $u^2 + 7u + 14 = 0$

$$u^2 + 7u = -14$$

$$u^2 + 7u + \left(\tfrac{7}{2}\right)^2 = -\tfrac{56}{4} + \tfrac{49}{4} \quad \text{Half of 7 is } \tfrac{7}{2}, \text{ and } \left(\tfrac{7}{2}\right)^2 = \tfrac{49}{4}.$$

$$\left(u + \tfrac{7}{2}\right)^2 = -\tfrac{7}{4}$$

$$u + \tfrac{7}{2} = \pm\sqrt{-\tfrac{7}{4}}$$

$$u = -\tfrac{7}{2} \pm \sqrt{-\tfrac{7}{4}}$$

The equation has no real solution.

**37.** $3x^2 - 6x = -9$

$$x^2 - 2x = -3$$

$$x^2 - 2x + (-1)^2 = -3 + 1 \quad \text{Half of } -2 \text{ is } -1, \text{ and } (-1)^2 = 1.$$

$$(x - 1)^2 = -2$$

$$x - 1 = \pm\sqrt{-2}$$

$$x = 1 \pm \sqrt{-2}$$

The equation has no real solution.

**39.** $2x^2 + 18x = -32$

$$x^2 + 9x = -16$$

$$x^2 + 9x + \left(\tfrac{9}{2}\right)^2 = -\tfrac{64}{4} + \tfrac{81}{4} \quad \text{Half of } -9 \text{ is } -\tfrac{9}{2}, \text{ and } \left(-\tfrac{9}{2}\right)^2 = \tfrac{81}{4}.$$

$$\left(x + \tfrac{9}{2}\right)^2 = \tfrac{17}{4}$$

$$x + \tfrac{9}{2} = \pm\sqrt{\tfrac{17}{4}}$$

$$x + \tfrac{9}{2} = \pm\frac{\sqrt{17}}{\sqrt{4}}$$

$$x + \tfrac{9}{2} = \pm\frac{\sqrt{17}}{2}$$

$$x = -\tfrac{9}{2} \pm \frac{\sqrt{17}}{2}$$

$$x = \frac{-9 \pm \sqrt{17}}{2}$$

The solutions are $\dfrac{-9 + \sqrt{17}}{2}$ and $\dfrac{-9 - \sqrt{17}}{2}$.

41.  $3x^2 + 4x + 5 = 0$

$$3x^2 + 4x = -5$$

$$x^2 + \frac{4}{3}x = -\frac{5}{3}$$

$$x^2 + \frac{4}{3}x + \left(\frac{2}{3}\right)^2 = -\frac{5}{3} + \frac{4}{9} \quad \text{Half of } \tfrac{4}{3} \text{ is } \tfrac{2}{3}, \text{ and } \left(\tfrac{2}{3}\right)^2 = \tfrac{4}{9}.$$

$$\left(x + \frac{2}{3}\right)^2 = -\frac{15}{9} + \frac{4}{9} = -\frac{11}{9}$$

$$x + \frac{2}{3} = \pm\sqrt{-\frac{11}{9}}$$

The equation has no real solution.

43.  $2y^2 + 3y - 1 = 0$

$$2y^2 + 3y = 1$$

$$y^2 + \frac{3}{2}y = \frac{1}{2}$$

$$y^2 + \frac{3}{2}y + \left(\frac{3}{4}\right)^2 = \frac{1}{2} + \frac{9}{16} \quad \text{Half of } \tfrac{3}{2} \text{ is } \tfrac{3}{4}, \text{ and } \left(\tfrac{3}{4}\right)^2 = \tfrac{9}{16}.$$

$$\left(y + \frac{3}{4}\right)^2 = \frac{8}{16} + \frac{9}{16} = \frac{17}{16}$$

$$y + \frac{3}{4} = \pm\sqrt{\frac{17}{16}}$$

$$y + \frac{3}{4} = \pm\frac{\sqrt{17}}{4}$$

$$y = -\frac{3}{4} \pm \frac{\sqrt{17}}{4}$$

$$y = \frac{-3 \pm \sqrt{17}}{4}$$

The solutions are $\dfrac{-3 + \sqrt{17}}{4}$ and $\dfrac{-3 - \sqrt{17}}{4}$.

**45.**     $6x^2 - 10x - 9 = 0$

$$6x^2 - 10x = 9$$

$$x^2 - \frac{10}{6}x = \frac{9}{6}$$

$$x^2 - \frac{5}{3}x = \frac{3}{2}$$

$$x^2 - \frac{5}{3}x + \left(-\frac{5}{6}\right)^2 = \frac{3}{2} + \frac{25}{36} \quad \text{Half of } -\frac{5}{3} \text{ is } -\frac{5}{6}, \text{ and } \left(-\frac{5}{6}\right)^2 = \frac{25}{36}.$$

$$\left(x - \frac{5}{6}\right)^2 = \frac{54}{36} + \frac{25}{36} = \frac{79}{36}$$

$$x - \frac{5}{6} = \pm\sqrt{\frac{79}{36}}$$

$$x - \frac{5}{6} = \pm\frac{\sqrt{79}}{6}$$

$$x = \frac{5}{6} \pm \frac{\sqrt{79}}{6}$$

$$x = \frac{5 \pm \sqrt{79}}{6}$$

The solutions are $\dfrac{5 + \sqrt{79}}{6}$ and $\dfrac{5 - \sqrt{79}}{6}$.

**47.**     $\frac{1}{3}x^2 + \frac{1}{3}x - 4 = 0$

$$3\left(\frac{1}{3}x^2 + \frac{1}{3}x - 4\right) = 3(0)$$

$$x^2 + x - 12 = 0$$

$$x^2 + x = 12$$

$$x^2 + x + \left(\frac{1}{2}\right)^2 = 12 + \frac{1}{4} \quad \text{Half of 1 is } \frac{1}{2}, \text{ and } \left(\frac{1}{2}\right)^2 = \frac{1}{4}.$$

$$\left(x + \frac{1}{2}\right)^2 = \frac{48}{4} + \frac{1}{4} = \frac{49}{4}$$

$$x + \frac{1}{2} = \pm\sqrt{\frac{49}{4}}$$

$$x + \frac{1}{2} = \pm\frac{7}{2}$$

$$x = -\frac{1}{2} \pm \frac{7}{2}$$

$$x = -\frac{1}{2} + \frac{7}{2} \Rightarrow x = 3$$

$$x = -\frac{1}{2} - \frac{7}{2} \Rightarrow x = -4$$

The solutions are 3 and $-4$.

**49.**     $\frac{1}{2}x^2 + x - 1 = 0$

$$\frac{1}{2}x^2 + x = 1$$

$$x^2 + 2x = 2$$

$$x^2 + 2x + 1^2 = 2 + 1 \quad \text{Half of 2 is 1, and } 1^2 = 1.$$

$$(x + 1)^2 = 3$$

$$x + 1 = \pm\sqrt{3}$$

$$x = -1 \pm \sqrt{3}$$

The solutions are $-1 + \sqrt{3}$ and $-1 - \sqrt{3}$.

**51.** (a)
$$x^2 - 11x = 0$$
$$x^2 - 11x + \left(-\frac{11}{2}\right)^2 = \frac{121}{4}$$
$$\left(x - \frac{11}{2}\right)^2 = \frac{121}{4}$$
$$x - \frac{11}{2} = \pm\sqrt{\frac{121}{4}}$$
$$x - \frac{11}{2} = \pm\frac{11}{2}$$
$$x = \frac{11}{2} \pm \frac{11}{2}$$
$$x = \frac{11}{2} + \frac{11}{2} = 11$$
$$x = \frac{11}{2} - \frac{11}{2} = 0$$

(b) $x^2 - 11x = 0$
$$x(x - 11) = 0$$
$$x = 0$$
$$x - 11 = 0 \Rightarrow x = 11$$
The solutions are 11 and 0.

**53.** (a)   $t^2 + 6t + 5 = 0$
$$t^2 + 6t = -5$$
$$t^2 + 6t + 3^2 = -5 + 9$$
$$(t + 3)^2 = 4$$
$$t + 3 = \pm\sqrt{4}$$
$$t + 3 = \pm 2$$
$$t = -3 \pm 2$$
$$t = -3 + 2 \Rightarrow t = -1$$
$$t = -3 - 2 \Rightarrow t = -5$$

(b)   $t^2 + 6t + 5 = 0$
$$(t + 5)(t + 1) = 0$$
$$t + 5 = 0 \Rightarrow t = -5$$
$$t + 1 = 0 \Rightarrow t = -1$$
The solutions are −1 and −5.

**55.** (a)   $x^2 + 5x + 6 = 0$
$$x^2 + 5x = -6$$
$$x^2 + 5x + \left(\frac{5}{2}\right)^2 = -6 + \frac{25}{4}$$
$$\left(x + \frac{5}{2}\right)^2 = \frac{1}{4}$$
$$x + \frac{5}{2} = \pm\sqrt{\frac{1}{4}} = \pm\frac{1}{2}$$
$$x = -\frac{5}{2} \pm \frac{1}{2}$$
$$x = -\frac{5}{2} + \frac{1}{2} \Rightarrow x = -2$$
$$x = -\frac{5}{2} - \frac{1}{2} \Rightarrow x = -3$$

(b)   $x^2 + 5x + 6 = 0$
$$(x + 2)(x + 3) = 0$$
$$x + 2 = 0 \Rightarrow x = -2$$
$$x + 3 = 0 \Rightarrow x = -3$$
The solutions are −2 and −3.

**57.** (a)   $2x^2 - 5x + 2 = 0$
$$2x^2 - 5x = -2$$
$$x^2 - \frac{5}{2}x = -1$$
$$x^2 - \frac{5}{2}x + \left(-\frac{5}{4}\right)^2 = -1 + \frac{25}{16}$$
$$\left(x - \frac{5}{4}\right)^2 = \frac{9}{16}$$
$$x - \frac{5}{4} = \pm\sqrt{\frac{9}{16}} = \pm\frac{3}{4}$$
$$x = \frac{5}{4} \pm \frac{3}{4}$$
$$x = \frac{5}{4} + \frac{3}{4} \Rightarrow x = 2$$
$$x = \frac{5}{4} - \frac{3}{4} \Rightarrow x = \frac{1}{2}$$

(b)   $2x^2 - 5x + 2 = 0$
$$(2x - 1)(x - 2) = 0$$
$$2x - 1 = 0 \Rightarrow 2x = 1 \Rightarrow x = \frac{1}{2}$$
$$x - 2 = 0 \Rightarrow x = 2$$
The solutions are 2 and $\frac{1}{2}$.

**59.** $2x^2 + 6x + 1 = 0$

$$2x^2 + 6x = -1$$

$$x^2 + 3x = -\frac{1}{2}$$

$$x^2 + 3x + \frac{9}{4} = -\frac{1}{2} + \frac{9}{4}$$

$$\left(x + \frac{3}{2}\right)^2 = \frac{7}{4}$$

$$x + \frac{3}{2} = \pm\sqrt{\frac{7}{4}}$$

$$x + \frac{3}{2} = \pm\frac{\sqrt{7}}{2}$$

$$x = \frac{-3 \pm \sqrt{7}}{2}$$

$$x = \frac{-3 - \sqrt{7}}{2} \approx -2.82$$

$$x = \frac{-3 + \sqrt{7}}{2} \approx -0.18$$

The solutions are approximately $-2.82$ and $-0.18$.

**61.** $3y^2 - y - 1 = 0$

$$3y^2 - y = 1$$

$$y^2 - \frac{1}{3}y = \frac{1}{3}$$

$$y^2 - \frac{1}{3}y + \left(\frac{1}{6}\right)^2 = \frac{1}{3} + \frac{1}{36}$$

$$\left(y - \frac{1}{6}\right)^2 = \frac{12}{36} + \frac{1}{36} = \frac{13}{36}$$

$$y - \frac{1}{6} = \pm\sqrt{\frac{13}{36}}$$

$$y - \frac{1}{6} = \pm\frac{\sqrt{13}}{6}$$

$$y = \frac{1}{6} \pm \frac{\sqrt{13}}{6}$$

$$y = \frac{1 \pm \sqrt{13}}{6}$$

$$y = \frac{1 + \sqrt{13}}{6} \Rightarrow y \approx 0.77$$

$$y = \frac{1 - \sqrt{13}}{6} \Rightarrow y \approx -0.43$$

The solutions are approximately $0.77$ and $-0.43$.

**63.** $2x^2 + 2x - 7 = 0$

$$2x^2 + 2x = 7$$

$$x^2 + x = \frac{7}{2}$$

$$x^2 + x + \frac{1}{4} = \frac{7}{2} + \frac{1}{4}$$

$$\left(x + \frac{1}{2}\right)^2 = \frac{15}{4}$$

$$x + \frac{1}{2} = \pm\sqrt{\frac{15}{4}}$$

$$x + \frac{1}{2} = \pm\frac{\sqrt{15}}{2}$$

$$x = -\frac{1}{2} \pm \frac{\sqrt{15}}{2}$$

$$x = \frac{-1 \pm \sqrt{15}}{2}$$

$$x = \frac{-1 - \sqrt{15}}{2} \approx -2.44$$

$$x = \frac{-1 + \sqrt{15}}{2} \approx 1.44$$

The solutions are approximately $-2.44$ and $1.44$.

**65.** $2x^2 + 24x + 10 = 0$

$x^2 + 12x + 5 = 0$

$x^2 + 12x = -5$

$x^2 + 12x + (6)^2 = -5 + 36$   Half of 12 is 6, and $(6)^2 = 36$.

$(x + 6)^2 = 31$

$x + 6 = \pm\sqrt{31}$

$x = -6 \pm \sqrt{31}$

$x = -6 + \sqrt{31} \approx -0.43$

$x = -6 - \sqrt{31} \approx -11.57$

The solutions are approximately $-0.43$ and $-11.57$.

**67.** $\dfrac{x}{2} + \dfrac{1}{x} = 2$

$2x\left(\dfrac{x}{2} + \dfrac{1}{x}\right) = 2x(2)$

$x^2 + 2 = 4x, \quad x \neq 0$

$x^2 - 4x + 2 = 0$

$x^2 - 4x = -2$

$x^2 - 4x + 4 = -2 + 4$

$(x - 2)^2 = 2$

$x - 2 = \pm\sqrt{2}$

$x = 2 \pm \sqrt{2}$

The solutions are $2 + \sqrt{2}$ and $2 - \sqrt{2}$.

**69.** $\dfrac{3}{x - 2} = 2x$

$(x - 2)\left(\dfrac{3}{x - 2}\right) = (x - 2)2x$

$3 = 2x^2 - 4x, \quad x \neq 2$

$\dfrac{3}{2} = x^2 - 2x$

$x^2 - 2x + (-1)^2 = \dfrac{3}{2} + 1$

$(x - 1)^2 = \dfrac{5}{2}$

$x - 1 = \pm\sqrt{\dfrac{5}{2}} = \pm\sqrt{\dfrac{10}{4}} = \pm\dfrac{\sqrt{10}}{2}$

$x = 1 \pm \dfrac{\sqrt{10}}{2} = \dfrac{2 \pm \sqrt{10}}{2}$

The solutions are $\dfrac{2 + \sqrt{10}}{2}$ and $\dfrac{2 - \sqrt{10}}{2}$.

**71.** $\sqrt{2x + 3} = x - 2$

$\left(\sqrt{2x + 3}\right)^2 = (x - 2)^2$

$2x + 3 = x^2 - 4x + 4$

$0 = x^2 - 6x + 1$

$-1 = x^2 - 6x$

$-1 + 9 = x^2 - 6x + 9$

$8 = (x - 3)^2$

$\pm\sqrt{8} = x - 3$

$\pm 2\sqrt{2} = x - 3$

$3 \pm 2\sqrt{2} = x$

The answer of $3 + 2\sqrt{2}$ checks, but the second answer is extraneous. (If $x = 3 - 2\sqrt{2}$, $x \approx 0.172$. Then $\sqrt{2x + 3} \approx 1.828$, but $x - 2 \approx -1.828$.) So, the only solution is $3 + 2\sqrt{2}$.

**73.** $2\sqrt{x - 1} = x - 4$

$\left(2\sqrt{x - 1}\right)^2 = (x - 4)^2$

$4(x - 1) = x^2 - 8x + 16$

$4x - 4 = x^2 - 8x + 16$

$0 = x^2 - 12x + 20$

$0 = (x - 10)(x - 2)$

$x - 10 = 0 \Rightarrow x = 10$

$x - 2 = 0 \Rightarrow x = 2$   Extraneous

The solution is 10.

**75.** $\sqrt{x^2 + 3} - 2\sqrt{x} = 0$

$$\sqrt{x^2 + 3} = 2\sqrt{x}$$

$$\left(\sqrt{x^2 + 3}\right)^2 = \left(2\sqrt{x}\right)^2$$

$$x^2 + 3 = 4x$$

$$x^2 - 4x + 3 = 0$$

$$(x - 3)(x - 1) = 0$$

$$x - 3 = 0 \Rightarrow x = 3$$

$$x - 1 = 0 \Rightarrow x = 1$$

The solutions are 3 and 1.

**77.**

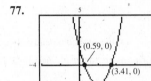

The $x$-intercepts appear to be approximately $(0.59, 0)$ and $(3.41, 0)$.

$$y = x^2 - 4x + 2$$

$$0 = x^2 - 4x + 2$$

$$x^2 - 4x = -2$$

$$x^2 - 4x + (-2)^2 = -2 + 4$$

$$(x - 2)^2 = 2$$

$$x - 2 = \pm\sqrt{2}$$

$$x = 2 \pm \sqrt{2}$$

$$x = 2 + \sqrt{2} \Rightarrow x \approx 3.41$$

$$x = 2 - \sqrt{2} \Rightarrow x \approx 0.59$$

These solutions verify the $x$-intercepts of approximately $(3.41, 0)$ and $(0.59, 0)$.

**79.**

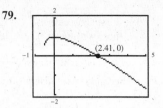

The $x$-intercept appears to be approximately $(2.41, 0)$.

$$y = \sqrt{2x + 1} - x$$

$$0 = \sqrt{2x + 1} - x$$

$$x = \sqrt{2x + 1}$$

$$x^2 = \left(\sqrt{2x + 1}\right)^2 = 2x + 1$$

$$x^2 - 2x = 1$$

$$x^2 - 2x + (-1)^2 = 1 + 1$$

$$(x - 1)^2 = 2$$

$$x - 1 = \pm\sqrt{2}$$

$$x = 1 \pm \sqrt{2}$$

$$x = 1 + \sqrt{2} \Rightarrow x \approx 2.41$$

$$x = 1 - \sqrt{2} \Rightarrow x \approx -0.41 \quad \text{Extraneous}$$

This verifies that the $x$-intercept is approximately $(2.41, 0)$.

**81.** *Verbal model:*    $\left(\boxed{\begin{array}{c}\text{First consecutive} \\ \text{positive integer}\end{array}}\right)^2 + \left(\boxed{\begin{array}{c}\text{Second consecutive} \\ \text{positive integer}\end{array}}\right)^2 = \boxed{85}$

*Labels:*    First consecutive positive integer $= n$

Second consecutive positive integer $= n + 1$

*Equation:*

$$n^2 + (n + 1)^2 = 85$$

$$n^2 + n^2 + 2n + 1 = 85$$

$$2n^2 + 2n - 84 = 0$$

$$n^2 + n - 42 = 0 \quad \text{Divide both sides by 2.}$$

$$(n + 7)(n - 6) = 0$$

$$n + 7 = 0 \Rightarrow n = -7 \text{ and } n + 1 = -6$$

$$n - 6 = 0 \Rightarrow n = 6 \quad \text{and } n + 1 = 7$$

The problem specifies that the integers are *positive*, so we discard the answers of $-7$ and $-6$. So, the two consecutive positive integers are 6 and 7. Note: The equation *could* be solved by completing the square, but factoring is an easier method for this equation.

**83.** $\frac{1}{2}(\text{Base})(\text{Height}) = \text{Area}$

$$\frac{1}{2}(x)(x + 3) = 12$$

$$x(x + 3) = 24$$

$$x^2 + 3x = 24$$

$$x^2 + 3x + \left(\frac{3}{2}\right)^2 = \frac{96}{4} + \frac{9}{4}$$

$$\left(x + \frac{3}{2}\right)^2 = \frac{105}{4}$$

$$x + \frac{3}{2} = \pm\sqrt{\frac{105}{4}}$$

$$x + \frac{3}{2} = \pm\sqrt{\frac{105}{2}}$$

$$x = -\frac{3}{2} \pm \frac{\sqrt{105}}{2}$$

$$x = \frac{-3 \pm \sqrt{105}}{2}$$

$$x = \frac{-3 + \sqrt{105}}{2} \approx 3.6$$

$$x = \frac{-3 - \sqrt{105}}{2} \approx -6.6 \quad \text{Discard negative solution.}$$

The base $x = \dfrac{-3 + \sqrt{105}}{2}$ centimeters or approximately 3.6 centimeters.

**85.** $(\text{Length})(\text{Width}) = \text{Area}$

$$x(x - 4) = 16$$

$$x^2 - 4x = 16$$

$$x^2 - 4x + (-2)^2 = 16 + 4$$

$$(x - 2)^2 = 20$$

$$x - 2 = \pm\sqrt{20} = \pm\sqrt{4}\sqrt{5} = \pm 2\sqrt{5}$$

$$x = 2 \pm 2\sqrt{5}$$

$$x = 2 + 2\sqrt{5} \Rightarrow x \approx 6.47$$

$$x = 2 - 2\sqrt{5} \Rightarrow x \approx -2.47 \quad \text{Discard negative solution.}$$

The length of the rectangle is $2 + 2\sqrt{5}$ mm, or approximately 6.47 mm.

**87.**

$$R = x\left(35 - \tfrac{1}{2}x\right), \quad 0 < x < 35$$

$$580.50 = x\left(35 - \tfrac{1}{2}x\right)$$

$$580.50 = 35x - \tfrac{1}{2}x^2$$

$$1161 = 70x - x^2$$

$$x^2 - 70x + 1161 = 0$$

$$x^2 - 70x = -1161$$

$$x^2 - 70x + (35)^2 = -1161 + 1225$$

$$(x - 35)^2 = 64$$

$$x - 35 = \pm\sqrt{64}$$

$$x - 35 = \pm 8$$

$$x = 35 \pm 8$$

$$x = 35 + 8 = 43$$

$$x = 35 - 8 = 27$$

The answer of 43 is extraneous because it is not in the domain, $0 < x < 35$. So, 27 vacuum cleaners must be sold.

**89.**

$$h = -16t^2 + 80t + 4$$

$$90 = -16t^2 + 80t + 4$$

$$16t^2 - 80t + 86 = 0$$

$$t^2 - 5t + \frac{86}{16} = 0$$

$$t^2 - 5t = -\frac{86}{16}$$

$$t^2 - 5t + \left(-\frac{5}{2}\right)^2 = -\frac{86}{16} + \frac{25}{4}$$

$$\left(t - \frac{5}{2}\right)^2 = -\frac{86}{16} + \frac{100}{16}$$

$$t - \frac{5}{2} = \pm\sqrt{\frac{14}{16}}$$

$$t - \frac{5}{2} = \pm\frac{\sqrt{14}}{4}$$

$$t = \frac{5}{2} \pm \frac{\sqrt{14}}{4}$$

$$t = \frac{10}{4} \pm \frac{\sqrt{14}}{4}$$

$$t = \frac{10 \pm \sqrt{14}}{4}$$

$$t = \frac{10 + \sqrt{14}}{4} \approx 3.44$$

$$t = \frac{10 - \sqrt{14}}{4} \approx 1.56$$

The object was at a height of 90 feet approximately 1.56 seconds after it was thrown and approximately 3.44 seconds after it was thrown.

**91. (a)**

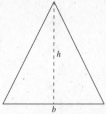

(b) Base + Height = 100

$$b + h = 100$$

$$b = 100 - h$$

(c)
$$\text{Area} = \tfrac{1}{2}(\text{Base})(\text{Height})$$

$$\text{Area} = \tfrac{1}{2}bh$$

$$\text{Area} = \tfrac{1}{2}(100 - h)(h)$$

$$1000 = \tfrac{1}{2}(100 - h)(h)$$

$$2(1000) = 2 \cdot \tfrac{1}{2}(100 - h)(h)$$

$$2000 = h(100 - h)$$

$$2000 = 100h - h^2$$

$$h^2 - 100h = -2000$$

$$h^2 - 100h + (50)^2 = -2000 + 2500$$

$$(x - 50)^2 = 500$$

$$x - 50 = \pm\sqrt{500}$$

$$x - 50 = \pm 10\sqrt{5}$$

$$x = 50 \pm 10\sqrt{5}$$

$$x = 50 \pm 10\sqrt{5} \Rightarrow x \approx 72.36$$

$$x = 50 - 10\sqrt{5} \Rightarrow x \approx 27.64$$

The height is greater than the base, so the height is $50 + 10\sqrt{5}$ centimeters or approximately 72.35 cm and the base is $50 - 10\sqrt{5}$ or approximately 27.64 cm.

**93.** $x^2 + \underline{\quad} + 144$

$$\sqrt{144} = 12 \ and \ (x + 12)^2 = x^2 + 24x + 144$$

The missing term is $24x$.

**95.** The solution is incorrect because

$$9x^2 - 4x + 4 \neq (3x - 2)^2.$$

$$(3x - 2)^2 = 9x^2 - 12x + 4$$

If the leading coefficient of a quadratic expression is not 1, you must divide each side of the equation by this coefficient before completing the square.

$$9x^2 - 4x - 2 = 0$$

$$x^2 - \frac{4}{9}x - \frac{2}{9} = 0$$

$$x^2 - \frac{4}{9}x = \frac{2}{9}$$

$$x^2 - \frac{4}{9}x + \left(\frac{2}{9}\right)^2 = \frac{2}{9} + \left(\frac{2}{9}\right)^2$$

$$\left(x - \frac{2}{9}\right)^2 = \frac{22}{81}$$

$$x - \frac{2}{9} = \pm\sqrt{\frac{22}{81}}$$

$$x - \frac{2}{9} = \pm\frac{\sqrt{22}}{9}$$

$$x = \frac{2 \pm \sqrt{22}}{9}$$

**97.** False. Any quadratic equation with real solutions can be solved by completing the square.

**99. (a)** $(0, 1)$

$$0 + 3(1) \overset{?}{=} 3$$

$$0 + 3 \overset{?}{=} 3$$

$$3 = 3$$

$$4(0) - (1) \overset{?}{=} -1$$

$$0 - 1 \overset{?}{=} -1$$

$$-1 = -1$$

$(0, 1)$ *is a solution.*

**(b)** $(3, 0)$

$$3 + 3(0) \overset{?}{=} 3$$

$$3 + 0 \overset{?}{=} 3$$

$$3 = 3$$

$$4(3) - (0) \overset{?}{=} -1$$

$$12 - 0 \overset{?}{=} -1$$

$$12 \neq -1$$

$(3, 0)$ *is not a solution.*

**101.** (a)  $(7, -13)$

$$-(7) - (-13) \overset{?}{=} 6$$

$$-7 + 13 \overset{?}{=} 6$$

$$6 = 6$$

$$-5(7) - 2(-13) \overset{?}{=} 3$$

$$-35 + 26 \overset{?}{=} 3$$

$$-9 \neq 3$$

$(7, -13)$ *is not* a solution.

(b)  $(3, -9)$

$$-(3) - (-9) \overset{?}{=} 6$$

$$-3 + 9 \overset{?}{=} 6$$

$$6 = 6$$

$$-5(3) - 2(-9) \overset{?}{=} 3$$

$$-15 + 18 \overset{?}{=} 3$$

$$3 = 3$$

$(3, -9)$ *is* a solution.

**103.**  $5x^2 = 65$

$$x^2 = \frac{65}{5}$$

$$x^2 = 13$$

$$x = \pm\sqrt{13}$$

The solutions are $\sqrt{13}$ and $-\sqrt{13}$.

**105.**  $u^2 + 8 = 0$

$$u^2 = -8$$

$$u = \pm\sqrt{-8}$$

The equation has no real solution.

**107.**  $9t^2 - 72 = 28$

$$9t^2 = 100$$

$$t^2 = \frac{100}{9}$$

$$t = \pm\sqrt{\frac{100}{9}}$$

$$t = \pm\frac{10}{3}$$

The solutions are $\dfrac{10}{3}$ and $-\dfrac{10}{3}$.

**109.**  $(x - 3)^2 = 24$

$$x - 3 = \pm\sqrt{24}$$

$$x - 3 = \pm2\sqrt{6}$$

$$x = 3 \pm 2\sqrt{6}$$

The solutions are $3 + 2\sqrt{6}$ and $3 - 2\sqrt{6}$.

# Section 10.3   Solution by the Quadratic Formula

**1.**  $3x^2 - x = 7$

$3x^2 - x - 7 = 0$

**3.**  $x^2 = 3 - 2x$

$x^2 + 2x - 3 = 0$

**5.**  $x(4 - x) = 10$

$$4x - x^2 = 10$$

$$-x^2 + 4x - 10 = 0$$

or

$$x^2 - 4x + 10 = 0$$

**7.**  $2x^2 - 3x - 1 = 0$

$a = 2, b = -3, c = -1$

$b^2 - 4ac = (-3)^2 - 4(2)(-1) = 9 + 8 = 17$

The *positive* discriminant indicates that the equation has *two* real solutions.

**9.**  $x^2 + 4x + 5 = 0$

$a = 1, b = 4, c = 5$

$b^2 - 4ac = 4^2 - 4(1)(5) = 16 - 20 = -4$

The *negative* discriminant indicates that the equation has *no* real solutions.

**11.** $x^2 + 6x + 1 = 0$

$a = 1, b = 6, c = 1$

$b^2 - 4ac = 6^2 - 4(1)(1) = 36 - 4 = 32$

The *positive* discriminant indicates that the equation has *two* real solutions.

**13.** $9x^2 - 12x + 4 = 0$

$a = 9, b = -12, c = 4$

$b^2 - 4ac = (-12)^2 - 4(9)(4) = 144 - 144 = 0$

The discriminant of *zero* indicates that the equation has *one* (repeated) real solution.

**15.** $x^2 - 3x - 18 = 0$

$a = 1, b = -3, x = -18$

$x = \dfrac{-(-3) \pm \sqrt{(-3)^2 - 4(1)(-18)}}{2(1)}$

$x = \dfrac{3 \pm \sqrt{9 + 72}}{2}$

$x = \dfrac{3 \pm \sqrt{81}}{2}$

$x = \dfrac{3 \pm 9}{2}$

$x = \dfrac{3 + 9}{2} = \dfrac{12}{2} = 6$

$x = \dfrac{3 - 9}{2} = \dfrac{-6}{2} = -3$

The solutions are 6 and $-3$.

**17.** $x^2 + 8x + 15 = 0$

$a = 1, b = 8, c = 15$

$x = \dfrac{-8 \pm \sqrt{8^2 - 4(1)(15)}}{2(1)}$

$x = \dfrac{-8 \pm \sqrt{64 - 60}}{2}$

$x = \dfrac{-8 \pm \sqrt{4}}{2}$

$x = \dfrac{-8 \pm 2}{2}$

$x = \dfrac{-8 + 2}{2} = \dfrac{-6}{2} = -3$

$x = \dfrac{-8 - 2}{2} = \dfrac{-10}{2} = -5$

The solutions are $-3$ and $-5$.

**19.** $t^2 - 5t + 6 = 0$

$a = 1, b = -5, c = 6$

$t = \dfrac{-(-5) \pm \sqrt{(-5)^2 - 4(1)(6)}}{2(1)}$

$t = \dfrac{5 \pm \sqrt{25 - 24}}{2}$

$t = \dfrac{5 \pm \sqrt{1}}{2}$

$t = \dfrac{5 \pm 1}{2}$

$t = \dfrac{5 + 1}{2} = \dfrac{6}{2} = 3$

$t = \dfrac{5 - 1}{2} = \dfrac{4}{2} = 2$

The solutions are 3 and 2.

**21.** $x^2 - 6x + 7 = 0$

$a = 1, b = -6, c = 7$

$x = \dfrac{-(-6) \pm \sqrt{(-6)^2 - 4(1)(7)}}{2(1)}$

$x = \dfrac{6 \pm \sqrt{36 - 28}}{2}$

$x = \dfrac{6 \pm \sqrt{8}}{2}$

$x = \dfrac{6 \pm 2\sqrt{2}}{2}$

$x = \dfrac{2(3 \pm \sqrt{2})}{2}$

$x = 3 \pm \sqrt{2}$

The solutions are $3 + \sqrt{2}$ and $3 - \sqrt{2}$.

**23.** $t^2 + 3t = 10$

$t^2 + 3t - 10 = 0$

$a = 1, b = 3, c = -10$

$t = \dfrac{-3 \pm \sqrt{3^2 - 4(1)(-10)}}{2(1)}$

$t = \dfrac{-3 \pm \sqrt{9 + 40}}{2}$

$t = \dfrac{-3 \pm \sqrt{49}}{2}$

$t = \dfrac{-3 \pm 7}{2}$

$t = \dfrac{-3 + 7}{2} = \dfrac{4}{2} = 2$

$t = \dfrac{-3 - 7}{2} = \dfrac{-10}{2} = -5$

The solutions are 2 and $-5$.

**25.**
$$-x^2 - 2x = -8$$
$$-x^2 - 2x + 8 = 0$$
$$a = -1, b = -2, c = 8$$

$$x = \frac{-(-2) \pm \sqrt{(-2)^2 - 4(-1)(8)}}{2(-1)}$$

$$x = \frac{2 \pm \sqrt{4 + 32}}{-2}$$

$$x = \frac{2 \pm \sqrt{36}}{-2}$$

$$x = \frac{2 \pm 6}{-2}$$

$$x = \frac{2 + 6}{-2} = \frac{8}{-2} = -4$$

$$x = \frac{2 - 6}{-2} = \frac{-4}{-2} = 2$$

The solutions are $-4$ and $2$.

**27.**
$$x^2 = 3x + 4$$
$$x^2 - 3x - 4 = 0$$
$$a = 1, b = -3, c = -4$$

$$x = \frac{-(-3) \pm \sqrt{(-3)^2 - 4(1)(-4)}}{2(1)}$$

$$x = \frac{3 \pm \sqrt{9 + 16}}{2}$$

$$x = \frac{3 \pm \sqrt{25}}{2}$$

$$x = \frac{3 \pm 5}{2}$$

$$x = \frac{3 + 5}{2} = \frac{8}{2} = 4$$

$$x = \frac{3 - 5}{2} = \frac{-2}{2} = -1$$

The solutions are $4$ and $-1$.

**29.**
$$4x^2 - 20x + 25 = 0$$
$$a = 4, b = -20, c = 25$$

$$x = \frac{-(-20) \pm \sqrt{20^2 - 4(4)(25)}}{2(4)}$$

$$x = \frac{20 \pm \sqrt{400 - 400}}{8}$$

$$x = \frac{20 \pm \sqrt{0}}{8}$$

$$x = \frac{20 \pm 0}{8}$$

$$x = \frac{20}{8} = \frac{5}{2}$$

The (repeated) solution is $\frac{5}{2}$.

**31.**
$$8x^2 - 10x + 3 = 0$$
$$a = 8, b = -10, c = 3$$

$$x = \frac{-(-10) \pm \sqrt{(-10)^2 - 4(8)(3)}}{2(8)}$$

$$x = \frac{10 \pm \sqrt{100 - 96}}{16}$$

$$x = \frac{10 \pm \sqrt{4}}{16}$$

$$x = \frac{10 \pm 2}{16}$$

$$x = \frac{10 + 2}{16} = \frac{12}{16} = \frac{3}{4}$$

$$x = \frac{10 - 2}{16} = \frac{8}{16} = \frac{1}{2}$$

The solutions are $\frac{3}{4}$ and $\frac{1}{2}$.

**33.** $10u^2 + 4u - 4 = 0$

$a = 10, b = 4, c = -4$

$$u = \frac{-4 \pm \sqrt{4^2 - 4(10)(-4)}}{2(10)}$$

$$u = \frac{-4 \pm \sqrt{16 + 160}}{20}$$

$$u = \frac{-4 \pm \sqrt{176}}{20}$$

$$u = \frac{-4 \pm 4\sqrt{11}}{20}$$

$$u = \frac{\cancel{4}\left(-1 \pm \sqrt{11}\right)}{\cancel{4}(5)}$$

$$u = \frac{-1 \pm \sqrt{11}}{5}$$

The solutions are $\dfrac{-1 + \sqrt{11}}{5}$ and $\dfrac{-1 - \sqrt{11}}{5}$.

**35.** $12x^2 - 12x - 3 = 0$

$a = 12, b = -12, c = -3$

$$x = \frac{-(-12) \pm \sqrt{(-12)^2 - 4(12)(-3)}}{2(12)}$$

$$x = \frac{12 \pm \sqrt{144 + 144}}{24}$$

$$x = \frac{12 \pm \sqrt{288}}{24}$$

$$x = \frac{12 \pm 12\sqrt{2}}{24}$$

$$x = \frac{\cancel{12}\left(1 \pm \sqrt{2}\right)}{\cancel{12}(2)}$$

$$x = \frac{1 \pm \sqrt{2}}{2}$$

The solutions are $\dfrac{1 + \sqrt{2}}{2}$ and $\dfrac{1 - \sqrt{2}}{2}$.

**37.** $3z^2 + 4z + 4 = 0$

$a = 3, b = 4, c = 4$

$$z = \frac{-4 \pm \sqrt{4^2 - 4(3)(4)}}{2(3)}$$

$$z = \frac{-4 \pm \sqrt{16 - 48}}{6}$$

$$z = \frac{-4 \pm \sqrt{-32}}{6}$$

Because $\sqrt{-32}$ is not a real number, the original equation has no real solution.

**39.** $\dfrac{1}{2}x^2 + 2x = 4$

$x^2 + 4x = 8$

$x^2 + 4x - 8 = 0$

$a = 1, b = 4, c = -8$

$$x = \frac{-4 \pm \sqrt{4^2 - 4(1)(-8)}}{2(1)}$$

$$x = \frac{-4 \pm \sqrt{16 + 32}}{2}$$

$$x = \frac{-4 \pm \sqrt{48}}{2}$$

$$x = \frac{-4 \pm 4\sqrt{3}}{2}$$

$$x = \frac{\cancel{2}\left(-2 \pm 2\sqrt{3}\right)}{\cancel{2}(1)}$$

$$x = -2 \pm 2\sqrt{3}$$

The solutions are $-2 + 2\sqrt{3}$ and $-2 - 2\sqrt{3}$.

**41.** $0.5x^2 - 0.8x + 0.3 = 0$

$a = 0.5, b = -0.8. c = 0.3$

$$x = \frac{-(-0.8) \pm \sqrt{(-0.8)^2 - 4(0.5)(0.3)}}{2(0.5)}$$

$$x = \frac{0.8 \pm \sqrt{0.64 - 0.6}}{1}$$

$$x = \frac{0.8 \pm \sqrt{0.04}}{1}$$

$$x = 0.8 \pm 0.2$$

$$x = 0.8 + 0.2 = 1$$

$$x = 0.8 - 0.2 = 0.6$$

The solutions are $1$ and $0.6$ $\left(\text{or } 1 \text{ and } \dfrac{3}{5}\right)$.

**43.** $0.2y^2 + y + 6 = 0$

$10\left(0.2y^2\right) + y + 6 = 10(0)$

$2y^2 + 10y + 60 = 0$

$y^2 + 5y + 30 = 0$

$a = 1, b = 5, c = 30$

$$y = \frac{-5 \pm \sqrt{5^2 - 4(1)(30)}}{2(3)}$$

$$y = \frac{-5 \pm \sqrt{25 - 120}}{6}$$

$$y = \frac{-5 \pm \sqrt{-95}}{6} \qquad \text{Not a real number.}$$

The equation has no real solution.

**45.** $0.36s^2 - 0.12s + 0.01 = 0$

$a = 0.36, b = -0.12, c = 0.01$

$$s = \frac{-(-0.12) \pm \sqrt{(-0.12)^2 - 4(0.36)(0.01)}}{2(0.36)}$$

$$s = \frac{0.12 \pm \sqrt{0.0144 - 0.0144}}{0.72}$$

$$s = \frac{0.12 \pm \sqrt{0}}{0.72}$$

$$s = \frac{0.12}{0.72}$$

$$s = \frac{1}{6}$$

The (repeated) solution is $\frac{1}{6}$ (or $0.1\overline{6}$).

**47.** $x^2 = 20$

$x = \pm\sqrt{20}$

$x = \pm2\sqrt{5}$

The solutions are $2\sqrt{5}$ and $-2\sqrt{5}$.

**49.** $y^2 - 8y = 0$

$y(y + 8) = 0$

$y = 0$

$y + 8 = 0 \Rightarrow y = -8$

The solutions are 0 and $-8$.

**51.** $2y(y - 12) + 3(y - 12) = 0$

$(y - 12)(2y + 3) = 0$

$y - 12 = 0 \Rightarrow y = 12$

$2y + 3 = 0 \Rightarrow 2y = -3 \Rightarrow y = -\frac{3}{2}$

The solutions are 12 and $-\frac{3}{2}$.

**53.** $(x - 3)^2 - 75 = 0$

$(x - 3)^2 = 75$

$x - 3 = \pm\sqrt{75}$

$x - 3 = \pm5\sqrt{3}$

$x = 3 \pm 5\sqrt{3}$

The solutions are $3 + 5\sqrt{3}$ and $3 - 5\sqrt{3}$.

**55.** $x^2 - 20x + 100 = 0$

$(x - 10)(x - 10) = 0$

$x - 10 = 0 \Rightarrow x = 10$

The solution is 10.

**57.** $-2x^2 + 6x + 1 = 0$

$a = -2, b = 6, c = 1$

$$x = \frac{-6 \pm \sqrt{6^2 - 4(-2)(1)}}{2(-2)}$$

$$x = \frac{-6 \pm \sqrt{36 + 8}}{-4}$$

$$x = \frac{-6 \pm \sqrt{44}}{-4}$$

$$x = \frac{-6 \pm 2\sqrt{11}}{-4}$$

$$x = \frac{3 \pm \sqrt{11}}{2}$$

The solutions are $\frac{3 + \sqrt{11}}{2}$ and $\frac{3 - \sqrt{11}}{2}$.

**59.** $10x^2 + x - 3 = 0$

$(5x + 3)(2x - 1) = 0$

$5x + 3 = 0 \Rightarrow 5x = -3 \Rightarrow x = -\frac{3}{5}$

$2x - 1 = 0 \Rightarrow 2x = 1 \Rightarrow x = \frac{1}{2}$

The solutions are $-\frac{3}{5}$ and $\frac{1}{2}$.

**61.** $3x^2 - 14x + 4 = 0$

$a = 3, b = -14, c = 4$

$$x = \frac{-(-14) \pm \sqrt{(-14)^2 - 4(3)(4)}}{2(3)}$$

$$x = \frac{14 \pm \sqrt{196 - 48}}{6} = \frac{14 \pm \sqrt{148}}{6} = \frac{14 \pm 2\sqrt{37}}{6}$$

$$x = \frac{2(7 \pm \sqrt{37})}{2(3)}$$

$$x = \frac{7 \pm \sqrt{37}}{3}$$

$$x = \frac{7 + \sqrt{37}}{3} \approx 4.361$$

$$x = \frac{7 - \sqrt{37}}{3} \approx 0.306$$

The solutions are *approximately* 4.361 and 0.306.

**63.** $-0.03x^2 + 2x - 0.5 = 0$

$0.03x^2 - 2x + 0.5 = 0$

$a = 0.03, b = -2, c = 0.5$

$x = \dfrac{-(-2) \pm \sqrt{(-2)^2 - 4(0.03)(0.5)}}{2(0.03)}$

$x = \dfrac{2 \pm \sqrt{4 - 0.06}}{0.06} = \dfrac{2 \pm \sqrt{3.94}}{0.06}$

$x = \dfrac{2 + \sqrt{3.94}}{0.06} \approx 66.416$

$x = \dfrac{2 - \sqrt{3.94}}{0.06} \approx 0.251$

The solutions are *approximately* 66.416 and 0.251.

Note: You could multiply both sides of the equation by 100 before determining the values of $a$, $b$, and $c$. Using the integer values, you will find $x = \dfrac{100 \pm 5\sqrt{394}}{3}$.

The decimal approximations are the same.

**65.** $\dfrac{x + 3}{2} - \dfrac{4}{x} = 2$

$2x\left(\dfrac{x + 3}{2} - \dfrac{4}{x}\right) = 2x(2)$

$x(x + 3) - 2(4) = 4x, \quad x \neq 0$

$x^2 + 3x - 8 = 4x$

$x^2 - x - 8 = 0$

$a = 1, b = -1, c = -8$

$x = \dfrac{-(-1) \pm \sqrt{(-1)^2 - 4(1)(-8)}}{2(1)}$

$x = \dfrac{1 \pm \sqrt{1 + 32}}{2}$

$x = \dfrac{1 \pm \sqrt{33}}{2}$

The solutions are $\dfrac{1 + \sqrt{33}}{2}$ and $\dfrac{1 - \sqrt{33}}{2}$.

**67.** $\sqrt{4x + 3} = x - 1$

$\left(\sqrt{4x + 3}\right)^2 = (x - 1)^2$

$4x + 3 = x^2 - 2x + 1$

$-x^2 + 6x + 2 = 0$

$x^2 - 6x - 2 = 0$

$a = 1, b = -6, c = -2$

$x = \dfrac{-(-6) \pm \sqrt{(-6)^2 - 4(1)(-2)}}{2(1)}$

$x = \dfrac{6 \pm \sqrt{36 + 8}}{2} = \dfrac{6 \pm \sqrt{44}}{2} = \dfrac{6 \pm 2\sqrt{11}}{2}$

$x = \dfrac{2(3 \pm \sqrt{11})}{2} = 3 \pm \sqrt{11}$

The answer of $3 + \sqrt{11}$ checks, but the second answer is extraneous. (If $x = 3 - \sqrt{11}$, $x \approx -0.317$. Then $\sqrt{4x + 3} \approx 1.317$ but $x - 1 \approx -1.317$.) So, the only solution is $3 + \sqrt{11}$.

**69.** *Verbal model:*     $\boxed{\text{Total time}} = \boxed{\text{Time up}} + \boxed{\text{Time Down}}$

*Labels:*

Rate going down $= x$          (miles per hour)

Rate going up $= x - 5$          (miles per hour)

Distance $= 20$          (miles)

Total time $= 4$          (hours)

Time up $= \dfrac{\text{Distance}}{\text{Rate}} = \dfrac{20}{x - 5}$          (hours)

Time down $= \dfrac{\text{Distance}}{\text{Rate}} = \dfrac{20}{x}$          (hours)

*Equation:*

$$4 = \frac{20}{x - 5} + \frac{20}{x}$$

$$4(x)(x - 5) = \left(\frac{20}{x - 5} + \frac{20}{x}\right)(x)(x - 5)$$

$$4x(x - 5) = 20x + 20(x - 5), \quad x \neq 0, x \neq 5$$

$$4x^2 - 20x = 20x + 20x - 100$$

$$4x^2 - 20x = 40x - 100$$

$$4x^2 - 60x + 100 = 0$$

$$x^2 - 15x + 25 = 0$$

$$x = \frac{15 \pm \sqrt{(-15)^2 - 4(1)(25)}}{2(1)}$$

$$x = \frac{15 \pm \sqrt{125}}{2}$$

$$x = \frac{15 \pm 5\sqrt{5}}{2}$$

$$x = \frac{15 + 5\sqrt{5}}{2} \Rightarrow x \approx 13.1$$

$$x = \frac{15 - 5\sqrt{5}}{2} \Rightarrow x \approx 1.9 \qquad \text{Extraneous}$$

The biker's speed coming down the trail is approximately 13.1 miles per hour. The other solution of approximately 1.9 miles per hour is excluded because the uphill rate $x - 5$ would be negative.

**71. (a)**
$$h = -16t^2 + 20t + 1465 \text{ and } h = 1465$$
$$1465 = -16t^2 + 20t + 1465$$
$$16t^2 - 20t = 0$$
$$4t(4t - 5) = 0$$
$$4t = 0 \Rightarrow t = 0$$
$$4t - 5 = 0 \Rightarrow 4t = 5 \Rightarrow t = \tfrac{5}{4}$$

The height is 1465 feet when $t = 0$ seconds and when $t = \tfrac{5}{4}$ second.

**(b)**
$$h = -16t^2 + 20t + 1465 \text{ and } h = 0$$
$$0 = -16t^2 + 20t + 1465$$
$$16t^2 - 20t - 1465 = 0$$
$$t = \frac{-(-20) \pm \sqrt{(-20)^2 - 4(16)(-1465)}}{2(16)}$$
$$t = \frac{20 \pm \sqrt{400 + 93,760}}{32}$$
$$t = \frac{20 \pm \sqrt{94,160}}{32}$$
$$t = \frac{20 \pm 4\sqrt{5885}}{32}$$
$$t = \frac{\cancel{4}\left(5 \pm \sqrt{5885}\right)}{\cancel{4}(8)}$$
$$t = \frac{5 \pm \sqrt{5885}}{8}$$
$$t = \frac{5 + \sqrt{5885}}{8} \Rightarrow t \approx 10.21$$
$$t = \frac{5 - \sqrt{5885}}{8} \Rightarrow t \approx -8.96 \qquad \text{Extraneous}$$

We discard the negative solution. The object strikes the ground in approximately 10.21 seconds.

**73.**   $(\text{Length})(\text{Width}) = \text{Area}$

$$(x + 6.3)(x) = 58.14$$

$$x^2 + 6.3x = 58.14$$

$$x^2 + 6.3x - 58.14 = 0$$

$$a = 1, b = 6.3, c = -58.14$$

$$x = \frac{-6.3 \pm \sqrt{6.3^2 - 4(1)(-58.14)}}{2(1)}$$

$$x = \frac{-6.3 \pm \sqrt{39.69 + 232.56}}{2}$$

$$x = \frac{-6.3 \pm \sqrt{272.25}}{2}$$

$$x = \frac{-6.3 \pm 16.5}{2}$$

$$x = \frac{-6.3 + 16.5}{2} \Rightarrow x = 5.1 \Rightarrow x + 6.3 = 11.4$$

$$x = \frac{-6.3 - 16.5}{2} \Rightarrow x = -11.4 \qquad \text{Discard negative solution.}$$

The dimensions are 5.1 inches by 11.4 inches.

**75.** Yes, the binomial factors could be multiplied together to put the equation into standard form, and then the Quadratic Formula could be used to solve the equation. No, the simplest method would be the factoring method because the left side of the equation is already factored and equal to 0.

**77.** The solution of the quadratic equation $ax^2 + bx + c = 0$ are $\dfrac{-b + \sqrt{b^2 - 4ac}}{2a}$ and $\dfrac{-b - \sqrt{b^2 - 4ac}}{2a}$.

The sum of the solutions:

$$\text{sum} = \frac{-b + \sqrt{b^2 - 4ac}}{2a} + \frac{-b - \sqrt{b^2 - 4ac}}{2a}$$

$$= \frac{-b + \sqrt{b^2 - 4ac} - b - \sqrt{b^2 - 4ac}}{2a} = \frac{-2b}{2a} = -\frac{b}{a}$$

The product of the solutions:

$$\text{product} = \frac{-b + \sqrt{b^2 - 4ac}}{2a} \cdot \frac{-b - \sqrt{b^2 - 4ac}}{2a}$$

$$= \frac{\left(-b + \sqrt{b^2 - 4ac}\right)\left(-b - \sqrt{b^2 - 4ac}\right)}{(2a)^2}$$

$$= \frac{b^2 + b\sqrt{b^2 - 4ac} - b\sqrt{b^2 - 4ac} - \left(\sqrt{b^2 - 4ac}\right)^2}{4a^2}$$

$$= \frac{b^2 - \left(b^2 - 4ac\right)}{4a^2}$$

$$= \frac{b^2 - b^2 + 4ac}{4a^2}$$

$$= \frac{4ac}{4a^2}$$

$$= \frac{c}{a}$$

**79.** $3 = 3(1)$

$27 = 3 \cdot 3 \cdot 3 = 3(9)$

Greatest common factor is 3.

**81.** $10z^2 = 2 \cdot 5 \cdot z \cdot z = 5z^2(2)$

$5z^3 = 5 \cdot z \cdot z \cdot z = 5z^2(z)$

Greatest common factor is $5z^2$.

**85.** $\qquad x^2 - 6x = 0$

$x^2 - 6x + (-3)^2 = 9$ Half of $-6$ is $-3$, and $(-3)^2 = 9$

$\qquad (x - 3)^2 = 9$

$\qquad x - 3 = \pm\sqrt{9}$

$\qquad x - 3 = \pm 3$

$\qquad x = 3 + 3 = 6$

$\qquad x = 3 - 3 = 0$

The solutions are 6 and 0.

**87.** $\qquad x^2 + 8x - 10 = 0$

$\qquad x^2 + 8x = 10$

$x^2 + 8x + (4)^2 = 10 + 16$ Half of 8 is 4, and $(4)^2 = 16$.

$\qquad (x + 4)^2 = 26$

$\qquad x + 4 = \pm\sqrt{26}$

$\qquad x = -4 \pm 26$

The solutions are $-4 + \sqrt{26}$ and $-4 - \sqrt{26}$.

**83.** $4(x + 1) = 2 \cdot 2(x + 1) = (x + 1)(4)$

$3(x + 1) = 3(x + 1) = (x + 1)(3)$

Greatest common factor is $x + 1$.

# Mid-Chapter Quiz for Chapter 10

**1.** $x^2 - 7x + 10 = 0$

$(x - 5)(x - 2) = 0$

$x - 5 = 0 \Rightarrow x = 5$

$x - 2 = 0 \Rightarrow x = 2$

The solutions are 5 and 2.

**2.** $x^2 = 400$

$x = \pm\sqrt{400}$

$x = \pm 20$

The solutions are 20 and $-20$.

**3.** $2x^2 + 9x - 35 = 0$

$(2x - 5)(x + 7) = 0$

$2x - 5 = 0 \Rightarrow 2x = 5 \Rightarrow x = \frac{5}{2}$

$x + 7 = 0 \Rightarrow x = -7$

The solutions are $\frac{5}{2}$ and $-7$.

**4.** $8x(x - 4) + 3(x - 4) = 0$

$(x - 4)(8x + 3) = 0$

$x - 4 = 0 \Rightarrow x = 4$

$8x + 3 = 0 \Rightarrow 8x = -3 \Rightarrow x = -\frac{3}{8}$

The solutions are 4 and $-\frac{3}{8}$.

**5.** $x^2 - 2500 = 0$

$x^2 = 2500$

$x = \pm\sqrt{2500}$

$x = \pm 50$

The solutions are 50 and $-50$.

**6.** $9(z - 4)^2 - 81 = 0$

$$9(z - 4)^2 = 81$$

$$(z - 4)^2 = 9$$

$$z - 4 = \pm\sqrt{9}$$

$$z - 4 = \pm 3$$

$$z = 4 \pm 3$$

$$z = 4 + 3 = 7$$

$$z = 4 - 3 = 1$$

The solutions are 7 and 1.

**7.** $y^2 + 6y - 11 = 0$

$$y^2 + 6y = 11$$

$$y^2 + 6y + 3^2 = 11 + 9$$

$$(y + 3)^2 = 20$$

$$y + 3 = \pm\sqrt{20}$$

$$y + 3 = \pm 2\sqrt{5}$$

$$y = -3 \pm 2\sqrt{5}$$

The solutions are $-3 + 2\sqrt{5}$ and $-3 - 2\sqrt{5}$.

**8.** $4u^2 + 12u - 1 = 0$

$$4u^2 + 12u = 1$$

$$u^2 + 3u = \frac{1}{4}$$

$$u^2 + 3u + \left(\frac{3}{2}\right)^2 = \frac{1}{4} + \frac{9}{4}$$

$$\left(u + \frac{3}{2}\right)^2 = \frac{10}{4}$$

$$u + \frac{3}{2} = \pm\sqrt{\frac{10}{4}}$$

$$u + \frac{3}{2} = \pm\frac{\sqrt{10}}{2}$$

$$u = -\frac{3}{2} \pm \frac{\sqrt{10}}{2}$$

$$u = \frac{-3 \pm \sqrt{10}}{2}$$

The solutions are $\dfrac{-3 + \sqrt{10}}{2}$ and $\dfrac{-3 - \sqrt{10}}{2}$.

**9.** $x^2 + 3x + 1 = 0$

$$a = 1, b = 3, c = 1$$

$$x = \frac{-3 \pm \sqrt{3^2 - 4(1)(1)}}{2(1)}$$

$$x = \frac{-3 \pm \sqrt{9 - 4}}{2}$$

$$x = \frac{-3 \pm \sqrt{5}}{2}$$

The solutions are $\dfrac{-3 + \sqrt{5}}{2}$ and $\dfrac{-3 - \sqrt{5}}{2}$.

**10.** $6x^2 - 8x - 20 = 0$

$$a = 6, b = -8, c = -20$$

$$x = \frac{-(-8) \pm \sqrt{(-8)^2 - 4(6)(-20)}}{2(6)}$$

$$x = \frac{8 \pm \sqrt{64 + 480}}{12}$$

$$x = \frac{8 \pm \sqrt{544}}{12}$$

$$x = \frac{8 \pm 4\sqrt{34}}{12}$$

$$x = \frac{\cancel{4}(2 \pm \sqrt{34})}{\cancel{4}(3)}$$

$$x = \frac{2 \pm \sqrt{34}}{3}$$

The solutions are $\dfrac{2 + \sqrt{34}}{3}$ and $\dfrac{2 - \sqrt{34}}{3}$.

**11.** $3x^2 - 48x = 96$

$$3x^2 - 48x - 96 = 0$$

$$a = 3, b = -48, c = -96$$

$$x = \frac{-(-48) \pm \sqrt{(-48)^2 - 4(3)(-96)}}{2(3)}$$

$$x = \frac{48 \pm \sqrt{2304 + 1152}}{6}$$

$$x = \frac{48 \pm \sqrt{3456}}{6}$$

$$x = \frac{48 \pm 24\sqrt{6}}{6}$$

$$x = \frac{\cancel{6}(8 \pm 4\sqrt{6})}{\cancel{6}(1)}$$

$$x = 8 \pm 4\sqrt{6}$$

The solutions are $8 + 4\sqrt{6}$ and $8 - 4\sqrt{6}$.

**12.** $5x = 3x^2 + 1$

$0 = 3x^2 - 5x + 1$

$x = \dfrac{-(-5) \pm \sqrt{(-5)^2 - 4(3)(1)}}{2(3)}$

$x = \dfrac{5 \pm \sqrt{25 - 12}}{6}$

$x = \dfrac{5 \pm \sqrt{13}}{6}$

The solutions are $\dfrac{5 + \sqrt{13}}{6}$ and $\dfrac{5 - \sqrt{13}}{6}$.

**13.** $x^2 + x + \frac{9}{4} = 0$

$a = 1, b = 1, c = \frac{9}{4}$

$b^2 - 4ac = 1^2 - 4(1)\left(\frac{9}{4}\right) = 1 - 9 = -8$

The *negative* discriminant indicates that the equation has *no* real solutions.

**14.** $y^2 - 7y - 1 = 0$

$a = 1, b = -7, c = 1$

$b^2 - 4ac = (-7)^2 - 4(1)(1) = 49 - 4 = 45$

The *positive* discriminant indicates that the equation has *two* real solutions.

**15.** $3x^2 - 4x - 4 = 0$

$a = 3, b = -4, c = -4$

$b^2 - 4ac = (-4)^2 - 4(3)(-4) = 16 + 48 = 64$

The *positive* discriminant indicates that the equation has *two* real solutions.

**16.** $9x^2 + 6x + 1 = 0$

$a = 9, b = 6, c = 1$

$b^2 - 4ac = 6^2 - 4(9)(1) = 36 - 36 = 0$

The discriminant of *0* indicates that the equation has *one* real solution.

**17.** $A = P(1 + r)^t$

$1669.54 = 1500(1 + r)^2$

$\dfrac{1669.54}{1500} = (1 + r)^2$

$1.113 \approx (1 + r)^2$

$\sqrt{1.113} \approx 1 + r$

$1.055 \approx 1 + r$

$0.055 \approx r$

The interest rate is approximately 5.5%.

**18.** $h = 343 - 16t^2$

$0 = 343 - 16t^2$  (Note: $h = 0$ when he finishes his dive.)

$16t^2 = 343$

$t^2 = \dfrac{343}{16}$

$t = \pm\sqrt{\dfrac{343}{16}}$

$t = \pm\dfrac{\sqrt{343}}{\sqrt{16}}$

$t = \pm\dfrac{7\sqrt{7}}{4}$

$t = \dfrac{7\sqrt{7}}{4}$   Discard the negative solution.

He was in the air for $\dfrac{7\sqrt{7}}{4}$ seconds, or approximately 4.63 seconds.

**19.**   $(\text{Length})(\text{Width}) = \text{Area}$

$$\left(x + 6\right)\left(\frac{1}{2}x\right) = 153.92$$

$$\frac{1}{2}x^2 + 3x = 153.92$$

$$0.5x^2 + 3x - 153.92 = 0$$

$$a = 0.5, b = 3, c = -153.92$$

$$x = \frac{-3 \pm \sqrt{(3)^2 - 4(0.5)(-153.92)}}{2(0.5)}$$

$$x = \frac{-3 \pm \sqrt{9 + 307.84}}{1}$$

$$x = -3 \pm \sqrt{316.84}$$

$$x = -3 + \sqrt{316.84} \Rightarrow x = 14.8$$

$$x = -3 - \sqrt{316.84} \Rightarrow x = -20.8 \quad \text{Extraneous}$$

If $x = 14.8$, then $x + 6 = 14.8 + 6 = 20.8$ and $\frac{1}{2}x = \frac{14.8}{2} = 7.4$.

The dimensions of the rectangle are 20.8 inches by 7.4 inches.

## Section 10.4   Graphing Quadratic Equations

**1.** Matches graph (c)

**2.** Matches graph (d)

**3.** Matches graph (f)

**4.** Matches graph (b)

**5.** Matches graph (a)

**6.** Matches graph (e)

**7.** $y = x^2 - 4x + 3$

   $a = 1$

   Since $a > 0$, the parabola opens *upward.*

**9.** $y = 6 + x - 2x^2$

   $y = -2x^2 + x + 6$

   $a = -2$

   Since $a < 0$, the parabola opens *downward.*

**11.** $y = 3 + x(3 - x)$

   $y = 3 + 3x - x^2$

   $y = -x^2 + 3x + 3$

   $a = -1$

   Since $a < 0$, the parabola opens *downward.*

**13.** $y = -(x + 1)^2 - 1$

   $y = -(x^2 + 2x + 1) - 1$

   $y = -x^2 - 2x - 1 - 1$

   $y = -x^2 - 2x - 2$

   $a = -1$

   Since $a < 0$, the parabola opens *downward.*

**15.** $y = 16 - x^2$

   *y-intercept:* $(\text{Let } x = 0.)$ $y = 16 - 0^2 = 16$

   The *y*-intercept is $(0, 16)$.

   *x-intercepts:* $(\text{Let } y = 0.)$ $0 = 16 - x^2$

   $0 = (4 + x)(4 - x)$

   $4 + x = 0 \Rightarrow x = -4$

   $4 - x = 0 \Rightarrow x = 4$

   The *x*-intercepts are $(-4, 0)$ and $(4, 0)$.

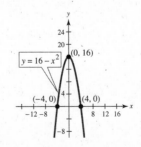

**17.** $y = x^2 - 2x$

*y-intercept:* $\left(\text{Let } x = 0.\right)$  $y = 0^2 - 2(0) = 0$

The *y*-intercept is $(0, 0)$.

*x-intercepts:* $\left(\text{Let } y = 0.\right)$  $0 = x^2 - 2x$

$0 = x(x - 2)$

$x = 0$

$x - 2 = 0 \Rightarrow x = 2$

The *x*-intercepts are $(0, 0)$ and $(2, 0)$.

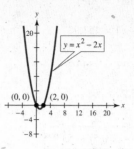

**19.** $y = x^2 - 4x + 3$

*y-intercept:* $\left(\text{Let } x = 0.\right)$  $y = 0^2 - 4(0) + 3 = 3$

The *y*-intercept is $(0, 3)$.

*x-intercepts:* $\left(\text{Let } y = 0.\right)$  $0 = x^2 - 4x + 3$

$0 = (x - 3)(x - 1)$

$x - 3 = 0 \Rightarrow x = 3$

$x - 1 = 0 \Rightarrow x = 1$

The *x*-intercepts are $(3, 0)$ and $(1, 0)$.

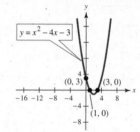

**21.** $y = 3x^2 + 4x - 4$

*y-intercept:* $\left(\text{Let } x = 0.\right)$  $y = 3(0)^2 + 4(0) - 4 = -4$

The *y*-intercept is $(0, -4)$.

*x-intercepts:* $\left(\text{Let } y = 0.\right)$  $0 = 3x^2 + 4x - 4$

$0 = (3x - 2)(x + 2)$

$3x - 2 = 0 \Rightarrow 3x = 2 \Rightarrow x = \frac{2}{3}$

$x + 2 = 0 \Rightarrow x = -2$

The *x*-intercepts are $\left(\frac{2}{3}, 0\right)$ and $(-2, 0)$.

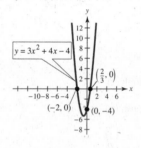

**23.** $y = \frac{1}{2}x^2 - 2x + 1$

*y-intercept:* $\left(\text{Let } x = 0.\right)$  $y = \frac{1}{2}(0)^2 - 2(0) + 1 = 1$

The *y*-intercept is $(0, 1)$.

*x-intercepts:* $\left(\text{Let } y = 0.\right)$  $0 = \frac{1}{2}x^2 - 2x + 1$

$x = \dfrac{2 \pm \sqrt{(-2)^2 - 4(1/2)(1)}}{2(1/2)}$

$x = \dfrac{2 \pm \sqrt{4 - 2}}{1}$

$x = 2 \pm \sqrt{2}$

The *x*-intercepts are $\left(2 + \sqrt{2}, 0\right)$ and $\left(2 - \sqrt{2}, 0\right)$.

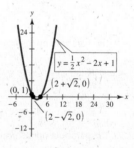

**25.** $y = -x^2 + 2$

$a = -1, b = 0$

$x = -\dfrac{b}{2a} = -\dfrac{0}{2(-1)} = 0$

$y = -0^2 + 2 = 2$

The vertex is located at $(0, 2)$.

**27.** $y = x^2 - 4x + 7$

$a = 1, b = -4$

$x = -\dfrac{b}{2a} = -\dfrac{-4}{2(1)} = -(-2) = 2$

$y = 2^2 - 4 \cdot 2 + 7 = 4 - 8 + 7 = 3$

The vertex is located at $(2, 3)$.

**29.** $y = 6 + 10x - x^2$

$y = -x^2 + 10x + 6$

$a = -1, b = 10$

$x = -\dfrac{b}{2a} = -\dfrac{10}{2(-1)} = 5$

$y = -(5)^2 + 10(5) + 6$

$y = -25 + 50 + 6 = 31$

The vertex is $(5, 31)$.

**31.** $y = x^2 + 5x - 3$

$a = 1, b = 5$

$x = -\dfrac{b}{2a} = -\dfrac{5}{2(1)} = -\dfrac{5}{2}$

$y = \left(-\dfrac{5}{2}\right)^2 + 5\left(-\dfrac{5}{2}\right) - 3$

$y = \dfrac{25}{4} - \dfrac{25}{2} - 3 = \dfrac{25}{4} - \dfrac{50}{4} - \dfrac{12}{4}$

$y = -\dfrac{37}{4}$

The vertex is $\left(-\dfrac{5}{2}, -\dfrac{37}{4}\right)$.

**33.** $y = 3x^2 + 6x - 8$

$a = 3,$ and $b = 6$

$x = -\dfrac{b}{2a} = -\dfrac{6}{2(3)} = -\dfrac{6}{6} = -1$

$y = 3(-1)^2 + 6(-1) - 8$

$y = 3 - 6 - 8 = -11$

The vertex is located at $(-1, -11)$.

**35.** $y = 8 - 9x - 3x^2$

$y = -3x^2 - 9x + 8$

$a = -3, b = -9$

$x = -\dfrac{b}{2a} = -\dfrac{-9}{2(-3)} = -\dfrac{9}{6}$

$x = -\dfrac{3}{2}$

$y = -3\left(-\dfrac{3}{2}\right)^2 - 9\left(-\dfrac{3}{2}\right) + 8$

$y = -3\left(\dfrac{9}{4}\right) + \dfrac{27}{2} + 8$

$y = -\dfrac{27}{4} + \dfrac{54}{4} + \dfrac{32}{4} = \dfrac{59}{4}$

The vertex is $\left(-\dfrac{3}{2}, \dfrac{59}{4}\right)$.

**37.** $y = x^2 - 1$

*Leading coefficient test:*     Since $a > 0$, the parabola opens upward.

*Vertex:* $(a = 1, b = 0)$     $x = -\dfrac{b}{2a} = -\dfrac{0}{2(1)} = 0$

$y = 0^2 - 1 = -1$

The vertex is located at $(0, -1)$.

*y-intercept:* $(\text{Let } x = 0.)$     $y = 0^2 - 1 = -1$

The *y*-intercept is $(0, -1)$.

*x-intercepts:* $(\text{Let } y = 0.)$     $0 = x^2 - 1$

$1 = x^2 \Rightarrow x = \pm\sqrt{1} \Rightarrow x = \pm 1$

The *x*-intercepts are $(1, 0)$ and $(-1, 0)$.

*Additional solution points:*

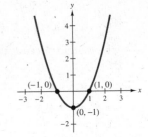

| $x$ | 2 | 3 | −2 | −3 |
|---|---|---|---|---|
| $y = x^2 - 1$ | 3 | 8 | 3 | 8 |
| Points | $(2, 3)$ | $(3, 8)$ | $(-2, 3)$ | $(-3, 8)$ |

**39.** $y = -x^2 + 1$

*Leading coefficient test:*     Since $a < 0$, the parabola opens downward.

*Vertex:* $(a = -1, b = 0)$     $x = -\dfrac{b}{2a} = \dfrac{-0}{2(-1)} = \dfrac{-0}{-2} = 0$

$y = -0^2 + 1 = 1$

The vertex is located at $(0, 1)$.

*y-intercept:* $(\text{Let } x = 0.)$     $y = -0^2 + 1 = -0 + 1 = 1$

The *y*-intercept is $(0, 1)$.

*x-intercepts:* $(\text{Let } y = 0.)$     $0 = -x^2 + 1$

$x^2 = 1$

$x = \pm\sqrt{1}$

$x = \pm 1$

The *x*-intercepts are $(1, 0)$ and $(-1, 0)$.

*Additional solution points:*

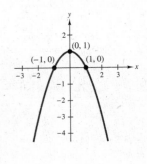

| $x$ | 2 | 3 | −2 | −3 |
|---|---|---|---|---|
| $y = -x^2 + 1$ | −3 | −8 | −3 | −8 |
| Points | $(2, -3)$ | $(3, -8)$ | $(-2, -3)$ | $(-3, -8)$ |

**41.** $y = x^2 - 4x$

*Leading coefficient test:*     Since $a > 0$, the parabola opens upward.

*Vertex:* $(a = 1, b = -4)$
$$x = -\frac{b}{2a} = -\frac{-4}{2(1)} = -(-2) = 2$$
$$y = 2^2 - 4 \cdot 2 = 4 - 8 = -4$$
The vertex is located at $(2, -4)$.

*y-intercept:* $(\text{Let } x = 0.)$     $y = 0^2 - 4 \cdot 0 = 0$

The *y*-intercept is $(0, 0)$.

*x-intercepts:* $(\text{Let } y = 0.)$
$$0 = x^2 - 4x \Rightarrow x(x - 4) = 0$$
$$x = 0$$
$$x - 4 = 0 \Rightarrow x = 4$$
The *x*-intercepts are $(0, 0)$ and $(4, 0)$.

*Additional solution points:*

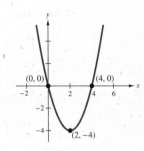

| $x$ | 5 | 3 | $-1$ | 1 |
|---|---|---|---|---|
| $y = x^2 - 4x$ | 5 | $-3$ | 5 | $-3$ |
| Points | $(5, 5)$ | $(3, -3)$ | $(-1, 5)$ | $(1, -3)$ |

**43.** $y = x^2 - 6x + 9$

*Leading coefficient test:*     Since $a > 0$, the parabola opens *upward*.

*Vertex:* $(a = 1, b = -6)$
$$x = -\frac{b}{2a} = -\frac{-6}{2(1)} = -(-3) = 3$$
$$y = (3)^2 - 6(3) + 9 = 9 - 18 + 9 = 0$$
The vertex is located at $(3, 0)$.

*y–intercept:* $(\text{Let } x = 0.)$     $y = 0^2 - 6(0) + 9 = 9$

The *y*-intercept is $(0, 9)$.

*x–intercept:* $(\text{Let } y = 0.)$
$$0 = x^2 - 6x + 9$$
$$0 = (x - 3)(x - 3)$$
$$x - 3 = 0 \Rightarrow x = 3$$
The *x*-intercept is $(3, 0)$.

*Additional solution points:*

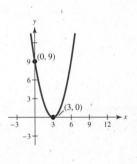

| $x$ | 1 | 2 | 4 | 5 |
|---|---|---|---|---|
| $y = x^2 - 6x + 9$ | 4 | 1 | 1 | 4 |
| Points | $(1, 4)$ | $(2, 1)$ | $(4, 1)$ | $(5, 4)$ |

**45.** $y = x^2 + 6x + 8$

*Leading coefficient test:*     Since $a > 0$, the parabola opens upward.

*Vertex:* $(a = 1, b = 6)$     $x = -\dfrac{b}{2a} = -\dfrac{6}{2(1)} = -3$

$$y = (-3)^2 + 6(-3) + 8 = 9 - 18 + 8 = -1$$

The vertex is located at $(-3, -1)$.

*y-intercept:* (Let $x = 0$.)     $y = 0^2 + 6 \cdot 0 + 8 = 8$; the $y$-intercept is $(0, 8)$.

*x-intercepts:* (Let $y = 0$.)     $0 = x^2 + 6x + 8 \Rightarrow (x + 4)(x + 2) = 0$

$$x + 4 = 0 \Rightarrow x = -4$$

$$x + 2 = 0 \Rightarrow x = -2$$

The $x$-intercepts are $(-4, 0)$ and $(-2, 0)$.

*Additional solution points:*

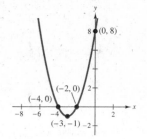

| $x$ | $-1$ | $-5$ | $-6$ |
|---|---|---|---|
| $y = x^2 + 6x + 8$ | 3 | 3 | 8 |
| Points | $(-1, 3)$ | $(-5, 3)$ | $(-6, 8)$ |

**47.** $y = -(x^2 + 2x - 3) = -x^2 - 2x + 3$

*Leading coefficient test:*     Since $a < 0$, the parabola opens downward.

*Vertex:* $(a = -1, b = -2)$     $x = -\dfrac{b}{2a} = -\dfrac{-2}{2(-1)} = -1$

$$y = -(-1)^2 - 2(-1) + 3 = -1 + 2 + 3 = 4$$

The vertex is located at $(-1, 4)$.

*y-intercept:* (Let $x = 0$.)     $y = -0^2 - 2 \cdot 0 + 3 = 3$; the $y$-intercept is $(0, 3)$.

*x-intercepts:* (Let $y = 0$.)     $0 = -x^2 - 2x + 3 \Rightarrow x^2 + 2x - 3 = 0 \Rightarrow (x + 3)(x - 1) = 0$

$$x + 3 = 0 \Rightarrow x = -3$$

$$x - 1 = 0 \Rightarrow x = 1$$

The $x$-intercepts are $(-3, 0)$ and $(1, 0)$.

*Additional solution points:*

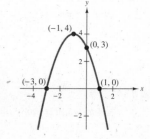

| $x$ | 2 | $-2$ | $-4$ |
|---|---|---|---|
| $y = -x^2 - 2x + 3$ | $-5$ | 3 | $-5$ |
| Points | $(2, -5)$ | $(-2, 3)$ | $(-4, -5)$ |

**49.** $y = x^2 + 4x + 8$

*Leading coefficient test:*  Since $a > 0$, the parabola opens *upward*.

*Vertex:* $\left(a = 1, b = 4\right)$

$$x = -\frac{b}{2a} = -\frac{4}{2(1)} = -2$$

$$y = (-2)^2 + 4(-2) + 8 = 4 - 8 + 8 = 4$$

The vertex is located at $(-2, 4)$.

*y–intercept:* $\left(\text{Let } x = 0.\right)$    $y = 0^2 + 4(0) + 8 = 8$

The $y$-intercept is $(0, 8)$.

*x–intercepts:* $\left(\text{Let } y = 0.\right)$    $0 = x^2 + 4x + 8$

$$x = \frac{-4 \pm \sqrt{4^2 - 4(1)(8)}}{2(1)} = \frac{-4 \pm \sqrt{16 - 32}}{2}$$

$$x = \frac{-4 \pm \sqrt{-16}}{2}$$

There are no $x$-intercepts.

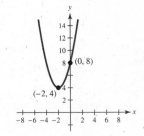

*Additional solution points:*

| $x$ | $-5$ | $-4$ | $-1$ | $1$ |
|---|---|---|---|---|
| $y = x^2 + 4x + 8$ | 13 | 8 | 5 | 13 |
| Points | $(-5, 13)$ | $(-4, 8)$ | $(-1, 5)$ | $(1, 13)$ |

**51.** $y = x^2 - 4x + 1$

*Leading coefficient test:*  Since $a > 0$, the parabola opens upward.

*Vertex:* $\left(a = 1, b = -4\right)$

$$x = -\frac{b}{2a} = -\frac{-4}{2(1)} = -(-2) = 2$$

$$y = 2^2 - 4(2) + 1 = 4 - 8 + 1 = -3$$

The vertex is located at $(2, -3)$.

*y-intercept:* $\left(\text{Let } x = 0.\right)$    $y = 0^2 - 4 \cdot 0 + 1 = 1$; the $y$-intercept is $(0, 1)$.

*x-intercepts:* $\left(\text{Let } y = 0.\right)$    $0 = x^2 - 4x + 1$

$$x = \frac{-(-4) \pm \sqrt{(-4)^2 - 4(1)(1)}}{2(1)} = \frac{4 \pm \sqrt{12}}{2}$$

$$x = \frac{4 \pm 2\sqrt{3}}{2} = 2 \pm \sqrt{3}$$

The $x$-intercepts are $\left(2 + \sqrt{3}, 0\right)$ and $\left(2 - \sqrt{3}, 0\right)$.

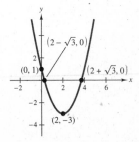

*Additional solution points:*

| $x$ | 4 | 5 | $-1$ |
|---|---|---|---|
| $y = x^2 - 4x + 1$ | 1 | 6 | 6 |
| Points | $(4, 1)$ | $(5, 6)$ | $(-1, 6)$ |

**53.** $y = 8 + 4x - 3x^2 = -3x^2 + 4x + 8$

*Leading coefficient test:*　　Since $a < 0$, the parabola opens downward.

*Vertex:* $(a = -3, b = 4)$

$$x = -\frac{b}{2a} = -\frac{4}{2(-3)} = \frac{4}{6} = \frac{2}{3}$$

$$y = -3\left(\frac{2}{3}\right)^2 + 4\left(\frac{2}{3}\right) + 8 = -\frac{4}{3} + \frac{8}{3} + \frac{24}{3} = \frac{28}{3}$$

The vertex is located at $\left(\frac{2}{3}, \frac{28}{3}\right)$.

*y-intercept:* (Let $x = 0$.)

$$y = -3(0)^2 + 4(0) + 8 = 8$$

The $y$-intercept is $(0, 8)$.

*x-intercepts:* (Let $y = 0$.)

$$0 = -3x^2 + 4x + 8$$

$$x = \frac{-4 \pm \sqrt{4^2 - 4(-3)(8)}}{2(-3)} = \frac{-4 \pm \sqrt{16 + 96}}{-6}$$

$$x = \frac{-4 \pm \sqrt{112}}{-6} = \frac{-4 \pm 4\sqrt{7}}{-6} = \frac{2 \pm 2\sqrt{7}}{3}$$

The $x$-intercepts are $\left(\frac{2 + 2\sqrt{7}}{3}, 0\right)$ and $\left(\frac{2 - 2\sqrt{7}}{3}, 0\right)$.

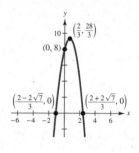

*Additional solution points:*

| $x$ | $-1$ | $1$ | $2$ | $3$ |
|---|---|---|---|---|
| $y = -3x^2 + 4x + 8$ | 1 | 9 | 4 | $-7$ |
| Points | $(-1, 1)$ | $(1, 9)$ | $(2, 4)$ | $(3, -7)$ |

**55.** $y = \dfrac{2}{3}x^2 - x - 3$

*Leading coefficient test:*    Since $a > 0$, the parabola opens upward.

*Vertex:* $\left(a = \dfrac{2}{3}, b = -1\right)$    $x = -\dfrac{-1}{2\left(\dfrac{2}{3}\right)} = \dfrac{1}{\left(\dfrac{4}{3}\right)} = \dfrac{3}{4}$

$$y = \dfrac{2}{3}\left(\dfrac{3}{4}\right)^2 - \dfrac{3}{4} - 3 = \dfrac{3}{8} - \dfrac{6}{8} - \dfrac{24}{8} = -\dfrac{27}{8}$$

The vertex is located at $\left(\dfrac{3}{4}, -\dfrac{27}{8}\right)$.

*y-intercept:* $\left(\text{Let } x = 0.\right)$    $y = \dfrac{2}{3}(0)^2 - (0) - 3 = -3$

The $y$-intercept is $(0, -3)$.

*x-intercepts:* $\left(\text{Let } y = 0.\right)$    $0 = \dfrac{2}{3}x^2 - x - 3 \Rightarrow 0 = 2x^2 - 3x - 9 \Rightarrow 0 = (2x + 3)(x - 3)$

$2x + 3 = 0 \Rightarrow 2x = -3 \Rightarrow x = -\dfrac{3}{2}$

$x - 3 = 0 \Rightarrow x = 3$

The $x$-intercepts are $\left(-\dfrac{3}{2}, 0\right)$ and $(3, 0)$.

*Additional solution points:*

| $x$ | $-6$ | $-3$ | 1.5 | 4.5 |
|---|---|---|---|---|
| $y = \dfrac{2}{3}x^2 - x - 3$ | 27 | 6 | $-3$ | 6 |
| Points | $(-6, 27)$ | $(-3, 6)$ | $(1.5, -3)$ | $(4.5, 6)$ |

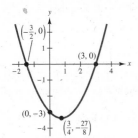

**57.**

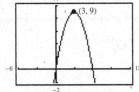

The vertex appears to be $(3, 9)$.

**59.**

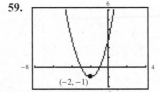

The vertex appears to be $(-2, -1)$.

**61.**

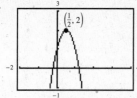

The vertex appears to be $\left(\frac{1}{2}, 2\right)$.

**63.**

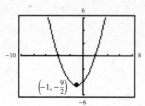

The vertex appears to be $\left(-1, -\frac{9}{2}\right)$.

**65.**

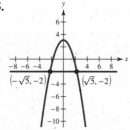

$y = -x^2 + 3$

$y = -2$

$-2 = -x^2 + 3$

$x^2 - 2 = 3$

$x^2 = 5$

$x = \pm\sqrt{5}$

The points of intersection are $\left(\sqrt{5}, -2\right)$ and $\left(-\sqrt{5}, -2\right)$.

**67.**

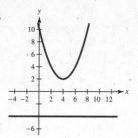

$y = \frac{1}{2}x^2 - 4x + 10$

$y = -4$

$\frac{1}{2}x^2 - 4x + 10 = -4$

$\frac{1}{2}x^2 - 4x + 14 = 0$

$x^2 - 8x + 28 = 0$

$x = \dfrac{-(-8) \pm \sqrt{(-8)^2 - 4(1)(28)}}{2(1)}$

$x = \dfrac{8 \pm \sqrt{64 - 112}}{2}$

$x = \dfrac{8 \pm \sqrt{-48}}{2}$

There is no real solution. The graphs have no point of intersection.

**69.**

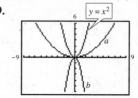

The graphs are all parabolas with the vertex at $(0, 0)$.

(a) The graph is wider than the graph of $y = x^2$. For the same $x$, $y$ is $\frac{1}{8}$ what it is on $y = x^2$.

(b) The parabola opens downward. The graph is not as wide as the graph of $y = x^2$. For the same $x$, $y$ is $-2$ times what it is on $y = x^2$.

**71.**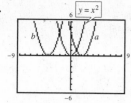

The graphs have the same shape but they are shifted left or right.

(a) The graph is shifted two units to the right of $y = x^2$; the vertex is $(2, 0)$.

(b) The graph is shifted four units to the left of $y = x^2$; the vertex is $(-4, 0)$.

**73.** $(-2, 0)$ and $(2, 0)$

$$y = \left[x - (-2)\right](x - 2)$$
$$y = (x + 2)(x - 2)$$
$$y = x^2 - 4$$
$$y = -x^2 + 4 \text{ or } y = 4 - x^4$$

**75.** $(-3, 0)$ and $(1, 0)$

$$y = \left[x - (-3)\right](x - 1)$$
$$y = (x + 3)(x - 1)$$
$$y = x^2 + 2x - 3$$
$$y = -x^2 - 2x + 3$$

**77.**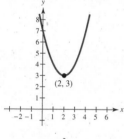

$$y = (x - 2)^2 + 3$$
$$y = x^2 - 4x + 4 + 3$$
$$y = x^2 - 4x + 7$$

Vertex: $x = -\dfrac{b}{2a} = -\dfrac{-4}{2(1)} = 2$

$$y = (2 - 2)^2 + 3 = 3$$

The vertex is located at $(2, 3)$.

The vertex is $(h, k)$ when the equation is in the form

$$y = (x - h)^2 + k.$$

**79.**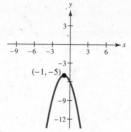

$$y = -(x + 1)^2 - 5$$
$$y = -\left(x^2 + 2x + 1\right) - 5$$
$$y = -x^2 - 2x - 6$$

Vertex: $x = -\dfrac{b}{2a} = -\dfrac{-2}{2(-1)} = -1$

$$y = -(-1 + 1)^2 - 5 = -5$$

The vertex is located at $(-1, -5)$.

The vertex is $(h, k)$ when the equation is in the form

$$y = (x - h)^2 + k.$$

**81.** (a)  When $x = 0$, $y = -0.1 \cdot 0^2 + 2(0) + 4 = 4$.

The ball was thrown from a height of 4 feet.

(b)  The maximum height is at the vertex of the parabola.

Vertex: $x = -\dfrac{b}{2a} = -\dfrac{2}{2(-0.1)} = -\dfrac{2}{-0.2} = 10$

$$y = -0.1(10)^2 + 2(10) + 4$$
$$y = -10 + 20 + 4 = 14$$

The vertex is $(10, 14)$, so the maximum height of the ball was 14 feet.

(c)  When the ball struck the ground, the height $y = 0$.

$$0 = -0.1x^2 + 2x + 4$$
$$10(0) = 10\left(-0.1x^2 + 2x + 4\right)$$
$$0 = -x^2 + 20x + 40$$
$$x^2 - 20x - 40 = 0$$

$$x = \frac{-(-20) \pm \sqrt{(-20)^2 - 4(1)(-40)}}{2(1)}$$

$$x = \frac{20 \pm \sqrt{560}}{2} = \frac{20 \pm 4\sqrt{35}}{2}$$

$$x = 10 \pm 2\sqrt{35} \quad \text{Choose the positive answer.}$$

$$x = 10 + 2\sqrt{35} \approx 21.8$$

The ball struck the ground approximately 21.8 feet from where it was thrown.

**83.** (a)  $y = -9.6t^2 + 624t + 22{,}360$

Vertex: $x = -\dfrac{b}{2a} = -\dfrac{624}{2(-9.6)} = -\dfrac{624}{-19.2} = 32.5$

$$y = -9.6(32.5)^2 + 624(32.5) + 22{,}360 = 32{,}500$$

The vertex is $(32.5, 32{,}500)$, so the maximum height of the plane is 32,500 feet.

(b)  $31{,}000 = -9.6t^2 + 624t + 22{,}360$

$$0 = -9.6t^2 + 624t - 8{,}640$$
$$0 = t^2 - 65t + 900$$
$$0 = (t - 20)(t - 45)$$
$$t - 20 = 0 \Rightarrow t = 20$$
$$t - 45 = 0 \Rightarrow t = 45$$

Weightlessness occurs from the time of 20 seconds until the time of 45 seconds.

**85.** (a) $2(\text{Length}) + 2(\text{Width}) = 36$

$$2x + 2(\text{Width}) = 36$$

$$2(\text{Width}) = 36 - 2x$$

$$\text{Width} = 18 - x$$

(b) $\text{Area} = (\text{Length})(\text{Width})$

$$\text{Area} = x(18 - x) = 18x - x^2$$

(c)

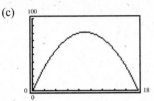

(d) The vertex of the graph appears to be located at $(9, 81)$. When the length $x = 9$, the width $18 - x = 9$. The rectangle of maximum area has dimensions 9 meters by 9 meters.

**87.** (a)

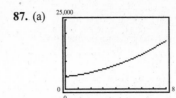

(b) The year in which there were 6000 stores appears to be 2002.

(c) $6000 = 146.29t^2 + 442.1t + 4471$

$$0 = 146.29t^2 + 442.1t - 1529$$

$$t = \frac{-442.1 \pm \sqrt{(442.1)^2 - 4(146.29)(-1529)}}{2(146.29)}$$

$$t \approx \frac{-442.1 \pm 1044.1}{292.58}$$

$$t \approx \frac{-442.1 + 1044.1}{292.58} \approx 2.1$$

$$t \approx \frac{-442.1 - 1044.1}{292.58} \approx -5.1$$

Choose the answer 2.1 that is in the domain. This verifies the answer that the year in which there were 6000 stores was 2002.

**89.** When the discriminant is positive, there are two real solutions for the equation and so there are two $x$-intercepts.

When the discriminant is 0, there is one real solution for the equation and so there is one $x$-intercept.

When the discriminant is negative, there are no real solutions for the equation and so there are no $x$-intercepts.

**91.** (a) $y = (x - 6)(x - 2) = x^2 - 8x + 12$

$a = 1, b = -8$

Vertex: $x = -\dfrac{b}{2a} = -\dfrac{-8}{2(1)} = -(-4) = 4$

$$y = (4)^2 - 8(4) + 12 = 16 - 32 + 12 = -4$$

So, the vertex is $(4, -4)$.

(b) A parabola with $x$-intercepts $(6, 0)$ and $(2, 0)$ has an equation of the form $y = a(x - 6)(x - 2) = a(x^2 - 8x + 12)$.

For the parabola in part (a) with vertex $(4, -4)$, $a = 1$.

Vertex: $(4, 4)$

The $y$-coordinate of this vertex is the opposite of the $y$-coordinate of the original, so $a = -1$.

$y = -(x - 6)(x - 2)$ or $y = -x^2 + 8x - 12$

Vertex: $(4, 8)$

The $y$-coordinate of this vertex is $-2$ times the $y$-coordinate of the original, so $a = -2$.

$y = -2(x - 6)(x - 2)$ or $y = -2x^2 + 16x - 24$

Vertex: $(4, -8)$

The $y$-coordinate of this vertex is 2 times the $y$-coordinate of the original, so $a = 2$.

$y = 2(x - 6)(x - 2)$ or $y = 2x^2 - 16x + 24$

**93.** $y - y_1 = m(x - x_1)$

$y - 9 = -3[x - (-4)]$

$y - 9 = -3(x + 4)$

$y - 9 = -3x - 12$

$y = -3x - 3$

**95.** $y - y_1 = m(x - x_1)$

$y - (-1) = -\frac{3}{5}[x - (-5)]$

$y + 1 = -\frac{3}{5}(x + 5)$

$y + 1 = -\frac{3}{5}x - 3$

$y = -\frac{3}{5}x - 4$

**97.** $b = 7$     $(2x + 9)(x - 1)$

$b = -7$     $(2x - 9)(x + 1)$

$b = -17$     $(2x + 1)(x - 9)$

$b = 17$     $(2x - 1)(x + 9)$

$b = -3$     $(2x + 3)(x - 3)$

$b = 3$     $(2x - 3)(x + 3)$

**99.** $b = 14$     $(5x + 4)(x + 2)$

$b = -14$     $(5x - 4)(x - 2)$

$b = 22$     $(5x + 2)(x + 4)$

$b = -22$     $(5x - 2)(x - 4)$

$b = 13$     $(5x + 8)(x + 1)$

$b = -13$     $(5x - 8)(x - 1)$

$b = 41$     $(5x + 1)(x + 8)$

$b = -41$     $(5x - 1)(x - 8)$

**101.** $3x^2 + 2x - 5 = 0$

$a = 3, b = 2, c = -5$

$b^2 - 4ac = 2^2 - 4(3)(-5) = 4 + 60 = 64$

The *positive* discriminant indicates that the equation has *two* real solutions.

**103.** $2x^2 - 4x + 2 = 0$

$a = 2, b = -4, c = 2$

$b^2 - 4ac = (-4)^2 - 4(2)(2) = 16 - 16 = 0$

The discriminant of *zero* indicates that the equation has *one* real solution.

## Section 10.5 · Applications of Quadratic Equations

**1.** *Verbal Model:*    $\boxed{\begin{array}{c}\text{First consecutive}\\\text{positive integer}\end{array}} \cdot \boxed{\begin{array}{c}\text{Second consecutive}\\\text{positive integer}\end{array}} = \boxed{132}$

*Labels:*    First consecutive positive integer $= n$

Second consecutive positive integer $= n + 1$

*Equation:*    $n(n + 1) = 132$

$n^2 + n = 132$

$n^2 + n - 132 = 0$

$(n + 12)(n - 11) = 0$

$n + 12 = 0 \Rightarrow n = -12$ and $n + 1 = -11$

$n - 11 = 0 \Rightarrow n = 11$ and $n + 1 = 12$

The problem specifies that these are positive integers, so we discard the negative answers. The two consecutive positive integers are 11 and 12.

**3.** *Verbal Model:* | First consecutive positive odd integer | $\cdot$ | Second consecutive positive odd integer | $=$ | 323 |

  *Labels:* First consecutive positive integer $= 2n + 1$

  Second consecutive positive integer $= 2n + 3$

  *Equation:*
$$(2n + 1)(2n + 3) = 323$$
$$4n^2 + 8n + 3 = 323$$
$$4n^2 + 8n - 320 = 0$$
$$n^2 + 2n - 80 = 0$$
$$(n + 10)(n - 8) = 0$$

$$n + 10 = 0 \Rightarrow n = -10 \Rightarrow 2n + 1 = -19 \text{ and } 2n + 3 = -17$$
$$n - 8 = 0 \Rightarrow n = 8 \Rightarrow 2n + 1 = 17 \text{ and } 2n + 3 = 19$$

The exercise specifies that these are positive integers, so we discard the negative solutions.
The two consecutive positive odd integers are 17 and 19.

**5.** *Verbal Model:* | First consecutive positive even integer | $\cdot$ | Second consecutive positive even integer | $=$ | 288 |

  *Labels:* First consecutive positive integer $= 2n$

  Second consecutive positive integer $= 2n + 2$

  *Equation:*
$$2n(2n + 2) = 288$$
$$4n^2 + 4n = 288$$
$$4n^2 + 4n - 288 = 0$$
$$n^2 + n - 72 = 0$$
$$(n + 9)(n - 8) = 0$$

$$n + 9 = 0 \Rightarrow n = -9 \Rightarrow 2n = -18 \text{ and } 2n + 2 = -16$$
$$n - 8 = 0 \Rightarrow n = 8 \Rightarrow 2n = 16 \text{ and } 2n + 2 = 18$$

The exercise specifies that these are positive integers, so we discard the negative solutions.
The two consecutive positive even integers are 16 and 18.

**7.** *Verbal Model:* $\left(\boxed{\begin{array}{c}\text{First consecutive}\\\text{positive integer}\end{array}}\right)^2$ $+$ $\left(\boxed{\begin{array}{c}\text{Second consecutive}\\\text{positive integer}\end{array}}\right)^2$ $=$ 113

  *Labels:* First consecutive positive integer $= n$

  Second consecutive positive integer $= n + 1$

  *Equation:*
$$n^2 + (n + 1)^2 = 113$$
$$n^2 + n^2 + 2n + 1 = 113$$
$$2n^2 + 2n + 1 = 113$$
$$2n^2 + 2n - 112 = 0$$
$$n^2 + n - 56 = 0 \qquad \text{Divide both sides by 2.}$$
$$(n + 8)(n - 7) = 0$$

$$n + 8 = 0 \Rightarrow n = -8 \text{ and } n + 1 = -7$$
$$n - 7 = 0 \Rightarrow n = 7 \text{ and } n + 1 = 8$$

The problem specifies that the integers are *positive*, so we discard the negative answers.
The two consecutive positive integers are 7 and 8.

**9.**  $h = h_0 - 16t^2$

$h_0 = 1600$ and $h = 0$

$0 = 1600 - 16t^2$

$0 = 16(100 - t^2)$

$0 = 16(10 + t)(10 - t)$

$16 \neq 0$

$10 + t = 0 \Rightarrow t = -10$

$10 - t = 0 \Rightarrow t = 10$

We choose the positive answer for this application. So, the time required for the object to fall to ground level is 10 seconds.

**11.**  $h = h_0 - 16t^2$

$h_0 = 1122$ and $h = 0$

$0 = 1122 - 16t^2$

$16t^2 = 1122$

$t^2 = \dfrac{1122}{16}$

$t = \pm\sqrt{\dfrac{1122}{16}}$

$t = \pm\dfrac{\sqrt{1122}}{4}$

$t \approx \pm8.37$

We choose the positive answer for this application. So, the time required for the object to fall to the ground is approximately 8.37 seconds.

**17.** *Formula:*  $(\text{Length})(\text{Width}) = \text{Area}$

$(2w)(w) = 50$

$2w^2 = 50$

$w^2 = 25$

$w = \pm\sqrt{25}$

$w = \pm5$

$w = 5$   Choose the positive answer.

$l = 2w = 10$

*Formula:* $2(\text{Length}) + 2(\text{Width}) = \text{Perimeter}$

$2 \cdot 10 + 2 \cdot 5 = P$

$20 + 10 = P$

$30 = P$

| Width | Length | Perimeter | Area |
|---|---|---|---|
| 5 ft | 10 ft | 30 ft | 50 ft$^2$ |

**13.**  $h = h_0 - 16t^2$

$h_0 = 555$ and $h = 0$

$0 = 555 - 16t^2$

$16t^2 = 555$

$t^2 = \dfrac{555}{16}$

$t = \pm\sqrt{\dfrac{555}{16}}$

$t \approx \pm5.89$

We choose the positive answer for this application. So, the time required for the object to fall to ground level is approximately 5.89 seconds.

**15.** *Formula:*  $2(\text{Length}) + 2(\text{Width}) = \text{Perimeter}$

$2l + 2(0.6l) = 64$

$2l + 1.2l = 64$

$3.2l = 64$

$l = 20$

$w = 0.6l = (0.6)(20) = 12$

*Formula:*  $(\text{Length})(\text{Width}) = \text{Area}$

$(20)(12) = A$

$240 = A$

| Width | Length | Perimeter | Area |
|---|---|---|---|
| 12 in. | 20 in. | 64 in. | 240 in.$^2$ |

**19.** *Formula:* $(\text{Length})(\text{Width}) = \text{Area}$

$$(l)\left(\tfrac{1}{4}l\right) = 100$$

$$\tfrac{1}{4}l^2 = 100$$

$$4\left(\tfrac{1}{4}l^2\right) = 4(100)$$

$$l^2 = 400$$

$$l = \pm\sqrt{400}$$

$$l = 20 \quad \text{Discard the negative answer.}$$

$$w = \tfrac{1}{4}l = \tfrac{1}{4}(20) = 5$$

*Formula:* $2(\text{Length}) + 2(\text{Width}) = \text{Perimeter}$

$$2(20) + 2(5) = P$$

$$40 + 10 = P$$

$$50 = P$$

| Width | Length | Perimeter | Area |
|---|---|---|---|
| 5 in. | 20 in. | 50 in. | 100 in$^2$ |

**21.** *Formula:* $2(\text{Length}) + 2(\text{Width}) = \text{Perimeter}$

$$2(w + 4) + 2(w) = 56$$

$$2w + 8 + 2w = 56$$

$$4w + 8 = 56$$

$$4w = 48$$

$$w = 12$$

$$l = w + 4 = 12 + 4 = 16$$

*Formula:* $(\text{Length})(\text{Width}) = \text{Area}$

$$(16)(12) = A$$

$$192 = A$$

| Width | Length | Perimeter | Area |
|---|---|---|---|
| 12 km | 16 km | 56 km | 192 km$^2$ |

**23.** *Formula:* $(\text{Length})(\text{Width}) = \text{Area}$

$$l(l - 10) = 75$$

$$l^2 - 10l = 75$$

$$l^2 - 10l - 75 = 0$$

$$(l - 15)(l + 5) = 0$$

$$l - 15 = 0 \Rightarrow l = 15 \text{ and } l - 10 = 5$$

$$l + 5 = 0 \Rightarrow l = -5 \qquad \text{Choose the positive answer.}$$

*Formula:* $2(\text{Length}) + 2(\text{Width}) = \text{Perimeter}$

$$2(15) + 2(5) = P$$

$$30 + 10 = P$$

$$40 = P$$

| Width | Length | Perimeter | Area |
|---|---|---|---|
| 5 m | 15 m | 40 m | 75 m$^2$ |

**25.** *Verbal Model:*  $\boxed{\text{Length of picture}} \cdot \boxed{\text{Width of picture}} = \boxed{\text{Area of picture}}$

*Labels:* 

Width of picture $= x$     (inches)

Length of picture $= x + 6$   (inches)

Area of picture $= 187$     (square inches)

*Equation:*

$$(x + 6)x = 187$$
$$x^2 + 6x = 187$$
$$x^2 + 6x - 187 = 0$$
$$(x + 17)(x - 11) = 0$$
$$x + 17 = 0 \Rightarrow x = -17 \text{ and } x + 6 = -11$$
$$x - 11 = 0 \Rightarrow x = 11 \text{ and } x + 6 = 17$$

We choose the positive solution for this application. The width of the picture is 11 inches and the length is 17 inches.

**27.** *Verbal Model:*  $\boxed{\text{Length of rectangle}} \cdot \boxed{\text{Width of rectangle}} = \boxed{\text{Area of rectangle}}$

*Labels:*

Area of rectangle $= 3750$     (square feet)

Width of rectangle $= w$     (feet)

Length of rectangle $= 175 - 2w$   (feet)

Note: $l + 2w = 175 \Rightarrow l = 175 - 2w$

*Equation:*

$$(175 - 2w)(w) = 3750$$
$$175w - 2w^2 = 3750$$
$$-2w^2 + 175w - 3750 = 0$$
$$2w^2 - 175w + 3750 = 0 \quad \text{Multiply both sides by } -1.$$
$$(2w - 75)(w - 50) = 0$$
$$2w - 75 = 0 \Rightarrow 2w = 75 \Rightarrow w = \tfrac{75}{2} \text{ and } 175 - 2w = 100$$
$$w - 50 = 0 \Rightarrow w = 50 \text{ and } 175 - 2w = 75$$

The storage area could be 37.5 by 100 ft *or* 50 ft by 75 ft.

**29.** *Verbal Model:*  $\boxed{\pi} \cdot \left(\boxed{\begin{array}{c}\text{Distance from station}\\\text{to farthest listener}\end{array}}\right)^{2} = \boxed{\begin{array}{c}\text{Area of}\\\text{broadcast range}\end{array}}$

*Labels:*

Distance from station to farthest listener $= x$     (miles)

Area of broadcast range $= 10,000$ square miles     (square miles)

*Equation:*

$$\pi x^2 = 10,000$$
$$x^2 = \frac{10,000}{\pi}$$
$$x = \pm\sqrt{\frac{10,000}{\pi}} \quad \text{Choose the positive answer.}$$
$$x = \sqrt{\frac{10,000}{\pi}}$$
$$x = \frac{100}{\sqrt{\pi}} = \frac{100\sqrt{\pi}}{\pi} \approx 56.4$$

The distance between the station and the listeners farthest from the station is approximately 56.4 miles.

**31.** *Verbal Model:*   $\dfrac{1}{2}$ $\boxed{\text{Base of triangle}}$ $\cdot$ $\boxed{\text{Height of triangle}}$ $=$ $\boxed{\text{Area of triangle}}$

*Labels:*   Base of triangle $= b$   (inches)

Height of triangle $= \dfrac{b}{3}$   (inches)

Area of picture $= 24$   (square inches)

*Equation:*   $\dfrac{1}{2}(b)\left(\dfrac{b}{3}\right) = 24$

$\dfrac{b^2}{6} = 24$

$b^2 = 144$

$b = \pm\sqrt{144}$

$b = \pm 12$

$b = 12 \Rightarrow \dfrac{b}{3} = 4$

$b = -12 \Rightarrow \dfrac{b}{3} = -4$

We choose the positive solution for this application. The triangle has a base of 12 inches and a height of 4 inches.

**33.**   $a^2 + b^2 = c^2$

$x^2 + 15^2 = 75^2$

$x^2 + 225 = 5625$

$x^2 = 5400$

$x = \pm\sqrt{5400}$

$x = \pm 30\sqrt{6}$

$x \approx 73.48$   Discard negative solution.

The distance from the boat to the dock is approximately 73.48 feet.

**35.** *Verbal Model:*   $\left(\boxed{\begin{array}{c}\text{Distance from pizza shop}\\ \text{to apartment complex}\end{array}}\right)^2$ $+$ $\left(\boxed{\begin{array}{c}\text{Distance from apartment}\\ \text{complex to furniture store}\end{array}}\right)^2$ $=$ $\left(\boxed{\begin{array}{c}\text{Distance from furniture}\\ \text{store to pizza shop}\end{array}}\right)^2$

*Labels:*   Distance from pizza shop to apartment complex $= x$   (miles)

Distance from apartment complex to furniture store $= 14 - x$   (miles)

Distance from furniture store to pizza shop $= 10$   (miles)

*Equation:*   $x^2 + (14 - x)^2 = 10^2$

$x^2 + 196 - 28x + x^2 = 100$

$2x^2 - 28x + 196 = 100$

$2x^2 - 28x + 96 = 0$

$x^2 - 14x + 48 = 0$

$(x - 6)(x - 8) = 0$

$x - 6 = 0 \Rightarrow x = 6$ and $14 - x = 8$

$x - 8 = 0 \Rightarrow x = 8$ and $14 - x = 6$

The distance from the pizza shop to the apartment complex is either 6 miles or 8 miles.

**37.** *Formula:*   $2l + 2w = p$

$$2l + 2w = 68$$

$$l + w = 34$$

$$l = 34 - w$$

*Formula:*   $a^2 + b^2 = c^2$

$$l^2 + w^2 = 26^2$$

$$(34 - w)^2 + w^2 = 26^2$$

$$1156 - 68w + w^2 + w^2 = 676$$

$$2w^2 - 68w + 1156 = 676$$

$$2w^2 - 68w + 480 = 0$$

$$w^2 - 34w + 240 = 0$$

$$(w - 24)(w - 10) = 0$$

$$w - 24 = 0 \Rightarrow w = 24 \text{ and } 34 - w = 10$$

$$w - 10 = 0 \Rightarrow w = 10 \text{ and } 34 - w = 24$$

The dimensions of the rectangle are 24 inches by 10 inches.

Note: You could use the quadratic formula instead of factoring to solve the equation.

**39.** *Verbal Model:*   | Area of upper rectangle | $+$ | Area of lower rectangle | $=$ | Total area |

*Labels:*   Area of upper rectangle $= x(x)$     (inches)

Area of lower rectangle $= x(x + 20)$   (inches)

Total area $= 2400$     (square inches)

*Equation:*   $x(x) + x(x + 20) = 2400$

$$x^2 + x^2 + 20x = 2400$$

$$2x^2 + 20x = 2400$$

$$2x^2 + 20x - 2400 = 0$$

$$x^2 + 10x - 1200 = 0$$

$$(x + 40)(x - 30) = 0$$

$$x + 40 = 0 \Rightarrow x = -40 \quad \text{Extraneous}$$

$$x - 30 = 0 \Rightarrow x = 30$$

We choose the positive answer for this application. The solution is $x = 30$ feet.

**41.** *Verbal Model:*   $\boxed{\text{Rate for faster person}} + \boxed{\text{Rate for slower person}} = \boxed{\text{Rate for both working together}}$

*Labels:*

Both people: time $= 4$   (hours)

$\text{rate} = \dfrac{1}{4}$   (tasks per hour)

Faster person: time $= x$   (hours)

$\text{rate} = \dfrac{1}{x}$   (tasks per hour)

Slower person: time $= x + 6$   (hours)

$\text{rate} = \dfrac{1}{(x+6)}$   (tasks per hour)

*Equation:*

$$\frac{1}{x} + \frac{1}{x+6} = \frac{1}{4}$$

$$4x(x+6)\left(\frac{1}{x}\right) + 4x(x+6)\left(\frac{1}{x+6}\right) = 4x(x+6)\left(\frac{1}{4}\right)$$

$$4(x+6) + 4x = x(x+6), \; x \neq 0, \; x \neq -6$$

$$4x + 24 + 4x = x^2 + 6x$$

$$8x + 24 = x^2 + 6x$$

$$0 = x^2 - 2x - 24$$

$$0 = (x-6)(x+4)$$

$$x - 6 = 0 \Rightarrow x = 6 \text{ and } x + 6 = 12$$

$$x + 4 = 0 \Rightarrow x = -4 \quad \text{Extraneous}$$

We choose the positive answer for $x$ in this application. The faster person could complete the job in 6 hours, and the slower person could complete the job in 12 hours.

**43.** *Verbal Model:* ⎡Rate for Combine A⎤ + ⎡Rate for Combine B⎤ = ⎡Rate of the two together⎤

*Labels:*    Time for Combine A to complete job $= x$    (hours)

Rate for Combine A $= \dfrac{1}{x}$    (jobs per hour)

Time for Combine B to complete job $= x + 2$    (hours)

Rate for Combine B $= \dfrac{1}{x + 2}$    (jobs per hour)

Time for the two combines to complete job $= 4$    (hours)

Rate for the two together $= \dfrac{1}{4}$    (jobs per hour)

*Equation:*

$$\frac{1}{x} + \frac{1}{x + 2} = \frac{1}{4}$$

$$4x(x + 2)\left(\frac{1}{x} + \frac{1}{x + 2}\right) = 4x(x + 2)\left(\frac{1}{4}\right)$$

$$4(x + 2)(1) + 4x(1) = x(x + 2), \quad x \neq 0, x \neq -2$$

$$4x + 8 + 4x = x^2 + 2x$$

$$8x + 8 = x^2 + 2x$$

$$0 = x^2 - 6x - 8$$

$a = 1, b = -6, c = -8$

$$x = \frac{-b \pm \sqrt{b^2 - 4ac}}{2a}$$

$$x = \frac{6 \pm \sqrt{(-6)^2 - 4(1)(-8)}}{2(1)}$$

$$x = \frac{6 \pm \sqrt{68}}{2}$$

$$x = \frac{6 + \sqrt{68}}{2} \approx 7.12 \Rightarrow x + 2 \approx 9.12$$

$$x = \frac{6 - \sqrt{68}}{2} \approx -1.12 \quad \text{Discard negative solution.}$$

Combine A would take approximately 7.12 hours and Combine B would take approximately 9.12 hours to harvest the field.

**45.** *Verbal Model:* $\boxed{\text{Final cost per person}} \cdot \boxed{\text{Final number of people}} = 100$

*Labels:*  Original number of people $= x$

Final number of people $= x + 5$

Original cost per person $= \dfrac{100}{x}$

Final cost per person $= \dfrac{100}{x} - 1$

*Equation:*  $\left(\dfrac{100}{x} - 1\right)(x + 5) = 100$

$\left(\dfrac{100 - x}{x}\right)(x + 5) = 100$

$(100 - x)(x + 5) = 100x, \; x \neq 0$

$100x + 500 - x^2 - 5x = 100x$

$-x^2 - 5x + 500 = 0$

$x^2 + 5x - 500 = 0$

$(x - 20)(x + 25) = 0$

$x - 20 = 0 \Rightarrow x = 20$ and $x + 5 = 25$

$x + 25 = 0 \Rightarrow x = -25$   Extraneous

We choose the positive solution for this application. So, 25 people are going to the game.

**47.** *Verbal Model:* $\boxed{\text{Final cost per person}} \cdot \boxed{\text{Number of bus riders}} = \boxed{\text{Total cost}}$

*Labels:*  Total cost $= 360$              (dollars)

Number of club members $= x$         (people)

Number of bus riders $= x + 6$        (people)

Original cost per person $= \dfrac{360}{x}$        (dollars per person)

Final cost per person $= \dfrac{360}{x} - 5$       (dollars per person)

*Equation:*  $\left(\dfrac{360}{x} - 5\right)(x + 6) = 360$

$\left(\dfrac{360 - 5x}{x}\right)(x + 6) = 360$

$(360 - 5x)(x + 6) = 360x, \;\; x \neq 0$

$360x + 2160 - 5x^2 - 30x = 360x$

$-5x^2 + 330x + 2160 = 360x$

$-5x^2 - 30x + 2160 = 0$

$x^2 + 6x - 432 = 0$

$(x + 24)(x - 18) = 0$

$x + 24 = 0 \Rightarrow x = -24$ and $x + 6 = -18$

$x - 18 = 0 \Rightarrow x = 18$ and $x + 6 = 24$

We choose the positive solution for this application. There are 24 people who are riding the bus to the science center.

**49.** *Verbal Model:*

$$\boxed{\begin{array}{c}\text{Travel time}\\\text{at faster speed}\end{array}} + \boxed{\begin{array}{c}\text{Travel time}\\\text{at slower speed}\end{array}} = \boxed{\begin{array}{c}\text{Total time}\\\text{for entire trip}\end{array}}$$

*Labels:*

| | |
|---|---|
| Distance at faster speed $= 200$ | (miles) |
| Faster speed $= x$ | (miles per hour) |
| Travel time at faster speed $= \dfrac{200}{x}$ | (hours) |
| Distance at slower speed $= 225$ | (miles) |
| Slower speed $= x - 5$ | (miles per hour) |
| Travel time at slower speed $= \dfrac{225}{x-5}$ | (hours) |
| Total time for entire trip $= 10$ | (hours) |

*Equation:*

$$\frac{200}{x} + \frac{225}{x-5} = 10$$

$$x(x-5)\left(\frac{200}{x}\right) + x(x-5)\left(\frac{225}{x-5}\right) = x(x-5)(10)$$

$$200(x-5) + 225x = 10x(x-5), \; x \neq 0, \, x \neq 5$$

$$200x - 1000 + 225x = 10x^2 - 50x$$

$$425x - 1000 = 10x^2 - 50x$$

$$0 = 10x^2 - 475x + 1000$$

$$0 = 2x^2 - 95x + 200 \qquad \text{Divide both sides by 5.}$$

$$x = \frac{-(-95) \pm \sqrt{(-95)^2 - 4(2)(200)}}{2(2)}$$

$$x = \frac{95 \pm \sqrt{7425}}{4}$$

$$x = \frac{95 \pm 15\sqrt{33}}{4}$$

$$x = \frac{95 + 15\sqrt{33}}{4} \approx 45.3 \text{ and } x - 5 \approx 40.3$$

$$x = \frac{95 - 15\sqrt{33}}{4} \approx 2.2 \text{ and } x - 5 \approx -2.8 \quad \text{Extraneous}$$

We choose the positive answers for *both* speeds. The two speeds are *approximately* 45.3 mi/hr and 40.3 mi/hr.

**51.**

$$A = P(1+r)^2$$

$$1123.60 = 1000(1+r)^2$$

$$\frac{1123.60}{1000} = \frac{1000(1+r)^2}{1000}$$

$$1.1236 = (1+r)^2$$

$$\pm\sqrt{1.1236} = 1 + r$$

$$\pm 1.06 = 1 + r$$

$$-1 \pm 1.06 = r$$

$$-1 + 1.06 = r \quad \text{Exclude negative solution.}$$

$$0.06 = r$$

The interest rate is 6%.

**53.**

$$A = P(1+r)^2$$

$$235.44 = 200(1+r)^2$$

$$\frac{235.44}{200} = \frac{200(1+r)^2}{200}$$

$$1.1772 = (1+r)^2$$

$$\pm\sqrt{1.1772} = 1 + r$$

$$\pm 1.085 \approx 1 + r$$

$$-1 \pm 1.085 \approx r$$

$$-1 + 1.085 \approx r \quad \text{Exclude negative solution.}$$

$$0.085 \approx r$$

The interest rate is approximately 8.5%.

**55.** (a)

(b) It appears that the year in which there were 36,000 movie screens in the U.S. was 2003.

(c) $36{,}000 = 63.86t^2 + 205.8t + 34{,}295$

$$0 = 63.86t^2 + 205.8t - 1705$$

$$t = \frac{-205.8 \pm \sqrt{(205.8)^2 - 4(63.86)(-1705)}}{2(63.86)}$$

$$t \approx \frac{-205.8 \pm 691.3}{127.72}$$

$$t \approx \frac{-205.8 + 691.3}{127.72} \approx 3.8$$

$$t \approx \frac{-205.8 - 691.3}{127.72} \approx -7.0$$

Choose the answer 3.8 that is in the domain. This verifies the answer that the year in which there were 36,000 movie screens in the United States was 2003.

**57.** To solve Exercise 17, you have to multiply the length and width. This product involves a squared term. The resulting equation is a quadratic equation.

**71.** $y = 2x^2 - 8x + 6 \Rightarrow a = 2, b = -8$

Vertex: $\quad x = -\dfrac{b}{2a} = -\dfrac{-8}{2(2)} = -\dfrac{-8}{4} = 2$

$$y = 2(2)^2 - 8(2) + 6 = 8 - 16 + 6 = -2$$

$(2, -2)$

# Section 10.6   Complex Numbers

**1.** $\sqrt{-4} = \sqrt{4(-1)} = \sqrt{4}\sqrt{-1} = 2i$

**3.** $-\sqrt{-144} = -\sqrt{144(-1)} = -\sqrt{144}\sqrt{-1} = -12i$

**5.** $\sqrt{\dfrac{-4}{25}} = \sqrt{\dfrac{4}{25}(-1)} = \sqrt{\dfrac{4}{25}}\sqrt{-1} = \dfrac{2}{5}i$

**7.** $\sqrt{-8} = \sqrt{4(2)(-1)} = \sqrt{4}\sqrt{2}\sqrt{-1} = 2\sqrt{2}i$

**9.** $\dfrac{\sqrt{-12}}{\sqrt{-3}} = \dfrac{\sqrt{4(3)(-1)}}{\sqrt{3(-1)}} = \dfrac{2\sqrt{3}i}{\sqrt{3}i} = 2$

**11.** $\sqrt{\dfrac{-18}{64}} = \sqrt{\dfrac{9(2)(-1)}{64}} = \dfrac{3\sqrt{2}}{8}i$

**59.** No. When you solve a quadratic equation with the Quadratic Formula, you simplify the radical $\sqrt{b^2 - 4ac}$. If the radical simplifies to a rational number, the equation has either one or two rational solutions, depending upon whether $b^2 - 4ac = 0$ or whether $b^2 - 4ac$ is a nonzero perfect square. If the radical simplifies to an irrational number, the equation has two irrational solutions. It is not possible for a quadratic equation to have a rational and an irrational solution.

**61.** The denominator $2x$ is zero when $x = 0$. So, the domain is all real values of $x$ such that $x \neq 0$.

**63.** The denominator $x^2 - 25 = (x + 5)(x - 5)$ is zero when $x = -5$ or $x = 5$. So, the domain is all real values of $x$ such that $x \neq -5$ and $x \neq 5$.

**65.** $\sqrt{4} = 2$

**67.** $\sqrt[3]{-64} = -4$

**69.** $y = x^2 + 5 \Rightarrow a = 1, b = 0$

Vertex: $\quad x = -\dfrac{b}{2a} = -\dfrac{0}{2(1)} = 0$

$$y = (0)^2 + 5 = 5$$

$(0, 5)$

**13.** $\sqrt{-16} + \sqrt{-36} = 4i + 6i = (4 + 6)i = 10i$

**15.** $\sqrt{-50} - \sqrt{-8} = \sqrt{25(2)(-1)} - \sqrt{(4)(2)(-1)}$

$$= 5\sqrt{2}i - 2\sqrt{2}i$$

$$= \left(5\sqrt{2} - 2\sqrt{2}\right)i$$

$$= 3\sqrt{2}i$$

**17.** $\sqrt{-75} + \sqrt{-27} = \sqrt{25(3)(-1)} + \sqrt{9(3)(-1)}$

$$= 5\sqrt{3}i + 3\sqrt{3}i$$

$$= \left(5\sqrt{3} + 3\sqrt{3}\right)i$$

$$= 8\sqrt{3}i$$

**19.** $\sqrt{-48} + \sqrt{-12} - \sqrt{-27} = \sqrt{16(3)(-1)} + \sqrt{4(3)(-1)} - \sqrt{9(3)(-1)}$

$$= 4\sqrt{3}i + 2\sqrt{3}i - 3\sqrt{3}i$$

$$= (4 + 2 - 3)\sqrt{3}i$$

$$= 3\sqrt{3}i$$

**21.** $\sqrt{-35}\sqrt{-35} = \sqrt{35(-1)}\sqrt{35(-1)}$

$$= \left(\sqrt{35}\,i\right)\left(\sqrt{35}\,i\right)$$

$$= 35 \cdot i^2$$

$$= 35(-1)$$

$$= -35$$

**23.** $\sqrt{-18}\sqrt{-3} = \left(3\sqrt{2}i\right)\left(\sqrt{3}i\right) = 3\sqrt{6} \cdot i^2 = -3\sqrt{6}$

**25.** $\sqrt{-3}\left(\sqrt{-3} + \sqrt{-4}\right) = \sqrt{3}i\left(\sqrt{3}i + 2i\right)$

$$= \left(\sqrt{3}i\right)^2 + 2\sqrt{3}i^2$$

$$= -3 - 2\sqrt{3}$$

**27.** $\sqrt{-5}\left(\sqrt{-16} - \sqrt{-10}\right) = \sqrt{5}i\left(4i - \sqrt{10}i\right)$

$$= 4\sqrt{5}i^2 - \sqrt{50}i^2$$

$$= -4\sqrt{5} + 5\sqrt{2}$$

**29.** $\sqrt{-2}\left(3 - \sqrt{-8}\right) = \sqrt{2}i\left(3 - 2\sqrt{2}i\right)$

$$= 3\sqrt{2}i - 2(2)i^2$$

$$= 3\sqrt{2}i + 4$$

$$= 4 + 3\sqrt{2}i$$

**31.** $\sqrt{64} + \sqrt{-25} \stackrel{?}{=} 8 + 5i$

$\sqrt{8^2} + \sqrt{5^2(-1)} \stackrel{?}{=} 8 + 5i$

$$8 + 5i = 8 + 5i$$

**33.** $\sqrt{27} - \sqrt{-8} \stackrel{?}{=} 3\sqrt{3} + 2\sqrt{2}i$

$\sqrt{3^2(3)} - \sqrt{2^2(2)(-1)} \stackrel{?}{=} 3\sqrt{3} + 2\sqrt{2}i$

$$3\sqrt{3} - 2\sqrt{2}i \neq 3\sqrt{3} + 2\sqrt{2}i$$

**35.** $-4 - \sqrt{-8} = a + bi$

$$-4 - 2\sqrt{2}i = a + bi$$

$$-4 = a \quad -2\sqrt{2} = b$$

**37.** $3a + \sqrt{-81} = 15 + 3bi$

$$3a + 9i = 15 + 3bi$$

$$3a = 15 \qquad 9 = 3b$$

$$a = 5 \qquad 3 = b$$

**39.** $8 + 2bi = 2a + \sqrt{-49}$

$$8 + 2bi = 2a + 7i$$

$$2a = 8 \qquad 2b = 7$$

$$a = 4 \qquad b = \frac{7}{2}$$

**41.** $(4 - 3i) + (6 + 7i) = (4 + 6) + (-3 + 7)i = 10 + 4i$

**43.** $(-4 - 7i) + (-10 - 33i) = (-4 - 10) + (-7 - 33)i$

$$= -14 - 40i$$

**45.** $13i - (14 - 7i) = (-14) + (13 + 7)i = -14 + 20i$

**47.** $(30 - i) - (18 + 6i) + 3 = (30 - 18 + 3) + (-1 - 6)i$

$$= 15 - 7i$$

**49.** $6 - (3 - 4i) + (4 + 2i) = (6 - 3 + 4) + \left[-(-4) + 2\right]i$

$$= 7 + 6i$$

**51.** $(15 - 3i) + \left(5 - \sqrt{-12}\right) - \sqrt{-81} = (15 - 3i) + \left(5 - 2\sqrt{3}i\right) - 9i$

$$= (15 + 5) + \left(-3 - 2\sqrt{3} - 9\right)i$$

$$= 20 + \left(-12 - 2\sqrt{3}\right)i$$

$$= 20 - \left(12 + 2\sqrt{3}\right)i$$

**53.** $(3i)(12i) = 36i^2 = -36$

**55.** $(3i)(-8i) = -24i^2 = -24(-1) = 24$

**57.** $(-6i)(-i)(6i) = 36i^3 = 36(-i) = -36i$

**59.** $5i(13 + 2i) = 65i + 10i^2 = 65i - 10 = -10 + 65i$

**61.** $(9 - 2i)(\sqrt{-4}) = (9 - 2i)(2i)$

$\qquad = 18i - 4i^2$

$\qquad = 18i + 4$

$\qquad = 4 + 18i$

**63.** $(-4 - 5i)(4 - 2i) = -16 + 8i - 20i + 10i^2$

$\qquad = -16 - 12i - 10$

$\qquad = -26 - 12i$

**65.** $(2 + 8i)(5 - 6i) = 10 - 12i + 40i - 48i^2$

$\qquad = 10 + 28i + 48$

$\qquad = 58 + 28i$

**67.** $(3 + 5i)(3 - 5i) = 9 - 15i + 15i - 25i^2$

$\qquad = 9 + 25$

$\qquad = 34$

**69.** $i^7 = (i^2)^3 i = (-1)^3 i = -i$

**71.** $i^{18} = (i^2)^9 = (-1)^9 = -1$

**73.** $i^9 = (i^2)^4 i = (-1)^4 i = i$

**75.** $(-i)^6 = i^6 = (i^2)^3 = (-1)^3 = -1$

**77.** $2 + i$, conjugate $= 2 - i$

$\qquad$ product $= (2 + i)(2 - i)$

$\qquad\qquad = 2^2 - i^2$

$\qquad\qquad = 4 + 1$

$\qquad\qquad = 5$

**79.** $-2 - 8i$, conjugate $= -2 + 8i$

$\qquad$ product $= (-2 - 8i)(-2 + 8i)$

$\qquad\qquad = (-2)^2 - (8i)^2$

$\qquad\qquad = 4 - 64i^2 = 4 + 64 = 68$

**81.** $5 - \sqrt{6}i$, conjugate $= 5 + \sqrt{6}i$

$\qquad$ product $= (5 - \sqrt{6}i)(5 + \sqrt{6}i)$

$\qquad\qquad = 5^2 - (\sqrt{6}i)^2$

$\qquad\qquad = 25 - 6i^2 = 25 + 6 = 31$

**83.** $10i$, conjugate $= -10i$

$\qquad$ product $= (10i)(-10i)$

$\qquad\qquad = -(10i)^2$

$\qquad\qquad = -100i^2$

$\qquad\qquad = 100$

**85.** $\dfrac{2 + i}{-5i} = \dfrac{2 + i}{-5i} \cdot \dfrac{5i}{5i}$

$\qquad = \dfrac{10i + 5i^2}{-25i^2}$

$\qquad = \dfrac{-5 + 10i}{25}$

$\qquad = -\dfrac{1}{5} + \dfrac{2}{5}i$

**87.** $\dfrac{6 + 7i}{8i} = \dfrac{6 + 7i}{8i} \cdot \dfrac{-8i}{-8i}$

$\qquad = \dfrac{-48i - 56i^2}{-64i^2}$

$\qquad = \dfrac{-48i - 56(-1)}{-64(-1)}$

$\qquad = \dfrac{-48i + 56}{64}$

$\qquad = \dfrac{-48}{64}i + \dfrac{56}{64}$

$\qquad = \dfrac{7}{8} - \dfrac{3}{4}i$

**89.** $\dfrac{-12}{2 + 7i} = \dfrac{-12}{2 + 7i} \cdot \dfrac{2 - 7i}{2 - 7i} = \dfrac{-12(2 - 7i)}{4 + 49}$

$\qquad = \dfrac{-12(2 - 7i)}{53}$

$\qquad = \dfrac{-24 + 84i}{53}$

$\qquad = -\dfrac{24}{53} + \dfrac{84}{53}i$

**91.** $\dfrac{3i}{5 + 2i} = \dfrac{3i}{5 + 2i} \cdot \dfrac{5 - 2i}{5 - 2i}$

$\qquad = \dfrac{15i - 6i^2}{5^2 + 2^2}$

$\qquad = \dfrac{6 + 15i}{29}$

$\qquad = \dfrac{6}{29} + \dfrac{15}{29}i$

**93.** $\dfrac{5 - i}{5 + i} = \dfrac{5 - i}{5 + i} \cdot \dfrac{5 - i}{5 - i}$

$\qquad = \dfrac{25 - 10i + i^2}{5^2 + 1}$

$\qquad = \dfrac{25 - 10i - 1}{26}$

$\qquad = \dfrac{24 - 10i}{26}$

$\qquad = \dfrac{12}{13} - \dfrac{5}{13}i$

**95.** $\dfrac{4 + 5i}{3 - 7i} = \dfrac{4 + 5i}{3 - 7i} \cdot \dfrac{3 + 7i}{3 + 7i}$

$\qquad = \dfrac{12 + 28i + 15i + 35i^2}{3^2 + 7^2}$

$\qquad = \dfrac{12 + 43i - 35}{58}$

$\qquad = \dfrac{-23 + 43i}{58}$

$\qquad = -\dfrac{23}{58} + \dfrac{43}{58}i$

**97.** (a) $x = -1 + 2i$

$\qquad (-1 + 2i)^2 + 2(-1 + 2i) + 5 \overset{?}{=} 0$

$\qquad 1 - 4i + 4i^2 - 2 + 4i + 5 \overset{?}{=} 0$

$\qquad\qquad 1 - 4 - 2 + 5 \overset{?}{=} 0$

$\qquad\qquad\qquad 0 = 0 \quad \text{Solution}$

(b) $x = -1 - 2i$

$\qquad (-1 - 2i)^2 + 2(-1 - 2i) + 5 \overset{?}{=} 0$

$\qquad 1 + 4i + 4i^2 - 2 - 4i + 5 \overset{?}{=} 0$

$\qquad\qquad 1 - 4 - 2 + 5 \overset{?}{=} 0$

$\qquad\qquad\qquad 0 = 0 \quad \text{Solution}$

**99.** (a) $x = -4$

$\qquad (-4)^3 + 4(-4)^2 + 9(-4) + 36 \overset{?}{=} 0$

$\qquad\qquad -64 + 64 - 36 + 36 \overset{?}{=} 0$

$\qquad\qquad\qquad 0 = 0 \quad \text{Solution}$

(b) $x = -3i$

$\qquad (-3i)^3 + 4(-3i)^2 + 9(-3i) + 36 \overset{?}{=} 0$

$\qquad -27t^3 + 36t^2 - 27i + 36 \overset{?}{=} 0$

$\qquad 27i - 36 - 27i + 36 \overset{?}{=} 0$

$\qquad\qquad\qquad 0 = 0 \quad \text{Solution}$

**101.** $z^2 = -36$

$\qquad z = \pm\sqrt{-36}$

$\qquad z = \pm 6i$

**103.** $x^2 + 32 = 0$

$\qquad x^2 = -32$

$\qquad x = \pm\sqrt{-32}$

$\qquad x = \pm\sqrt{16(2)(-1)}$

$\qquad x = \pm 4\sqrt{2}i$

**105.** $9u^2 + 17 = 0$

$\qquad 9u^2 = -17$

$\qquad u = \pm\sqrt{-\dfrac{17}{9}}$

$\qquad\quad = \pm\dfrac{\sqrt{17}}{3}i$

**107.** $(t - 3)^2 = -25$

$\qquad t - 3 = \pm\sqrt{-25}$

$\qquad\quad t = 3 \pm 5i$

**109.** $(2x - 5)^2 = -54$

$\qquad 2x - 5 = \pm\sqrt{-54}$

$\qquad\quad 2x = 5 \pm 3\sqrt{6}i$

$\qquad\quad x = \dfrac{5}{2} \pm \dfrac{3\sqrt{6}}{2}i$

**111.** $9(x + 6)^2 = -121$

$\qquad (x + 6)^2 = \dfrac{-121}{9}$

$\qquad x + 6 = \pm\sqrt{\dfrac{-121}{9}}$

$\qquad\quad x = -6 \pm \dfrac{11}{3}i$

**113.** $2(9x - 4)^2 + 50 = 0$

$\qquad 2(9x - 4)^2 = -50$

$\qquad (9x - 4)^2 = -25$

$\qquad 9x - 4 = \pm\sqrt{-25}$

$\qquad 9x - 4 = \pm 5i$

$\qquad\quad 9x = 4 \pm 5i$

$\qquad\quad x = \dfrac{4 \pm 5i}{9}$

$\qquad\quad x = \dfrac{4}{9} \pm \dfrac{5}{9}i$

**115.**
$$x^2 + 10 = 6x$$
$$x^2 - 6x + 10 = 0$$
$$x^2 - 6x = -10$$
$$x^2 - 6x + 9 = -10 + 9 \quad \text{Half of } -6 \text{ is } -3, \text{ and } (-3)^2 = 9.$$
$$(x - 3)^2 = -1$$
$$x - 3 = \pm\sqrt{-1}$$
$$x = 3 \pm i$$

**117.** $-x^2 + x - 1 = 0$
$$x^2 - x + 1 = 0$$
$$x^2 - x + \frac{1}{4} = -1 + \frac{1}{4} \quad \text{Half of } -1 \text{ is } -\frac{1}{2}, \text{ and } \left(-\frac{1}{2}\right)^2 = \frac{1}{4}.$$
$$\left(x - \frac{1}{2}\right)^2 = -\frac{3}{4}$$
$$x - \frac{1}{2} = \pm\sqrt{-\frac{3}{4}}$$
$$x = \frac{1}{2} \pm \frac{\sqrt{3}}{2}i$$
$$x \approx 0.50 + 0.87i$$
$$x \approx 0.50 - 0.87i$$

**119.** $4z^2 - 3z + 2 = 0$
$$z^2 - \frac{3}{4}z = \frac{-2}{4}$$
$$z^2 - \frac{3}{4}z + \frac{9}{64} = \frac{-1}{2} + \frac{9}{64} \quad \text{Half of } -\frac{3}{4} \text{ is } -\frac{3}{8}, \text{ and } \left(-\frac{3}{8}\right)^2 = \frac{9}{64}.$$
$$\left(z - \frac{3}{8}\right)^2 = \frac{-23}{64}$$
$$z - \frac{3}{8} = \pm\sqrt{\frac{-23}{64}}$$
$$z = \frac{3}{8} \pm \frac{\sqrt{-23}}{8}$$
$$z = \frac{3}{8} \pm \frac{\sqrt{23}}{8}i$$
$$z \approx 0.38 + 0.60i$$
$$z \approx 0.38 - 0.60i$$

**121.** $x^2 - 8x + 19 = 0$

$$x = \frac{-(-8) \pm \sqrt{(-8)^2 - 4(1)(19)}}{2(1)}$$

$$x = \frac{8 \pm \sqrt{64 - 76}}{2}$$

$$x = \frac{8 \pm \sqrt{-12}}{2}$$

$$x = \frac{8 \pm 2\sqrt{3}i}{2}$$

$$x = 4 \pm \sqrt{3}i$$

**123.** $2x^2 + 3x + 3 = 0$

$$x = \frac{-3 \pm \sqrt{3^2 - 4(2)(3)}}{2(2)}$$

$$x = \frac{-3 \pm \sqrt{9 - 24}}{4}$$

$$x = \frac{-3 \pm \sqrt{-15}}{4}$$

$$x = \frac{-3 \pm \sqrt{15}i}{4}$$

$$x = -\frac{3}{4} \pm \frac{\sqrt{15}}{4}i$$

**125.** $5x^2 - 5x + 7 = 0$

$$a = 5, b = -5, c = 7$$

$$x = \frac{-(-5) \pm \sqrt{(-5)^2 - 4(5)(7)}}{2(5)}$$

$$x = \frac{5 \pm \sqrt{25 - 140}}{10}$$

$$x = \frac{5 \pm \sqrt{-115}}{10}$$

$$x = \frac{5 \pm \sqrt{115}i}{10}$$

$$x = \frac{5}{10} \pm \frac{\sqrt{115}}{10}i$$

$$x = \frac{1}{2} \pm \frac{\sqrt{115}}{10}i$$

**127. (a)**

$$\left(\frac{-5 + 5\sqrt{3}i}{2}\right)^3 = \left(\frac{-5}{2} + \frac{5}{2}\sqrt{3}i\right)^2\left(\frac{-5}{2} + \frac{5}{2}\sqrt{3}i\right)$$

$$= \left(\frac{25}{4} - \frac{25}{2}\sqrt{3}i + \frac{25}{4}(3)i^2\right)\left(\frac{-5}{2} + \frac{5}{2}\sqrt{3}i\right)$$

$$= \left(\frac{25}{4} - \frac{25}{2}\sqrt{3}i - \frac{75}{4}\right)\left(\frac{-5}{2} + \frac{5}{2}\sqrt{3}i\right)$$

$$= \left(\frac{-50}{4} - \frac{25}{2}\sqrt{3}i\right)\left(\frac{-5}{2} + \frac{5}{2}\sqrt{3}i\right)$$

$$= \left(\frac{-25}{2} - \frac{25}{2}\sqrt{3}i\right)\left(\frac{-5}{2} + \frac{5}{2}\sqrt{3}i\right)$$

$$= \frac{125}{4} - \frac{125}{4}\sqrt{3}i + \frac{125}{4}\sqrt{3}i - \frac{125}{4}(3)i^2$$

$$= \frac{125}{4} + \frac{375}{4}$$

$$= \frac{500}{4} = 125$$

**(b)** Use same method as part (a).

$$\left(\frac{-5 - 5\sqrt{3}i}{2}\right)^3 = 125$$

**129.** (a) $1, \dfrac{-1 + \sqrt{3}i}{2}, \dfrac{-1 - \sqrt{3}i}{2}$

(b) $2, \dfrac{-2 + 2\sqrt{3}i}{2} = -1 + \sqrt{3}i, \dfrac{-2 - 2\sqrt{3}i}{2} = -1 - \sqrt{3}i$

(c) $4, \dfrac{-4 + 4\sqrt{3}i}{2} = -2 + 2\sqrt{3}i, \dfrac{-4 - 4\sqrt{3}i}{2} = -2 - 2\sqrt{3}i$

**131.** $(a + bi) + (a - bi) = (a + a) + (b - b)i$
$$= 2a + 0i$$
$$= 2a$$

**133.** $(a + bi) - (a - bi) = (a - a) + (b + b)i$
$$= 0 + 2bi$$
$$= 2bi$$

**135.** The imaginary numbers were not written in $i$-form before they were multiplied.
$$\sqrt{-3}\sqrt{-3} = \left(\sqrt{3}i\right)\left(\sqrt{3}i\right) = 3i^2 = -3$$

**137.** $(x + i)(x - i) = x^2 - ix + ix - i^2$
$$= x^2 - (-1)$$
$$= x^2 + 1$$

So, the polynomial $x^2 + 1$ can be factored using complex numbers:
$$x^2 + 1 = (x + i)(x - i)$$

**139.** (a) The complex number $a + bi$ is a real number when $a$ is a real number and $b = 0$.

(b) The complex number $a + bi$ is an imaginary number when $a$ and $b$ are real numbers and $b \neq 0$.

(c) The complex number $a + bi$ is a pure imaginary number when $a = 0$ and $b$ is a real number, $b \neq 0$.

**141.** Standard form: $13x^2 - x + 7$

Degree: 2

Leading coefficient: 13

**143.** Standard form: $5x^4 - 2x^3 + 3x + 7$

Degree: 4

Leading coefficient: 5

**145.** $16 - \sqrt{t} = 7$
$$-\sqrt{t} = -9$$
$$\left(-\sqrt{t}\right)^2 = (-9)^2$$
$$t = 81$$

**147.** $\sqrt{x - 4} = 6$
$$\left(\sqrt{x - 4}\right)^2 = 6^2$$
$$x - 4 = 36$$
$$x = 40$$

**149.** $\sqrt{2x - 15} - 2 = 3$
$$\sqrt{2x - 15} = 5$$
$$\left(\sqrt{2x - 15}\right)^2 = 5^2$$
$$2x - 15 = 25$$
$$2x = 40$$
$$x = 20$$

**151.** $\sqrt{x + 2} = \sqrt{5x - 7}$
$$\left(\sqrt{x + 2}\right)^2 = \left(\sqrt{5x - 7}\right)^2$$
$$x + 2 = 5x - 7$$
$$-4x + 2 = -7$$
$$-4x = -9$$
$$x = \dfrac{-9}{-4}$$
$$x = \dfrac{9}{4}$$

# Section 10.7   Relations, Functions, and Graphs

**1.** Domain: $\{-4, 1, 2, 4\}$

Range: $\{-3, 2, 3, 5\}$

**3.** Domain: $\{-9, \frac{1}{2}, 2\}$

Range: $\{-10, 0, 16\}$

**5.** Domain: $\{-1, 1, 5, 8\}$

Range: $\{-7, -2, 3, 4\}$

**7.** Yes. No first component has two different second components, so this relation *is* a function.

9. No. Since one of the first components, –2, is paired with two second components, this relation is *not* a function.

11. Yes. No first component has two different second components, so this relation *is* a function.

13. No. One first component, 0, is paired with two second components, so the relation is *not* a function.

15. No. Since first components, CBS and NBC, are paired with three second components, this relation is *not* a function.

17. Yes. No first component has two different second components, so this relation *is* a function.

19. Yes. No first component has two different second components, so the relation *is* a function.

21. No. Two first components, 1 and 3, are paired with two second components, so the relation is *not* a function.

23. Yes. No first component has two different second components, so the relation *is* a function.

25. Yes. No first component has two different second components, so this relation *is* a function.

27. Yes. This graph indicates that *y is* a function of *x* because *no* vertical line intersects more than one point on the graph.

29. Yes. This graph indicates that *y is* a function of *x* because *no* vertical line intersects more than one point on the graph.

31. Yes. This graph indicates that *y is* a function of *x* because *no* vertical line intersects more than one point on the graph.

33. No. This graph indicates that *y* is *not* a function of *x*. Some vertical lines intersect the graph at more than one point.

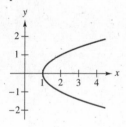

35. No. This graph indicates that *y* is *not* a function of *x* because some vertical lines intersect the graph at more than one point.

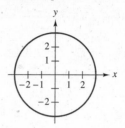

37. (a) $f(2) = \frac{1}{2}(2) = 1$

 (b) $f(5) = \frac{1}{2}(5) = \frac{5}{2}$

 (c) $f(-4) = \frac{1}{2}(-4) = -2$

 (d) $f\left(-\frac{2}{3}\right) = \frac{1}{2}\left(-\frac{2}{3}\right) = -\frac{1}{3}$

39. (a) $f(0) = 2(0) - 1 = -1$

 (b) $f(3) = 2(3) - 1 = 5$

 (c) $f(-3) = 2(-3) - 1 = -7$

 (d) $f\left(-\frac{1}{2}\right) = 2\left(-\frac{1}{2}\right) - 1 = -1 - 1 = -2$

41. (a) $h(200) = \frac{1}{4}(200) - 1 = 49$

 (b) $h(-12) = \frac{1}{4}(-12) - 1 = -4$

 (c) $h(8) = \frac{1}{4}(8) - 1 = 1$

 (d) $h\left(-\frac{5}{2}\right) = \frac{1}{4}\left(-\frac{5}{2}\right) - 1 = -\frac{13}{8}$

43. (a) $f(-4) = \frac{1}{2}(-4)^2 = \frac{1}{2}(16) = 8$

 (b) $f(4) = \frac{1}{2}(4)^2 = \frac{1}{2}(16) = 8$

 (c) $f(0) = \frac{1}{2}(0)^2 = \frac{1}{2}(0) = 0$

 (d) $f(2) = \frac{1}{2}(2)^2 = \frac{1}{2}(4) = 2$

45. (a) $f(1) = 4(1)^2 + 2 = 4 \cdot 1 + 2 = 6$

 (b) $f(-1) = 4(-1)^2 + 2 = 4 \cdot 1 + 2 = 6$

 (c) $f(-4) = 4(-4)^2 + 2 = 4 \cdot 16 + 2 = 66$

 (d) $f\left(-\frac{3}{2}\right) = 4\left(-\frac{3}{2}\right)^2 + 2 = 4 \cdot \frac{9}{4} + 2 = 11$

**47.** (a)  $g(0) = 2(0)^2 - 3(0) + 1 = 0 - 0 + 1 = 1$

(b)  $g(-2) = 2(-2)^2 - 3(-2) + 1 = 8 + 6 + 1 = 15$

(c)  $g(1) = 2(1)^2 - 3(1) + 1 = 2 - 3 + 1 = 0$

(d)  $g\left(\frac{1}{2}\right) = 2\left(\frac{1}{2}\right)^2 - 3\left(\frac{1}{2}\right) + 1 = \frac{1}{2} - \frac{3}{2} + 1 = 0$

**49.** (a)  $g(2) = |2 + 2| = |4| = 4$

(b)  $g(-2) = |-2 + 2| = |0| = 0$

(c)  $g(10) = |10 + 2| = |12| = 12$

(d)  $g\left(-\frac{5}{2}\right) = \left|-\frac{5}{2} + 2\right| = \left|-\frac{1}{2}\right| = \frac{1}{2}$

**51.** (a)  $h(0) = 0^3 - 1 = -1$

(b)  $h(1) = 1^3 - 1 = 0$

(c)  $h(3) = 3^3 - 1 = 26$

(d)  $h\left(\frac{1}{2}\right) = \left(\frac{1}{2}\right)^3 - 1 = \frac{1}{8} - 1 = -\frac{7}{8}$

**53.**  $D = \{0, 1, 2, 3, 4\}$

**55.**  $D = \{-8, -6, 2, 5, 12\}$

**57.**  $D = \{-5, -4, -3, -2, -1\}$

**59.** The domain is the set of all real numbers $r$ such that $r > 0$.

**61.** (a)  $f(10) = 20 - 0.5(10)$     $f(15) = 20 - 0.5(15)$

$\qquad\quad = 20 - 5$ $\qquad\qquad\quad = 20 - 7.5$

$\qquad\quad = 15$ $\qquad\qquad\qquad = 12.5$

(b) As the price increases, the demand decreases.

**63.** (a)  $d(2) = 50(2) = 100$

When $t = 2$ hours, the car will travel 100 miles.

(b)  $d(4) = 50(4) = 200$

When $t = 4$ hours, the car will travel 200 miles.

(c)  $d(10) = 50(10) = 500$

When $t = 10$ hours, the car will travel 500 miles.

**65.** Yes, the high school enrollment is a function of the year.

**67.**  $f(2001) = 15,000,000$

**69.**  $P = 4s$

Yes, $P$ *is* a function of $s$. If you make a table of values where $s > 0$, no first component will have two different second components.

**71.** (a) Yes, $L$ is a function of $t$.

(b) The range is approximately $9 \le L \le 15$.

**73.** Yes. There are many correct examples
Here are two:

| Domain | Range |
|--------|-------|
| 1 | 4 |
| 2 | 5 |
| 3 | 6 |

$\{(2, 5), (4, 8), (4, 3)\}$

**75.** Yes, the number of elements in the domain can be greater than the number of elements in the range. Here are two examples:

$f: \{(1, 5), (2, 5), (3, 5), (4, 5)\}$

$f(x) = 10$

**77.** *Verbal Model:*   $\boxed{5}$ = $\boxed{\text{what percent of } 35}$

*Label:*      Unknown percent $($in decimal form$)$ = $p$

*Equation:*      $5 = p(35)$

$\qquad\qquad\quad \frac{5}{35} = p$

$\qquad\qquad\quad 0.143... \approx p$

$\qquad\qquad\quad$ 5 is approximately 14.3% of 35.

**79.** *Verbal Model:*   $\boxed{215}$ = $\boxed{125\% \text{ of what number}}$

*Label:*      Unknown number = $b$

*Equation:*      $215 = 1.25b$

$\qquad\qquad\quad \frac{215}{1.25} = b$

$\qquad\qquad\quad 172 = b$

$\qquad\qquad\quad$ 215 is 125% of 172.

**81.** $\sqrt{-4} + \sqrt{-100} = 2i + 10i = 12i$

**83.** $(3 - 8i) + (8 - 4i) = (3 + 8) + (-8 - 4)i = 11 - 12i$

**85.** $(9i)(i) = 9i^2 = 9(-1) = -9$

**87.** $2i(-5 + 8i) = -10i + 16i^2$
$$= -10i + 16(-1)$$
$$= -16 - 10i$$

## Review Exercises for Chapter 10

**1.** $x^2 + 10x = 0$
$x(x + 10) = 0$
$x = 0$
$x + 10 = 0 \Rightarrow x = -10$
The solutions are 0 and −10.

**3.** $x^2 - 9x + 18 = 0$
$(x - 6)(x - 3) = 0$
$x - 6 = 0 \Rightarrow x = 6$
$x - 3 = 0 \Rightarrow x = 3$

**5.** $8z^2 - 32 = 0$
$8(z^2 - 4) = 0$
$8(z - 2)(z + 2) = 0$
$8 \neq 0$
$z - 2 = 0 \Rightarrow z = 2$
$z + 2 = 0 \Rightarrow z = -2$
The solutions are −2 and 2.

**7.** $x^2 = 49$
$x = \pm\sqrt{49}$
$x = \pm 7$
The solutions are 7 and −7.

**9.** $y^2 - 18 = 0$
$y^2 = 18$
$y = \pm\sqrt{18}$
$y = \pm 3\sqrt{2}$
The solutions are $3\sqrt{2}$ and $-3\sqrt{2}$.

**11.** $x^2 - 48 = 0$
$x^2 = 48$
$x = \pm\sqrt{48}$
$x = \pm 4\sqrt{3}$
The solutions are $4\sqrt{3}$ and $-4\sqrt{3}$.

**13.** $(x - 5)^2 = 3$
$x - 5 = \pm\sqrt{3}$
$x = 5 \pm\sqrt{3}$
The solutions are $5 + \sqrt{3}$ and $5 - \sqrt{3}$.

**15.** $(x - 2)^2 - 6 = 0$
$(x - 2)^2 = 6$
$x - 2 = \pm\sqrt{6}$
$x = 2 \pm\sqrt{6}$
The solutions are $2 + \sqrt{6}$ and $2 - \sqrt{6}$.

**17.** $8(x - 4)^2 - 32 = 0$
$8(x - 4)^2 = 32$
$(x - 4)^2 = 4$
$x - 4 = \pm\sqrt{4}$
$x - 4 = \pm 2$
$x = 4 \pm 2$
$x = 4 + 2 = 6$
$x = 4 - 2 = 2$
The solutions are 6 and 2.

**19.** $2(x + 4)^2 - 16 = 0$
$2(x + 4)^2 = 16$
$(x + 4)^2 = 8$
$x + 4 = \pm\sqrt{8}$
$x + 4 = \pm 2\sqrt{2}$
$x = -4 \pm 2\sqrt{2}$
The solutions are $-4 + 2\sqrt{2}$ and $-4 - 2\sqrt{2}$.

**21.** $9(x - 7)^2 - 25 = 25$

$$9(x - 7)^2 = 50$$

$$(x - 7)^2 = \frac{50}{9}$$

$$x - 7 = \pm\sqrt{\frac{50}{9}}$$

$$x - 7 = \pm\frac{5\sqrt{2}}{3}$$

$$x = 7 \pm \frac{5\sqrt{2}}{3}$$

$$x = \frac{21}{3} \pm \frac{5\sqrt{2}}{3} \Rightarrow x = \frac{21 \pm 5\sqrt{2}}{3}$$

The solutions are $\dfrac{21 + 5\sqrt{2}}{3}$ and $\dfrac{21 - 5\sqrt{2}}{3}$.

**23.** $x^2 + 12x + \boxed{36}$

Half of 12 is 6, and $6^2 = 36$.

$$x^2 + 12x + 36 = (x + 6)^2$$

**25.** $t^2 - 15t + \boxed{\dfrac{225}{4}}$

Half of $-15$ is $-\frac{15}{2}$, and $\left(-\frac{15}{2}\right)^2 = \frac{225}{4}$.

$$t^2 - 15t + \tfrac{225}{4} = \left(t - \tfrac{15}{2}\right)^2$$

**27.** $x^2 - 6x - 1 = 0$

$$x^2 - 6x = 1$$

$$x^2 - 6x + 9 = 1 + 9 \quad \text{Half of } -6 \text{ is } -3 \text{ and } (-3)^2 = 9.$$

$$(x - 3)^2 = 10$$

$$x - 3 = \pm\sqrt{10}$$

$$x = 3 \pm \sqrt{10}$$

The solutions are $3 + \sqrt{10}$ and $3 - \sqrt{10}$.

**29.** $x^2 - x - 1 = 0$

$$x^2 - x = 1$$

$$x^2 - x + \frac{1}{4} = 1 + \frac{1}{4} \quad \text{Half of } -1 \text{ is } -\frac{1}{2} \text{ and } \left(-\frac{1}{2}\right)^2 = \frac{1}{4}.$$

$$\left(x - \frac{1}{2}\right)^2 = \frac{5}{4}$$

$$x - \frac{1}{2} = \pm\sqrt{\frac{5}{4}}$$

$$x - \frac{1}{2} = \pm\frac{\sqrt{5}}{\sqrt{4}}$$

$$x - \frac{1}{2} = \pm\frac{\sqrt{5}}{2}$$

$$x = \frac{1}{2} \pm \frac{\sqrt{5}}{2}$$

$$x = \frac{1 \pm \sqrt{5}}{2}$$

The solutions are $\dfrac{1 + \sqrt{5}}{2}$ and $\dfrac{1 - \sqrt{5}}{2}$.

**31.** $2y^2 + 10y + 5 = 0$

$$2y^2 + 10y = -5$$

$$y^2 + 5y = -\frac{5}{2}$$

$$y^2 + 5y + \frac{25}{4} = -\frac{5}{2} + \frac{25}{4} \quad \text{Half of 5 is } \frac{5}{2} \text{ and } \left(\frac{5}{2}\right)^2 = \frac{25}{4}.$$

$$\left(y + \frac{5}{2}\right)^2 = -\frac{10}{4} + \frac{25}{4}$$

$$\left(y + \frac{5}{2}\right)^2 = \frac{15}{4}$$

$$y + \frac{5}{2} = \pm\sqrt{\frac{15}{4}}$$

$$y + \frac{5}{2} = \pm\frac{\sqrt{15}}{2}$$

$$y = -\frac{5}{2} \pm \frac{\sqrt{15}}{2}$$

$$y = \frac{-5 \pm \sqrt{15}}{2}$$

The solutions are $\dfrac{-5 + \sqrt{15}}{2}$ and $\dfrac{-5 - \sqrt{15}}{2}$.

**33.** $4x^2 - 2x - 1 = 0$

$$x^2 - \frac{1}{2}x - \frac{1}{4} = 0$$

$$x^2 - \frac{1}{2}x = \frac{1}{4}$$

$$x^2 - \frac{1}{2}x + \left(\frac{1}{4}\right)^2 = \frac{1}{4} + \frac{1}{16} \quad \text{Half of } -\frac{1}{2} \text{ is } -\frac{1}{4} \text{ and } \left(-\frac{1}{4}\right)^2 = \frac{1}{16}.$$

$$\left(x - \frac{1}{4}\right)^2 = \frac{5}{16}$$

$$x - \frac{1}{4} = \pm\sqrt{\frac{5}{16}}$$

$$x - \frac{1}{4} = \pm\frac{\sqrt{5}}{4}$$

$$x = \frac{1}{4} \pm \frac{\sqrt{5}}{4}$$

$$x = \frac{1 \pm \sqrt{5}}{4}$$

The solutions are $\dfrac{1 + \sqrt{5}}{4}$ and $\dfrac{1 - \sqrt{5}}{4}$.

**35.** $y^2 + y - 42 = 0$

$a = 1, b = 1, c = -42$

$$y = \frac{-1 \pm \sqrt{1^2 - 4(1)(-42)}}{2(1)}$$

$$y = \frac{-1 \pm \sqrt{169}}{2}$$

$$y = \frac{-1 \pm 13}{2}$$

$$y = \frac{-1 + 13}{2} = \frac{12}{2} = 6$$

$$y = \frac{-1 - 13}{2} = \frac{-14}{2} = -7$$

The solutions are 6 and $-7$.

**37.** $c^2 - 6c + 5 = 0$

$$c = \frac{-(-6) \pm \sqrt{(-6)^2 - 4(1)(5)}}{2(1)}$$

$$c = \frac{6 \pm \sqrt{16}}{2}$$

$$c = \frac{6 \pm 4}{2}$$

$$c = \frac{6 + 4}{2} = \frac{10}{2} = 5$$

$$c = \frac{6 - 4}{2} = \frac{2}{2} = 1$$

The solutions are 1 and 5.

**39.** $-c^2 + 6c - 6 = 0$

$$c = \frac{-6 \pm \sqrt{6^2 - 4(-1)(-6)}}{2(-1)}$$

$$c = \frac{-6 \pm \sqrt{36 - 24}}{-2}$$

$$c = \frac{-6 \pm \sqrt{12}}{-2}$$

$$c = \frac{-6 \pm 2\sqrt{3}}{-2}$$

$$c = \frac{\cancel{2}\left(3 \pm \sqrt{3}\right)}{\cancel{2}(1)}$$

$$c = 3 \pm \sqrt{3}$$

The solutions are $3 + \sqrt{3}$ and $3 - \sqrt{3}$.

**41.** $2y^2 + y - 42 = 0$

$a = 2, b = 1, c = -42$

$$y = \frac{-1 \pm \sqrt{1^2 - 4(2)(-42)}}{2(2)}$$

$$y = \frac{-1 \pm \sqrt{337}}{4}$$

The solutions are $\dfrac{-1 + \sqrt{337}}{4}$ and $\dfrac{-1 - \sqrt{337}}{4}$.

**43.** $9x^2 + 30x + 25 = 0$

$$x = \frac{-30 \pm \sqrt{30^2 - 4(9)(25)}}{2(9)}$$

$$x = \frac{-30 \pm \sqrt{900 - 900}}{18}$$

$$x = \frac{-30 \pm \sqrt{0}}{18}$$

$$x = \frac{-30 \pm 0}{18}$$

$$x = \frac{-30}{18} = -\frac{5}{3}$$

The (repeated) solution is $-\dfrac{5}{3}$.

**45.** $v^2 = 250$

$v^2 - 250 = 0$

$a = 1, b = 0, c = -250$

$$x = \frac{-0 \pm \sqrt{0^2 - 4(1)(250)}}{2(1)}$$

$$x = \frac{0 \pm \sqrt{0 + 1000}}{2}$$

$$x = \frac{0 \pm \sqrt{1000}}{2}$$

$$x = \frac{0 \pm 10\sqrt{10}}{2}$$

$$x = \frac{\pm 10\sqrt{10}}{2}$$

$$x = \pm 5\sqrt{10}$$

The solutions are $5\sqrt{10}$ and $-5\sqrt{10}$.

**47.** $0.3t^2 - 2t + 1 = 0$

$a = 0.3, b = -2, c = 1$

$$t = \frac{-(-2) \pm \sqrt{(-2)^2 - 4(0.3)(1)}}{2(0.3)}$$

$$t = \frac{2 \pm \sqrt{2.8}}{0.6}$$

$$t = \frac{2 + \sqrt{2.8}}{0.6} \approx 6.12$$

$$t = \frac{2 - \sqrt{2.8}}{0.6} \approx 0.54$$

The solutions are *approximately* 6.12 and 0.54.

Note: If you multiply both sides of the original equation by 10 to obtain $3t^2 - 20t + 10 = 0$, the solutions will

be in the *equivalent* form $t = \dfrac{10 \pm \sqrt{70}}{3}$.

**49.** $0.7x^2 - 0.14x + 0.007 = 0$

$a = 0.7, b = -0.14, c = 0.007$

$$x = \frac{0.14 \pm \sqrt{(-0.14)^2 - 4(0.7)(0.007)}}{2(0.7)}$$

$$x = \frac{0.14 \pm \sqrt{0}}{1.4}$$

$$x = \frac{0.14}{1.4} = 0.1$$

The (repeated) solution is 0.1.

**51.**
$$\frac{1}{x} + \frac{1}{x+1} = \frac{1}{2}$$

$$2x(x+1)\left(\frac{1}{x}\right) + 2x(x+1)\left(\frac{1}{x+1}\right) = 2x(x+1)\left(\frac{1}{2}\right)$$

$$2(x+1) + 2x = x(x+1), \; x \neq 0, x \neq -1$$

$$2x + 2 + 2x = x^2 + x$$

$$4x + 2 = x^2 + x$$

$$0 = x^2 - 3x - 2$$

$$x = \frac{-(-3) \pm \sqrt{(-3)^2 - 4(1)(-2)}}{2(1)}$$

$$x = \frac{3 \pm \sqrt{17}}{2}$$

The solutions are $\dfrac{3 + \sqrt{17}}{2}$ and $\dfrac{3 - \sqrt{17}}{2}$.

**53.** $\sqrt{2x + 5} = x - 3$

$$\left(\sqrt{2x+5}\right)^2 = (x-3)^2$$

$$2x + 5 = x^2 - 6x + 9$$

$$0 = x^2 - 8x + 4$$

$$x = \frac{8 \pm \sqrt{(-8)^2 - 4(1)(4)}}{2(1)}$$

$$x = \frac{8 \pm \sqrt{48}}{2}$$

$$x = \frac{8 \pm 4\sqrt{3}}{2}$$

$$x = 4 \pm 2\sqrt{3}$$

The solution is $4 + 2\sqrt{3}$. The other answer is extraneous.

**55.** $y = x^2 - 9x + 3 \Rightarrow a = 1$

$a > 0 \Rightarrow$ The parabola opens upward.

**57.** $y = 6 - 5x - 7x^2 \Rightarrow a = -7$

$a < 0 \Rightarrow$ The parabola opens downward.

**59.** $y = 3 - (x + 4)^2$

$y = 3 - (x^2 + 8x + 16)$

$y = 3 - x^2 - 8x - 16$

$y = -x^2 - 8x - 13 \Rightarrow a = -1$

$a < 0 \Rightarrow$ The parabola opens downward.

**61.** $y = x^2 - 2x + 1$

*Leading coefficient test:*

Since $a > 0$, the parabola opens upward.

*Vertex:* $(a = 1, b = -2)$

$$x = -\frac{b}{2a} = -\frac{-2}{2(1)} = 1$$

$$y = 1^2 - 2(1) + 1 = 1 - 2 + 1 = 0$$

The vertex is located at $(1, 0)$.

*y-intercept:* (Let $x = 0$.)

$$y = 0^2 - 2(0) + 1 = 1$$

The $y$-intercept is $(0, 1)$.

*x-intercept:* (Let $y = 0$.)

$$0 = x^2 - 2x + 1 \Rightarrow (x - 1)^2 = 0 \Rightarrow x - 1 = 0$$

$$x - 1 = 0 \Rightarrow x = 1$$

The $x$-intercept is $(1, 0)$.

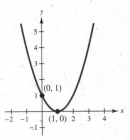

*Additional solution points:*

| $x$ | $-2$ | $-1$ | $2$ | $3$ |
|---|---|---|---|---|
| $y = x^2 - 2x + 1$ | 9 | 4 | 1 | 4 |
| Points | $(-2, 9)$ | $(-1, 4)$ | $(2, 1)$ | $(3, 4)$ |

**63.** $y = -x^2 + 4x - 3$

*Leading coefficient test:*

Since $a < 0$, the parabola opens downward.

*Vertex:* $(a = -1, b = 4)$

$$x = -\frac{b}{2a} = -\frac{4}{2(-1)} = -(-2) = 2$$

$$y = -(2)^2 + 4(2) - 3 = 1$$

The vertex is located at $(2, 1)$.

*y-intercept:* (Let $x = 0$.)

$$y = -0^2 + 4(0) - 3 = -3;$$ the $y$-intercept is $(0, -3)$.

*x-intercept:* (Let $y = 0$.)

$$0 = -x^2 + 4x - 3 = x^2 - 4x + 3 = (x - 3)(x - 1)$$

$$x - 3 = 0 \Rightarrow x = 3$$

$$x - 1 = 0 \Rightarrow x = 1$$

The $x$-intercepts are $(3, 0)$ and $(1, 0)$.

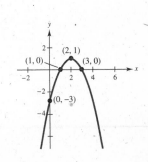

*Additional solution points:*

| $x$ | $-1$ | $4$ | $5$ |
|---|---|---|---|
| $y = -x^2 + 4x - 3$ | $-8$ | $-3$ | $-8$ |
| Points | $(-1, -8)$ | $(4, -3)$ | $(5, -8)$ |

**65.** $y = -x^2 + 3x$

*Leading coefficient test:* Since $a < 0$, the parabola opens downward.

*Vertex:* $(a = -1, b = 3)$

$$x = -\frac{b}{2a} = -\frac{3}{2(-1)} = -\left(-\frac{3}{2}\right) = \frac{3}{2}$$

$$y = -\left(\frac{3}{2}\right)^2 + 3\left(\frac{3}{2}\right) = -\frac{9}{4} + \frac{9}{2} = \frac{9}{4}$$

The vertex is located at $\left(\frac{3}{2}, \frac{9}{4}\right)$.

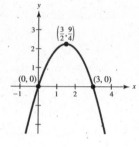

*y-intercept:* (Let $x = 0$.) $\quad y = -0^2 + 3 \cdot 0 = 0$; the $y$-intercept is $(0, 0)$.

*x-intercepts:* (Let $y = 0$.) $\quad 0 = -x^2 + 3x \Rightarrow x = -x(x - 3)$

$-x = 0 \Rightarrow x = 0$

$x - 3 = 0 \Rightarrow x = 3$

The $x$-intercepts are $(0, 0)$ and $(3, 0)$.

*Additional solution points:*

| $x$ | 4 | 5 | $-2$ | $-1$ |
|---|---|---|---|---|
| $y = -x^2 + 3x$ | $-4$ | $-10$ | $-10$ | $-4$ |
| Points | $(4, -4)$ | $(5, -10)$ | $(-2, -10)$ | $(-1, -4)$ |

**67.** $y = \frac{1}{4}\left(4x^2 - 4x + 3\right)$ or $y = x^2 - x + \frac{3}{4}$

*Leading coefficient test:* Since $a > 0$, the parabola opens upward.

*Vertex:* $(a = 1, b = -1)$

$$x = -\frac{b}{2a} = -\frac{-1}{2(1)} = \frac{1}{2}$$

$$y = \left(\frac{1}{2}\right)^2 - \left(\frac{1}{2}\right) + \frac{3}{4} = \frac{1}{4} - \frac{1}{2} + \frac{3}{4} = \frac{1}{2}$$

The vertex is located at $\left(\frac{1}{2}, \frac{1}{2}\right)$.

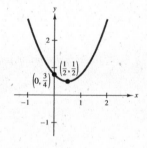

*y-intercept:* (Let $x = 0$.) $\quad y = 0^2 - 0 + \frac{3}{4} = \frac{3}{4}$; the $y$-intercept is $\left(0, \frac{3}{4}\right)$.

*x-intercepts:* (Let $y = 0$.) $\quad 0 = x^2 - x + \frac{3}{4}$ or $0 = 4x^2 - 4x + 3$

$$x = \frac{-(-4) \pm \sqrt{(-4)^2 - 4(4)(3)}}{2(4)} = \frac{4 \pm \sqrt{-32}}{8}$$

There are no $x$-intercepts.

*Additional solution points:*

| $x$ | 2 | 1 | $-1$ |
|---|---|---|---|
| $y = \frac{1}{4}(4x - 4x + 3)$ | 2.75 | 0.75 | 2.75 |
| Points | $(2, 2.75)$ | $(1, 0.75)$ | $(-1, 2.75)$ |

**69.** $y = 2x^2 + 4x + 5$

*Leading coefficient test:*    Since $a > 0$, the parabola opens upward.

*Vertex:* $(a = 2, b = 4)$

$$x = -\frac{b}{2a} = -\frac{4}{2(2)} = -1$$

$$y = 2(-1)^2 + 4(-1) + 5 = 2 - 4 + 5 = 3$$

The vertex is located at $(-1, 3)$.

*y-intercept:* (Let $x = 0$.)    $y = 2 \cdot 0^2 + 4 \cdot 0 + 5 = 5$; the y-intercept is $(0, 5)$.

*x-intercepts:* (Let $y = 0$.)    $0 = 2x^2 + 4x + 5$

$$x = \frac{-4 \pm \sqrt{4^2 - 4(2)(5)}}{2(2)} = \frac{-4 \pm \sqrt{-24}}{4}$$

There are no x-intercepts.

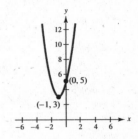

*Additional solution points:*

| $x$ | 1 | $-2$ | $-3$ |
|---|---|---|---|
| $y = 2x^2 + 4x + 5$ | 11 | 5 | 11 |
| Points | $(1, 11)$ | $(-2, 5)$ | $(-3, 11)$ |

**71.** $y = -\left(3x^2 - 4x - 2\right) = -3x^2 + 4x + 2$

*Leading coefficient test:*    Since $a < 0$, the parabola opens downward.

*Vertex:* $(a = -3, b = 4)$

$$x = -\frac{b}{2a} = -\frac{4}{2(-3)} = \frac{2}{3}$$

$$y = -3\left(\frac{2}{3}\right)^2 + 4\left(\frac{2}{3}\right) + 2 = -\frac{4}{3} + \frac{8}{3} + 2 = \frac{10}{3}$$

The vertex is located at $\left(\frac{2}{3}, \frac{10}{3}\right)$.

*y-intercept:* (Let $x = 0$.)    $y = -3(0)^2 + 4(0) + 2 = 2$ The y-intercept is $(0, 2)$.

*x-intercepts:* (Let $y = 0$.)    $0 = -3x^2 + 4x + 2$

$$x = \frac{-4 \pm \sqrt{(4)^2 - 4(-3)(2)}}{2(-3)} = \frac{-4 \pm \sqrt{16 + 24}}{-6}$$

$$x = \frac{-4 \pm \sqrt{40}}{-6} = \frac{-4 \pm 2\sqrt{10}}{-6} = \frac{2 \pm \sqrt{10}}{3}$$

The x-intercepts are $\left(\frac{2 + \sqrt{10}}{3}, 0\right)$ and $\left(\frac{2 + \sqrt{10}}{3}, 0\right)$.

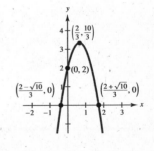

*Additional solution points:*

| $x$ | $-2$ | $-1$ | 1 | 2 |
|---|---|---|---|---|
| $y = -3x^2 + 4x + 2$ | $-18$ | $-5$ | 3 | $-2$ |
| Points | $(-2, -18)$ | $(-1, -5)$ | $(1, 3)$ | $(2, -2)$ |

**73.**

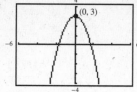

The vertex appears to be $(0, 3)$.

**75.**

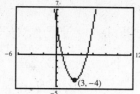

The vertex appears to be $(3, -4)$.

**77.** *Verbal model:* First consecutive positive integer $\cdot$ Second consecutive positive integer $=$ 240

*Labels:*  First consecutive positive integer $= n$

Second consecutive positive integer $= n + 1$

*Equation:*
$$n(n + 1) = 240$$
$$n^2 + n = 240$$
$$n^2 + n - 240 = 0$$
$$(n + 16)(n - 15) = 0$$
$$n + 16 = 0 \Rightarrow n = -16 \text{ and } n + 1 = -15$$
$$n - 15 = 0 \Rightarrow n = 15 \text{ and } n + 1 = 16$$

The problem specifies that these are *positive* integers, so we discard the negative answers. The two consecutive positive integers are 15 and 16.

**79.** $2332.80 = 2000(1 + r)^2$

$$\frac{2332.80}{2000} = (1 + r)^2$$
$$1.1664 = (1 + r)^2$$
$$\sqrt{1.1664} = 1 + r$$
$$1.08 = 1 + r$$
$$0.08 = r$$

The interest rate is 8%.

**81.**  $h = 48 - 16t^2$ and $h = 0$

$$0 = 48 - 16t^2$$
$$16t^2 = 48$$
$$t^2 = 3$$
$$t = \pm\sqrt{3}$$
$$t \approx \pm 1.73$$

We choose the positive answer for this application. The object strikes the ground in approximately 1.73 seconds. The object was dropped from a height of 48 feet.

**83.** $(\text{Length})(\text{Width}) = \text{Area}$

$$(x + 10)(x - 4) = 32$$
$$x^2 + 6x - 40 = 32$$
$$x^2 + 6x - 72 = 0$$
$$(x + 12)(x - 6) = 0$$
$$x + 12 = 0 \Rightarrow x = -12 \quad \text{Extraneous}$$
$$x - 6 = 0 \Rightarrow x = 6$$

So $x = 6$ centimeters.

**85.** Area of large rectangle: $l \cdot w = (2x + 28)(x)$

Area of each small rectangle: $l \cdot w = (x)(x)$

Total area: $(2x + 28)(x) + 2(x)(x) = 1800$

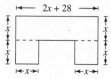

$(2x + 28)(x) + 2(x)(x) = 1800$

$2x^2 + 28x + 2x^2 = 1800$

$4x^2 + 28x - 1800 = 0$

$x^2 + 7x - 450 = 0$

$$x = \frac{-7 \pm \sqrt{7^2 - 4(1)(-450)}}{2(1)}$$

$$x = \frac{-7 \pm \sqrt{49 + 1800}}{2}$$

$$x = \frac{-7 \pm \sqrt{1849}}{2}$$

$$x = \frac{-7 \pm 43}{2}$$

$$x = \frac{-7 + 43}{2} = \frac{36}{2} = 18$$

$$x = \frac{-7 - 43}{2} = \frac{-50}{2} = -25$$

We choose the positive answer for this application. The length $x$ is 18 meters.

**87.** *Formula:* $\frac{1}{2}(\text{Base})(\text{Height}) = \text{Area}$

$$\frac{1}{2}(x)\left(\frac{3}{2}x\right) = 60$$

$$\frac{3}{4}x^2 = 60$$

$$x^2 = 80$$

$$x = \pm\sqrt{80}$$

$$x = \pm 4\sqrt{5} \qquad \text{Discard negative solution.}$$

$$x = 4\sqrt{5} \approx 8.94$$

$$\frac{3}{2}x = \frac{3}{2}\left(4\sqrt{5}\right) = 6\sqrt{5} \approx 13.42$$

The base of the triangle is approximately 8.94 inches and the height is approximately 13.42 inches.

**89.** Formula: $2l + 2w = 94$

$$l + w = 47$$

$$l = 47 - w$$

Formula: $\qquad a^2 + b^2 = c^2$

$$l^2 + w^2 = 37^2$$

$$(47 - w)^2 + w^2 = 1369$$

$$2209 - 94w + w^2 + w^2 = 1369$$

$$2w^2 - 94w + 840 = 0$$

$$w^2 - 47w + 420 = 0$$

$$w = \frac{-(-47) \pm \sqrt{(-47)^2 - 4(1)(420)}}{2(1)}$$

$$w = \frac{47 \pm \sqrt{2209 - 1680}}{2}$$

$$w = \frac{47 \pm \sqrt{529}}{2}$$

$$w = \frac{47 \pm 23}{2}$$

$$w = \frac{47 + 23}{2} = 35 \text{ and } 47 - w = 12$$

$$w = \frac{47 - 23}{2} = 12 \text{ and } 47 - w = 35$$

The dimensions of the driveway are 12 feet by 35 feet.

**91.** *Verbal Model:* $\boxed{\begin{array}{c}\text{Rate of}\\\text{faster person}\end{array}} + \boxed{\begin{array}{c}\text{Rate of}\\\text{slower person}\end{array}} = \boxed{\begin{array}{c}\text{Rate of two}\\\text{working together}\end{array}}$

*Labels:* Both people: time = 10 (hours); rate = $\dfrac{1}{10}$(task per hour)

Faster person: time = $x$ (hours); rate = $\dfrac{1}{x}$(task per hour)

Slower person: time = $x + 4$ (hours); rate = $\dfrac{1}{x+4}$(task per hour)

*Equation:*

$$\frac{1}{x} + \frac{1}{x+4} = \frac{1}{10}$$

$$10x(x+4)\left(\frac{1}{x}\right) + 10x(x+4)\left(\frac{1}{x+4}\right) = 10x(x+4)\left(\frac{1}{10}\right)$$

$$10(x+4) + 10x = x(x+4), \ x \neq 0, \ x \neq -4$$

$$10x + 40 + 10x = x^2 + 4x$$

$$20x + 40 = x^2 + 4x$$

$$0 = x^2 - 16x - 40$$

$$x = \frac{-(-16) \pm \sqrt{(-16)^2 - 4(1)(-40)}}{2(1)}$$

$$x = \frac{16 \pm \sqrt{416}}{2}$$

$$x = \frac{16 \pm 4\sqrt{26}}{2}$$

$$x = 8 \pm 2\sqrt{26}$$

$$x = 8 + 2\sqrt{26} \qquad \text{Choose the positive answer.}$$

$$x \approx 18.2 \text{ and } x + 4 \approx 22.2$$

The faster person could complete the task in approximately 18.2 hours and the slower person could complete the task in approximately 22.2 hours.

**93.** *Verbal Model:* $\boxed{\begin{array}{c}\text{Final cost}\\\text{per person}\end{array}} \cdot \boxed{\begin{array}{c}\text{Final number}\\\text{of people}\end{array}} = 72$

*Labels:* Original number of people = $x$

Final number of people = $x + 3$

Original cost per person = $\dfrac{72}{x}$

Final cost per person = $\dfrac{72}{x} - 1.20$

*Equation:*

$$\left(\frac{72}{x} - 1.20\right)(x + 3) = 72$$

$$\left(\frac{72 - 1.20x}{x}\right)(x + 3) = 72$$

$$(72 - 1.20x)(x + 3) = 72x, \ x \neq 0$$

$$72x + 216 - 1.20x^2 - 3.60x = 72x$$

$$-1.20x^2 - 3.60x + 216 = 0$$

$$x^2 + 3x - 180 = 0$$

$$(x - 12)(x + 15) = 0$$

$$x - 12 = 0 \Rightarrow x = 12 \text{ and } x + 3 = 15$$

$$x + 15 = 0 \Rightarrow x = -15 \qquad \text{Extraneous}$$

We choose the positive solution for this application. So, 15 people are going to the game.

**95.** *Verbal model:*   $\boxed{\begin{array}{c}\text{Travel time at}\\\text{slower speed}\end{array}} + \boxed{\begin{array}{c}\text{Travel time at}\\\text{faster speed}\end{array}} = \boxed{\begin{array}{c}\text{Total time}\\\text{for entire trip}\end{array}}$

*Labels:*

Distance at slower speed $= 165$ (miles)

Rate at slower speed $= x$ (miles per hour)

Time at slower speed $= \dfrac{165}{x}$ (hours)

Distance at faster speed $= 300$ (miles)

Rate at faster speed $= x + 5$ (miles per hour)

Time at faster speed $= \dfrac{300}{x + 5}$ (hours)

Total time for entire trip $= 8$ (hours)

*Equation:*

$$\frac{165}{x} + \frac{300}{x + 5} = 8$$

$$x(x + 5)\left(\frac{165}{x} + \frac{300}{x + 5}\right) = (x)(x + 5)(8)$$

$$165(x + 5) + 300(x) = 8x(x + 5), \; x \neq 0, \; x \neq -5$$

$$165x + 825 + 300x = 8x^2 + 40x$$

$$465x + 825 = 8x^2 + 40x$$

$$-8x^2 + 425x + 825 = 0$$

$$8x^2 - 425x - 825 = 0$$

$$x = \frac{425 \pm \sqrt{(-425)^2 - 4(8)(-825)}}{2(8)}$$

$$x = \frac{425 \pm \sqrt{207.025}}{16}$$

$$x = \frac{425 \pm 455}{16}$$

$$x = \frac{425 + 455}{16} \Rightarrow x = 55 \text{ and } x + 5 = 60$$

$$x = \frac{425 - 455}{16} \Rightarrow x = -1.875$$

We choose the positive solution. The two average speeds of the train are 55 miles per hour and 60 miles per hour.

**97.** $y = -\dfrac{1}{10}x^2 + 3x + 3$

(a) When the ball is thrown, $x = 0$.

$$y = -\frac{1}{10}(0)^2 + 3(0) + 3 = 3$$

The ball is thrown from a height of 3 feet.

(b) The maximum height occurs at the vertex.

$$x = -\frac{b}{2a} = -\frac{3}{2\left(-\frac{1}{10}\right)} = \frac{3}{\left(\frac{1}{5}\right)} = 15$$

$$y = -\frac{1}{10}(15)^2 + 3(15) + 3 = -22.5 + 45 + 3 = 25.5$$

The maximum height is 25.5 feet.

(c) When the ball strikes the ground, $y = 0$.

$$0 = -\frac{1}{10}x^2 + 3x + 3$$

$$10(0) = 10\left(-\frac{1}{10}x^2 + 3x + 3\right)$$

$$0 = -x^2 + 30x + 30$$

$$x = \frac{-30 \pm \sqrt{30^2 - 4(-1)(30)}}{2(-1)}$$

$$x = \frac{-30 \pm \sqrt{1020}}{-2}$$

$$x = \frac{-30 \pm 2\sqrt{255}}{-2}$$

$$x = 15 \pm \sqrt{255}$$

$$x = 15 + \sqrt{255} \Rightarrow x \approx 31.0$$

$$x = 15 - \sqrt{255} \Rightarrow x \approx -1.0 \qquad \text{Extraneous}$$

The ball strikes the ground approximately 31.0 feet from the child.

**99.** (a)

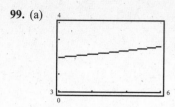

(b) It appears that the year in which there were $2.5 billion in sales was 2005.

(c) $2.5 = 0.0125t^2 + 0.103t + 1.58$

$0 = 0.0125t^2 + 0.103t - 0.92$

$$t = \frac{-0.103 \pm \sqrt{(0.103)^2 - 4(0.0125)(-0.92)}}{2(0.0125)}$$

$$t \approx \frac{-0.103 \pm 0.238}{0.025}$$

$$t \approx \frac{-0.103 + 0.238}{0.025} \approx 5.4$$

$$t \approx \frac{-0.103 - 0.238}{0.025} \approx -13.6$$

The answer in the domain is 5.4. This verifies the year 2005.

**101.** $\sqrt{-81} = \sqrt{9^2(-1)} = 9i$

**103.** $\sqrt{-5} = \sqrt{5(-1)} = \sqrt{5}i$

**105.** $\dfrac{\sqrt{-50}}{\sqrt{-2}} = \dfrac{\sqrt{25(2)(-1)}}{\sqrt{2(-1)}}$

$= \dfrac{5\sqrt{2}i}{\sqrt{2}i}$

$= \dfrac{5\cancel{\sqrt{2}i}}{\cancel{\sqrt{2}i}}$

$= 5$

**107.** $\sqrt{-81} + \sqrt{-36} = 9i + 6i = 15i$

**109.** $\sqrt{-121} - \sqrt{-84} = 11i - 2\sqrt{21}i = \left(11 - 2\sqrt{21}\right)i$

**111.** $\sqrt{-5}\sqrt{-5} = i\sqrt{5} \cdot i\sqrt{5} = i^2 \cdot 5 = -5$

**113.** $\sqrt{-10}\left(\sqrt{-4} - \sqrt{-7}\right) = i\sqrt{10}\left(2i - i\sqrt{7}\right)$

$= 2i^2\sqrt{10} - i^2\sqrt{70}$

$= -2\sqrt{10} + \sqrt{70}$

**115.** $12 - 5i = (a + 2) + (b - 1)i$

$12 = a + 2 \qquad -5 = b - 1$

$10 = a \qquad\qquad -4 = b$

**117.** $\sqrt{-49} + 4 = a + bi$

$7i + 4 = a + bi$

$4 + 7i = a + bi$

$a = 4 \quad b = 7$

**119.** $(-4 + 5i) - (-12 + 8i) = (-4 + 12) + (5 - 8)i$

$= 8 - 3i$

**121.** $(3 - 5i) + (7 + 12i) = (3 + 7) + (-5 + 12)i$

$= 10 + 7i$

**123.** $(4 - 3i)(4 + 3i) = 4^2 - (3i)^2 = 16 + 9 = 25$

**125.** $(6 - 5i)^2 = 6^2 - 2(6)(5i) + (5i)^2$

$= 36 - 60i - 25$

$= 11 - 60i$

**127.** $\dfrac{7}{3i} = \dfrac{7}{3i} \cdot \dfrac{-i}{-i} = \dfrac{-7i}{-3i^2} = \dfrac{-7i}{3} = -\dfrac{7}{3}i$

**129.** $\dfrac{4i}{2 - 8i} = \dfrac{4i}{2 - 8i} \cdot \dfrac{2 + 8i}{2 + 8i}$

$= \dfrac{8i + 32i^2}{2^2 - (8i)^2}$

$= \dfrac{8i - 32}{4 + 64}$

$= \dfrac{8i - 32}{68}$

$= \dfrac{-8 + 2i}{17}$

$= -\dfrac{8}{17} + \dfrac{2}{17}i$

**131.** $\dfrac{3 - 5i}{6 + i} = \dfrac{3 - 5i}{6 + i} \cdot \dfrac{6 - i}{6 - i}$

$= \dfrac{18 - 3i - 30i + 5i^2}{6^2 - i^2}$

$= \dfrac{18 - 33i - 5}{36 + 1}$

$= \dfrac{13 - 33i}{37}$

$= \dfrac{13}{37} - \dfrac{33}{37}i$

**133.** $z^2 = -121$

$z = \pm\sqrt{-121}$

$z = \pm 11i$

**135.** $y^2 + 50 = 0$

$$y^2 = -50$$
$$y = \pm\sqrt{-50}$$
$$y = \pm\sqrt{25 \cdot 2}\,i$$
$$y = \pm 5\sqrt{2}\,i$$

**137.** $(y + 4)^2 + 18 = 0$

$$(y + 4)^2 = -18$$
$$y + 4 = \pm\sqrt{-18}$$
$$y + 4 = \pm\sqrt{9 \cdot 2}\,i$$
$$y = -4 \pm 3\sqrt{2}\,i$$

**139.** $x^2 - 2x + 26 = 0$

$$x^2 - 2x = -26$$
$$x^2 - 2x + 1 = -26 + 1 \qquad \text{Half of } -2 \text{ is } -1, \text{ and } (-1)^2 = 1.$$
$$(x - 1)^2 = -25$$
$$x - 1 = \pm\sqrt{-25}$$
$$x - 1 = \pm 5i$$
$$x = 1 \pm 5i$$

**141.** $x^2 - 3x + 3 = 0$

$$x^2 - 3x = -3$$
$$x^2 - 3x + \frac{9}{4} = -3 + \frac{9}{4} \qquad \text{Half of } -3 \text{ is } -\frac{3}{2}, \text{ and } \left(-\frac{3}{2}\right)^2 = \frac{9}{4}.$$
$$\left(x - \frac{3}{2}\right)^2 = -\frac{12}{4} + \frac{9}{4}$$
$$\left(x - \frac{3}{2}\right)^2 = -\frac{3}{4}$$
$$x - \frac{3}{2} = \pm\sqrt{-\frac{3}{4}}$$
$$x - \frac{3}{2} = \pm\frac{\sqrt{3}}{2}\,i$$
$$x = \frac{3}{2} \pm \frac{\sqrt{3}}{2}\,i$$
$$x = \frac{3}{2} \pm \frac{\sqrt{3}}{2}\,i = 1.50 \pm 0.87i$$

**143.** $x^2 + 6x + 13 = 0$

$$x = \frac{-6 \pm \sqrt{6^2 - 4(1)(13)}}{2(1)}$$
$$x = \frac{-6 \pm \sqrt{36 - 52}}{2}$$
$$x = \frac{-6 \pm \sqrt{-16}}{2}$$
$$x = \frac{-6 \pm 4i}{2}$$
$$x = -3 \pm 2i$$

**145.** $3z^2 - 3z + \dfrac{49}{64} = 0$

$$64\left(3z^2 - 3z + \frac{49}{64}\right) = 64(0)$$
$$192z^2 - 192z + 49 = 0$$
$$x = \frac{192 \pm \sqrt{192^2 - 4(192)(49)}}{2(192)}$$
$$x = \frac{192 \pm \sqrt{36864 - 37632}}{384}$$
$$x = \frac{192 \pm \sqrt{-768}}{384}$$
$$x = \frac{192 \pm 16\sqrt{3}\,i}{384}$$
$$x = \frac{1}{2} \pm \frac{\sqrt{3}}{24}\,i$$

**147.** Domain: $\{-2, 3, 5, 8\}$

Range: $\{1, 3, 7, 8\}$

**149.** Domain: $\{-2, 2, 4\}$

Range: $\{-3, 0, 3, 4\}$

**151.** Yes. No first component has two different second components, so this relation is a function.

**153.** No, the relation is not a function because the input value 2 has two different output values.

**155.** Yes. No first component has two different second components, so this relation *is* a function.

**157.** Yes, this graph indicates that $y$ is a function of $x$ because no vertical line intersects more than one point on the graph.

**159.** No. This graph indicates that $y$ is not a function of $x$. Some vertical lines intersect the graph at more than one point.

**161.** Yes, this graph indicates that $y$ is a function of $x$ because no vertical line intersects more than one point on the graph.

**163.** (a) $f(-1) = \frac{3}{4}(-1) = -\frac{3}{4}$

(b) $f(4) = \frac{3}{4}(4) = 3$

(c) $f(10) = \frac{3}{4}(10) = \frac{30}{4} = \frac{15}{2}$

(d) $f\left(-\frac{4}{3}\right) = \frac{3}{4}\left(-\frac{4}{3}\right) = \frac{-12}{12} = -1$

# Chapter Test for Chapter 10

**1.** $x^2 - 400 = 0$

$x^2 = 400$

$x = \pm\sqrt{400}$

$x = \pm 20$

The solutions are 20 and −20.

**165.** (a) $g(0) = -16(0)^2 + 64 = 0 + 64 = 64$

(b) $g\left(\frac{1}{4}\right) = -16\left(\frac{1}{4}\right)^2 + 64 = -1 + 64 = 63$

(c) $g(1) = -16(1)^2 + 64 = -16 + 64 = 48$

(d) $g(2) = -16(2)^2 + 64 = -64 + 64 = 0$

**167.** (a) $f(0) = |2(0) + 3| = |3| = 3$

(b) $f(5) = |2(5) + 3| = |13| = 13$

(c) $f(-4) = |2(-4) + 3| = |-5| = 5$

(d) $f\left(-\frac{3}{2}\right) = \left|2\left(-\frac{3}{2}\right) + 3\right| = |0| = 0$

**169.** $f(p) = 40 - 0.2p$

(a) $f(10) = 40 - 0.2(10) = 40 - 2 = 38$

(b) $f(50) = 40 - 0.2(50) = 40 - 10 = 30$

(c) $f(100) = 40 - 0.2(100) = 40 - 20 = 20$

**171.** $D = \{1, 2, 3, 4, 5\}$

**173.** $D = \{-2, 0, 3, 4, 7\}$

**2.** $(x + 4)^2 + 100 = 0$

$(x + 4)^2 = -100$

$x + 4 = \pm\sqrt{-100}$

$x + 4 = \pm 10i$

$x = -4 \pm 10i$

The solutions are $-4 + 10i$ and $-4 - 10i$.

**3.** $t^2 - 6t + 11 = 0$

$t^2 - 6t = -11$

$t^2 - 6t + 9 = -11 + 9$   Half of $-6$ is $-3$, and $(-3)^2 = 9$.

$(t - 3)^2 = -2$

$t - 3 = \pm\sqrt{-2}$

$t - 3 = \pm\sqrt{2}i$

$t = 3 \pm \sqrt{2}i$

The solutions are $3 + \sqrt{2}i$ and $3 - \sqrt{2}i$.

**4.** $3z^2 + 9z + 5 = 0$

$3z^2 + 9z = -5$

$z^2 + 3z = -\dfrac{5}{3}$

$z^2 + 3z + \dfrac{9}{4} = -\dfrac{5}{3} + \dfrac{9}{4}$   Half of 3 is $\dfrac{3}{2}$, $\left(\dfrac{3}{2}\right)^2 = \dfrac{9}{4}$.

$\left(z + \dfrac{3}{2}\right)^2 = -\dfrac{20}{12} + \dfrac{27}{12}$

$\left(z + \dfrac{3}{2}\right)^2 = \dfrac{7}{12}$

$z + \dfrac{3}{2} = \pm\sqrt{\dfrac{7}{12}}$

$z + \dfrac{3}{2} = \pm\dfrac{\sqrt{7}}{\sqrt{12}} \cdot \dfrac{\sqrt{3}}{\sqrt{3}}$

$z + \dfrac{3}{2} = \pm\dfrac{\sqrt{21}}{\sqrt{36}}$

$z + \dfrac{3}{2} = \pm\dfrac{\sqrt{21}}{6}$

$z = -\dfrac{3}{2} \pm \dfrac{\sqrt{21}}{6}$

$z = -\dfrac{9}{6} \pm \dfrac{\sqrt{21}}{6}$

$z = \dfrac{-9 \pm \sqrt{21}}{6}$

The solutions are $\dfrac{-9 + \sqrt{21}}{6}$ and $\dfrac{-9 - \sqrt{21}}{6}$.

**5.** $x^2 - 2x + 3 = 0$

$a = 1, b = -2, c = 3$

$x = \dfrac{-(-2) \pm \sqrt{(-2)^2 - 4(1)(3)}}{2(1)}$

$x = \dfrac{2 \pm \sqrt{4 - 12}}{2}$

$x = \dfrac{2 \pm \sqrt{-8}}{2}$

$x = \dfrac{2 \pm 2\sqrt{2}i}{2}$

$x = \dfrac{\cancel{2}\left(1 \pm \sqrt{2}i\right)}{\cancel{2}(1)}$

$x = 1 \pm \sqrt{2}i$

The solutions are $1 + \sqrt{2}i$ and $1 - \sqrt{2}i$.

**6.** $2u^2 + 4u + 1 = 0$

$a = 2, b = 4, c = 1$

$u = \dfrac{-4 \pm \sqrt{4^2 - 4(2)(1)}}{2(2)}$

$u = \dfrac{-4 \pm \sqrt{8}}{4}$

$u = \dfrac{-4 \pm 2\sqrt{2}}{4}$

$u = \dfrac{\cancel{2}\left(-2 \pm \sqrt{2}\right)}{\cancel{2}(2)}$

$u = \dfrac{-2 \pm \sqrt{2}}{2}$

The solutions are $\dfrac{-2 + \sqrt{2}}{2}$ and $\dfrac{-2 - \sqrt{2}}{2}$.

7. $\dfrac{1}{x+1} - \dfrac{1}{x-2} = 1$

$$(x+1)(x-2)\left(\dfrac{1}{x+1}\right) - (x+1)(x-2)\left(\dfrac{1}{x-2}\right) = (x+1)(x-2)(1)$$

$$(x-2) - (x+1) = (x+1)(x-2), x \neq -1, x \neq 2$$

$$x - 2 - x - 1 = x^2 - x - 2$$

$$-3 = x^2 - x - 2$$

$$0 = x^2 - x + 1$$

$$x = \dfrac{-(-1) \pm \sqrt{(-1)^2 - 4(1)(1)}}{2(1)}$$

$$x = \dfrac{1 \pm \sqrt{-3}}{2} \Rightarrow x = \dfrac{1 \pm \sqrt{3}i}{2}$$

The solutions are $\dfrac{1 + \sqrt{3}i}{2}$ and $\dfrac{1 - \sqrt{3}i}{2}$.

8. $\sqrt{2x} = x - 1$

$$\left(\sqrt{2x}\right)^2 = (x-1)^2$$

$$2x = x^2 - 2x + 1$$

$$0 = x^2 - 4x + 1$$

$$x = \dfrac{-(-4) \pm \sqrt{(-4)^2 - 4(1)(1)}}{2(1)}$$

$$x = \dfrac{4 \pm \sqrt{12}}{2}$$

$$x = \dfrac{4 \pm 2\sqrt{3}}{2}$$

$$x = \dfrac{\cancel{2}(2 \pm \sqrt{3})}{\cancel{2}}$$

$$x = 2 \pm \sqrt{3}$$

The answer of $2 + \sqrt{3}$ checks, but the second answer is extraneous. (If $x = 2 - \sqrt{3}$, $x \approx 0.27$. Then $\sqrt{2x} \approx 0.73$, but $x - 1 \approx -0.73$.) So, the only solution is $2 + \sqrt{3}$.

9. $y = -2x^2 - 7 \Rightarrow a = -2, b = 0$

$a < 0 \Rightarrow$ the parabola opens *downward*.

Vertex: $x = -\dfrac{b}{2a} = -\dfrac{0}{2(-2)} = 0$

$$y = -2(0)^2 - 7 = -7$$

$$(0, -7)$$

The vertex is located at $(0, -7)$.

10. $y = 5 - 2x - x^2$

$y = -x^2 - 2x + 5 \Rightarrow a = -1, b = -2$

$a < 0 \Rightarrow$ the parabola opens *downward*.

Vertex: $x = -\dfrac{b}{2a} = -\dfrac{-2}{2(-1)} = -1$

$$y = -(-1)^2 - 2(-1) + 5 = 6$$

$$(-1, 6)$$

The vertex is located at $(-1, 6)$.

11. $y = (x-2)^2 + 3$

$y = x^2 - 4x + 4 + 3$

$y = x^2 - 4x + 7 \Rightarrow a = 1, b = -4$

$a > 0 \Rightarrow$ the parabola opens *upward*.

Vertex: $x = -\dfrac{b}{2a} = -\dfrac{-4}{2(1)} = 2$

$$y = (2-2)^2 + 3 = 3$$

$$(2, 3)$$

The vertex is located at $(2, 3)$.

**12.** *Leading coefficient test:*

Since $a > 0$, the parabola opens upward.

*Vertex:* $(a = 1, b = -8)$

$x = -\dfrac{b}{2a} = -\dfrac{-8}{2(1)} = -\dfrac{-8}{2} = -(-4) = 4$

$y = 4^2 - 8(4) + 12 = 16 - 32 + 12 = -4$

The vertex is located at $(4, -4)$.

*y-intercept:* $(\text{Let } x = 0.)$   $y = 0^2 - 4(0) + 12 = 12$  The *y*-intercept is $(0, 12)$.

*x-intercepts:* $(\text{Let } y = 0.)$

$0 = x^2 - 8x + 12 = (x - 6)(x - 2)$

$x - 6 = 0 \Rightarrow x = 6$

$x - 2 = 0 \Rightarrow x = 2$

The *x*-intercepts are $(6, 0)$ and $(2, 0)$.

*Additional solution points:*

| $x$ | 1 | 3 | 5 | 7 |
|---|---|---|---|---|
| $y - x^2 \cdot 8x + 12$ | 5 | $-3$ | $-3$ | 5 |
| Points | $(1, 5)$ | $(3, -3)$ | $(5, -3)$ | $(7, 5)$ |

**13.** $y = -x^2 - 4x$

*Leading coefficient test:*

Since $a < 0$, the parabola opens *downward*.

*Vertex:* $(a = -1, b = -4)$

$x = -\dfrac{b}{2a} = -\dfrac{-4}{2(-1)} = -2$

$y = -(-2)^2 - 4(-2) = -4 + 8 = 4$

The vertex is located at $(-2, 4)$.

*y-intercept:* $(\text{Let } x = 0.)$   $y = -0^2 - 4(0) = 0$

The *y*-intercept is $(0, 0)$.

*x-intercepts:* $(\text{Let } y = 0.)$

$0 = -x^2 - 4x$

$0 = -x(x + 4)$

$-x = 0 \Rightarrow x = 0$

$x + 4 = 0 \Rightarrow x = -4$

The *x*-intercepts are $(0, 0)$ and $(-4, 0)$.

*Additional solution points:*

| $x$ | $-5$ | $-2$ | $-1$ | 1 |
|---|---|---|---|---|
| $y = -x^2 - 4x$ | $-5$ | 4 | 3 | $-5$ |
| Points | $(-5, -5)$ | $(-2, 4)$ | $(-1, 3)$ | $(1, -5)$ |

**14.** $(2 + 3i) - \sqrt{-25} = 2 + 3i - 5i$

$= 2 - 2i$

**15.** $(2 + 9i)^2 = 2^2 + 2(2)(9i) + (9i)^2$

$= 4 + 36i + 81i^2$

$= 4 + 36i - 81$

$= -77 + 36i$

**16.** $\sqrt{-16}(1 + \sqrt{4}) = 4i(1 + 2i)$

$= 4i + 8i^2$

$= -8 + 4i$

**17.** $(3 - 8i)(4 + i) = 12 + 3i - 32i - 8i^2$

$= 12 - 29i + 8$

$= 20 - 29i$

**18.** $\dfrac{5-2i}{3+i} = \dfrac{5-2i}{3+i} \cdot \dfrac{3-i}{3-i}$

$= \dfrac{15 - 5i - 6i + 2i^2}{3^2 - i^2}$

$= \dfrac{15 - 11i - 2}{9 - (-1)}$

$= \dfrac{13 - 11i}{10}$

$= \dfrac{13}{10} - \dfrac{11}{10}i$

**19.** No. Some first components, 0 and 1, have two different second components, so the table does not represent $y$ as a function of $x$.

**20.** Yes. It passes the Vertical Line Test. No vertical line intersects the graph at more than one point.

**21.** (a) $f(0) = 0^3 - 2(0)^2$

$\qquad = 0 - 2(0)$

$\qquad = 0$

(b) $f(2) = 2^3 - 2(2)^2$

$\qquad = 8 - 2(4)$

$\qquad = 8 - 8$

$\qquad = 0$

(c) $f(-2) = (-2)^3 - 2(-2)^2$

$\qquad = -8 - 2(4)$

$\qquad = -8 - 8$

$\qquad = -16$

(d) $f\left(\dfrac{1}{2}\right) = \left(\dfrac{1}{2}\right)^3 - 2\left(\dfrac{1}{2}\right)^2$

$\qquad = \dfrac{1}{8} - 2\left(\dfrac{1}{4}\right)$

$\qquad = \dfrac{1}{8} - \dfrac{1}{2}$

$\qquad = \dfrac{1}{8} - \dfrac{4}{8}$

$\qquad = -\dfrac{3}{8}$

**22.** *Formula:*   $\boxed{\begin{array}{c}\text{Length of}\\\text{rectangle}\end{array}} \cdot \boxed{\begin{array}{c}\text{Width of}\\\text{rectangle}\end{array}} = \boxed{\begin{array}{c}\text{Area of}\\\text{rectangle}\end{array}}$

*Labels:*   Width of rectangle $= x$        (inches)

Length of rectangle $= x + 10$     (inches)

Area of rectangle $= 96$        (square inches)

*Equation:*        $(x + 10)(x) = 96$

$x^2 + 10x = 96$

$x^2 + 10x - 96 = 0$

$(x + 16)(x - 6) = 0$

$x + 16 = 0 \Rightarrow x = -16$ and $x + 10 = -6$

$x - 6 = 0 \Rightarrow x = 6$ and $x + 10 = 16$

We choose the positive solutions for this application. The dimensions of the rectangle are 6 inches by 16 inches.

**23.** *Verbal Model:*   $\boxed{\text{Rate of faster person}} + \boxed{\text{Rate of slower person}} = \boxed{\text{Rate of two persons together}}$

*Labels:*   Two persons: time = 6 (hours), rate = $\dfrac{1}{6}$ (lawn per hour)

Faster person: time $x$ (hours), rate = $\dfrac{1}{x}$ (lawn per hour)

Slower person: time = $x + 5$ (hours), rate = $1/(x + 5)$ (lawn per hour)

*Equation:*

$$\frac{1}{x} + \frac{1}{x+5} = \frac{1}{6}$$

$$6x(x+5)\left(\frac{1}{x}\right) + 6x(x+5)\left(\frac{1}{x+5}\right) = 6x(x+5)\left(\frac{1}{6}\right)$$

$$6(x+5) + 6x = x(x+5), \ x \neq 0, \ x \neq -5$$

$$6x + 30 + 6x = x^2 + 5x$$

$$12x + 30 = x^2 + 5x$$

$$0 = x^2 - 7x - 30$$

$$0 = (x-10)(x+3)$$

$$x - 10 = 0 \Rightarrow x = 10 \text{ and } x + 5 = 15$$

$$x + 3 = 0 \Rightarrow x = -3 \quad \text{Discard the negative answer.}$$

The faster person needed 10 hours to mow the lawn, and the slower person needed 15 hours.

# Appendix A  Introduction to Graphing Calculators

**1.** $y = -3x$

*Keystrokes:*

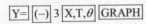

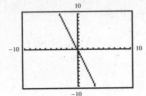

**3.** $y = \frac{3}{4}x - 6$

*Keystrokes:*

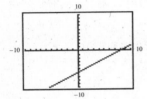

**5.** $y = \frac{1}{2}x^2$

*Keystrokes:*

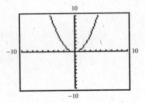

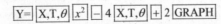

**7.** $y = x^2 - 4x + 2$

*Keystrokes:*

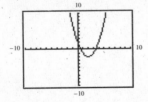

**9.** $y = |x - 5|$

*Keystrokes:*

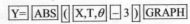

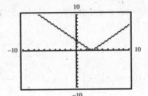

**11.** $y = |x^2 - 4|$

*Keystrokes:*

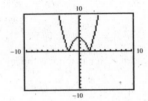

**13.** $y = 27x + 100$

*Keystrokes:*

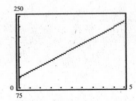

**15.** $y = 0.001x^2 + 0.5x$

*Keystrokes:*

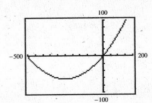

**17.** $y = 15 + |x - 12|$

*Keystrokes:*

Y= 15 + ABS (( X,T,$\theta$ − 12 )) GRAPH

Xmin = 4
Xmax = 20
Xscl = 1
Ymin = 14
Ymax = 22
Yscl = 1

**19.** $y = -15 + |x + 12|$

*Keystrokes:*

Y= (−) 15 + ABS (( X,T,$\theta$ + 12 )) GRAPH

Xmin = −20
Xmax = −4
Xscl = 1
Ymin = −16
Ymax = −8
Yscl = 1

**21.** $y = -4$, $y = -|x|$

*Keystrokes:* $y_1$: Y= (−) 4 ENTER

$y_2$: (−) ABS X,T,$\theta$ GRAPH

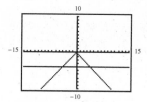

Triangle

**23.** $y = |x| - 8$, $y = -|x| + 8$

*Keystrokes:* $y_1$: Y= ABS X,T,$\theta$ − 8 ENTER

$y_2$: (−) ABS X,T,$\theta$ + 8 GRAPH

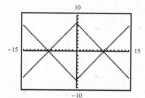

Square

**25.** $y_1 = 2x + (x + 1)$

$y_2 = (2x + x) + 1$

*Keystrokes:*

$y_1$: Y= 2 X,T,$\theta$ + (( X,T,$\theta$ + 1 )) ENTER

$y_2$: (( 2 X,T,$\theta$ + X,T,$\theta$ )) + 1 GRAPH

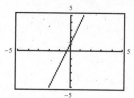

Associative Property of Addition

**27.** $y_1 = 2\left(\frac{1}{2}\right)$

$y_2 = 2$

*Keystrokes:* $y_1$: Y= 2 (( 1 ÷ 2 )) ENTER

$y_2$: 1                    GRAPH

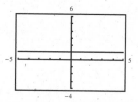

Multiplicative Inverse Property

**29.** $y = 9 - x^2$

*Keystrokes:* Y= 9 − X,T,$\theta$ $x^2$ GRAPH

Trace to $x$-intercepts: $(-3, 0)$ and $(3, 0)$

Trace to $y$-intercept: $(0, 9)$

**31.** $y = 6 - |x + 2|$

*Keystrokes:*

Y= 6 − ABS (( X,T,$\theta$ + 2 )) GRAPH

Trace to $x$-intercepts: $(-8, 0)$ and $(4, 0)$

Trace to $y$-intercept: $(0, 4)$

**33.** $y = 2x - 5$

*Keystrokes:* Y= 2 X,T,$\theta$ − 5 GRAPH

Trace to $x$-intercept: $\left(\frac{5}{2}, 0\right)$

Trace to $y$-intercept: $(0, -5)$

**35.** $y = x^2 + 1.5x - 1$

*Keystrokes:*

[Y=] [X,T,$\theta$] [$x^2$] [+] 1.5 [X,T,$\theta$] [−] 1 [GRAPH]

Trace to *x*-intercepts: $(-2, 0)$ and $\left(\frac{1}{2}, 0\right)$

Trace to *y*-intercept: $(0, -1)$

**37.** *Keystrokes:*

$y_1$: [Y=] 0.5 [X,T,$\theta$] [$x^2$] [−] 5.06 [X,T,$\theta$] [+] 110.3 [ENTER]

$y_2$: [(−)] 0.221 [X,T,$\theta$] [$x^2$] [+] 5.88 [X,T,$\theta$] [+] 75.8 [GRAPH]

Set window as indicated in problem.

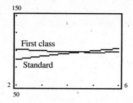